The Nature and Tectonic Significance of Fault Zone Weakening

Special Publication reviewing procedures

The Society makes every effort to ensure that the scientific and production quality of its books matches that of its journals. Since 1997, all book proposals have been refereed by specialist reviewers as well as by the Society's Publications Committee. If the referees identify weaknesses in the proposal, these must be addressed before the proposal is accepted.

Once the book is accepted, the Society has a team of series editors (listed above) who ensure that the volume editors follow strict guidelines on refereeing and quality control. We insist that individual papers can only be accepted after satisfactory review by two independent referees. The questions on the review forms are similar to those for *Journal of the Geological Society*. The referees' forms and comments must be available to the Society's series editors on request.

Although many of the books result from meetings, the editors are expected to commission papers that were not presented at the meeting to ensure that the book provides a balanced coverage of the subject. Being accepted for presentation at the meeting does not guarantee inclusion in the book.

Geological Society Special Publications are included in the ISI Science Citation Index, but they do not have an impact factor, the latter being applicable only to journals.

More information about submitting a proposal and producing a Special Publication can be found on the Society's web site: www.geolsoc.org.uk

GEOLOGICAL SOCIETY SPECIAL PUBLICATION No. 186

The Nature and Tectonic Significance of Fault Zone Weakening

EDITED BY

R.E. HOLDSWORTH
University of Durham, UK

R.A. STRACHAN
Oxford Brookes University, UK

J.F. MAGLOUGHLIN
Colorado State University, USA

and

R.J. KNIPE
University of Leeds, UK

2001
Published by
The Geological Society
London

THE GEOLOGICAL SOCIETY

The Geological Society of London was founded in 1807 and is the oldest geological society in the world. It received its Royal Charter in 1825 for the purpose of 'investigating the mineral structure of the Earth' and is now Britain's national society for geology.

Both a learned society and a professional body, the Geological Society is recognized by the Department of Trade and Industry (DTI) as the chartering authority for geoscience, able to award Chartered Geologist status upon appropriately qualified Fellows. The Society has a membership of 9099, of whom about 1500 live outside the UK.

Fellowship of the Society is open to persons holding a recognized honours degree in geology or a cognate subject, or not less than six years' relevant experience in geology or a cognate subject. A Fellow with a minimum of five years' relevant postgraduate experience in the practice of geology may apply for chartered status. Successful applicants are entitled to use the designatory postnominal CGeol (Chartered Geologist). Fellows of the Society may use the letters FGS. Other grades of membership are available to members not yet qualifying for Fellowship.

The Society has its own Publishing House based in Bath, UK. It produces the Society's international journals, books and maps, and is the European distributor for publications of the American Association of Petroleum Geologists (AAPG), the Society for Sedimentary Geology (SEPM) and the Geological Society of America (GSA). Members of the Society can buy books at considerable discounts. The Publishing House has an online bookshop (http://bookshop.geolsoc.org.uk).

Further information on Society membership may be obtained from the Membership Services Manager, The Geological Society, Burlington House, Piccadilly, London W1V 0JU (E-mail: enquiries@geolsoc.org.uk; tel: +44 (0) 207 434 9944).

The Society's Web Site can be found at http://www.geolsoc.org.uk/. The Society is a Registered Charity, number 210161.

Published by The Geological Society from:
The Geological Society Publishing House
Unit 7, Brassmill Enterprise Centre
Brassmill Lane
Bath BA1 3JN, UK
(*Orders*: Tel. +44 (0)1225 445046
Fax +44 (0)1225 442836)
Online bookshop: http://bookshop.geolsoc.org.uk

British Library Cataloguing in Publication Data
A catalogue record for this book is available from the British Library.

ISBN 1-86239-090-8

Typeset and Printed by The Alden Press, Oxford, UK.

Distributors

USA
AAPG Bookstore
PO Box 979
Tulsa
OK 74101-0979
USA
Orders: Tel. +1 918 584-2555
Fax +1 918 560-2652
E-mail *bookstore@aapg.org*

Australia
Australian Mineral Foundation Bookshop
63 Conyngham Street
Glenside
South Australia 5065
Australia
Orders: Tel. +61 88 379-0444
Fax +61 88 379-4634
E-mail *bookshop@amf.com.au*

India
Affiliated East-West Press PVT Ltd
G-1/16 Ansari Road, Daryaganj,
New Delhi 110 002
India
Orders: Tel. +91 11 327-9113
Fax +91 11 326-0538
E-mail *affiliat@nda.vsnl.net.in*

Japan
Kanda Book Trading Co.
Cityhouse Tama 204
Tsurumaki 1-3-10
Tama-shi
Tokyo 206-0034
Japan
Orders: Tel. +81 (0)423 57-7650
Fax +81 (0)423 57-7651

Contents

It is recommended that reference to all or part of this book should be made in one of the following ways:

HOLDSWORTH, R.E., STRACHAN, R.A., MAGLOUGHLIN, J.F. & KNIPE, R.J. (eds) 2001. *The Nature and Tectonic Significance of Fault Zone Weakening*. Geological Society, London, Special Publications, **186**.

FURLONG, K.P., SHEAFFER, S.D. & MALSERVISI, R. 2001. Thermo-rheological controls on deformation within oceanic transforms. *In:* HOLDSWORTH, R.E., STRACHAN, R.A., MAGLOUGHLIN, J.F. & KNIPE, R.J. (eds) *The Nature and Tectonic Significance of Fault Zone Weakening*. Geological Society, London, Special Publications, **186,** 65–84.

Preface

Many faults appear to form persistent zones of weakness that fundamentally influence the distribution, architecture and kinematic patterns of crustal-scale deformation and associated geological processes in both continental and oceanic regions. To date, however, our understanding of the mechanisms that lead to changes in fault zone rheology, their many geological consequences and the larger-scale implications that they may have for lithosphere dynamics are still poorly understood. This publication contains 18 papers written by an international group of Earth Scientists based around a central theme of the causes and consequences of fault zone weakening in both continental and oceanic regions.

The opening paper (**Rutter *et al.***) presents a basic review and overview of the causes and consequences of fault zone weakness during crustal deformation. The papers that follow are grouped into four sections. In the first section, *Insights from Neotectonic Settings, Deformation Experiments and Modelling Studies*, the issues of fault strength and rheology are explored using earthquake focal mechanisms (**Townend & Zoback**), direct analysis of fault core from an active low-angle detachment (**Kopf**), experimental deformation studies (**Main *et al.***) and numerical modelling (**Furlong *et al.***).

In the second section, *Insights from Natural Fault Rocks,* the nature and significance of clay-mineral transformations are examined (**Warr & Cox, Yan *et al.***), together with detailed case studies illustrating the use of microfractures in the analysis of reactivated fault zones (**Mitra & Ismat**) and metamorphic/microstructural evidence for episodic weakening and hardening in a deep crustal shear zone (**Steffen *et al.***).

In the third section, *Geometric Controls and Fault System Evolution*, the fundamental influence of factors such as fault size, connectivity, position and orientation upon strain localization and fault growth is examined (**Walsh *et al.***). The following papers are concerned with the nature and origin of asymmetric arrays of shear surfaces in natural fault zones (**Wojtal**) and a quantitative study of the way in which pre-existing basement heterogeneities influence brittle fracture zone development during reactivation (**Beacom *et al.***).

The final section, *Insights from Lithosphere-to Crustal-Scale Fault Zones***,** presents a series of case studies in which issues related to long-term reactivation of faults/shear zones and weakening are examined on various scales. Examples are drawn from: Idaho, USA (**Tikoff *et al.***); Central Argentina (**Simpson *et al.***); the Southern Appalachians, USA (**Hatcher**); the Apennine–Maghrebide system, Italy (**Tavarnelli *et al.***); and Sweden (**Talbot**). The final paper in the volume (**Handy *et al.***) presents a synthesis and review of the relationships between fault zones and melting in the continental crust, and how the presence of molten material and its subsequent crystallization may lead to profound changes in crustal strength.

The volume derives from a conference held in March 2000 at Burlington House, London under the joint auspices of the Tectonic Studies Group (Geological Society of London, UK), the Structural Geology & Tectonics Division (Geological Society of America) and InterRidge.

Bob Holdsworth, Durham, UK
Rob Strachan, Oxford, UK
Jerry Magloughlin, Colorado, USA
Rob Knipe, Leeds, UK

Acknowledgements

The editors would like to thank the following colleagues and friends who kindly donated their valuable time and expertise to help with the reviewing of papers submitted to this volume:

Mark Allen
Ian Alsop
Torgeir Andersen
Andy Barnicoat
Donna Blackman
Al Bolton
Joe Cann
Massimo Cocco
Darrel Cowan
Mike Curtis
George Davis
Allen Dennis
Mike Edwards
Dan Faulkner
David Ferrill
Quentin Fisher
Haakon Fossen
Laurel Goodwin
John Grocott
Mark Handy
Achim Kopf
Geoff Lloyd
Ken McCaffrey
Alessandro Michetti
Brendan Murphy
Tim Needham
Kieran O'Hara
Bob Pankhurst
Don Pecor
Gerald Roberts
Ernie Rutter
Peter Sammonds
Roger Searle
Rick Sibson
Chris Talbot
Enrico Tavarnelli
Rob Twiss
John Walsh
Laurence Warr
John Wheeler
Chris Wibberley
Bob Wintsch
Mike Watkeys

We would like to especially thank Roger Searle who was a co-covenor of the conference at Burlington House. We are once again grateful to the staff at both Burlington House and the Geological Society Publishing House, especially Angharad Hills, whose efforts were invaluable in ensuring a successful conference and the rapid production of this volume.

Bob Holdsworth would like to dedicate this volume to his son Ronan who passed away 23 November 2000.

The nature and tectonic significance of fault-zone weakening: an introduction

E.H. RUTTER[1], R.E. HOLDSWORTH[2] & R.J. KNIPE[3]

[1]*Rock Deformation Laboratory, Earth Sciences Department, University of Manchester, Manchester M13 9PL, UK (e-mail: ernie.rutter@man.ac.uk)*

[2]*Reactivation Research Group, Department of Geological Sciences, University of Durham, Durham DH1 3LE, UK*

[3]*Rock Deformation Research, Earth Sciences Department, University of Leeds, Leeds LS2 9JT, UK*

Abstract: Fault zones control the location, architecture and evolution of a broad range of geological features, act as conduits for the focused migration of economically important fluids and, as most seismicity is associated with active faults, they also constitute one of the most important global geological hazards. In general, the repeated localization of displacements along faults and shear zones, often over very long time scales, strongly suggests that they are weak relative to their surrounding wall rocks. Geophysical observations from plate boundary faults such as the San Andreas fault additionally suggest that this fault zone is weak in an absolute sense, although this remains a controversial issue. Our understanding of fault-zone structure and mechanical behaviour derive from three main sources of information: (1) studies of natural fault zones and their deformation products (fault rocks); (2) seismological and neotectonic studies of currently active natural fault systems; (3) laboratory-based deformation experiments using rocks or rock-analogue materials. These provide us with a basic understanding of brittle faulting in the upper crust of the Earth where the stress state is limited by the frictional strength of networks of faults under the prevailing fluid-pressure conditions. Under the long-term loading conditions typical of geological fault zones, poorly understood phenomena such as subcritical crack growth in fracture process zones are likely to be of major importance in controlling both fault growth and strength. Grain-size reduction in highly strained fault rocks produced in the plastic–viscous and deeper parts of frictional regime can lead to changes in deformation mechanisms and relative weakening that can account for the localization of deformation and repeated reactivation of crustal faults. Our understanding the interactions between deformation mechanisms, metamorphic processes and the flow of chemically active fluids is a key area for future study. An improved understanding of how fault- or shear-zone linkages, strength and microstructure evolve over large changes in finite strain will ultimately lead to the development of geologically more realistic numerical models of lithosphere deformation that incorporate displacements concentrated into narrow, weaker fault zones.

In continental and oceanic regions, the deformation of the Earth's crust (and lithosphere) is characteristically heterogeneous, with most displacements being localized into linked systems of faults and shear zones. In both intraplate and plate margin settings, these approximately planar or tabular deformation zones influence strongly the location, architecture and evolution of a broad range of geological features, including rift basins, orogenic belts and transcurrent fault systems. Many fault zones are known to act as conduits for the focused migration of fluids and clearly play a central role in determining the location, modes of transport and emplacement of economically important hydrocarbon reservoirs, hydrothermal mineral deposits and igneous intrusions. In addition, most active seismicity is associated with displacements along fault zones, which therefore represent one of the most important global geological hazards.

Fault-zone structure and mechanical behaviour

In the upper, seismogenic part of the crust, deformation in fault zones occurs by frictional

From: HOLDSWORTH, R.E., STRACHAN, R.A., MAGLOUGHLIN, J.F. & KNIPE, R.J. (eds) 2001. *The Nature and Tectonic Significance of Fault Zone Weakening*. Geological Society, London, Special Publications, **186**, 1–11. 0305-8719/01/$15.00

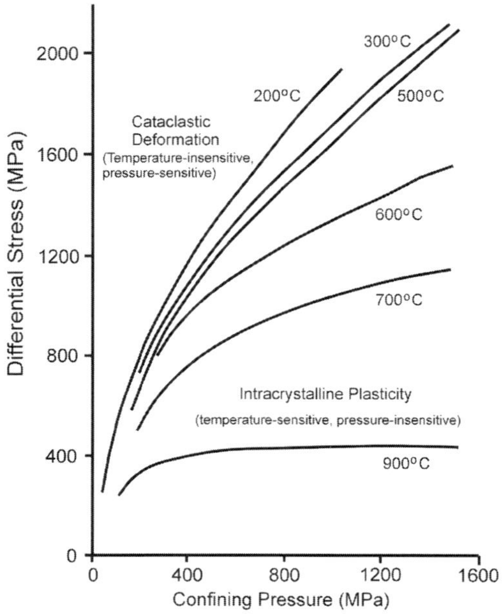

Fig. 1. Simplified representation of the data of Tullis & Yund (1977) for the ultimate strength of Westerly granite at a strain rate of $c. 10^{-5}\,s^{-1}$; frictional strength behaves in much the same way. The figure shows that strength in the brittle regime (up to 300 °C) is insensitive to temperature, but very sensitive to confining pressure, giving way at higher temperatures to an increased temperature sensitivity, but a reduction in pressure sensitivity as intracrystalline plastic processes begin to dominate.

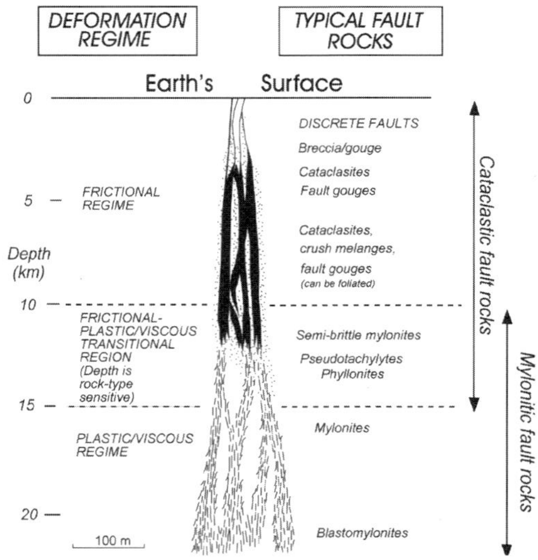

Fig. 2. A widely accepted conceptual model for the way that the character of a crustal fault zone might vary with depth (based on Sibson 1977). Narrow, brittle–frictional faults with a range of possible cataclastic fault rock products pass with increasing depth into foliated mylonitic fault rocks in which intracrystalline plastic and diffusion-accommodated viscous flow processes progresssively dominate. The transitional region between the upper and lower flow regimes is expected to correspond to a crustal strength maximum. The horizontal scale length is greatly exaggerated relative to the vertical. Although the fault zone is shown broadening with depth, this may not necessarily occur, depending on rock type.

processes in which the deformation mechanisms involve brittle fracture and frictional sliding. Except in initially porous rocks, these processes lead to dilatancy. Thus, the strength of brittle faults increases with effective pressure and hence depth of burial (Fig. 1; Byerlee 1978; Paterson 1978; Sibson 1983). Recurrence of movements on localized faults usually points to the fault zone being weaker than the stress required to form a fresh fault in the surrounding protolith. This degree of weakening after fault initiation is probably due to some combination of the formation of fragmentary rock products that are more porous and less cohesive than the protolith (thereby allowing enhanced fluid–rock interaction) coupled with the development of a foliated fabric, which may involve local concentration of clay minerals. Whether the fault motion is steady or seismogenic depends on whether the fault zones display transiently velocity-strengthening or -weakening characteristics (Scholz 1990, Scholz, 1998). This characteristic is sensitive both to fault rock type and local conditions (slip velocity, effective pressure, temperature).

At greater crustal depths and hence higher temperatures, brittle–frictional faults pass downward into shear zones (e.g. Fig. 2) and the regime changes to one of viscous flow in which a range of non-frictional, thermally activated deformation mechanisms are involved to produce crystal plasticity and diffusional creep (Sibson 1977; Tullis & Yund 1977; Schmid & Handy 1991). In the region separating these two regimes, a frictional–viscous (or sometimes called brittle–ductile) transition is likely to coincide with a strength maximum in the lithosphere, based on the findings of laboratory experiments and seismological studies (e.g. Sibson 1977, Sibson, 1983). If the deformation becomes isovolumetric, the stable continuation of localized flow demands that the material inside the fault zone be weaker than that outside. It is clear, however, that even high-temperature shearing can be accompanied by some dilatancy, which may be crucial to explain the ability of deep shear zones to transport fluids, both aqueous and melts (Bruhn *et al.* 2000). Even at the very high pressures in the deeper parts of subduction zones (400–600 km depth), localized deformation is indicated by the occurrence of earthquakes whose first-motion patterns indicate shear faulting. These too demand a dramatic weakening process that is either isovolumetric or compactive in nature, so that there is no requirement for work to be done against the enormous effective pressures at such depths (Kirby *et al.* 1996).

The repeated localization of displacements along existing faults and shear zones over a wide depth range, on either geologically short (<1 Ma; fault recurrence) or long (>1 Ma; fault reactivation) time scales, is likely to be largely determined by their geometric and internal rheological evolution. This is significant in continental regions where the resistance of the buoyant quartzofeldspathic crust to subduction means that once major fault zones have formed, they have the potential to be preserved for very long periods of geological time. The widespread recognition of reactivated faults in continental deformation zones and regions of inversion tectonics seems to confirm this suggestion (e.g. Butler *et al.* 1997; Holdsworth *et al.* 1997) and is manifested by the diffuse character of continental seismicity especially in collision zones. Although the initial localization of deformation can occur in both strain-hardening and strain-softening deformation regimes (e.g. Griggs & Handin 1960; Cobbold 1977), stable fault zones must be weak relative to their surrounding wall rocks. This would account for processes such as reactivation and recurrence, and might also help to explain why apparently large displacements can be accommodated along faults in mechanically unfavourable orientations, notably low-angle extensional detachments in both oceanic and continental settings (e.g. Lister & Davis 1989; Cann *et al.* 1997). It is also important to distinguish between the constitutive or material behaviour of the rocks in the fault zone and the overall system behaviour, which may be affected by additional boundary conditions, particularly during brittle deformation (see Hobbs *et al.* 1990). Our improving understanding of the scaling relationships between fault attributes such as length, width, magnitude of displacement and interconnectivity suggests that these factors will additionally influence the processes of fault growth and reactivation (e.g. Cowie & Scholz 1992; Sornette *et al.* 1993; Walsh *et al.* 2001).

Our understanding of fault-zone structure and mechanical behaviour derives from three main sources of information: (1) studies of natural fault zones and their deformation products (fault rocks); (2) seismological and neotectonic studies of currently active natural fault systems; (3) laboratory-based deformation experiments using rocks or rock-analogue materials. Natural fault zones generally preserve fault rocks whose composition and microstructure can be used to gain insights into the nature and evolution of deformation mechanisms and the rheological behaviour of fault zones under a wide range of pressure and temperature conditions (e.g. Handy 1989; Snoke *et al.* 1998). The traditional model for crustal-scale fault zones (Fig. 2) suggests that an interlinked network of brittle faults and cataclastic fault rocks connects directly at depth into a broader, anastomosing system of viscous shear zones with mylonitic fault rocks (Sibson 1977; Schmid & Handy 1991). In natural fault zones, the situation is made more complex by the fact that rocks are compositionally heterogeneous on all scales, with adjacent rock units responding in very different ways to the imposed deformation under a given set of environmental conditions (e.g. Handy 1990). Furthermore, as strain accumulates, deformational and associated metamorphic processes often profoundly modify fault rock mineralogy and microstructure in ways that may lead to significant changes in rheological behaviour and mechanical strength (e.g. Schmid & Handy 1991). Recent case studies suggest that such changes may be particularly important in the presence of a chemically active fluid phase and that this can lead to profound changes in fault-zone strength, together with the location and character of the frictional–viscous transition (e.g. Imber *et al.* 1997; Stewart *et al.* 2000; Handy *et al.* 2001).

The textural evolution and distribution of deformation within most fault zones can be determined by the operation of up to six interrelated factors that can be conveniently subdivided into lithological and environmental controls (Fig. 3). The importance of the six factors will change in both space and time as the fault system evolves and accumulates displacement, leading to a heterogeneous distribution of fault rocks, textures and deformation histories (e.g. see reviews by Handy 1989; Schmid & Handy 1991; Holdsworth *et al.* 2001). However, faults and shear zones tend to develop as self-organizing deformation systems on all scales (e.g. Handy 1990, 1990; Sornette *et al.* 1990, Sornette *et al.*, 1993). Strains become increasingly localized to form interconnected, narrow displacement zones (faults, shear zones) that surround elongate lenses of less highly

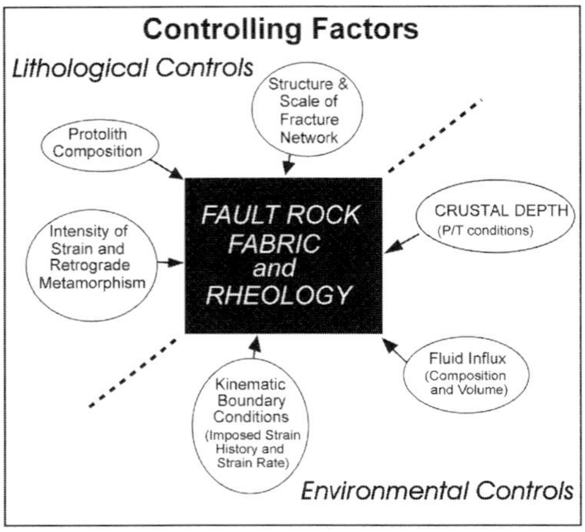

Fig. 3. Schematic summary illustrating that fault rock fabric and rheology depend on the linked operation of six controlling factors (after Holdsworth *et al.* 2001). Three of them depend on lithological factors and the remainder are externally imposed (following the approach of Knipe (1989)).

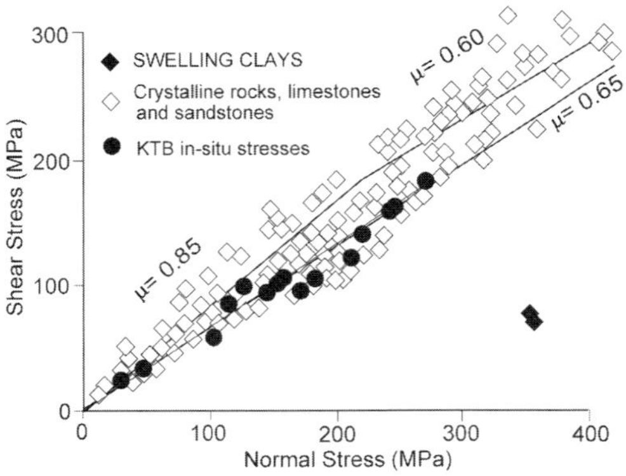

Fig. 4. Measurements of the resistance to frictional sliding in a wide range of rock types (where μ is the coefficient of frictional sliding), compiled by Byerlee (1978). These data have been used widely as a basis for generalization about the mechanical behaviour of upper-crustal rocks. The data of Brudy *et al.* (1997) for the state of stress in the German Deep Continental Borehole (KTB) have been transformed to the same coordinate frame to show their consistency with Byerlee's generalization.

deformed material. After initiation, this configuration is mechanically stable and is thought to allow rock systems to deform over long time scales by heterogeneous, but effectively steady-state flow in which the strain response of the entire system will be controlled by the kinematic behaviour of the interconnected fault- or shear-zone network. Therefore, in mature fault- or shear-zone systems, it is likely that the rheological properties and evolution of the fault rocks in the interconnected, highest strain fault- or shear-zone strands will ultimately control the behaviour of the whole system (Handy 1990).

Fault rock rheology in the brittle–frictional regime

Byerlee's 'rule' and intraplate faulting

Our understanding of brittle, upper-crustal faulting is largely based on the results of laboratory faulting and friction experiments carried out on a great many rock types. These show that frictional strength increases rather rapidly with effective normal stress, at a rate corresponding to a friction coefficient between about 0.5 and 0.9. Byerlee (1978) pointed out that, to a first approximation, the friction coefficient is not strongly dependent on rock type, so that for simple modelling of crustal fault behaviour, we have a useful rule of thumb that friction is independent of rock type (Byerlee's 'rule'). Although this is a useful approximation, it must be remembered that the frictional strength of various rocks may in fact vary substantially (note the spread implied by the low-pressure data in Fig. 4; there are relatively fewer high-stress data). However, a range of observed geometrical relationships or natural fault structures are broadly consistent with Byerlee friction (Sibson 1994). The utility of the approximation is enhanced by the fact that, whereas cataclastic–frictional deformation is very sensitive to variations in total pressure or pore pressure, it is much less sensitive to variations in temperature, at least up to *c.* 300 °C (Fig. 1; Blanpied *et al.* 1998). Because of the trade-off between deformation rate and temperature that exists for most materials, frictional behaviour is also relatively insensitive to large changes in strain rate. Thus it has become common to discuss the strength of the upper crust in terms of a generalized friction law for all rock types, independent of temperature and strain rate. This approach has been used successfully to account for borehole stress measurements made in intraplate regions in a variety of tectonic stress regimes (Townend & Zoback 2000).

There are a number of instances in which it has been possible to test Byerlee's 'rule' in present-day intraplate localities (e.g. the German Deep Continental Borehole, KTB). Great care was taken to acquire a dataset that would describe the stress difference down to a depth of *c.* 9 km (Fig. 4; Brudy *et al.* 1997). These data show that *in situ* stresses are indeed consistent with laboratory friction data. The long-term persistence of unrelaxed, near-sliding stresses also supports the idea that upper-crustal rheology is almost rate independent. When *in situ* stresses are consistent with laboratory friction data, we would say that the crust is relatively strong, or is as strong as it can be. For such a 'strong' crust, the orientation of potentially active faults should be *c.* 30–45° to the local orientation of maximum principal stress, so that high values of resolved shear stress act along the fault. By comparison with laboratory data again, we would expect that the stresses necessary to initiate a new fault, rather than inducing continued slip on an old one, would be even higher. It should be noted, however, that the regional principal stress difference required to cause slip on unfavourably oriented faults (e.g. low-angle detachments where σ_1 stress trajectories are vertical) would also be greater than the resistance to the formation of a fresh fault, for a range of unfavourable fault orientations.

Seismological data tend to yield relatively small stress drops associated with earthquakes ($\Delta\tau$<10 MPa in almost all cases). If this were to represent total stress drop on faults, it would imply that fault strength is much lower than implied by Byerlee friction under hydrostatic fluid-pressure conditions. However, laboratory data again showed that, provided seismogenic slip is analogous to the stick–slip behaviour observed in laboratory experiments, the seismogenic stress drop is only a small (*c.* 10%) fraction of the total shear stress on the fault (e.g. McGarr 1999).

The San Andreas plate boundary fault

In contrast to the cosy image of fault behaviour in intraplate situations, two key observations from the San Andreas plate boundary fault suggest that, even in the shallow elastic frictional regime where deformation can be seismogenic, this fault may be anomously weak, both in a relative and in an absolute sense. This begs the question of whether plate boundary faults, which must cut right through the lithosphere in a fairly narrow zone, are fundamentally different in some way from smaller intraplate faults. First, heat-flow data progressively accumulated since the late 1960s (Lachenbruch & Sass 1973; Lachenbruch & Sass, 1980, Lachenbruch & Sass, 1992) seem to suggest that slip in the San Andreas fault is not generating frictional heat at a rate that would be expected of a fault displaying 'Byerlee' friction under hydrostatic fluid-pressure conditions. Second, estimates of the orientation of the maximum principal stress in the country rocks on either side of the San Andreas fault suggest that it lies nearly normal to the fault trace (Mount & Suppe 1987; Zoback *et al.* 1987), implying very low values of resolved shear stress along the fault, even if the differential stresses in the crust around the fault are consistent with Byerlee friction.

A range of explanations have been proposed to account for the inferred weakness of the San Andreas fault. These include the presence of supposed low-friction materials (e.g. smectite-bearing fault gouge or certain serpentine minerals) in the fault zone (Morrow *et al.* 1992; Moore *et al.* 1996), maintenance of high fluid pressure in the fault-zone core (Byerlee 1990; Rice 1992; Chester *et al.* 1993) relative to lower pressures outside, or a range of dynamic mechanisms (e.g. frictional melting, thermal fluid pressurization) for the production of transiently low friction during slip events (e.g. Sibson 1980; Melosh 1996). All of these explanations pose difficulties that must be addressed. Furthermore, Scholz (2000) has now called into question the validity of the interpretations of the data that suggest either that there is no heat flow anomaly or that the maximum principal stress is at a high angle to the fault trace, and instead proposes that the San Andreas fault is not anomalously weak at all. This point of view is disputed by Zoback (2000), but it is important that such a level of scrutiny be applied to the San Andreas problem, because it underlines our present lack of clear understanding of whether there are fundamental differences in crustal mechanics in different plate tectonic settings. It is worth noting, however, that there is also no evidence for frictional heat generation along the seismic thrust interface at fast subduction zones, a region responsible for >90% of the global seismic moment release; this restricts shear stresses on these thrusts to <20 MPa (e.g. see Wang *et al.* 1995).

Fluids, faults and detachments

Natural fault zones exhibit complex internal architectures composed of clusters and networks of small faults and fractures surrounding larger slip surfaces (Engelder 1974; Knipe *et al.* 1998). The behaviour of fluids and strength characteristics in these zones depend upon the distribution, evolution and connectivity of the fault rocks with different properties in the network (Caine & Foster 1999; Haneberg *et al.*, 1999). Sealed volumes bounded by faults and complex baffles between pressure compartments are possible within the fault zone (Byerlee 1993). Evaluation of the dynamic flow behaviour of these networks is important to mineral and hydrocarbon reserves and the associated operating plumbing systems. This has an impact on the effective stress and therefore on the strength behaviour of the faults (Wong & Zhu 1999). The important contribution made by low-displacement fault or fracture networks that act as conduits for fluid flow is increasingly recognized (Sibson 1996, Sibson, 2000), but their role is still poorly constrained for many deformation environments. Faulted geothermal, mineral and hydrocarbon systems provide a rich source of data for understanding fault behaviour at comparatively shallow crustal depths.

The fault zones that underlie accretionary prisms developed at convergent plate margins provide another example of an apparently weak fault zone able to concentrate deformation (Moore *et al.* 1995). In some cases, these are associated with fluid overpressures and focused fluid flow; in other cases they are not. These fault zones provide an opportunity for evaluating complex fluid flow and mechanical behaviour where concentrated slip takes place. An understanding of such zones needs to incorporate knowledge concerning the deformation behaviour of weak to poorly lithified granular aggregates (Jones & Addis 1985; Jones 1994; Bolton & Maltman 1997). The International Ocean Drilling Programme (IODP) is poised to provide new information on these fault zones via a new initiative for deep drilling of the active margins from a specially designed research ship.

The issue of the orientation of major extensional dip-slip faults relative to the orientation of principal stresses pertains particularly to the problem of understanding relatively flat-lying detachment faults. These are seen in exhumed orogenic belts, in extended continental terranes (e.g. Lister & Davis 1989) and also adjacent to the inside corners of some ridge–transform boundaries in oceanic settings (e.g. Cann *et al.* 1997). In continental settings, the commonly observed juxtaposition of brittle deformation in the hanging wall with mylonitic rocks in the footwall seems to imply large displacements. Present-day seismicity patterns in regions of continental extension are dominated by high-angle normal faulting (i.e. dips >30°; e.g. Jackson & White 1989), so are flat-lying detachment faults naturally aseismic? Does their very existence and orientation imply that they are anomalously weak in the same way as inferred for the San Andreas fault or do they form and reactivate under highly anomalous stress trajectory systems?

Fracture mechanisms and subcritical crack growth

A range of classical grain-scale fracture mechanisms have been identified based on the results of laboratory deformation experiments carried out on crystalline materials (e.g. Atkinson 1982). This approach generally assumes, however, that cracks propagate at a critical value of a parameter such as the stress intensity factor (K_c, or fracture toughness), which requires essentially linear elastic behaviour. It is increasingly apparent, however, that such assumptions are not appropriate for most geological faults where loading occurs over long time periods, often at elevated temperatures and in the presence of chemically active fluids. This is thought to give rise to a range of time-dependent behaviours in fracture process zones around crack tips known collectively as *subcritical crack growth*, in which slow fracture occurs when $K \ll K_c$ (Atkinson 1982, Atkinson, 1987). The mechanisms and controls of this behaviour (long recognized in ceramics and glasses) are still very poorly understood in geological systems, but they are very likely to influence strongly long-term crustal shear stress levels. The recent recognition of significant amounts of grain-scale low-temperature crystal plasticity and diffusion mass transfer adjacent to shallow crustal brittle faults (e.g. Lloyd & Knipe 1992) reinforces the view that subcritical crack growth is likely to be a major influence upon both the growth and strength of natural fault zones.

Fault rock rheology in the plastic–viscous regime

Plastic deformation that demands elevated temperatures may localize into shear zones ranging in thickness from a few millimetres to kilometre scale. Despite the general increase in deformability that accompanies elevated temperature, even the rocks of the lower continental crust and upper mantle can be carried about as large, relatively undeformed blocks between localized shear zones (e.g. Rutter & Brodie 1992). Thus, slip on interconnected, localized zones appears to be the dominant process during lithosphere-scale deformation, even in continental orogens. In many cases, major shear zones, once localized, appear to be relatively weak features, as they undergo repeated episodes of reactivation during successive cycles of crustal deformation, particularly in the continents where the crust persists over aeons owing to its resistance to being subducted (Butler *et al.* 1997; Holdsworth *et al.* 1997, Holdsworth *et al.*, 2001). Many shear zones also act as fluid channelways, localizing hydrothermal or magmatic activity, and these factors may additionally contribute to their weakness relative to their surroundings.

A number of mechanisms can be identified whereby, in rocks at high temperature, flow localization can be made stable (i.e. not spread laterally into the protolith). Several such processes have been identified through experimental studies, or have been inferred to operate from textural and petrographic studies on naturally deformed rocks. These processes are as follows.

(1) *Metamorphic overpressure and embrittlement.* Dehydration reactions during prograde metamorphism evolve high-pressure water, which can cause weakening and embrittlement if undrained. This has been demonstrated experimentally in several systems (e.g. Raleigh & Paterson 1965; Heard & Rubey 1966; Ko *et al.* 1995). If the system is drained, then the availability of collapsible porosity created during reaction can be expected to produce transient weakening during the progress of the reaction (Rutter & Brodie 1995). This latter effect has not been documented in nature, because crystallization of product phases repairs the damage done during shearing. It has also not yet been demonstrated in the laboratory, owing to the difficulties of performing the requisite experiments.

(2) Geometric softening. Intracrystalline plastic flow generally leads to the formation of crystallographic preferred orientation. If 'easy slip' orientations become aligned with the shear plane, a relative weakening may result (Schmid *et al.* 1987).

(3) Grain-size reduction and the onset of diffusional mechanisms. High-strain zones are often characterized by dramatic tectonic grain-size reduction. Sufficient grain refinement may favour a switch in deformation mechanism to a grain-size-sensitive, diffusion-dominated process, particularly if a hydrous fluid phase is present (e.g. Brodie & Rutter 1987; Hoogerduijn Strating & Vissers 1991). If this is characterized by a marked sensitivity of flow stress to strain rate, weakening will result. Grain refinement may arise from extreme cataclasis, dynamic recrystallization, or neocrystallization of at least transiently fine-grained products of a metamorphic reaction or polymorphic transformation. The last process has been invoked to explain the occurrence of deep focus (400–670 km depth) earthquakes. This depth range coincides with that necessary for the transformation of orthorhombic olivine to a denser spinel structure (Burnley *et al.* 1991; Kirby *et al.* 1996).

(4) Cyclic dynamic recrystallization. This may result in weakening through some combination of the effects of restoration of a low dislocation density and the formation of preferred crystallographic orientation (e.g. in feldspars, which work harden very rapidly; Tullis & Yund 1985). The sweeping of grain boundaries through the volume of a rock may cause the defect chemistry of grain interiors to equilibrate with the pore fluid much more rapidly than solid-state diffusion will allow, perhaps facilitating hydrolytic weakening in minerals such as quartz (Rutter & Brodie 1995).

Much of the above understanding of plastic–viscous weakening processes comes from before-and-after studies of naturally sheared rocks and their adjacent protoliths, or from experimental deformation of either undeformed rocks or mature fault rocks. The evolution of strength through the microstructural and metamorphic changes that accompany localization have been difficult to track, owing to the large strain ranges over which they can occur. Little has yet been done in the direction of including a strain-dependent term in constitutive flow laws. However, recent applications of high-strain extension testing and torsion testing offer the promise of being able to follow strength evolution over large changes in finite strain (Rutter 1998, Rutter, 1999).

Conclusions

Understanding the behaviour of localized faults and shear zones at all depths (and orientations) in the lithosphere is clearly a central issue in tectonics, yet our understanding remains rather piecemeal. A combination of mechanical experiments, field-based and microstructural study of fault rocks, analogue and numerical modelling, and seismological studies has resulted in a basic understanding of brittle faulting in the upper crust of the Earth. Here the stress state is limited by the frictional strength of networks of faults under the prevailing fluid-pressure conditions. The behaviour of brittle fault rocks and the structure of fault zones must largely determine their ability to produce earthquakes, yet our capability to predict such events is limited. This may in part reflect our poor understanding of the behaviour of faults and stress under long-term loading conditions, where grain-scale mechanisms related to subcritical crack growth in fracture process zones may play a major role in controlling both fault growth and strength. Radical new directions in research are planned and include the San Andreas Fault Observatory at Depth (SAFOD), the establishment of which is contingent upon the funding of a deep drilling programme (Dalton 1999). Such studies are essential if we are to establish whether plate boundary faults really can be weaker than their intraplate counterparts and, if so, by what means: trapped fluid overpressures or faults possessing anomalous (non-'Byerlee') friction properties or some other process(es).

Our understanding of the mechanics of deeper-crustal, higher-temperature shear zones is built around before-and-after observations of the sheared rock and its protolith, whether on naturally or experimentally deformed samples. Highly strained fault rocks produced in the plastic–viscous regime, and in the deeper parts of the frictional regime, are often characterized by marked grain-size reduction. The results of studies on naturally produced fault or shear zones and from rock deformation experiments have identified a number of processes that cause grain refinement, and recognize that these can lead to relative weakening and hence favour the observed localization of flow and, following periods of quiescence, reactivation. For the future, few attempts have yet been made experimentally to study the evolution of mechanical properties and microstructure over large ranges of strain, not least owing to the difficulties of carrying out such experiments. Our understanding of deformation when it is accompanied by metamorphic reactions, so evident from microstructural study of naturally deformed rocks, remains far from perfect. In natural fault rocks, much syntectonic damage is thought to be repaired by post-tectonic microstructural readjustments while the rock remains at high temperature. Crucially, the mechanism(s) of enhanced hydraulic conductivity of high-temperature shear zones, long clear from field evidence, remains largely in the area of speculation, and will be a key focus and challenge for future experimental work.

It may be particularly important to examine the nature and role of weakening mechanisms in faults and shear zones where they cut through the main load-bearing regions of the lithosphere (Holdsworth *et al.* 2001). In continental regions, most experimentally derived rheological strength profiles predict that these regions lie within the upper mantle and/or the mid–upper crust close to the frictional–viscous transition (Molnar 1992; Kohlstedt *et al.*, 1995). It is uncertain whether such models are even approximately correct in both intraplate and plate boundary settings. Nevertheless, some of the apparent differences between faults in these settings could also be a function of differences in the location, nature and importance of the main load-bearing regions in the lithosphere.

Finally, it is important to emphasize that most discussion and modelling of large-scale lithosphere rheology is based on presumed 'steady-state' flow of a lithosphere that deforms homogeneously, a view that is incompatible with the heterogeneous deformation observed on all scales by geologists in nature. The development of a better understanding of how fault- or shear-zone linkages, strength and microstructure evolve over large changes in finite strain or displacement is a vital next step if we are to model the behaviour of lithosphere where deformation in concentrated into narrow, weaker zones.

The authors would like to thank all the participants of the London conference for their inputs and discussion during the meeting, particularly M. Handy, who wrote an excellent summary of the proceedings. We would also like to thank R. Sibson for his insightful review of this manuscript, although the authors remain responsible for the views expressed here.

References

ATKINSON, B.K. 1982. Subcritical crack-propagation in rocks: theory, experimental results and applications. *Journal of Structural Geology*, **4**, 41–56.

ATKINSON, B.K. *Fracture Mechanics of Rock*. Academic Press, London.

BLANPIED, M.L., MARONE, C.J., LOCKNER, D.A., BYERLEE, J.D. & KING, D.P. 1998. Quantitative measure of the variation in fault rheology due to fluid–rock interactions. *Journal of Geophysical Research*, **103**, 9691–9712.

BOLTON, A. & MALTMAN, A.J. 1997. Fluid flow pathways in actively deforming sediments, the role of pore fluid pressure and volume change. *In*: HENDRY, J., CAREY, P., PARNELL, J., RUFFELL, A. & WORDEN, R. (eds) *Geofluids II*. Queen's University, Belfast, 356–359.

BRODIE, K.H. & RUTTER, E.H. 1987. The role of transiently fine grained reaction products in syntectonic metamorphism: natural and experimental examples. *Canadian Journal of Earth Sciences*, **24**, 554–564.

BRUDY, M., ZOBACK, M.D., FUCHS, K., RUMMEL, F. & BAUMGARTNER, J. 1997. Estimation of the complete stress tensor to 8 km depth in the KTB scientific drill holes; implications for crustal strength. *Journal of Geophysical Research*, **102**, 18453–18475.

BRUHN, D., GROEBNER, N. & KOHLSTEDT, D.L. 2000. An interconnected network of core-forming melts produced by shear deformation. *Nature*, **403**, 883–886.

BURNLEY, P.C., GREEN, H.W. & PRIOR, D.J. 1991. Faulting associated with the olivine to spinel phase transformation in Mg_2GeSiO_4 and its implications for deep focus earthquakes. *Journal of Geophysical Research*, **96**, 425–443.

BUTLER, R.W.H., HOLDSWORTH, R.E. & LLOYD, G.E. 1997. The role of basement reactivation in continental deformation. *Journal of the Geological Society, London*, **154**, 69–71.

BYERLEE, J.D. 1978. Friction of rocks. *Pure and Applied Geophysics*, **116**, 615–626.

BYERLEE, J.D. 1990. Friction, overpressure and fault-normal compression. *Geophysical Research Letters*, **17**, 2109–2112.

BYERLEE, J.D. 1993. Model for episodic flow of high pressure water in fault zones before earthquakes. *Geology*, **21**, 303–306.

CAINE, J.S. & FOSTER, C.B. 1999. Fault zone architecture and fluid flow: insights from field data and numerical modelling. *In*: HANEBERG, W.C., MOZLEY, P.S., MOORE, J.C. & GOODWIN, L.B. (eds) *Faults and Subsurface Fluid Flow in the Shallow Crust*. Geophysical Monograph, American Geophysical Union, **113**, 101–127.

CANN, J.R., BLACKMAN, D.K., SMITH, D.K. & 6 OTHERS 1997. Corrugated slip surfaces formed at ridge–transform intersections on the Mid-Atlantic Ridge. *Nature*, **385**, 329–332..

CHESTER, F.M., EVANS, J.P. & BIEGEL, R.L. 1993. Internal structure and weakening mechanisms of the San Andreas fault. *Journal of Geophysical Research*, **98**, 771–786.

COBBOLD, P.R. 1977. Description and origin of banded deformation structures, II Rheology and growth of banded perturbations. *Canadian Journal of Earth Sciences*, **14**, 2510–2523.

COWIE, P. & SCHOLZ, C.H. 1992. Growth of faults by accumulation of seismic slip. *Journal of Geophysical Research*, **97**, 11085–11095.

DALTON, R. 1999. Go-ahead for San Andreas drilling project. *Nature*, **401**, 5.

ENGELDER, J.T. 1974. Cataclasis and the generation of fault gouge. *Geological Society of America Bulletin*, **85**, 1515–1522.

GRIGGS, D.T. & HANDIN, J. 1960. Observations on fracture and a hypothesis of earthquakes. *In*: GRIGGS, D. & HANDIN, J. (eds) *Rock Deformation*. Geological Society of America, Memoirs, **79**, 347–364.

HANDY, M.R. 1989. Deformation regimes and the rheological evolution of fault zones in the lithosphere: the effects of pressure, temperature, grainsize and time. *Tectonophysics*, **163**, 119–152.

HANDY, M.R. 1990. The solid-state flow of polymineralic rocks. *Journal of Geophysical Research*, **95**, 8647–8661.

HANDY, M.R. 1994. The energetics of steady state heterogeneous shear in mylonitic rock. *Material Science and Engineering*, **A175**, 261–272.

HANDY, M.R., MULCH, R., ROSENAU, M. & ROSENBERG, C.R. 2001. The role of transcurrent shear zones as melt conduits and reactors and as agents of weakening in the continental crust. *In*: HOLDSWORTH, R.E., STRACHAN, R.A., MAGLOUGHLIN, J.F. & KNIPE, R.J. (eds) *The Nature and Significance of Fault Zone Weakening*. Geological Society, London, Special Publications **186**, 303–330.

HANEBERG, W.C., MOZLEY, P.S., MOORE, J.C. & GOODWIN, L.B. (eds) *Faults and Subsurface Fluid Flow in the Shallow Crust*. Geophysical Monograph, American Geophysical Union **113**.

HEARD, H.C. & RUBEY, W.W. 1966. Tectonic implications of gypsum dehydration. *Geological Society of America Bulletin*, **77**, 741–760.

HOBBS, B.E., MÜHLHAUS, H.-B. & ORD, A. 1990. Instability, softening and localization of deformation. *In*: KNIPE, R.J. & RUTTER, E.H. (eds) *Deformation Mechanisms, Rheology and Tectonics*. Geological Society, London, Special Publications, **54**, 143–165.

HOLDSWORTH, R.E., BUTLER, C.A. & ROBERTS, A.M. 1997. The recognition of reactivation during continental deformation. *Journal of the Geological Society, London*, **154**, 73–78.

HOLDSWORTH, R.E., STEWART, M., IMBER, J. & STRACHAN, R.A. 2001. The structure and rheological evolution of reactivated continental fault zones: a review and case study. *In*: MILLER, J.A., HOLDSWORTH, R.E., BUICK, I.S. & HAND, M. (eds) *Continental Reactivation and Reworking*. Geological Society, London, Special Publications **184**, 115–137.

HOOGERDUIJN STRATING, E.H. & VISSERS, R.L.M. 1991. Dehydration-induced fracturing of eclogite-facies peridotites, implications for the mechanical behaviour of subducting lithosphere. *Tectonophysics*, **200**, 187–198.

IMBER, J., HOLDSWORTH, R.E., BUTLER, C.A. & LLOYD, G.E. 1997. Fault zone weakening processes along the reactivated Outer Hebrides Fault Zone, Scotland. *Journal of the Geological Society, London*, **154**, 105–109.

JACKSON, J.A. & WHITE, N.J. 1989. Normal faulting in the upper continental crust: observations from regions of active extension. *Journal of Structural Geology*, **11**, 15–36.

JONES, M.E. 1994. Mechanical principles of sediment deformation. *In*: MALTMAN, A.J. (ed.) *Geological Deformation of Sediments*. Chapman & Hall, London, 37–71.

JONES, M.E. & ADDIS, M.A. 1985. The application of stress path and critical state to the analysis of sediment deformation. *Journal of Structural Geology*, **8**, 575–580.

KIRBY, S.H., STEIN, S., OKAL, E.A. & RUBIE, D.H. 1996. Metastable mantle phase transformations and deep earthquakes in subducting oceanic lithosphere. *Reviews of Geophysics*, **34**, 261–306.

KNIPE, R.J. 1989. Deformation mechanisms—recognition from natural tectonites. *Journal of Structural Geology*, **11**, 127–146.

KNIPE, R.J., JONES, G. & FISHER, Q.J. 1998. Faulting, fault seal and fluid flow in hydrocarbon reservoirs: an introduction. *In*: JONES, G., FISHER, Q.J. & KNIPE, R.J. (eds) *Faulting, Fault Sealing and Fluid Flow*. Geological Society, London, Special Publications **147**, vii–xxi.

KO, S.-C., OLGAARD, D.L. & BRIEGEL, U. 1995. The transition from weakening to strengthening in dehydrating gypsum: evolution of excess pore pressures. *Geophysical Research Letters*, **22**, 1009–1012.

KOHLSTEDT, D.L., EVANS, B. & MACKWELL, S.J. 1995. Strength of the lithosphere: constraints imposed by laboratory experiments. *Journal of Geophysical Research*, **100**, 17587–17602.

LACHENBRUCH, A.H. & SASS, J.H. 1973. Thermo-mechanical aspects of the San Andreas. *In*: NUR, A. (ed.) *Proceedings of the Conference on the Tectonic Problems of the San Andreas Fault System*. Stanford University Press, Palo Alto, CA, 192–205.

LACHENBRUCH, A.H. & SASS, J.H. 1980. Heat flow and energetics of the San Andreas fault zone. *Journal of Geophysical Research*, **85**, 6185–6222.

LACHENBRUCH, A.H. & SASS, J.H. 1992. Heat flow from Cajon Pass, fault strength and tectonic implications. *Journal of Geophysical Research*, **97**, 4995–5015.

LISTER, G.S. & DAVIS, G.R. 1989. The origin of metamorphic core complexes and detachment faults formed during Tertiary continental extension in the northern Colorado River region, U.S.A. *Journal of Structural Geology*, **11**, 65–94.

LLOYD, G.E. & KNIPE, R.J. 1992. Deformation mechanisms accommodating faulting of quartzite under upper crustal conditions. *Journal of Structural Geology*, **14**, 127–143.

MCGARR, A. 1999. On relating apparent stress to the stress causing earthquake fault slip. *Journal of Geophysical Research*, **104**, 3003–3011.

MELOSH, H.J. 1996. Dynamical weakening of faults by acoustic fluidization. *Nature*, **379**, 601–606.

MOLNAR, P. 1992. Brace–Goetze strength profiles, the partitioning of strike-slip and thrust faulting at zones of oblique convergence and the stress–heat flow paradox of the San Andreas Fault. *In*: EVANS, B., WONG, T.-F. (eds) *Fault Mechanics and Transport Properties in Rocks*. Academic Press, London, 359–435.

MOORE, D.E., LOCKNER, D.A., SUMMERS, R., SHENGLI, M. & BYERLEE, J.D. 1996. Strength of chrysotile-serpentinite gouge under hydrothermal conditions: can it explain a weak San Andreas fault? *Geology*, **24**, 1041–1044.

MOORE, J.C., SHIPLEY, T.H., GOLDBERG, D. & 25 OTHERS 1995. Abnormal fluid pressures and fault zone dilation in the Barbados accretionary prism: evidence from logging while drilling. *Geology*, **23**, 605–608.

MORROW, C., RADLEY, B. & BYERLEE, J.D. 1992. Frictional strength and the effective pressure law for montmorillonite and illite clays. *In*: EVANS, B. & WONG, T.-F. (eds) *Fault Mechanics and Transport Properties of Rocks*. Academic Press, London, 69–88.

MOUNT, V.S. & SUPPE, J. 1987. State of stress near the San Andreas fault—implications for wrench tectonics. *Geology*, **15**, 1143–1146.

PATERSON, M.S. 1978. *Experimental Rock Deformation—The Brittle Field*. Springer, Berlin.

RALEIGH, C.B. & PATERSON, M.S. 1965. Experimental deformation of serpentinite and its tectonic implications. *Journal of Geophysical Research*, **70**, 3965–3985.

RICE, J.R. 1992. Fault stress states, pore pressure distributions and the weakness of the San Andreas fault. *In*: EVANS, B. & WONG, T.-F. (eds) *Fault Mechanics and Transport Properties of Rocks*. Academic Press, London, 475–504.

RUTTER, E.H. 1998. Use of extension testing to investigate the influence of finite strain on the rheological behaviour of marble. *Journal of Structural Geology*, **20**, 243–254.

RUTTER, E.H. 1999. On the relationship between the formation of shear zones and the form of the flow law for rocks undergoing dynamic recrystallization. *Tectonophysics*, **303**, 147–158.

RUTTER, E.H. & BRODIE, K.H. 1992. The rheology of the lower continental crust. *In*: FOUNTAIN, D., ARCULUS, R. & KAY, R. (eds) *Continental Lower Crust, Developments in Geodynamics*. Elsevier, Amsterdam, 201–268.

RUTTER, E.H. & BRODIE, K.H. 1995. Mechanistic interactions between deformation and metamorphism. *Geological Journal*, **30**, 27–240.

SCHMID, S.M. & HANDY, M.R. 1991. Towards a genetic classification of fault rocks: geological usage and tectonophysical implications. *In*: MÜLLER, D.W., MCKENZIE, J.A. & WEISSERT, H. (eds) *Controversies in Modern Geology*. Academic Press, London, 339–361.

SCHMID, S.M., PANOZZO, R. & BAUER, S. 1987. Simple shear experiments on calcite rocks: rheology and microfabric. *Journal of Structural Geology*, **9**, 747–778.

SCHOLZ, C.H. *Mechanics of Earthquakes and Faulting*. Cambridge University Press, Cambridge.

SCHOLZ, C.H. 1998. Earthquakes and friction laws. *Nature*, **391**, 37–42.

SCHOLZ, C.H. 2000. Evidence for a strong San Andreas fault. *Geology*, **28**, 163–166.

SIBSON, R.H. 1977. Fault rocks and fault mechanisms. *Journal of the Geological Society, London*, **133**, 191–213.

SIBSON, R.H. 1980. Power dissipation and stress levels on faults in the upper crust. *Journal of Geophysical Research*, **85**, 6239–6247.

SIBSON, R.H. 1983. Continental fault structure and the shallow earthquake source. *Journal of the Geological Society, London*, **140**, 741–767.

SIBSON, R.H. 1994. An assessment of field evidence for Byerlee friction. *Pure and Applied Geophysics*, **142**, 645–662.

SIBSON, R.H. 1996. Structural permeability of fluid-driven fault–fracture meshes. *Journal of the Structural Geology*, **18**, 655–660.

SIBSON, R.H. 2000. Fluid involvement in normal faulting. *Journal of Geodynamics*, **29**, 469–499.

SNOKE, A.W., TULLIS, J. & TODD, V.R. 1998. *Fault-related Rocks: a Photographic Atlas*. Princeton University Press, Princeton, NJ.

SORNETTE, D., DAVY, P. & SORNETTE, A. 1990. Structuration of the lithosphere in plate tectonics as a self-organised critical phenomenon. *Journal of Geophysical Research*, **95**, 17353–17361.

SORNETTE, A., DAVY, P. & SORNETTE, D. 1993. Fault growth in brittle–ductile experiments and the mechanics of continental collisions. *Journal of Geophysical Research*, **98**, 12111–12139.

STEWART, M., HOLDSWORTH, R.E. & STRACHAN, R.A. 2000. Deformation processes and weakening mechanisms within the frictional–viscous transition zone of major crustal-scale faults: insights from the Great Glen Fault Zone, Scotland. *Journal of Structural Geology*, **22**, 543–560.

TOWNEND, J. & ZOBACK, M.D. 2000. How faulting keeps the crust strong. *Geology*, **28**, 399–402.

TULLIS, J. & YUND, R.A. 1977. Experimental deformation of dry Westerly granite. *Journal of Geophysical Research*, **82**, 5705–5718.

TULLIS, J. & YUND, R.A. 1985. Dynamic recrystallization of feldspar: a mechanism for ductile shear zone formation. *Geology*, **13**, 238–241.

WALSH, J.J., CHILDS, C., MEYER, V., & 9 OTHERS, 2001. Geometric controls on the evolution of normal fault systems. *In*: HOLDSWORTH, R.E., STRACHAN, R.A., MAGLOUGHLIN, J.F. & KNIPE, R.J. (eds) *The Nature and Significance of Fault Zone Weakening*. Geological Society, London, Special Publications, **186**, 157–170.

WANG, K., MULDER, T., ROGERS, G.C. & HYNDMAN, R.D. 1995. Case for very low coupling stress on the Cascadia subduction fault. *Journal of Geophysical Research*, **100**, 12907–12918.

WONG, T.-F. & ZHU, W. 1999. Brittle faulting and permeability evolution: hydromechanical measurement, microstructural observations and network modeling. *In*: HANEBERG, W.C., MOZLEY, P.S., MOORE, J.C. & GOODWIN, L.B. (eds) *Faults and Subsurface Fluid Flow in the Shallow Crust*. Geophysical Monograph, American Geophysical Union **113**, 88–99.

ZOBACK, M.D. 2000. Strength of the San Andreas. *Nature*, **405**, 31–32.

ZOBACK, M.D., ZOBACK, M.L., MOUNT, V.S. & 10 OTHERS 1987. New evidence on the state of stress on the San Andreas fault system. *Science*, **238**, 1105–1111.

Implications of earthquake focal mechanisms for the frictional strength of the San Andreas fault system

JOHN TOWNEND & MARK D. ZOBACK

Department of Geophysics, Stanford University, Stanford, CA 94305-2215, USA

(e-mail: jtownend@stanford.edu)

Abstract: Analysis of stress orientation data from earthquake focal plane mechanisms adjacent to the San Andreas fault in the San Francisco Bay area and throughout southern California indicates that the San Andreas fault has low frictional strength. In both regions, available stress orientation data indicate low levels of shear stress on planes parallel to the San Andreas fault. In the San Francisco Bay area, focal plane mechanisms from within 5 km of the San Andreas and Calaveras fault zones indicate a direction of maximum horizontal compression nearly orthogonal to both subvertical, right-lateral strike-slip faults, a result consistent with those obtained previously from studies of aftershocks of the 1989 Loma Prieta earthquake. In southern California, the direction of maximum horizontal stress near the San Andreas fault is nearly everywhere at a high angle to it, similarly indicating that the fault has low frictional strength. Thus, along these two major sections of the San Andreas fault (which produced great earthquakes in southern California in 1857 and central and northern California in 1906), the frictional strength of the fault is much lower than expected for virtually any common rock type if near-hydrostatic pore pressure exists at depth, and so low as to produce no discernible shear-heating anomaly.

Our findings in southern California are in marked contrast to recent suggestions by Hardebeck & Hauksson that stress orientations rotate systematically within *c.* 25 km of the fault, which prompted a high frictional strength model of the San Andreas fault. As we utilize the same stress data and inversion technique as Hardebeck & Hauksson, we interpret the difference in our findings as being related to the way in which we group focal plane mechanisms to find the best-fitting stress tensor. We suggest that the Hardebeck & Hauksson gridding scheme may not be consistent with the requisite *a priori* assumption of stress homogeneity for each set of earthquakes.

Finally, we find no evidence of regional stress changes associated with the occurrence of the 1992 M7.4 Landers earthquake, again in apparent contradiction with the findings of Hardebeck & Hauksson.

Although it is well known that the San Andreas fault (SAF) system (Fig. 1) has been the locus of prolonged, localized deformation in the lithosphere separating the Pacific and North American plates for several million years, the level of shear stress required to cause the major faults of the San Andreas system to slip during earthquakes in the seismogenic upper *c.* 15 km of the crust remains controversial (Lachenbruch & McGarr 1990). Specifically, although laboratory experiments systematically reveal coefficients of friction of between 0.6 and 1.0 (Byerlee 1978), consistent with stress levels measured to depths of several kilometres in intraplate regions (Townend & Zoback 2000) two lines of evidence suggest that coefficients of friction, μ, along the major faults of the San Andreas system are substantially lower. First, heat flow data reveal that the surface trace of the SAF is not associated with the marked, positive thermal anomaly expected if the coefficient of friction were any higher than 0.2 (Brune *et al.* 1969; Lachenbruch & Sass 1980, 1992; Lachenbruch & McGarr 1990). Second, measurements of principal stress directions along the SAF suggest that the angle between the greatest compressive stress direction and the fault plane almost invariably exceeds the directions expected for $0.6 \leqslant \mu \leqslant 1.0$ ('hydrostatic Byerlee's law'; Brace & Kohlstedt 1980; Townend & Zoback 2000) and, in fact, locally exceeds 80° at several locations (Mount & Suppe 1987; Zoback *et al.* 1987; Jones 1988; Oppenheimer *et al.* 1988; Zoback & Beroza 1993). Such high angles of compression imply friction coefficients of <0.1 (Lachenbruch &

From: HOLDSWORTH, R.E., STRACHAN, R.A., MAGLOUGHLIN, J.F. & KNIPE, R.J. (eds) 2001. *The Nature and Tectonic Significance of Fault Zone Weakening*. Geological Society, London, Special Publications, **186**, 13–21. 0305-8719/01/$15.00

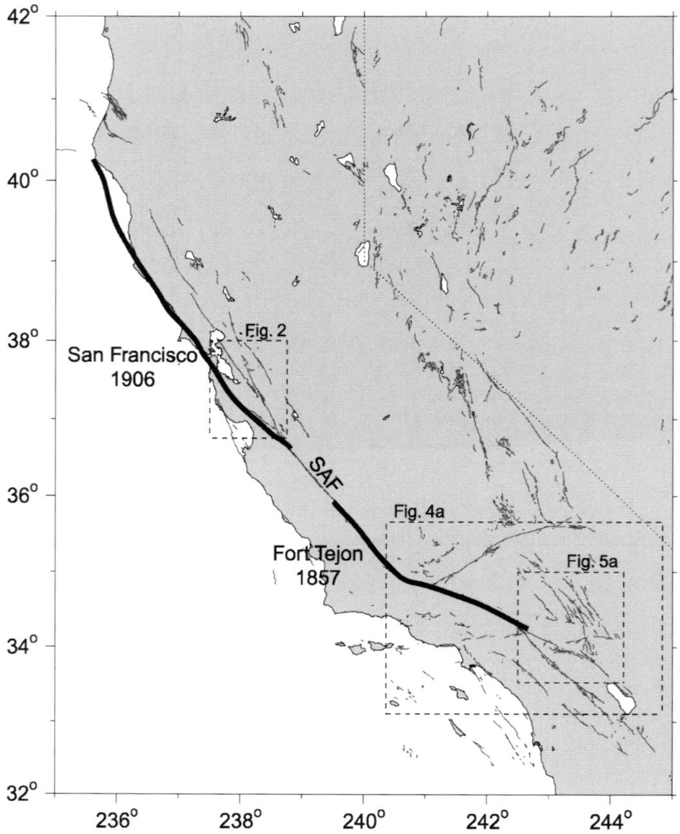

Fig. 1. Map of California illustrating the San Andreas fault (SAF) and other major faults. The thick lines along sections of the San Andreas fault indicate those portions of the fault that ruptured during the 1857 Fort Tejon and 1906 San Francisco earthquakes. The areas corresponding to Figs. 2, 4a and 5a are shown by the dashed boxes.

Sass 1992) or high fluid pressures within the fault zone (Rice 1992).

Hardebeck & Hauksson (1999) recently suggested that whereas the far-field axis of greatest horizontal compression, $S_{H\max}$, is at a high angle to the average strike of the SAF throughout much of southern California, the stress tensor rotates systematically adjacent to the fault so that the relevant near-field angle is *c.* 45°. Further, it was suggested that this behaviour is particularly pronounced in the vicinity of the 'Big Bend' area near Fort Tejon. This led Scholz (2000) to postulate a mechanical model of the SAF in which the fault is strong and stress rotation is a direct consequence of frictionless slip below the locked zone.

In this paper we use well-located earthquake focal mechanisms to infer principal stress orientations in the vicinity of the San Andreas fault system in the San Francisco Bay area and in southern California (Fig. 1). Our primary aim is to determine the orientation of $S_{H\max}$ close to the SAF and adjacent faults, where 'close' denotes a distance less than the thickness of seismogenic crust (*c.* 10–15 km). In the San Francisco Bay area we have compiled previous workers' results and obtained an additional stress orientation result by focal mechanism inversion. With the analysis of the southern California data our motivation is to repeat the analysis of Hardebeck & Hauksson (1999) with the specific objective of investigating the influence of their gridding scheme. In other words, we wish to address the method with which they grouped focal mechanism data to determine stress orientations.

The fundamental basis for inverting a group of fault slip observations or focal mechanisms to infer stress orientations is the assumption that slip in each event occurred within a uniform stress field (Angelier 1979, 1984; Gephart & Forsyth 1984; Michael 1984, 1987). Hardebeck

& Hauksson (1999) used rectangular 'boxes' when selecting focal mechanisms for inversion to examine stress orientation as a function of distance from the San Andreas fault. The dimensions of these boxes varied greatly, from as little as 25 km and 2 km in the local strike-parallel and strike-perpendicular directions, respectively, to more than 150 km and 60 km. Along the Fort Tejon profile, on which the model of Scholz (2000) was subsequently based, the strike-parallel length of the boxes is 80 km: consequently, focal mechanisms separated by as much as *c.* 80 km were combined into boxes for inversion. Given that the SAF exhibits a *c.* 25° change of strike in the Fort Tejon area and intersects the left-lateral Garlock fault, stress is unlikely to be homogeneous on 80 km scales, and it may be inappropriate to group such widely separated focal mechanisms. In effect, the gridding scheme used by Hardebeck & Hauksson may not be consistent with the *a priori* assumption of stress homogeneity within each group of earthquakes.

Finally, using the southern California dataset we wish to examine the possibility of stress changes induced by the 1992m M7.4 Landers earthquake. Hardebeck & Hauksson (1999) reported an apparent clockwise stress rotation of as much as 40° caused by this earthquake. If that were the case, it would indicate that the stress changes associated with this earthquake were of comparable magnitude to tectonic stress levels. As average earthquake stress drops (determined seismologically) of 1–10 MPa (Kanamori & Anderson 1975) are small relative to crustal stress levels consistent with hydrostatic Byerlee's law, any such stress rotations would imply that the crust, as well as major faults, had low frictional strength.

San Francisco Bay area

Four major SSE–NNW-striking faults, the San Gregorio, San Andreas, Hayward and Calaveras faults (Fig. 2) accommodate varying amounts of right-lateral slip and relative plate motion in the San Francisco Bay area. San Francisco Bay itself, bounded to the SW and NE by the San Francisco Peninsula Ranges and Diablo Ranges–East Bay Hills, respectively, is relatively aseismic, but microseismicity levels are high along each of the principal fault strands, and the region has experienced several major historic earthquakes. Notable 20th-century events include the 1906 M7.8 San Francisco and 1989 M6.9 Loma Prieta earthquakes on the SAF and the 1984 M6.2 Morgan Hill earthquake on the Calaveras fault.

Previous studies using focal mechanisms of aftershocks following the Morgan Hill and Loma Prieta events have revealed stress orientations in which the angles between $S_{H\max}$ and the local fault strike are 80–90° (Oppenheimer *et al.* 1988; Zoback & Beroza 1993). We have supplemented these two results with one other obtained for a suite of 28 earthquakes that occurred on the southern San Francisco Peninsula (the area encompassed by the box along the SAF west of San Jose in Fig. 2) between 1970 and 1999 (Zoback *et al.* 1999). These earthquakes are a subset of those shown in cross-section in Fig. 3a: they occurred on either side of the quiescent SAF and exhibited a mixture of reverse and strike-slip focal mechanisms. We have used the inversion algorithm of Gephart & Forsyth (1984) and Gephart (1990*a*, *b*) to determine an optimal stress tensor that minimizes the one-norm misfit between the observed and calculated shear tractions on either of the two nodal planes in each focal mechanism. Not unexpectedly, the inversion indicates a strike-slip–reverse faulting regime in which the intermediate and least principal stresses have approximately the same magnitude. The best-fitting solution is illustrated in Fig. 3b: the trend and plunge of S_1, which is $S_{H\max}$ in this case, are 221° and 6°, respectively. S_2 and S_3 (the vertical stress) trend 130° and 334°, and plunge 14° and 75°, respectively. The average misfit of the optimal solution for the 28 focal mechanisms is 4.9°. It is clear from these results that near fault--normal compressive stress exists in the seismogenic crust immediately adjacent to the SAF.

This finding is similar to that of Oppenheimer *et al.* (1988), who studied aftershocks of the 1984 Morgan Hill earthquake. A precise relocation of these events is shown in Fig. 3c (Schaff, pers. comm.). Thrust events on planes parallel to the Calaveras fault and strike-slip events on north–south-trending faults are observed adjacent to the right-lateral, strike-slip Calaveras fault. As in the case of the events along the SAF, Oppenheimer *et al.* (1988) found a best-fitting stress tensor for which $S_{H\max}$ was essentially perpendicular to the trend of the Calaveras fault (Fig. 3d). These results, combined with those of Zoback & Beroza (1993) for the diverse aftershock focal plane mechanisms of the Loma Prieta earthquake (Fig. 2), make a convincing case that $S_{H\max}$ orientations are consistently at an angle of >80° to the local strike of the San Andreas or Calaveras faults. Importantly, the stress orientation estimates in each case are indicative of conditions within 5–

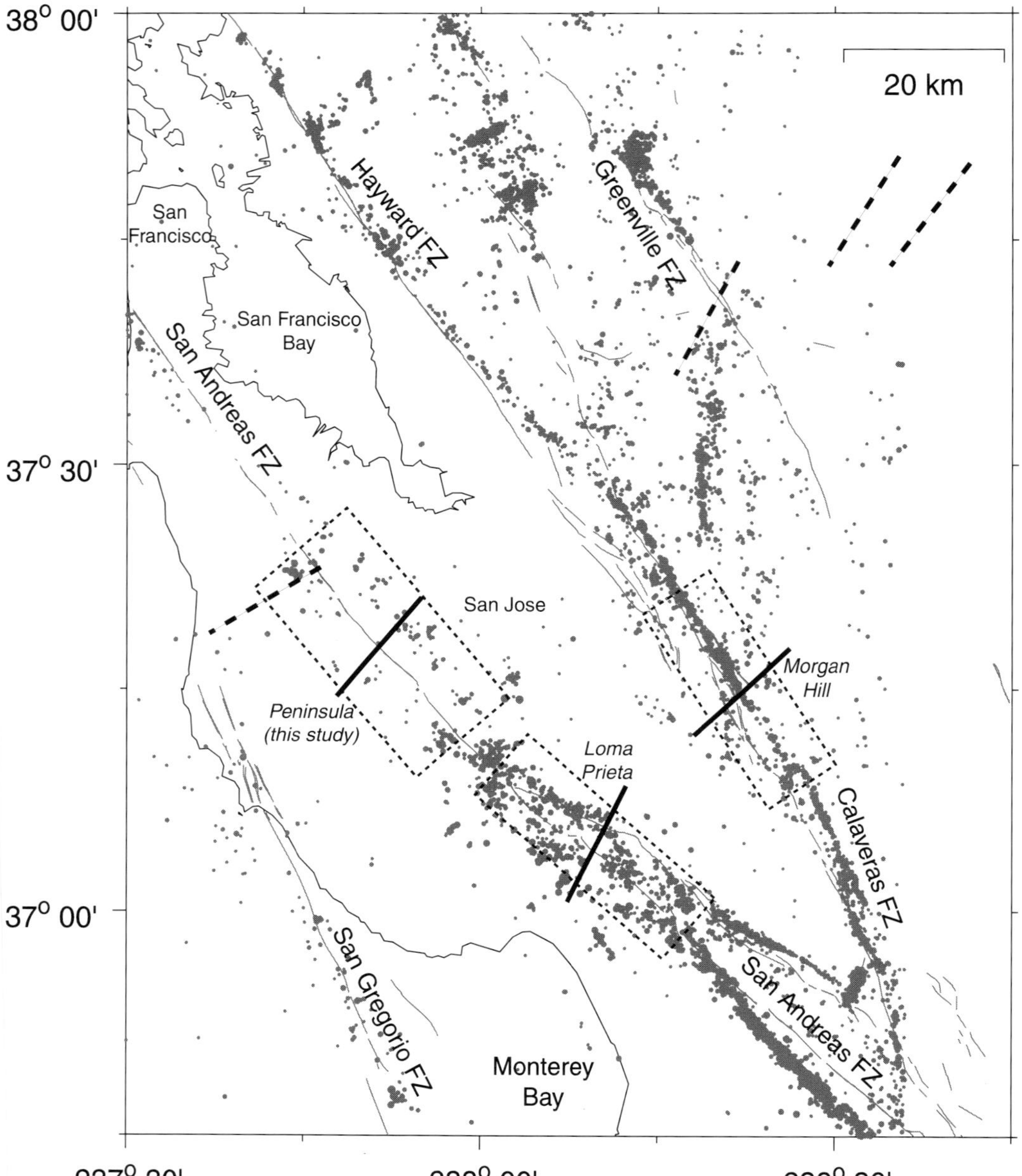

Fig. 2. Map of the San Francisco Bay area showing the orientation of $S_{H\max}$ obtained by focal mechanism inversion (continuous lines) and borehole breakout analysis (dashed lines). The dotted rectangles centred on the three focal mechanism results illustrate the spatial distribution of focal mechanisms used in the 1984 Morgan Hill (Oppenheimer *et al.* 1988), 1991 Loma Prieta (Zoback & Beroza 1993), and southern San Francisco Peninsula (this study) stress inversions. The dots indicate seismicity larger than M2 between 1965 and 2000. Historic and Holocene fault data courtesy of the California Department of Mines and Geology.

10 km of the respective faults, thus indicating fault-normal compressive stress within the crustal volumes adjacent to the seismogenic sections of the San Andreas and Calaveras faults.

Four $S_{H\max}$ estimates obtained from borehole breakout data and shown in Fig. 2 reinforce the picture of fault normal compression in the San Francisco Bay area. That both the San Andreas

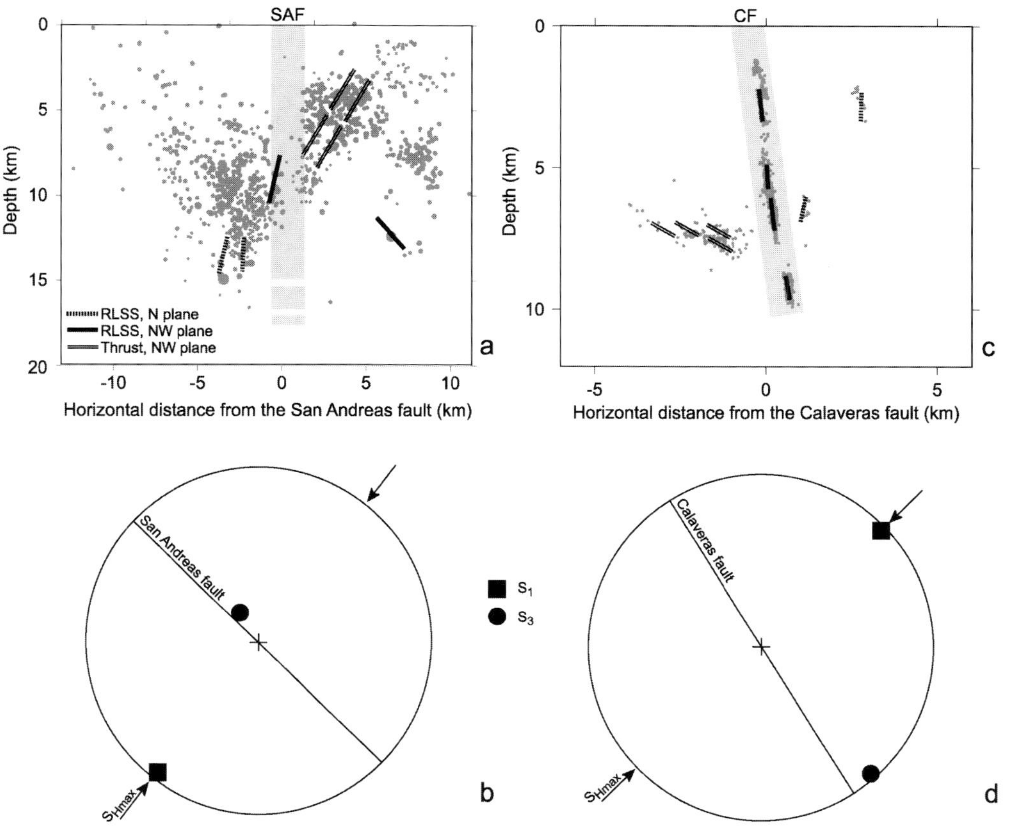

Fig. 3. (**a**) Cross-section across the San Andreas fault on the southern San Francisco Peninsula, west of San Jose, illustrating the distribution of near-fault seismicity (M1.5 and greater). The data have been projected onto a plane orthogonal to the SAF. RLSS, right lateral strike-slip. Data and geological interpretations from Zoback *et al.* (1999). (**b**) Optimal stress tensor obtained by inversion of focal mechanisms for 28 earthquakes larger than M3; the locations of the earthquakes lie within the upper rectangle centred on the San Andreas fault in Fig. 2. S_1 and S_3 are axes of greatest and least compressive stress, respectively. Data from Zoback *et al.* (1999). (**c**) Cross-section across the Calaveras fault (CF) in the vicinity of the 1984 Morgan Hill earthquake. Hypocentral data and fault interpretation courtesy of Schaff (pers. comm.). (**d**) Preferred stress tensor obtained by Oppenheimer *et al.* (1988) for the Morgan Hill aftershock sequence.

and Calaveras faults are able to slip under near-field fault-normal compression indicates that they are very weak with respect to hydrostatic Byerlee's law.

Stress in southern California

As alluded to above, the gridding method of Hardebeck & Hauksson (1999) results in widely separated focal mechanisms being combined for stress inversion. The essential limitation of this method is that it takes no account of either structural trends (other than the average strike of the SAF) or data clustering. To determine whether their grid design adversely affects stress inversion results in southern California, we have opted to use a recursive quadtree algorithm to grid the data. Starting from a single square box encompassing all the earthquake locations, the algorithm operates by dividing the box into quarters and continuing recursively within each quarter until there are fewer than n_{max} earthquake locations in each box or the box reaches a minimum dimension of x_{min}. The latter condition is a modification of the standard quadtree algorithm that we have incorporated to avoid boxes that are smaller than errors in the earthquake locations. The resultant grid comprises an irregular mesh of square boxes that are smaller and more densely spaced in areas containing many earthquakes. In practice, we impose an additional condition that restricts the stress inversion to those boxes containing more than n_{min} events. We feel that this gridding scheme is geologically reasonable because it combines earthquakes within contiguous crustal volumes of limited extent.

To compare our results directly to those of Hardebeck & Hauksson, we have used the same dataset and stress inversion method. The dataset contains first-motion focal mechanisms for *c*. 49 000 earthquakes recorded by the southern California seismic network between 1981 and 1999 and relocated using a three-dimensional velocity model (Hardebeck & Hauksson 1999; Hauksson 2000). We use the Michael (1984, 1987) stress inversion method with equal weighting for both nodal planes.

Figure 4a illustrates the results for southern California obtained by inverting the entire dataset after gridding with the parameters $n_{min} = 30$, $n_{max} = 100$, and $x_{min} = 5$ km. It should be noted that this procedure produces square boxes of *c*. 5.6 km width in regions of dense seismicity, such as the vicinities of the Northridge and Landers earthquakes (denoted by N and L, respectively) and boxes of 5.6–22 km width along the SAF. Within the Mojave Desert, the triangular fault-bounded region in the middle of Fig. 4, the average box size is *c*. 44 km, reflecting the relatively sparse seismicity. We are confident that the gridding technique is robust because the $S_{H\max}$ orientations in adjacent boxes are very similar despite each actual focal mechanism inversion being independent of those surrounding it.

In general, $S_{H\max}$ trends slightly east of north and is oriented at a high angle to the fault strike of each major fault strand illustrated. Figures 4b and c illustrate the near-field stress orientation data (those data points closer than 5 km to the fault trace) along the San Andreas fault. The mean and standard deviation of the angle between $S_{H\max}$ and the local fault strike are 64° and 14°, respectively, indicating that the near-field angle is significantly different at the 95% level of confidence from that suggested by Hardebeck & Hauksson (1999) for some of their profiles and later utilized by Scholz (2000) to argue for a strong San Andreas fault. This result agrees, however, with that of Jones (1988), who found that the mean angle exceeded 60°. Contrary to Hardebeck & Hauksson (1999), we do not observe a consistent pattern of stress rotation close to either the SAF or the other major faults.

We have also investigated the effect of the 1992 Landers earthquake on the local stress field. Using a relatively simple approach, we have inverted the focal mechanisms before and after this event separately to determine whether or not any detectable stress rotation is evident (Fig. 5a). The nature of the quadtree algorithm means that the grids corresponding to different datasets are not necessarily the same, implying that the pre- and post-earthquake stress inversions are made at different locations. We have compared the two datasets only at locations <5 km apart. The comparisons were made separately inside and outside the zones of pronounced aftershock activity (defined by areas of aftershock activity occurring within 30 days of the main shock). This distinction is made on the presumption that the occurrence of aftershocks (the 'aftershock zone') implies a possibly significant stress perturbation, whereas a lack of aftershock activity implies no significant earthquake-induced change. Thus, the latter case constitutes a control zone in which we expect to observe no rotation if the technique is appropriate.

It is qualitatively apparent from Fig. 5a that there is little systematic difference in the pre- and post-Landers stress orientations. This observation is substantiated in Fig. 5b and c, in which the differences in orientation in the control zone (mean −0.8°; SD 12.2°; 46 observations) and aftershock zone (mean 1.4°; SD 12.7°; 19 observations) are compared. Neither of these samples is significantly different from zero at the 95% level of confidence. Both standard deviations are similar to those obtained for the southern California dataset as a whole, and suggest that small earthquake-induced rotations are unlikely to be observed given the stress uncertainties.

Discussion

The data from both the San Francisco Bay area and southern California reveal high angles between the local fault strike and the direction of maximum horizontal compression. Indeed, this observation applies both to the SAF and associated faults, and suggests that weakness is characteristic of these portions of the North America–Pacific plate boundary.

The discrepancies between our southern California results and those of Hardebeck & Hauksson (1999) must stem from the different gridding methods used, as the dataset and focal mechanism inversion procedures were the same. By using an iterative gridding algorithm that incorporates spatial variations in seismicity, it has been possible to obtain a high-resolution image of near-fault stress orientations (where justified by the data) without needlessly com-

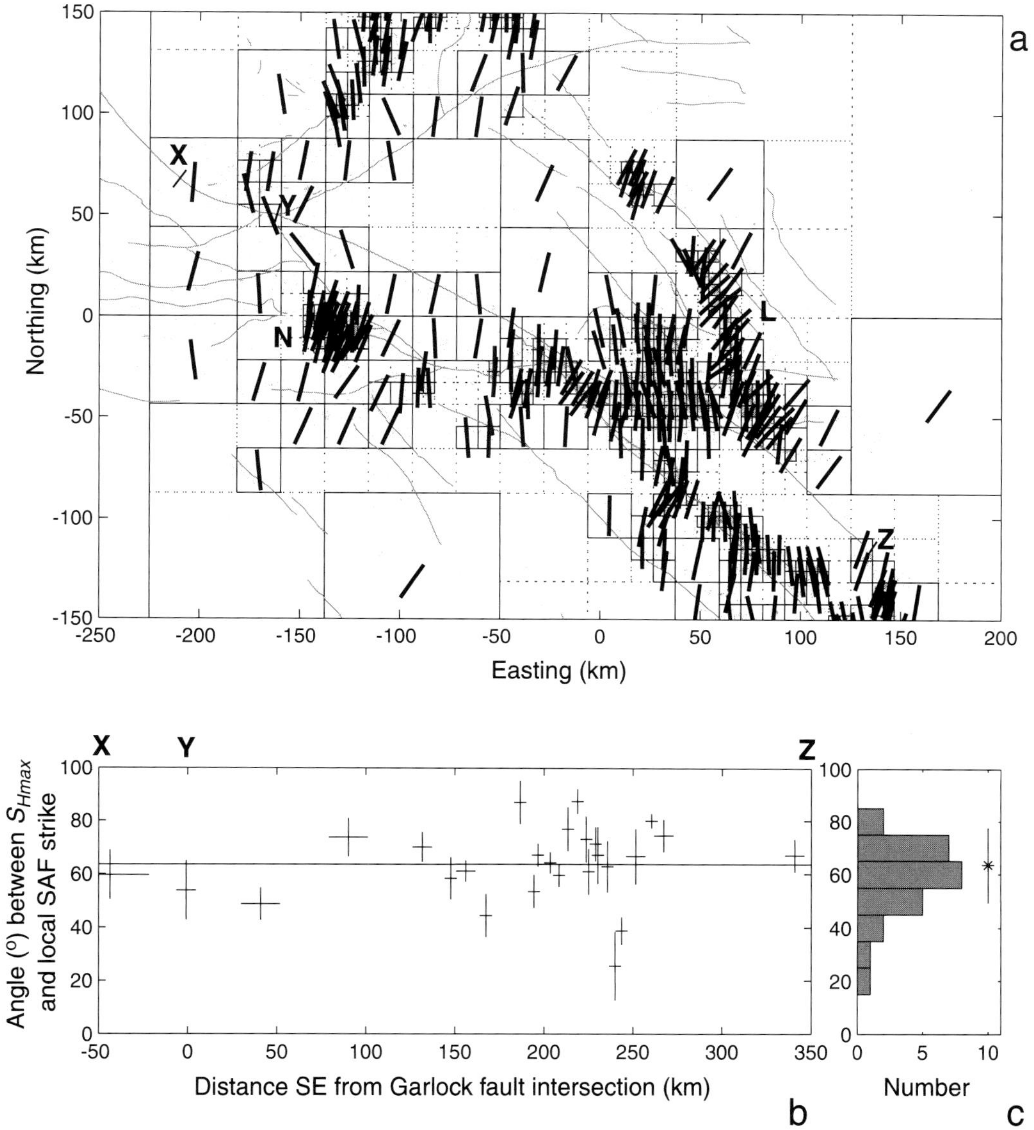

Fig. 4. (**a**) $S_{H\max}$ orientation results obtained using the quadtree gridding algorithm and the Michael (1994, 1987) stress inversion routine for the 1981–1999 M1+ southern California dataset utilized by Hardebeck & Hauksson (1999). Boxes containing <30 focal mechanisms are dashed, and no inversion was made at those locations. L, N, areas of seismicity associated with the 1992 Landers and 1994 Northridge earthquakes, respectively; X, Y, Z, trace of the San Andreas fault profile illustrated in (**b**). (**b**) Along-fault profile (X–Y–Z) of the angle between near-field $S_{H\max}$ estimates (lying within 5 km of the San Andreas fault trace) and the local fault strike. The horizontal error bars indicate the width of the box used for each inversion, and the vertical error bars indicate the 95% confidence bounds estimated by bootstrap analysis. The mean of all estimates (64°) is indicated by the continuous horizontal line. (**c**) Histogram of the data shown in (**b**). The mean and standard deviation (14°) are indicated by the star and vertical error bars, respectively.

promising the assumption of local stress homogeneity. Our results are incompatible with the model of Scholz (2000), which predicts symmetric rotation about the SAF over lateral distances of ± 20 km. We have particularly strong reservations about the applicability of a 2D plane-strain model to a segment of the SAF that is clearly curved and, moreover, intersects

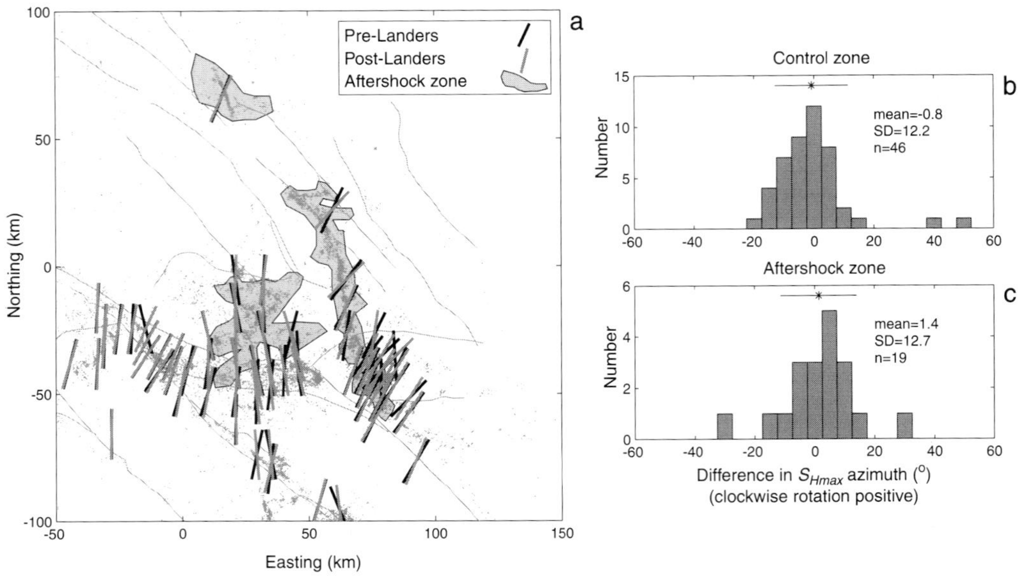

Fig. 5. (**a**) $S_{H\max}$ orientations before (black) and after (grey) the Landers 1992 earthquake. The gridding parameters for each inversion are the same as in Fig. 4a, but the resultant grids are different before and after the earthquake, and are not shown. The light grey shaded regions indicate regions of aftershocks (see text for details). (**b**) Histogram of differences in pre- and post-Landers stress orientations outside the zones of aftershock activity. The values plotted are the post-Landers orientation minus the pre-Landers orientation; clockwise rotations are positive. The mean (−0.8°) and standard deviation (12.2°) of the 46 data points are illustrated by the star and horizontal error bar, respectively. (**c**) Histogram of differences in pre- and post-Landers stress orientations within the aftershock zone. No significant change in stress orientation is observed in either the control zone or the aftershock zone.

another large fault. It is noteworthy that the only data presented by Hardebeck & Hauksson (1999) that are satisfactorily fitted by the Scholz model are those from the complex 'Big Bend' area.

The significance of our results is that any postulated stress rotation must occur within a zone no more than several kilometres wide, centred on the fault zone, or otherwise it would be detectable. Any such zone must be narrow, on theoretical grounds, or the fluid pressure would overcome the least principal stress and hydrofracture the surrounding crust (Rice 1992). As pointed out by Scholz (2000), the 20–30 km wide zone of stress rotation suggested by Hardebeck & Hauksson (1999) is inconsistent with the Rice (1992) model of a high-pressure fault core. Although our data do not possess sufficient resolution to observe stress rotations on the scale of *c.* 1 km, it is clear that within ± 5 km of the SAF $S_{H\max}$ is at a much higher angle to the fault than is consistent with conventional Coulomb faulting and hydrostatic Byerlee's law. In other words, these data clearly illustrate that the fault has low frictional strength.

Conclusions

We have compiled stress orientation data in the San Francisco Bay area and reanalysed stress orientations in southern California using focal mechanism data. In the San Francisco Bay area, focal mechanism stress inversions indicate that the maximum horizontal compressive stress is oriented almost orthogonally to the strikes of the San Andreas and Calaveras faults. Using the same dataset and inversion method as Hardebeck & Hauksson (1999) we have demonstrated that their postulated near-fault stress rotation may be an artefact of combining widely separated focal mechanisms into single inversions. Using a geologically reasonable and mathematically more sophisticated gridding algorithm that takes into account spatial clustering of seismicity, we have been able to obtain stress orientations while justifying to the extent that the data warrant the necessary assumption of stress homogeneity within contiguous crustal blocks. Furthermore, applying this technique to the 1992 Landers earthquake reveals no consistent differences in stress orientation before and after the main shock.

Overall, available stress orientation results from the San Francisco Bay area and southern California strongly indicate low levels of shear stress on planes parallel to the San Andreas fault. In the San Francisco Bay area, several data are available within ± 5 km of the main fault zone that indicate a direction of maximum horizontal compression nearly orthogonal to the San Andreas and Calaveras faults, and the fault that produced the Loma Prieta earthquake. In southern California, the direction of maximum horizontal stress near the San Andreas fault is generally at a high angle to it (averaging >60°), similarly indicating that the fault has low frictional strength.

We are grateful to D. Schaff and G. Beroza of Stanford University for making their Calaveras fault earthquake relocation data available, S. Prejean of Stanford University for valuable suggestions, M. L. Zoback for the San Francisco Peninsula seismicity data, and to D. Cowan and P. Sammonds for considerate reviews. Fault data courtesy of the California Department of Mines and Geology. This work was supported by the National Science Foundation (award 96-14267) and an Arco Stanford Graduate Fellowship (to J.T.).

References

ANGELIER, J. 1979. Determination of the mean principal directions of stresses for a given fault population. *Tectonophysics*, **56**, T17–T26.

ANGELIER, J. 1984. Tectonic analysis of fault slip data sets. *Journal of Geophysical Research*, **89**, 5835–5848.

BRACE, W.F. & KOHLSTEDT, D. 1980. Limits on lithospheric stress imposed by laboratory measurements. *Journal of Geophysical Research*, **85**, 6248–6252.

BRUNE, J.N., HENYEY, T.L. & ROY, R.F. 1969. Heat flow, stress, and rate of slip along the San Andreas fault, California. *Journal of Geophysical Research*, **74**, 3821–3827.

BYERLEE, J.D. 1978. Friction of rocks. *Pure and Applied Geophysics*, **116**, 615–626.

GEPHART, J.W. 1990*a*. Stress and the direction of slip on fault planes. *Tectonics*, **9**, 845–858.

GEPHART, J.W. 1990*b*. FMSI: A Fortran program for inverting fault/slickenside and earthquake focal mechanism data to obtain the regional stress tensor. *Computers and Geosciences*, **16**, 953–989.

GEPHART, J.W. & FORSYTH, D.W. 1984. An improved method for determining the regional stress tensor using earthquake focal mechanism data: application to the San Fernando earthquake sequence. *Journal of Geophysical Research*, **89**, 9305–9320.

HARDEBECK, J.L. & HAUKSSON, E. 1999. Role of fluids in faulting inferred from stress field signatures. *Science*, **285**, 236–239.

HAUKSSON, E. 2000. Crustal structure and seismicity distributions adjacent to the Pacific and North America plate boundary in southern California. *Journal of Geophysical Research*, **105**, 13875–13903.

JONES, L.M. 1988. Focal mechanisms and the state of stress on the San Andreas fault in southern California. *Journal of Geophysical Research*, **93**, 8869–8891.

KANAMORI, H. & ANDERSON, D.L. 1975. Theoretical basis of some empirical relations in seismology. *Bulletin of the Seismological Society of America*, **65**, 1073–1095.

LACHENBRUCH, A.H. & MCGARR, A. 1990. Stress and heat flow. *US Geological Survey Professional Paper*, **1515**, 261–277.

LACHENBRUCH, A.H. & SASS, J.H. 1980. Heat flow and energetics of the San Andreas fault zone. *Journal of Geophysical Research*, **85**, 6185–6223.

LACHENBRUCH, A.H. & SASS, J.H. 1992. Heat flow from the Cajon Pass, fault strength, and tectonic implications. *Journal of Geophysical Research*, **97**, 4995–5015.

MICHAEL, A.J. 1984. Determination of stress from slip data: faults and folds. *Journal of Geophysical Research*, **89**, 11517–11526.

MICHAEL, A.J. 1987. Use of focal mechanisms to determine stress: a control study. *Journal of Geophysical Research*, **92**, 357–368.

MOUNT, V.S. & SUPPE, J. 1987. State of stress near the San Andreas fault: implications for wrench tectonics. *Geology*, **15**, 1143–1146.

OPPENHEIMER, D.H., REASENBERG, P.A. & SIMPSON, R.W. 1988. Fault plane solutions for the 1984 Morgan Hill, California, earthquake sequence: evidence for the state of stress on the Calaveras fault. *Journal of Geophysical Research*, **93**, 9007–9026.

RICE, J.R. 1992. Fault stress states, pore pressure distributions, and the weakness of the San Andreas fault. *In*: EVANS, B. & WONG, T.F. (eds) *Fault Mechanics and Transport Properties of Rocks*. Academic Press, New York, 475–503.

SCHOLZ, C.H. 2000. Evidence for a strong San Andreas fault. *Geology*, **28**, 163–166.

TOWNEND, J. & ZOBACK, M.D. 2000. How faulting keeps the crust strong. *Geology*, **28**, 399–402.

ZOBACK, M.D. & BEROZA, G.C. 1993. Evidence for near-frictionless faulting in the 1989 (M6.9) Loma Prieta, California, earthquake and its aftershocks. *Geology*, **21**, 181–185.

ZOBACK, M.D., ZOBACK, M.L., MOUNT, V.S. & 20 OTHERS 1987. New evidence for the state of stress of the San Andreas fault system. *Science*, **238**, 1105–1111.

ZOBACK, M.L., JACHENS, R.C. & OLSON, J.A. 1999. Abrupt along-strike change in tectonic style: San Andreas fault zone, San Francisco Peninsula. *Journal of Geophysical Research*, **104**, 10719–10742.

Permeability variation across an active low-angle detachment fault, western Woodlark Basin (ODP Leg 180), and its implication for fault activation

ACHIM KOPF
GEOMAR, Wischhofstrasse 1–3, 24148 Kiel, Germany
Present address: Scripps Institution of Oceanography, 9500, Gilman Drive, La Jolla, CA 92093-0220, USA (e-mail: akopf@ucsd.edu)

Abstract: In the western Woodlark Basin, off Papua New Guinea, the variation from continental rifting to sea-floor spreading has profound effects on the mechanical response of the lithosphere. The extension is well expressed in a seismically active, shallow-dipping detachment fault. Recent Ocean Drilling Program drilling (ODP Leg 180) in the area obtained cores from the hanging wall, detachment fault gouge, and footwall, of which samples underwent permeability testing in the laboratory. Permeability variation was found to be critically dependent on (1) flow direction, i.e. fabric anisotropy of the rocks, and (2) deformational structures in the hanging wall to the fault. Regarding the first, a slight but distinct increase in permeability has been recorded parallel to the fabric (compared with flow normal to this, as indicated by anisotropy indices of $k_{\text{horizontal}}/k_{\text{vertical}}$ of >1.7). This phenomenon appears most profound directly above fault zones in the hanging-wall block, which are interpreted to represent splays to the main detachment fault plane at depth. Here, shear-enhanced compaction seals fluid flow to the sea floor, so that conductive flow parallel to the fault plane is favoured (in general one order of magnitude higher). The fault gouge, mainly consisting of highly serpentinized basalt and chlorite, exhibits an increase in permeability relative to the clay- and siltstones of the hanging-wall block. In the metamorphic basalt from the tectonic footwall, permeability decreases again by three orders of magnitude (k is *c.* 6e−17 to 5e−18 m^2). Consequently, the detachment and synthetic splays related to it are zones of enhanced fluid migration in the fault plane direction. Fluid overpressure, and hence fault activity, is suggested to be triggered by seal of the top of the fault zone, owing to both shear fabrics and cementation processes.

The lateral variation from active continental rifting to sea-floor spreading within a small region makes the western Woodlark Basin an attractive area to investigate the mechanics of lithospheric extension (Abers 1991). Earthquake source parameters and seismic reflection data indicate that low-angle (*c.* 25–30°) normal faulting is active in the region of incipient continental separation (Taylor *et al.* 1995; 1996; Mutter *et al.* 1996; Abers *et al.* 1997). A low-angle normal fault emerges along the northern flank of Moresby Seamount, a continental crustal block with greenschist metamorphic basement, which to its north has a down-flexed pre-rift sedimentary basin (i.e. the tectonic hanging wall to the fault).

The main objective behind a permeability study across a low-angle detachment is to investigate the phenomenon of very low friction that allows such faults to develop. Detachments with low angles of dip were previously thought to be in contradiction to simple fault theories, or be of secondary origin (i.e. reactivation after footwall uplift or erosion, then rotating to a low angle after fault movement ceased), but have since been demonstrated to be seismogenic (Abers 1991). Normal faulting on such detachment surfaces requires the fault to be extremely weak, almost frictionless, an assumption that is necessary for models found in the literature in which low-angle detachment faulting is an essential mechanism of large-scale strain accommodation (e.g. Wernicke & Burchfiel 1982). Structural, textural, and geochemical arguments suggest that movement under low resolved shear stresses, both in brittle and ductile regimes, may be partly or totally the result of pore fluid overpressure within the fault zone (e.g. Axen 1992). Accompanying phenomena are hydrogeological fracturing or dilatant microstructures, elevated

From: HOLDSWORTH, R.E., STRACHAN, R.A., MAGLOUGHLIN, J.F. & KNIPE, R.J. (eds) 2001. *The Nature and Tectonic Significance of Fault Zone Weakening*. Geological Society, London, Special Publications, **186**, 23–41. 0305-8719/01/$15.00

fluid flow rates (e.g. Brown *et al.* 1994a; Fisher *et al.* 1996; Screaton *et al.* 1997; Takizawa & Ogawa 1999) and mineral precipitation in the hanging wall of the fault (Roller *et al.* 2001), formation of (chlorite) microbreccia layers, threshold pressure gradients, and low-permeability mylonites just below the fault zone (Axen 1992).

One possible explanation for the mechanism by which friction might be effectively reduced on low-angle fault plane may be a shear-enhanced permeability contrast with the surrounding crust (by opening cracks more quickly than precipitation can heal them; see discussion by Labaume *et al.* (1997) of scaly fabric formation and effect on permeability). This would allow pore pressures as high as fault-normal compressive stresses within the fault zone, but relatively low with distance into the adjacent crust (Axen 1992; Rice 1992). A second possibility is the sealing of fault planes as a result of clay mineral alignment, which has been shown to be crucial for décollement physical properties (Kimura *et al.* 1997; Labaume *et al.* 1997; Zhu & Wong 1997). A third possible explanation for weakness are fluid–rock reactions forming phyllosilicates in the fault zone that are particularly weak because of their well-developed fabrics (Wintsch *et al.* 1995; Moore 1997). As a fourth hypothesis, the rotation of principal-stress orientations consistent with low-angle faulting has been put forward (Wills & Buck 1997). Testing of the various hypotheses, including permeability and pore pressures anomalies, presence of weak phyllosilicates, and/or fluid–rock reactions affecting deformation (constitutive response, frictional stability, long-term fault strength; e.g. Hickman *et al.* 1993; Barton *et al.* 1995; Sibson 2000) is beyond the scope of this study. However, the study was inspired by these ideas, and the understanding of the permeability variation over the active Woodlark detachment will help in the critical assessment of some of these hypotheses.

Recent deep-sea drilling (Ocean Drilling Program (ODP) Leg 180; Taylor *et al.* 1999) recovered drill cores from 11 locations, including a suite of samples through the hanging wall, fault zone, and footwall of the detachment. In a synthetic profile from three drillsites, a succession of clays and claystones, the serpentinized fault gouge, and metamorphic basalts have been studied to examine permeability variation and its relation to faulting. Key objectives of the study focused on the effect of anisotropy on permeability, according to, for example, state of consolidation, sedimentary fabrics and physical properties (Bolton *et al.* 1999; Clennell *et al.* 1999; Dewhurst *et al.* 1999*b*). These measurements are interpreted in concert with fabric observations, such as hydrofracturing and subsequent fracture seal mechanisms. For this purpose, pairs of minicores from the same piece of core were drilled vertically and parallel to the fabric, and then examined. The effect of anisotropy and its development are related to fault activity and formation of enhanced fluid pressures.

Geological setting

Active extensional processes, laterally varying from continental rifting to sea-floor spreading in the western Woodlark Basin, off Papua New Guinea, have been suggested to be active during the last 6 Ma (Fig. 1). A westward-propagating spreading centre opened the Woodlark Basin, with its current spreading tip close to 9.8° S, 151.7° E. In the west, extension is accommodated by continental rifting, with associated full and half grabens, and metamorphic core complexes (such as the Moresby Seamount; see Fig. 2). Earthquake source parameters and seismic reflection data reveal a seismically active, major low-angle normal fault (e.g. Abers 1991; Abers & Roecker 1991). The spreading history of the western Woodlark Basin is well constrained by the wealth of geophysical survey data acquired during preparation of the drill cruise (e.g. bathymetry, acoustic imagery, magnetization and gravity maps, and seismic reflection profiles; see Taylor *et al.* 1995; 1996; Goodliffe *et al.* 1997). The upper-crustal architecture of the rifting region illustrates the presence of low-angle normal faults (Mutter *et al.* 1996; Taylor *et al.* 1996).

A schematic cross-section (Fig. 2a) based on a series of parallel, north–south-trending seismic lines allows identification of the main features. These are: (1) the Pocklington Rise to the south; a continuation of Papuan Peninsula mainland; (2) the Moresby Seamount at its northernmost continuation; (3) the Woodlark Rise to the north. The last is separated from the metamorphic basement by the low-angle detachment fault. As a consequence of the northern block slipping downwards, a half-graben basin opened, which has subsequently been filled with turbidites. The area of interest to this study is the vicinity of the metamorphic core complex of the Moresby Seamount, which was subject to strong uplift during exhumation facilitated by fault movement. In Fig. 2a, the relative tectonostratigraphic positions of the drillsites are shown (see below). The seismic section illustrates the

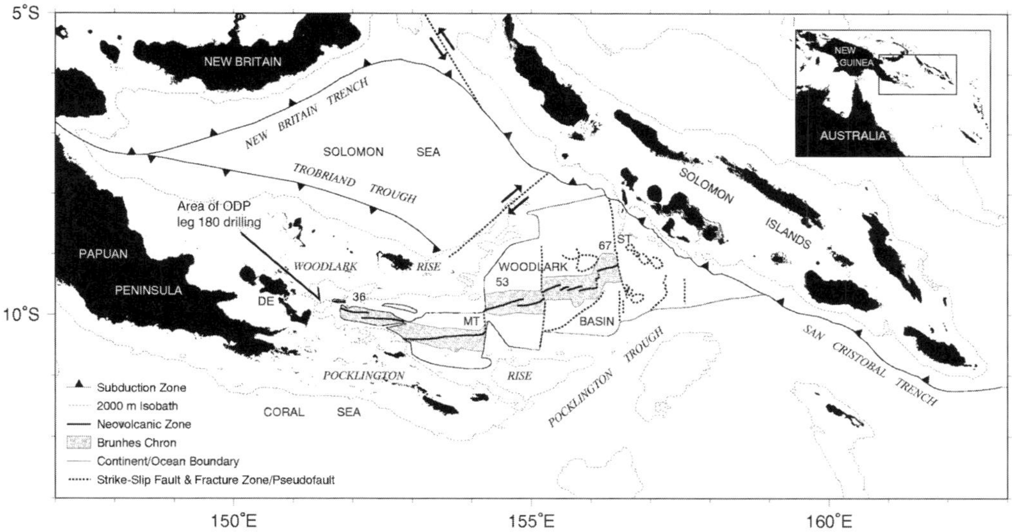

Fig. 1. Major physiographic features and active plate boundaries of the Woodlark Basin region. The stippled area encloses oceanic crust formed during the Brunhes Chron at spreading rates labelled in mm a^{-1}. MT, Moresby transform fault; ST, Simbo transform fault; DE, D'Entrecasteaux Islands. The inset shows the wider region of the Woodlark Basin (after Taylor *et al.* 1999).

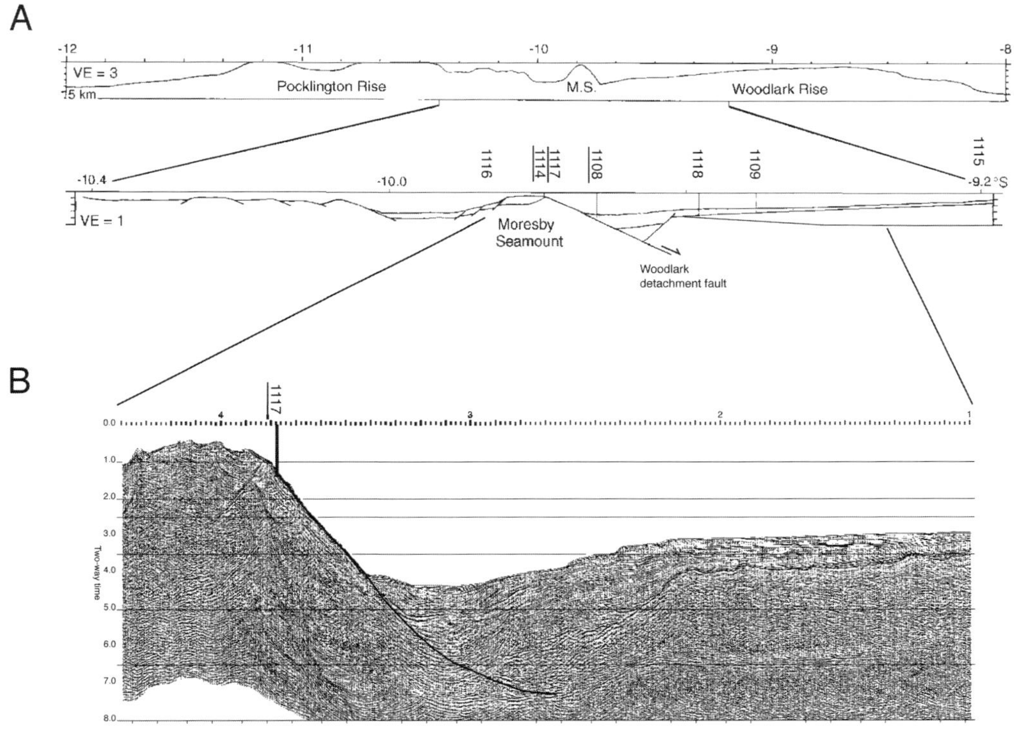

Fig. 2. (**a**) Schematic north–south-trending cross-section over the study area, as taken from interpreted seismic reflection profiles (after Taylor *et al.* 1999). The enlarged middle part of the section also shows the projected relative locations of Sites 1114, 1117, and 1108 (from south to north) across the Moresby Seamount (underlined) and the adjacent detachment fault. (**b**) The enlarged lowermost part is a multi-channel seismic (MCS) cross-section across Site 1117 having penetrated the fault gouge where exposed to the sea floor after slip of the hanging-wall block (from Taylor *et al.* 1999).

trace of the detachment fault plane and the overlying onlapping synrift sediments (Fig. 2b).

Results from drilling

The locations of the three drillsites of interest in this study are shown in the schematic cross-section in Fig. 2a. The principal lithologies at each locality are, from north to south: rift basin sediments (Site 1108), serpentinized gouge from a structurally high part of the detachment fault plane (Site 1117), and the metamorphic footwall on Moresby Seamount (Site 1114).

Site 1108 is located in the seismically active region of incipient continental separation 2.5 km ahead of the Woodlark Basin spreading centre. Here, a continental fault block (Moresby Seamount) forms the footwall to a low-angle normal fault imaged to 9 km depth that dips 25–30° beneath a 3.2 km deep, asymmetric rift basin with over 2 km of sediment fill (Fig. 2, bottom). Site 1108 aimed to drill through *c*. 1 km of the rift basin fill, the low-angle normal fault zone, and into the footwall metamorphic rocks. Unfortunately, as a result of hole instability (possibly caused by overpressure) and the presence of higher hydrocarbons, only 485 m of sediment was drilled (see Taylor *et al.* (1999) for details). The rocks recovered comprised nannofossil ooze, talus from the nearby Moresby Seamount at the top *c*. 60 m of the hole, and a succession of more or less uniform siliciclastic deposits (terrigenous turbiditic sandstones, siltstones, claystones) underneath. Ages increase from 1.67–1.75 Ma at 82.8 mbsf (metres below sea floor) to <3.35 Ma at the base. Sedimentation rates increase down section, from 325 m Ma^{-1} at 1.7–2 Ma to 425 m Ma^{-1} at 3–3.2 Ma (Taylor *et al.* 1999). Abundant evidence for brittle deformation, characterized by rotated bedding dips up to 35°, low-angle shearing, brecciation and ubiquitous slickensides, is mainly observed in the fine-grained rocks. Two fault zones are seen, one at *c*. 165–190 mbsf, and one below 350 mbsf (see Fig. 3). Scaly fabrics are observed, and the shear sense, when recognizable, indicates normal displacement in an extensional fault zone. The greatest frequency of fractures and faults is concentrated in the upper fault zone, where porosities are anomalously high. Temperature measurements suggest an average thermal gradient of 100°C km^{-1} to 390 mbsf, the maximum value found so far in the area (Taylor *et al.* 1999).

Unfortunately, the main detachment could not be penetrated owing to hole instability. However, the two fault zones penetrated at shallower depth are interpreted as being splays of the main detachment fault, because of both their normal sense of shear (Taylor *et al.* 1999) and the composition of vein cements (Kopf, unpublished data). Rock from the detachment zone, however, was recovered at Hole 1117A further up the fault plane in a shallow position (see Taylor *et al.* 1999). These light-coloured rocks show rudimentary scaly fabrics, and consist of mainly talc, chlorite, serpentine, and calcite, consistent with alteration of the underlying, deformed metamorphic basement. The fault gouge has a porosity of 30%, and an unconfined compressive strength of up to 90 kPa (Taylor *et al.* 1999). Both the high (in fault gouge terms) pore volume and the low strength are indicative of the weakness of the material. The footwall to the detachment was successfully recovered immediately beneath the detachment (in the deeper part of Hole 1117A), and also at Hole 1114A on Moresby Seamount (see Fig. 2, and Taylor *et al.* 1999). This footwall comprises metamorphic diabase of different degrees of alteration and deformation, mylonitic rock fragments, and gabbros.

As a consequence of the unsuccessful attempt to penetrate hanging wall, detachment, and footwall in one hole (1108B), a synthetic profile was compiled for the purposes of this study (Fig. 3). Beneath the cores of Site 1108 in the upper part of the hanging-wall basin fill, the detachment would have been expected at a depth of *c*. 1 km below the sea floor. Consequently, 1000 m were added to the real depth of recovery of the samples from Site 1117 (see Table 1). Also, the footwall samples recovered from Site 1114 were plotted at a depth below 1 km bsf, so that the permeability trends with depth can be illustrated as a 'continuous' profile. A justification for such a correlation and adjustment of the Site 1114 and 1117 specimens down to 'artificial' depths of >1 km bsf is in strong disagreement with the widely accepted decrease of k with depth. With respect to Site 1117 fault gouge, however, an overpressured fault would display the same effective stress values along its entire plane. Although the gouge was recovered in close proximity to the sea floor (and hence was altered by seawater circulation), corresponding rocks at kilometre depth may have been altered to a similar degree, given the evidence for enhanced flow of deep-seated fluids (e.g. healed veins and hydrofractures; Fig. 4). Regarding the metadiabase recovered at Site 1114 and 1117, it can be safely ruled out that this rock would undergo stronger compaction when buried to a depth of 1–2 km bsf.

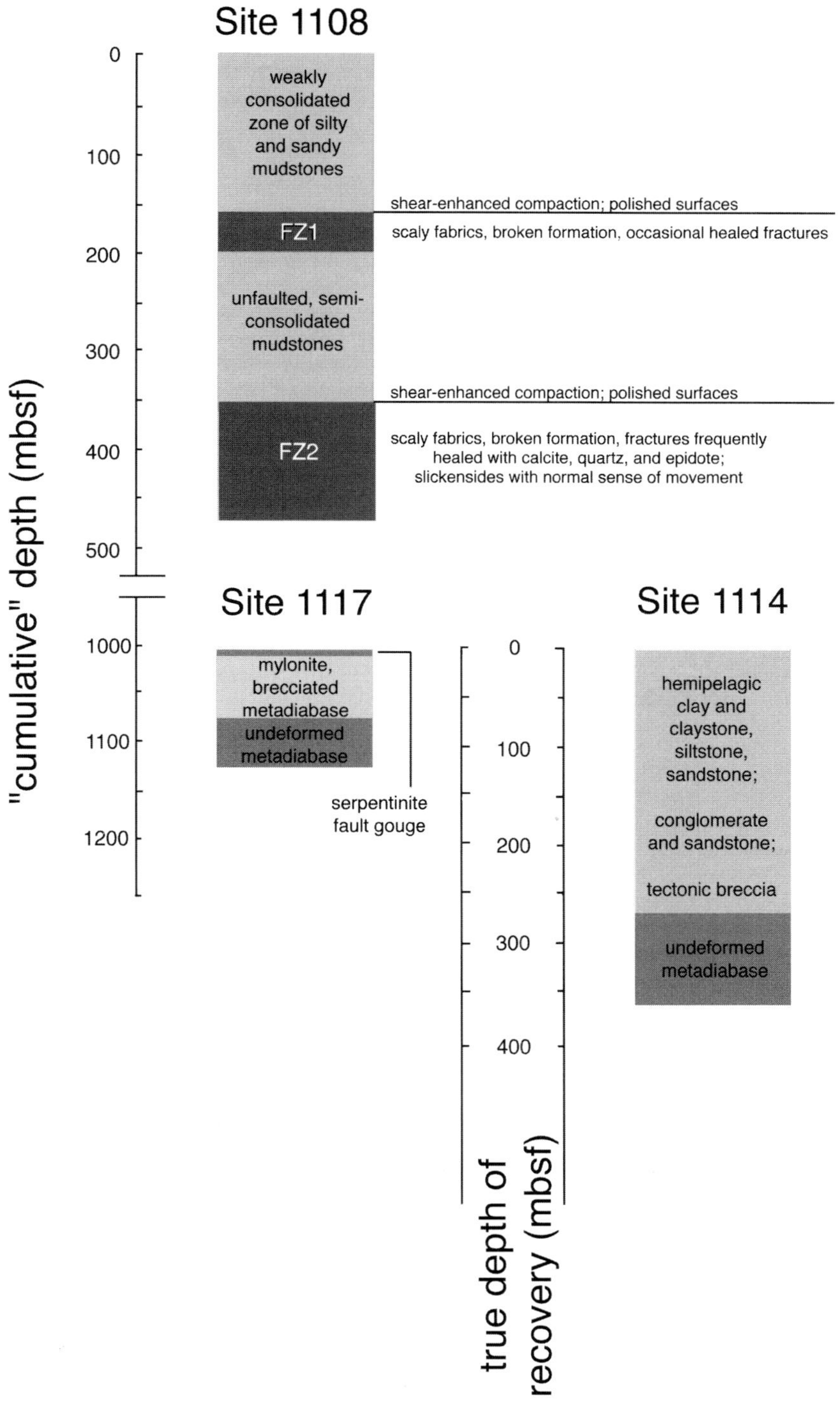

Fig. 3. Lithological profile and main structural observations for the synthetic profile compiled from rocks recovered at Sites 1108, 1114, and 1117. It should be noted that rocks from Sites 1114 and 1117 are projected into a position near the proposed depth of the detachment and footwall at the location of Site 1108 (left depth axis). Depth of recovery below the sea floor is given for reference.

Table 1. *Sample data*

Sample ID	Depth (mbsf)	Sample length (mm)	Type of lithology	Grain size (% <10 μm)	Type of sample	Bulk density (g cm^{-3})	Grain density (g cm^{-3})
Hanging wall to detachment fault							
180-1108B-1R-02, 0–4 cm	1	30	hemipelagic clay	–	soft sediment	1.44	2.71
180-1108B-1R-02, 6–8 cm	1.06	28	hemipelagic clay	42	soft sediment	1.44	2.71
180-1108B-08R-CC, 5–7 cm	63.43	20	silt	32	soft sediment	2.01	2.71
180-1108B-08R-CC, 9–12 cm	63.47	28	silt	–	soft sediment	2.01	2.71
180-1108B-10R-01, 105–107 cm	82.95	24	silty sand	–	soft sediment	2.02	2.7
180-1108B-10R-01, 118–121 cm	83.08	30	silty sand	40	soft sediment	2.02	2.7
180-1108B-15R-02, 29–31 cm	130.92	21	clayey/silty sandstone	40	minicore	1.97	2.73
180-1108B-15R-02, 42–44 cm	131.05	33	clayey/silty sandstone	–	minicore	1.97	2.73
180-1108B-16R-01, 19–21 cm	139.59	23	silty claystone	35	minicore	2.11	2.74
180-1108B-16R-01, 21–27 cm	139.64	17	silty claystone	–	minicore	2.11	2.74
180-1108B-17R-01, 90–92 cm	149.9	26	clayey siltstone	–	minicore	1.82	2.67
180-1108B-17R-01, 94–96 cm	149.94	26	clayey siltstone	41	minicore	1.82	2.67
180-1108B-18R-01, 117–118 cm	159.75	31	siltstone	–	minicore	1.85	2.71
180-1108B-18R-01, 115–117 cm	159.73	28	siltstone	38	minicore	1.85	2.71
180-1108B-20R-CC, 16–18 cm	180.74	18	sandstone	33	minicore	2.15	2.7
180-1108B-20R-CC, 19–21 cm	180.71	20	sandstone	–	minicore	2.15	2.7
180-1108B-23R-02, 12–14 cm	208.42	28	silty claystone	–	minicore	2.12	2.7
180-1108B-23R-02, 18–20 cm	208.48	26	silty claystone	–	minicore	2.12	2.7
180-1108B-27R-02, 139–141 cm	247.79	26	silty claystone	49	minicore	2.17	2.74
180-1108B-27R-02, 135–139 cm	247.75	22	silty claystone	–	minicore	2.17	2.74
180-1108B-34R-02, 55–57 cm	313.89	25	clayey siltstone	–	minicore	2.33	2.74
180-1108B-34R-02, 57–59 cm	313.91	24	clayey siltstone	38	minicore	2.33	2.74
180-1108B-37R-02, 15–17 cm	342.25	30	sandy claystone	40	minicore	2.29	2.72
180-1108B-37R-02, 12–14 cm	342.22	25	sandy claystone	–	minicore	2.29	2.72
180-1108B-39R-02, 50–52 cm	361.76	22	siltstone	–	minicore	2.08	2.66
180-1108B-39R-02, 41–43 cm	361.67	33	siltstone	–	minicore	2.08	2.66
180-1108B-41R-02, 25–27 cm	381.15	22	clayey siltstone	39	minicore	2.12	2.72
180-1108B-41R-02, 28–30 cm	381.18	26	clayey siltstone	–	minicore	2.12	2.72
180-1108B-45R-02, 4–6 cm	418.95	22	silty claystone	–	minicore	2.28	2.71
180-1108B-45R-02, 13–15 cm	419.04	27	silty claystone	38	minicore	2.28	2.71
180-1108B-48R-01, 20–22 cm	446.9	18	sandy siltstone	–	minicore	2.25	2.69
180-1108B-48R-01, 50–51 cm	447.2	19	sandy siltstone	29	minicore	2.25	2.69
180-1108B-51R-01, 117–119 cm	476.67	24	clayey siltstone	37	minicore	2.22	2.71
180-1108B-51R-01, 60–62 cm	476.1	23	clayey siltstone	–	minicore	2.22	2.71
Detachment fault zone							
180-1117A-01R-02, 108–110 cm	2.58	38	fault gouge	–	soft sediment	2.28	2.79
180-1117A-01R-02, 118–120 cm	2.68	38	fault gouge	39	soft sediment	2.28	2.79
180-1117A-01R-02, 135–137 cm	2.85	34	fault gouge	45	soft sediment	2.2	2.77
180-1117A-01R-02, 144–146 cm	2.94	35	fault gouge	–	soft sediment	2.2	2.77
Footwall to detachment fault							
180-1114A-32R-01, 144–146 cm	296.84	22	brecciated metadiabase	–	minicore	2.56	2.73
180-1114A-32R-01, 141–143 cm	296.81	27	brecciated metadiabase	–	minicore	2.56	2.73
180-1117A-09R-01, 16–18 cm	66.76	22	massive metadiabase	–	minicore	2.76	2.82
180-1117A-09R-01, 27–28 cm	66.87	27	massive metadiabase	–	minicore	2.76	2.82

Porosity (%)	Void ratio	Direction of flow	Hydraulic conductivity (m s^{-1})	Intrinsic permeability (m^2)	anisotropy index $k_{horizontal}/k_{vertical}$ (dimensionless)	Number of repeat measurements	Comments
75.3	3.05	vertical	6.79602E−05	6.79602E−12		8	
75.3	3.05	horizontal	1.77114E−05	1.77114E−12		11	
41.7	0.72	horizontal	1.801E−05	1.801E−12		9	
41.7	0.72	vertical	1.72239E−05	1.72239E−12		7	
40.2	0.69	horizontal	1.51244E−05	1.51244E−12		6	
40.2	0.69	vertical	2.30846E−05	2.30846E−12		4	
44.4	0.8	horizontal	1.07861E−05	1.07861E−12		11	
44.4	0.8	vertical	7.34826E−06	7.34826E−13		12	
36.6	0.58	horizontal	2.52736E−07	2.52736E−14		6	
36.6	0.58	vertical	1.65174E−08	1.65174E−15		2	
50.8	1.03	vertical	3.60199E−08	3.60199E−15		3	
50.8	1.03	horizontal	1.51244E−07	1.51244E−14		4	
50.6	1.02	vertical	1.52239E−07	1.52239E−14		1	
50.6	1.02	horizontal	7.69154E−07	7.69154E−14		3	
36.1	0.56	horizontal	1.75124E−06	1.75124E−13		2	
36.1	0.56	vertical	2.03085E−06	2.03085E−13		3	
34.9	0.54	vertical	9.87065E−07	9.87065E−14		2	
34.9	0.54	horizontal	1.31244E−06	1.31244E−13	1.329637097	9	horizontal fissures
33.3	0.5	vertical	6.14925E−07	6.14925E−14		2	
33.3	0.5	horizontal	5.42289E−07	5.42289E−14	0.881877023	3	
23.7	0.31	horizontal	1.88557E−08	1.88557E−15		4	
23.7	0.31	vertical	3.25373E−08	3.25373E−15	0.579510703	2	
21.1	0.27	horizontal	4.15622E−07	4.15622E−14		1	
21.1	0.27	vertical	6.29552E−07	6.29552E−14	0.660186502	2	
35.7	0.56	horizontal	1.04876E−07	1.04876E−14		2	
35.7	0.56	vertical	7.52239E−08	7.52239E−15	1.394179894	4	
35.3	0.47	horizontal	1.33134E−07	1.33134E−14		4	
35.3	0.47	vertical	9.35821E−08	9.35821E−15	1.422647528	2	
25.2	0.34	horizontal	1.08458E−07	1.08458E−14		5	fractured
25.2	0.34	vertical	9.89055E−08	9.89055E−15	1.096579477	2	
27.8	0.39	horizontal	6.47935E−08	6.47935E−15		4	
27.8	0.39	vertical	3.78289E−08	3.78289E−15	1.712807091	2	
29	0.41	horizontal	7.1227E−08	7.12269E−15		1	
29	0.41	vertical	4.0995E−08	4.0995E−15	1.737451456	2	
28.7	0.4	horizontal	6.12935E−07	6.12935E−14		3	scaly fabrics in some firmer clay and cloritic parts of the sample
28.7	0.4	vertical	3.42289E−07	3.42289E−14	1.790697674	7	
32.3	0.48	horizontal	7.46269E−07	7.46269E−14		2	
32.3	0.48	vertical	4.0995E−07	4.0995E−14	1.82038835	4	
9.8	0.11	horizontal	5.8607E−10	5.8607E−17		2	rock pervasively fractured (secondary porosity)
9.8	0.11	vertical	4.36816E−10	4.36816E−17	1.341685649	2	rock pervasively fractured (secondary porosity)
2.95	0.03	horizontal	6.0995E−11	6.0995E−18		1	
2.95	0.03	vertical	5.23085E−11	5.23085E−18	1.166064295	1	

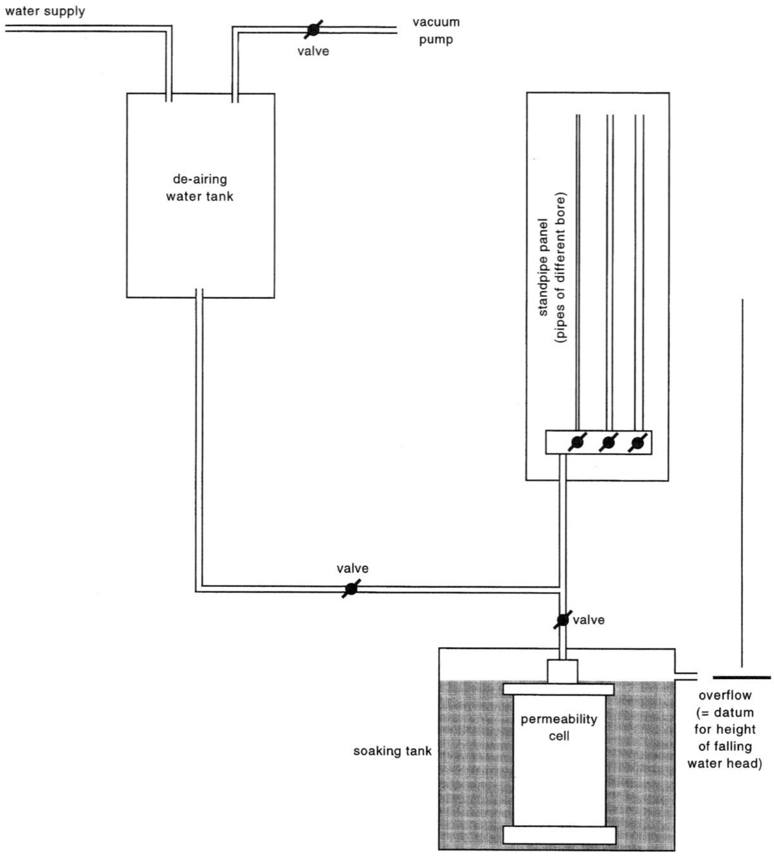

Fig. 4. ELE falling-head permeameter set-up used in the shipboard laboratory during ODP Leg 180. (See text for explanation.)

Permeability tests

Background

Hydraulic conductivity, K (in m s^{-1}), was directly determined as velocity of the migration of a fluid phase through a sediment of known flow rate, following Darcy's law:

$$K = Q/Ai$$

where Q is flow rate (m^3 s^{-1}), A is cross-sectional area (m^2), and i is hydraulic gradient (dimensionless).

The intrinsic permeability, k (in m^2), is related to the hydraulic conductivity, K, by accounting for the unit weight and viscosity (η) of the pore water (see Narasimhan 1982):

$$k = K\eta/\gamma_w$$

where γ_w is the density (or specific weight) of the fluid. K is also known as Darcy's coefficient of permeability, being a function of both the sediment and pore fluid properties. For pure water as a fluid, the conversion factor between K and k has been found to be *c.* 10^{-7} (Olson & Daniel 1981). As Darcy's law is linear for small hydraulic gradients, but non-linear for high hydraulic gradients (which are often needed to determine low hydraulic conductivities encountered with fine-grained sediments), different methods have been applied to measure permeability in these materials (see review by Tavenas *et al.* 1983*a*). Falling-head and constant-head permeameters are traditionally used for more permeable rock and sediment, whereas flow pumps are often used for lower-permeability materials (see Aiban & Znidarcic 1989). Nevertheless, falling-head set-ups can be used to measure K in fine-grained sediments as well, given that a closed, saturated system is used (as is necessary for all permeability tests) and tests are conducted over long periods of time (including a number of repeat runs). Another impetus for using a falling-head kit was its relative ease of assembly in the laboratory of R.V. *Joides Resolution*.

Sampling strategy and methods

A series of 21 pairs of undisturbed minicores (2.5 cm across; variable length, as given in Table 1) were collected from the three drillsites described above. The necessity for a synthetic profile emerged from the unsuccessful 'deep' hole at Site 1108, where hole instability and the occurrence of thermogenic gas stopped the drilling operations at an early stage (Taylor *et al.* 1999). Nevertheless, owing to successful recovery of both fault gouge (Site 1117) and footwall rock (i.e. basement at Sites 1114 and 1117), a vertical profile was synthesized (see below). The minicores were taken as pairs in directions parallel to the primary core axis and perpendicular to it. For each pair, the individual cores were taken within tens of centimetres depth, so that they originated from the same piece of rock (see Table 1), hence avoiding lithological or fabric diversity. This allows the distinction between horizontal (i.e. subparallel to the detachment and the dominating rock fabric) and vertical flow. Owing to intense faulting of the entire succession drilled, core recovery was very low so that the samples chosen for permeability testing are limited in size. Nevertheless, a representative set of intact cylindrical samples (both taken in plastic tubes in the case of soft sediments and drilled as minicores) were obtained from the recovered material.

A modified ELE falling-head permeability apparatus (type EL25-0605) was used to determine hydraulic conductivity at ambient pressures (see Fig. 4). The tests were carried out during the cruise so that unconsolidated to semi-lithified deposits did not suffer fabric alteration as a result of evaporation of pore water during transport. Testing principles and procedures closely followed ASTM 2434 (ASTM 1989) and ASTM D5084 (ASTM 1990) procedures (see also Pane *et al.* 1983; Jose *et al.* 1989), but modified slightly to suit the requirements and restrictions of the shipboard laboratory, and accommodate for the low permeability of the more consolidated claystones at depth. The apparatus measures the hydraulic gradient and quantity of water flowing through a sample using a varying head of water. The panel of pipes of different bore was chosen longer than in the original ELE set-up, resulting in a higher head of water (Fig. 4). Consequently, a wider range of possible permeabilities could be determined. The undisturbed sample was loaded into the permeability cell, which was equipped with a sample holder made of hardened silicone rubber adhesive sealant. Samples went into a cylindrical hole of 2.5 cm diameter, and were sealed against the silicone inset with silicone gel. Soft sediments were kept in the plastic plugs (i.e. the silicone gel was spread on the outside of the plug) whereas the outside of the rock minicores was protected with parafilm so that no silicone gel could enter the pore space and hinder water flow through the sample. Given their similar mantling (hardened silicone sealant, plastic tube, parafilm), the confining pressures, and hence permeability, can be compared between the samples of different physical properties. The sample was placed in the permeability cell together with a loofah sponge on top (to fill the upper part of the spacious cell) and immersed in a soaking tank filled with water to isolate the specimen from air during the tests. Before testing, the entire system was bled to remove air, and a vacuum applied to the cell to ensure full saturation of the sample. Depending on the grain size and consistency of the sample, one of three standpipes of different bore was chosen to allow maximum accuracy during logging of the falling head. Water flow through the sample was then recorded as a function of both the water head and time. For permeability, k (in m^2), it follows that

$$k = 2.3026al \ln(h_1/h_2)/\Delta tA$$

where a is volume of fluid, l is length of the sample, A is area of the sample, h_1 and h_2 are head before and after testing, respectively, and Δt is time elapsed during test.

Each sample was tested at least twice, some of them more often, to assess variation in the results. The number of repeat measurements is reported in Table 1. A low number of repeat measurements indicates that the first permeability runs revealed very similar results (high reproducibility); permeability was found to be more variable in the upper part of the succession (i.e. for the more permeable samples), so that up to 12 repeat measurements were carried out (Table 1).

Results

Fabric studies. Petrographic study of thin sections of mudstone (mostly Site 1108, hanging wall to the detachment) and metadiabase (Site 1117, footwall to the detachment) revealed

significant development of a secondary fracture porosity. These fractures were up to 1 mm wide, but generally ranged from 100 to 400 μm. The wider structures are probably a combination of stress relief of naturally evolved fissures, whereas the narrower ones (especially when not part of a fracture network) appear to be natural. The majority of these features are parallel or at low angles to the dominant sedimentary fabric (i.e. subhorizontal). Hence, such fracture porosity favours permeability anisotropy if the fracture network is interconnected, and most specifically enhanced fluid flow in subhorizontal direction (see below). Very rarely, the fractures and microfissures have been sealed again. Authigenic vein fills, both calcareous and siliceous, are usually restricted to distinct intervals near highly deformed zones. They include fault zone 2 at Site 1108 as well as the mylonitized interval beneath the detachment fault gouge at Site 1117 (see Fig. 3).
A typical example of a consolidated mudstone fabric intersected by a calcite vein is shown in Fig. 5a. Here, the homogeneous nature of the matrix can be observed in strong contrast to the fracture with its sharp boundary. In contrast, some mudstones are not only fractured, but show microveins and fissures along their grain boundaries away from the main fracture (Fig. 5b). Such secondary porosity probably results from pore fluid overpressure hindering contact between the particles. If this secondary pore space is not filled by authigenic precipitates, such pore space may enhance fluid flow significantly. At Site 1108, healed fractures are most commonly found in the fault zones 1 (only one sample) and 2 (abundant fractures), but not in the consolidated mudstones above or below (Fig. 3). Consequently, the sealing of fractures is interpreted as a direct consequence of conductive fluid flow in these fault zones. With increasing distance from the faulted zone, fluid entrainment becomes negligible as a result of low porosity and high tortuosity (e.g. Clennell 1997), as observed in thin section.

In the mylonitized rocks in the detachment fault zone (Site 1117, see Fig. 3), the foliated fabrics are often characterized by anastomosing fracture networks filled with calcite (Fig. 5a–d), quartz (Fig. 5e) or, less often, epidote. These fractures are filled with either micritic or blocky calcite, and can show twinning to accommodate local differential stress (Rowe & Rutter 1990). In places, the plastically deformed foliation fractures have suffered a later brittle deformation event, sometimes leading to extensional en echelon arrangement (Fig. 5e). Owing to continued regional extension and possibly elevated pore pressures, pervasive fractures developed, some of which were later filled with calcite. The later calcite vein shown in Fig. 5d shows closely spaced twins. The complex multi-phase deformational history is beyond the scope of this study; the reader is referred to Roller *et al.* (2001).

Physical properties. As outlined in Table 1 and in the drilling results from ODP Leg 180 (Taylor *et al.* 1999, Chapter 3), some physical properties are routinely determined immediately after core recovery. Regarding the porosity, the hanging-wall mudstones show a decrease from *c.* 45% at a depth of *c.* 150 mbsf to *c.* 25–30% near 450 mbsf (Fig. 6a; Taylor *et al.* 1999). Given that near sea-floor porosities of marine sediments range between 65 and 80% (e.g. Brückmann 1989), this trend suggests either erosion of overlying sediments, or that processes other than compaction reduce the pore volume. There is no direct evidence for the first from drilling, whereas petrography supports the latter, with precipitation being one mechanism of pore-space reduction (at least in the pores in the vicinity of fractures; see above and Fig. 5a and b). In the fault zone, porosity drops to an average 30% (fault gouge) and <10% (mylonite) at the detachment and in the underlying metadiabase.

As porosity variations often do not directly correlate with permeability (Dewhurst *et al.* 1999*b*), but with grain-size distribution (Dewhurst *et al.* 1999*a*), grain size was analysed for a number of deflocculated specimens by simple settling tests in Atterberg cylinders. The only aim was to separate the fraction smaller than 10 μm, so that each test was interrupted after having decanted this portion, and weighing the two fractions (<10 μm and >10 μm) after drying. The results reveal that porosity excursions with depth are only sometimes mirrored by variation in clay fraction (Table 1). Although some of the samples show relatively high clay contents, they still have high pore volumes, and hence their permeabilities range at the upper end of the spectrum of this study (e.g. sample 1108B-10R-01; Table 1). From this, it appears that the maximum grain size (i.e. sand in sample 1108B-37R-02; Table 1) is the crucial factor controlling hydraulic conductivity, even if by volume more clay size particles than sand size particles occur. This is in agreement with recent systematic studies on London clay (Dewhurst *et al.* 1999*a*), where it was demonstrated that the largest pores contribute most profoundly to hydraulic conductivity

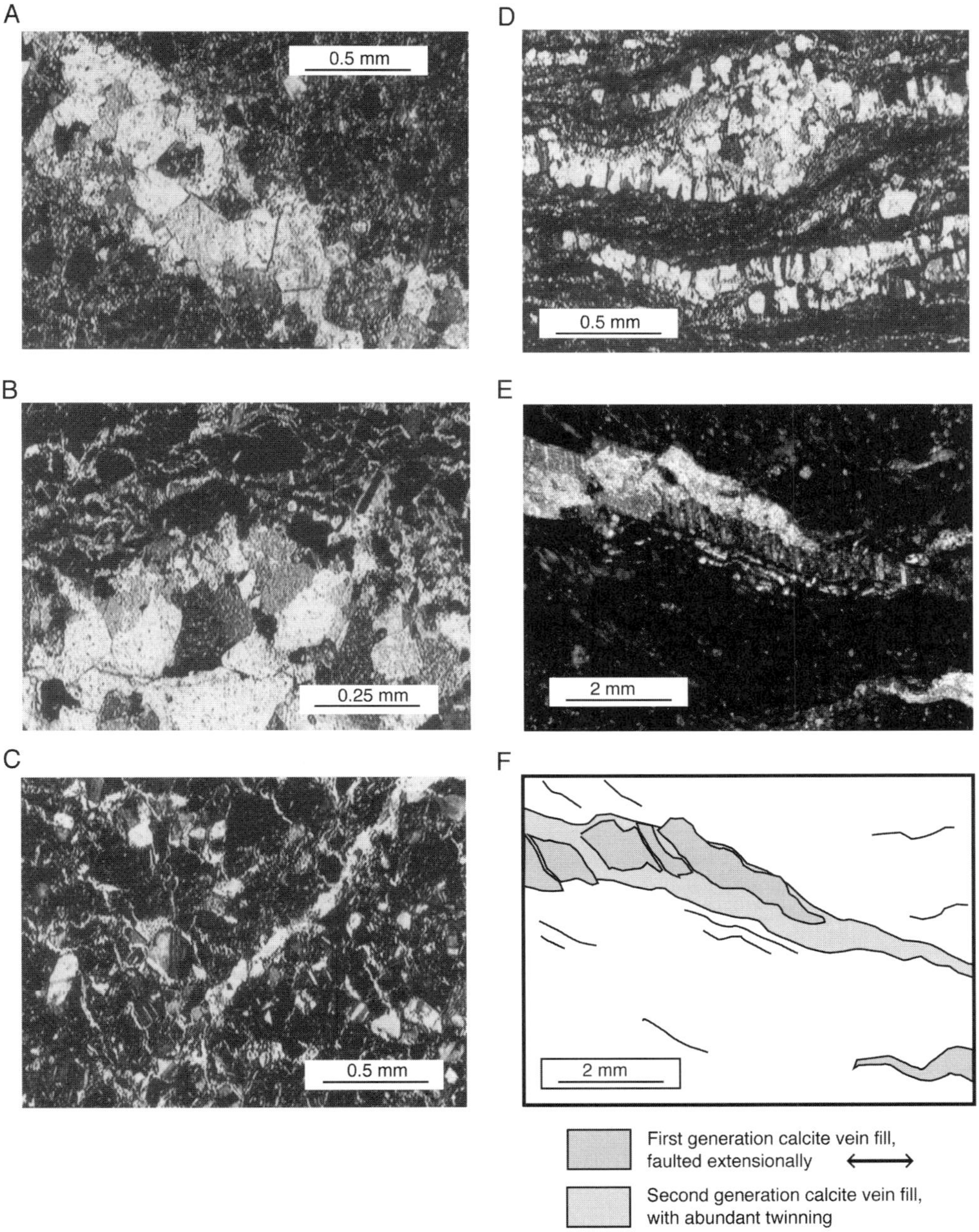

Fig. 5. (**a**) Mudstone from Hole 1108B, 428 mbsf, showing very homogeneous fabric of fine-grained clay with a fracture filled with authigenic calcite. (**b, c**) Silty mudstone from 1108B, 371 mbsf, with the terminal part of a thin, calcite-filled fracture. (Note that perpendicular to the fracture, fissures along grain boundaries occur frequently.) (**d**) Mylonite from Hole 1117A, 57 mbsf, with fractures parallel to the foliation. Fractures are either chloritized or cemented with blocky calcite. (**e**) Mylonite from Hole 1117A, 67 mbsf, indicating reactivation of fractures or shear bands. An earlier vein filled with calcite cement is separated by extensional stress; the resulting second generation fracture is now filled with twinned calcite. All micrographs taken with crossed Nicols. (**f**) Sketch of (**e**).

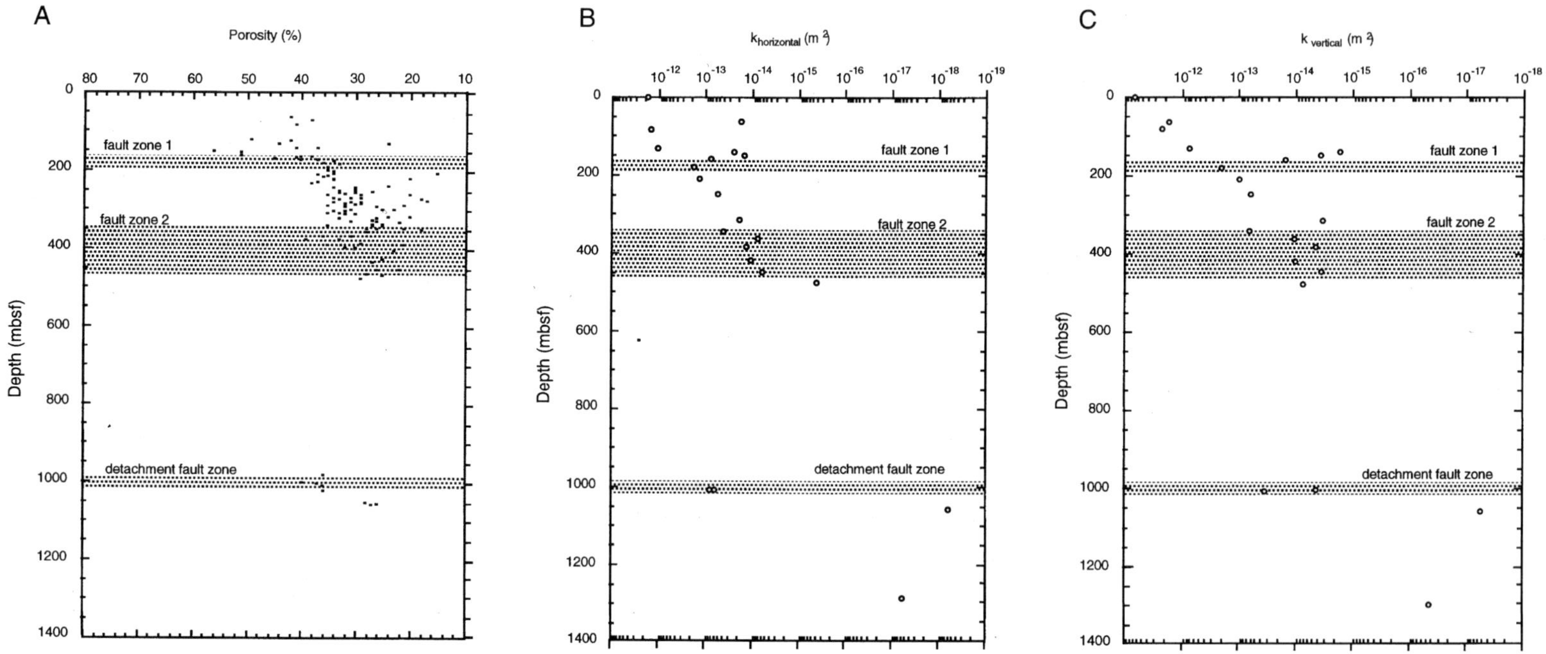

Fig. 6. (**a**) Results from porosity measurements from ODP Leg 180 cores v. depth (Taylor *et al.* 1999), and from permeability data of tested specimens recovered from Sites 1108, 1114, and 1117 v. depth, presented for horizontal (**b**) and vertical flow (**c**). It should be noted that there is no real depth scale in the lower part of the figure (below 500 mbsf), as samples from Sites 1114 and 1117 have been projected to depth (see text for justification).

of a mudstone, so that low clay contents lead to a disproportional increase in permeability. If, however, shear deformation allows a well-compacted fabric to develop (as is seen immediately above the fault splays in FZ1 and FZ2; Fig. 3), permeabilities reach local minima (in the context of results from this study; see Table 1).

Next to data obtained from permeability tests, wet bulk and grain densities as well as porosities are reported in Table 1. Values of the latter were determined on discrete samples after splitting the core as part of the physical property measurement routines on board. The data shown represent the shipboard result closest to the location of the permeability specimen from this study. The void ratio is calculated from the ODP physical properties data, being the volume of voids (i.e. the porosity) v. the volume of grains. Void ratio as well as porosity show a roughly log–linear relationship with depth, despite considerable scatter (see Fig. 6a and Table 1). These variations in lithology attributes only partly account for the significant change in porosity, so that Taylor *et al.* (1999) proposed erosion to explain the apparent undercompaction in the upper part of the hole (see Athy function as a reference (Athy 1930); Fig. 6a).

The permeability data show three main results. First, the hydraulic conductivity in vertical direction is generally slightly lower than the horizontal, fabric-parallel flow (Table 1). Second, a general downhole decrease in permeability is seen. Third, despite the lithology of the hanging-wall succession (i.e. the Site 1108 cores) being fairly homogeneous, two fault zones (FZ1 and FZ2 in Figs 3 and 6) have a profound effect on permeability. This is outlined in the upper part of Fig. 6, where distinct excursions of an otherwise exponential downhole trend can be found. At the topmost edge of the two fault zones at *c.* 165 and *c.* 350 mbsf, respectively (see above, and Taylor *et al.* 1999), a local permeability minimum is reached, before permeability increases as a result of fracturing and scaly fabrics in the fault zone. Below the fault zones, the overall log–linear decrease of permeability with depth continues. The second profound change in this trend is the high permeability of the fault gouge samples (Fig. 6, projected to *c.* 1000 mbsf). It should be noted here that the results from the Sites 1114 and 1117 are plotted in their 'artificial' (not recovered) depth according to the synthetic profile (see captions of Figs 3 and 6). It is, however, important to recognize the alignment of the permeability of the footwall metadiabase with the exponential trend defined in the hanging-wall section, illustrating the generally low porosity of the hanging-wall mudstones as a result of precipitation processes (see discussion by Taylor *et al.* (1999) of porosity data and possible erosion at Site 1108). The increase in permeability within the detachment fault may have several explanations, not all of which may relate to tectonic events. In particular, alteration on the sea floor at Site 1117 may have changed and weakened the deformation fabrics, thus triggering mineral reactions and serpentinization. In addition, thin-section study suggests volume increase and fracturing accompanying serpentinization. As alteration of the surrounding footwall rocks is less profound compared with the fault gouge, post-deformational fluid processes appear to have occurred in localized, fractured intervals. Although the gouge samples studied had been recovered near the sea floor (and hence have probably been a subject to sea-water alteration), fault gouge buried at a depth of 1 km or more may well be affected by migrating (possibly) overpressured fluids (see Discussion).

Discussion

The discussion is split into two parts, the first of which addresses problems associated with mudstone permeability measurements and the validity of the ELE set-up for the measurement of low-permeability rocks, whereas the second considers the geological significance of the findings for fault development and activity.

Difficulties in assessment of the permeability of mudstones

As has been pointed out by previous workers (e.g. Neuzil 1994; Dewhurst *et al.* 1999*b*), the limited number of mudstone permeability data vary over almost ten orders of magnitude. This demonstrates the lack of quantitative understanding of fluid flow through well-characterized muds and mudstones. Dewhurst *et al.* (1999*a*) have shown that not only porosity, but also grain-size variation and surface area within a mudrock cause permeability variations of several orders of magnitude. Also, it has been shown that faults and fractures play a considerable role in altering mudstone permeability on a temporary basis. Fluid overpressures can produce hydrofracture leading to enhanced flow (e.g. Brown *et al.* 1994*a*; Roberts & Nunn 1995). However, the extent to which this fracturing affects long-term permeability is almost impossible to predict, because secondary porosity varies on a short-term basis, and more so when fluid migration and sealing (precipitation in veins) occur. In the first part of this discussion, hydrofracture development and other mechanisms producing secondary pore space are only mentioned as one effect which caused permeabilities to be high enough to be measured with the ELE set-up. The geological and physical implications of such hydrofracturing are now discussed.

It has been argued that a falling-head system does not allow permeability determination of very fine-grained material, mostly as Darcy's flow law becomes non-linear under large hydraulic gradients conditions (Tavenas *et al.* 1983*a*, *b*; Silva *et al.* 1989; Moran *et al.* 1995). However, the limitations of the system become a concern only with the results from the footwall rock, as both the altered, chloritized fault gouge material in the detachment zone as well as the predominantly silty sediments and rocks of the hanging-wall block require permeabilities well within the range that the ELE system was designed for. Also, the results obtained are in good agreement with permeabilities determined in a backpressured constant-rate flow pump system outlined elsewhere, for example by Bolton *et al.* (1999). Those workers tested undisturbed whole round core samples, which range in the same order of magnitude (i.e. 10^{-17}–10^{-18} m^2 for the hanging-wall claystones; Bolton *et al.* 2001). Such results also correspond well to accreted clay- and siltstones across a prominent reverse fault at the Cascadian margin. There, different techniques were applied to measure permeability, using (1) an oedometer system (Moran *et al.* 1995), (2) a constant-rate flow pump apparatus (Brown 1995), and (3) an *in situ* drill-string PACKER (Screaton *et al.* 1995; 1997). Values ranged from 10^{-11} to 10^{-18} m^2 for the claystones (methods (1) and (2)), and from 10^{-11} to 10^{-16} m^2 for the fault zone (methods (2) and (3)). The studies off Cascadia were carried out independently and corroborate the results found in this study. Similarly, such comparative studies yielded comparable results for clays recovered from backthrust faults in the Mediterranean Ridge (Kopf *et al.* 1998). Older, more consolidated rocks have undergone permeability testing in an apparatus very similar to the one in this study (McLatchie *et al.* 1958). These argillaceous Cretaceous shales yielded 10^{-11} to 10^{-14} m^2 (Young *et al.* 1964), i.e. within the range of the Woodlark mudstones.

Regarding the basement rocks, however, the ELE system may well have been inappropriate to determine the permeability reliably. Despite this, the results are in broad agreement with permeability data for other metamorphic or basement rocks (10^{-14}–10^{-18} m^2 for unaltered basaltic crust; Matthess & Ubell 1983). For young oceanic crust with active hydrothermal fluid circulation (e.g. the Costa Rica Rift), Becker (1996) has measured bulk *in situ* permeabilities between 10^{-13} and 10^{-14} m^2. In summary, there may remain some uncertainty regarding a potential overestimate in permeability in the deepest part of the succession (particularly the metadiabase of the footwall), but none the less there is a relative change in permeability of several orders of magnitude when crossing major faults in the hanging wall and at the detachment. The implications of these results, as well as the anisotropy owing to the fabric, are the subject of the second part of the discussion.

Effect of permeability on low-angle normal fault activity

Distinct variation of permeability as a function of sedimentary fabrics has long been known from studies of remoulded or natural sediments (e.g. Young *et al.* 1964; Schultheiss & Gunn 1985; Arch & Maltman 1990). Generally, flow parallel to the fabric (often subhorizontal when compaction dominates deformational processes) exceeded that perpendicular to it. In a systematic study, Dewhurst *et al.* (1996) compared permeability in compaction-induced and shear fabrics. Starting from remoulded, isotropic samples, uniaxial consolidation alone caused very little anisotropy in permeability. This result is in agreement with naturally compacted samples, where despite grain alignment, permeability anisotropy is negligible (Taylor & Leonard 1990). This somewhat counter-intuitive finding has previously been attributed to the silt fraction, because pure clays exhibit moderate k anisotropy when compacted uniaxially (e.g. Al-Tabbaa & Wood 1987), most evidently as a result of change in tortuosity (Clennell 1997). When sheared, however, clay size particles align and cause barrier formation, which profoundly affects hydraulic conductivity. Parallel to these shear-enhanced fabrics, conductive fluid migration may be several orders of magnitudes higher than perpendicular to it (Dewhurst *et al.* 1996; Faulkner & Rutter 1998). This phenomenon is known from main fault zones, such as décollements in an accretionary prism, and has been widely accepted (e.g. Screaton *et al.* 1990; Brown 1995; Kimura *et al.* 1997). In particular, in case studies of the décollement of the Costa Rican forearc wedge or the Cascadia accretionary complex, the immediate top to the actual fault zone has been described as having undergone enhanced compaction and particle alignment (e.g. Kimura *et al.* 1997; Labaume *et al.* 1997). In addition, permeability variation over a prominent frontal thrust at the Hydrate Ridge, Cascadia, reaches its minimum just above the thrust, and higher values are recorded in the highly conductive fault zone (Brown 1995; Screaton *et al.* 1995). Abundant scaly fabric development has been found at the uppermost contact between the two fault zones and their hanging-wall sediments at Site 1108. The hanging-wall edge of the fault zones is characterized by polished bands of clay-rich material, suggesting localized shear, which caused enhanced compaction. Consequently, discrete minicore samples from these intervals revealed minimum permeabilities with respect to the data of this study (Fig. 6). The disintegrated fabrics of faulted intervals themselves have obviously been subject to hydrofracturing and scaly fabric formation, so that permeabilities in the fault zones (FZ1 and FZ2; Fig. 6) increase about two orders of magnitude compared with their hanging-wall edge. It should be noted here that these values probably underestimate the absolute change, because the minicores were taken in the most intact parts of the fault zones, and not in intervals where fabric degradation or destruction is most intense. Thus, with respect to the definition by Labaume *et al.* (1997), they originate from the deformed, fractured lenses of the interval of scaly fabric formation, but not from the contact of deformed and undeformed lenses (along which samples are likely to part as a result of former shear localization).

Hydrofracture formation and its implications for geological sediment deformation have been the subject of a great body of work by numerous workers, for example, Behrmann (1991), Brown *et al.* (1994a) and discussion by Fischer & Engelder (1994) (see also reply by Brown *et al.* 1994*b*), most of which cannot be repeated here. Also, the significance of shear zones and its importance in plumbing marine fine-grained sediments has been pointed out repeatedly (e.g. Maltman 1988). In general, major shear surfaces are capable of acting as seals to migrating fluids (in a subvertical direction), and hence result in the build-up of overpressures for the above-mentioned reasons. These fluid pressures have a crucial control on episodic fluid expulsion (e.g. Yassir & Bell 1994; Roberts & Nunn 1995), and are also believed to allow, or at least trigger, fault movement while approaching lithostatic pressure (Moore *et al.* 1995; Moore & Tobin 1997). Sibson (2000) has shown that activation of low-angle normal faults is facilitated only if pore pressures are high in the fault zone, and the host rock to the fault zone retains sufficient tensile strength. Especially in areas where horizontal shear bands and veining are abundant, fluid pressures equal lithostatic loading (as at the Barbados proto-décollement thrust; fig. 4 of Moore & Tobin 1997). Fault dilation has been proposed to be the direct consequence (Moore *et al.* 1995). The effective stresses at the Barbados accretionary prism form a log–linear relationship with *in situ* permeabilities measured (Fisher *et al.* 1996). Those results range between 10^{-13} and 10^{-15} m^2 at a depth of *c.* 500 mbsf, and thus commensurate with those in this study.

Applied to the western Woodlark Basin, the load of the overburden synrift basin fill can be overcome by excess fluid pressures built up by the low-permeability units that immediately overlie the fault zones. Fluid pressures near lithostatic then enable the hanging-wall block to slide downdip along the detachment plane. As has been shown previously, closure of rock joints first occurs elastically (e.g. Brown & Scholz 1986), but subsequently plastic deformation and precipitation processes may play a crucial role. The results from healed fractures in the semi-consolidated clayey sediments at Site 1108 provide evidence that precipitation of authigenic calcite (and, to a lesser extent, epidote and quartz) may have the primary control on effective stress and fluid pressure variation. After closure of fractures, pore pressure rises until failure occurs, with fluids then migrating along both reactivated fractures or new joints and particle or aggregate boundaries (Fig. 4b and c). Similarly, mylonitic rocks of the detachment fault zone are affected by such deformation (Fig. 4d and e). Thereafter, precipitation healing is probably triggered by the flux of warm fluids generated at depth near the tip of the Woodlark spreading centre some 2 km eastward (Fig. 1). The continuing activity of the Woodlark detachment, as well as its status as a fluid conduit, is best reflected by the porosity decrease owing to cementation reactions. Soft fault gouge, although recovered at <20 mbsf at Site 1117 (Fig. 3), has low water contents of 15–30% (Taylor *et al.* 1999) and a volumetrically stiff response. Undercompaction and fluid overpressures are suggested, and hydrothermal alteration may also contribute to weakening of the fault zone. As known from other detachment faults (Vrolijk & van der Pluijm 2000), such fault zone fabrics may have undergone degradation as a result of mineral reactions (e.g. smectite–illite transformation), causing not only weakening, but also an overall increase in hydraulic conductivity. For at least the upper part of the Woodlark detachment, mineral reactions, namely chloritization and serpentinization (as evidenced from X-ray diffraction and petrography), severely affect the strength of the rock. The lubricating effect of serpentine-bearing layers in, for example, the Akapnou Forest detachment (Troodos ophiolite, Cyprus; see MacLeod 1999) has been demonstrated by field observations and shear strength experiments on simulated fault gouge (Moore *et al.* 1997).

Although mylonites provide evidence for a phase of plastic deformation during evolution of the Woodlark detachment (Fig. 4d and e), present-day seismicity (e.g. Abers *et al.* 1997) and multi-phase fracturing suggest episodic brittle failure and seal to be the dominant deformation mechanisms. The overall weakness of the gouge layer, as reflected by a relatively high porosity of 30% and an unconfined compressive strength not higher than 90 kPa, may provide an additional trigger for fault activity.

Conclusions

This paper reports a series of hydraulic conductivity tests on intact discrete minicore samples from deep-sea drilling into an active rift basin off Papua New Guinea. The main findings include the following.

i. Mudstone permeability across the Woodlark Basin detachment fault varies over several orders of magnitude, with its minimum values immediately at the base of hanging walls to major fault zones.
ii. Anisotropy of permeability is a consequence of shear-enhanced compaction, i.e. fault rock fabric and layering at the hanging-wall edge of the fault zone, and fabric degradation within the fault zone.
iii. Low subvertical fluid conductivity may cause overpressures exceeding lithostatic, leading to hydrofracture, and both temporal fluid expulsion and fault movement.
iv. Multi-phase mineralization of the hydrofractures favours cycles of fault or fracture seal by authigenic precipitates, fluid pressure build-up, brittle failure and fault movement, followed by a phase of enhanced fluid flow before precipitation occurs again.
v. Along the detachment fault plane *sensu stricto*, i.e. the uppermost part of the underlying core complex, weakening is a function of a number of processes, namely mineral reactions (forming serpentine, talc, and chlorite), fluid migration, shear deformation and fabric degradation.

Many thanks are due to L. Screaton for discussion of physical properties, and her magnanimity and good humour during cruise ODP Leg 180. J. Behrmann generously provided the equipment for the permeability tests. A. Bolton and D. Faulkner are thanked for their detailed, helpful suggestions and critical reviews of the manuscript. Proponents, scientists, shipboard technicians, and crew of ODP Leg 180 are acknowledged for having made things happen. Special thanks go to the drilling engineers and drillers for an excellent job recovering rock from this highly disturbed region. The author is supported by a stipendium through BASF AG, Germany.

References

ABERS, G.A. 1991. Possible seismogenic shallow-dipping normal faults in the Woodlark–D'Entrecasteaux extensional province Papua New Guinea. *Geology*, **19**, 1205–1208.

ABERS, G.A. & ROECKER, S.W. 1991. Deep structure of an arc–continent collision: earthquake relocation and inversion for upper mantle P and S wave velocities beneath Papua New Guinea. *Journal of Geophysical Research*, **96**, 6379–6401.

ABERS, G.A., MUTTER, C.Z. & FANG, J. 1997. Earthquakes and normal faults in the Woodlark–D'Entrecasteaux rift system Papua New Guinea. *Journal of Geophysical Research*, **102**, 15301–15317.

AIBAN, S.A. & ZNIDARCIC, D. 1989. Evaluation of the flow pump and constant head techniques for permeability measurement. *Géotechnique*, **39**, 655–666.

AL-TABBAA, A. & WOOD, D.M. 1987. Some measurements of the permeability of kaolin. *Géotechnique*, **37**, 499–503.

ARCH, J. & MALTMAN, A. 1990. Anisotropy permeability and tortuosity in deformed wet sediments. *Journal of Geophysical Research*, **95**, 9035–9046.

ASTM 2434, 1989. Standard test method for permeability of granular soils (constant head). *American Society for Testing and Materials, Annual Handbook*, **04.08**, 278–282.

ASTM D5084, 1990. Standard test method for measurement of hydraulic conductivity of saturated porous materials using a flexible wall permeameter. *American Society for Testing and Materials. Annual Handbook*, **04.02**, 151–158.

ATHY, L.F. 1930. Density, porosity, and compaction of sedimentary rocks. *AAPG Bulletin*, **14**, 1–23.

AXEN, G.J. 1992. Pore pressure, stress increase, and fault weakening in low-angle normal faulting. *Journal of Geophysical Research*, **97**, 8979–8991.

BARTON, C.A., ZOBACK, M.D. & MOOS, D. 1995. Fluid flow along potentially active faults in crystalline rock. *Geology*, **23**, 683–686.

BECKER, K. 1996. Permeability measurements in Hole 896A and implications for the lateral variability of upper crustal permeability at Sites 504B and 896. *In*: Alt, J.C. *et al.* (eds) *Proceedings of the Ocean Drilling Program, Scientific Results*, **148**. Ocean Drilling Program, College Station, TX, 353–362.

BOLTON, A.J., CLENNELL, M.B. & MALTMAN, A.J. 1999. Nonlinear stress dependence of permeability: a mechanism for episodic fluid flow in accretionary wedges. *Geology*, **27**, 239–242.

BOLTON, A.J., MALTMAN, A.J. & FISHER, Q.J. 2001. Anisotropic permeability and bimodal pore size distributions of fine grained marine sediments. *Marine and Petroleum Geology*, in press.

BROWN, K.M. 1995. The variation of the hydraulic conductivity structure of an overpressured thrust zone with effective stress. *In*: CARSON, B., WESTBROOK, G.K., MUSGRAVE, R.J. & SUESS, E. (eds) *Proceedings of the Ocean Drilling Program, Scientific Results*, **146**. Ocean Drilling Program, College Station, TX, 281–290.

BROWN, K.M., BEKINS, B., CLENNELL, M.B., DEWHURST, D. & WESTBROOK, G.K. 1994*a*. Heterogeneous hydrofracture development and accretionary fault dynamics. *Geology*, **22**, 259–262.

BROWN, K.M., BEKINS, B., CLENNELL, M.B., DEWHURST, D. & WESTBROOK, G.K. 1994*b*. Reply. *Geology*, **22**, 1053–1054.

BROWN, S.R. & SCHOLZ, C.H. 1986. Closure of rock joints. *Journal of Geophysical Research*, **91**, 4939–4948.

BRÜCKMANN, W. 1989. *Typische Kompaktionsabläufe mariner Sedimente und ihre Modifikation in einem rezenten Akkretionskeil (Barbados Ridge)*. Tübinger Geowissenschaftliche Abhandlungen, **A5**.

CLENNELL, M.B. 1997. Tortuosity: a guide through the maze. *In*: LOVELL, M.A. & HARVEY, P.K. (eds) *Developments in Petrophysics*. Geological Society, London, Special Publications, **122**, 299–344.

CLENNELL, M.B., DEWHURST, D.N., BROWN, K.M. & WESTBROOK, G.K. 1999. Permeability anisotropy of consolidated clays. *In*: APLIN, A.C., FLEET, A.J. & MACQUAKER, J.H.S. (eds) *Muds and Mudstones: Physical and Fluid Flow Properties*. Geological Society, London, Special Publications, **158**, 79–96.

DEWHURST, D.N., BROWN, K.M., CLENNELL, M.B. & WESTBROOK, G.K. 1996. A comparison of the fabric and permeability anisotropy of consolidated and sheared silty clay. *Engineering Geology*, **42**, 253–267.

DEWHURST, D.N., APLIN, A.C. & SARDA, J.P. 1999. Influence of clay fraction on pore-scale properties and hydraulic conductivity of experimentally compacted mudstones. *Journal of Geophysical Research*, **104**, 29261–29274.

DEWHURST, D.N., APLIN, A.C. & YANG, Y. 1999*b*. Permeability and fluid flow in natural mudstones. *In*: APLIN, A.C., FLEET, A.J. & MACQUAKER, J.H.S. (eds) *Muds and Mudstones: Physical and Fluid Flow Properties*. Geological Society, London, Special Publications, **158**, 23–43.

FAULKNER, D., & RUTTER E.H. 1998. The gas permeability of clay-bearing fault gouge at 20 °C. *In*: JONES, G., FISHER, Q.J. & KNIPE, R.J. (eds) *Faulting, Fault Sealing and Fluid Flow in Hydrocarbon Reservoirs*. Geological Society, London, Special Publications, **147**, 147–156.

FISCHER, M.P. & ENGELDER, T. 1994. Discussion to Brown et al., 1994. *Geology*, **22**, 1052–1054.

FISHER, A.T., ZWART, G. & ODP Leg 156 Scientific Party 1996. Relation between permeability and effective stress along a plate-boundary fault, Barbados accretionary complex. *Geology*, **24**, 307–310.

GOODLIFFE, A., TAYLOR, B., MARTINEZ, F., HEY, R., MAEDA, K. & OHNO, K. 1997. Synchronous reorientation of the Woodlark Basin spreading center. *Earth and Planetary Science Letters*, **146**, 233–242.

HICKMAN, S., SIBSON, R. & BRUHN, R. (eds) 1993. *Proceedings of Workshop LXIII: the Mechanical Involvement of Fluids in Faulting*. US Geological Survey, Open File Report, 94-228.

JOSE, U.V., BHAT, S.T., NAYAK, B.U. 1989. Influence of salinity on permeability characteristics of marine sediments. *Marine Geotechnology*, **8**, 249–258.

KIMURA, G., SILVER, E.E., BLUM, P. & Shipboard Scientific Party Leg 170, 1997. *Proceedings of the Ocean Drilling Program, Initial Reports*, **170**. Ocean Drilling Program, College Station, TX.

KOPF, A., CLENNELL, M.B. & CAMERLENGHI, A. 1998. Variations in sediment physical properties and permeability of mud volcano deposits from Napoli dome and adjacent mud volcano. *In*: ROBERTSON, A.H.F. & EMEIS, K.-C. (eds) *Proceedings of the Ocean Drilling Program, Scientific Results*, **160**. Ocean Drilling Program, College Station, TX, 625–644.

LABAUME, P., MALTMAN, A.J., BOLTON, A., TESSIER, D., OGAWA, Y & TAKIZAWA, S 1997. Scaly fabrics in sheared clays from the dĕcollement zone of the Barbados accretionary prism. *In*: SHIPLEY, T.H., OGAWA, Y. & BLUM, P. (eds) *Proceedings of the Ocean Drilling Program, Scientific Results*, **156**. Ocean Drilling Program, College Station, TX, 59–77.

MACLEOD, C.J. 1999. Moho-level detachment faulting in the Troodos ophiolite. *Geophysical Research Abstracts*, **1**, 188.

MALTMAN, A.J. 1988. The importance of shear zones in naturally deformed wet sediments. *Tectonophysics*, **145**, 163–175.

MATTHESS, G. & UBELL, K. *Lehrbuch der Hydrogeologie: Allgemeine Hydrogeologie—Grundwasserhaushalt, Band 1*. Bornträger, Berlin.

MCLATCHIE, A.S., HEMSTOCK, R.A. & YOUNG, J.W. 1958. The effective compressibility of reservoir rock and its effects on permeability. *Journal of Petroleum Technology*, **10**, 49–51.

MOORE, D.E. 1997. *Mineralogical and Microstructural Investigations of Core Samples from the Vicinity of the Great Sumatran Fault, Indonesia.* US Geological Survey, Open File Report.

MOORE, J.C. & Shipboard Party ODP Leg 156, 1995. Abnormal fluid pressures and fault-zone dilation in the Barbados accretionary prism: evidence from logging while drilling. *Geology*, **23**, 605–608.

MOORE, J.C., TOBIN, H. 1997. Estimated fluid pressures of the Barbados accretionary prism and adjacent sediments. *In*: SHIPLEY, T.H., OGAWA, Y. & BLUM, P. (eds) *Proceedings of the Ocean Drilling Program, Scientific Results*, **156**. 229–238. Ocean Drilling Program, College Station, TX.

MOORE, D. E., LOCKNER, D. A., SHENGLI, M., SUMMERS, R. & BYERLEE, J. D. 1997. Strengths of serpentinite gouges at elevated temperatures. *Journal of Geophysical Research*, **102**, 14787–14801.

MORAN, K., GRAY, W.G.D & JARRETT, C.A. 1995. Permeability and stress history of sediment from the Cascadia Margin. *In*: CARSON, B., WESTBROOK, G.K., MUSGRAVE, R.J. & SUESS, E. (eds) *Proceedings of the Ocean Drilling Program, Scientific Results*, **146**. Ocean Drilling Program, College Station, TX, 275–280.

MUTTER, J.C., MUTTER, C.Z. & FANG, J. 1996. Analogies to oceanic behaviour in the continental breakup of the western Woodlark Basin. *Nature*, **380**, 333–336.

NARASIMHAN, T.N. 1982. Physics of saturated–unsaturated subsurface flow. *In*: Narasimhan, T.N. (ed.) *Recent Trends in Hydrogeology*. Geological Society of America, Special Papers, **189**, 2–23.

NEUZIL, C.E. 1994. How permeable are clays and shales? *Water Resources Research*, **30**, 145–150.

OLSON, R.E. & DANIEL, D.E. 1981. Measurement of the hydraulic conductivity of fine grained soils. *In*: ZIMMIE, T.F. & RIGGS, C.O. (eds) *Permeability and Groundwater Contaminant Transport, STP 746*. American Society for Testing of Materials, Philadelphia, PA, 18–64.

PANE, V., CROCE, P., ZNIDARIC, D., KO, H.Y., OLSEN, H.W. & SHIFFMAN, R.L. 1983. Effects of consolidation on permeability measurements for soft clay. *Géotechnique*, **33**, 67–72.

RICE, J.R. 1992. Fault stress states, pore pressure distributions, and the weakness of the San Andreas fault. *In*: EVANS, B. & WONG, T.-F. (eds) *Fault Mechanics and Transport Properties of Rock. International Geophysics Series* **51**. Academic Press, New York.

ROBERTS, S.J. & NUNN, J.A. 1995. Episodic fluid expulsion from geopressured sediments. *Marine and Petroleum Geology*, **12**, 195–204.

ROLLER, S., BEHRMANN, J.H. & KOPF, A. 2001. Deformation fabrics of faulted rocks and syntectonic stress estimates from the active Woodlark Basin detachment zone. *In*: WILSON, R.C.L, WHITMARSH, R.B., TAYLOR, B. & FROITZHEIM, N. (eds) *Non-Volcanic Rifting of Continental Margins: a Comparison of Evidence from Land and Sea*. Geological Society, London, Special Publications, **187**, XX–XXX.

ROWE, K.J. & RUTTER, E.H. 1990. Paleostress estimation using calcite twinning: experimental calibration and application to nature. *Journal of Structural Geology*, **12**, 1, 1–17.

SCHULTHEISS, P.J. & GUNN, D.E. 1985. *The Permeability and Consolidation of Deep-sea Sediments.* Institute of Oceanographic Sciences, Report, **201**.

SCREATON, E.J., CARSON, B. & LENNON, G.P. 1995. *In situ* permeability tests at Site 892: characteristics of a hydrogeologically active fault zone on the Oregon accretionary prism. *In*: CARSON, B., WESTBROOK, G.K., MUSGRAVE, R.J. & SUESS, E. (eds) *Proceedings of the Ocean Drilling Program, Scientific Results*, **146**. Ocean Drilling Program, College Station, TX, 291–298.

SCREATON, E.J., FISHER, A.T., CARSON, B. & BECKER, K. 1997. Barbados Ridge hydrogeological tests: implications for fluid migration along an active décollement. *Geology*, **25**, 239–242.

SCREATON, E.J., WUTHRICH, D.R. & DREISS, S.J. 1990. Permeabilities, fluid pressures, and flow rates in the Barbados accretionary complex. *Journal of Geophysical Research*, **95**, 8997–9007.

SIBSON, R.H. 2000. Fluid involvement in normal faulting. *Journal of Geodynamics*, **29**, 469–499.

SILVA, A.J., HETHERMAN, J.R. & CALNAN, D.I. 1989. Low-gradient permeability testing of fine-grained marine sediments. *In*: ZIMMIE, T.F. & RIGGS, C.O. (eds) *Permeability and Groundwater Contaminant Transport*. American Society for Testing and Materials, **STP746**, 121–136.

TAKIZAWA, S. & OGAWA, Y. 1999. Dilatant clayey microstructure in the Barbados décollement zone. *Journal of Structural Geology*, **21**, 117–122.

TAVENAS, F., LEBLOND, P., JEAN, P. & LEROUEIL, S. 1983. The permeability of natural soft clays. Part 1: Methods of labororatory measurement. *Canadian Geotechnical Journal*, **20**, 629–644.

TAVENAS, F., LEBLOND, P., JEAN, P. & LEROUEIL, S. 1983. The permeability of natural soft clays. Part 2: Permeability characteristics. *Canadian Geotechnical Journal*, **20**, 645–660.

TAYLOR, B., GOODLIFFE, A., MARTINEZ, F. & HEY, R. 1995. A new view of continental rifting and initial seafloor spreading. *Nature*, **374**, 534–537.

TAYLOR, B., HUCHON, P., KLAUS, A. *et al.* 1999. *Proceedings of the Ocean Drilling Program, Scientific Results*, **180**. Ocean Drilling Program, College Station, TX.

TAYLOR, B., MUTTER, C., GOODLIFFE, A. & FANG, J. 1996. Active continental extension: the Woodlark Basin. *JOI/USSAC Newsletter*, **9**, 1–4.

TAYLOR, E. & LEONARD, J. 1990. Sediment consolidation and permeability at the Barbados forearc. *In*: MOORE, J.C. & MASCLE, A. (eds) *Proceedings of the Ocean Drilling Program, Scientific Results*, **110**. Ocean Drilling Program, College Station, TX, 289–308.

VROLIJK, P. & VAN DER PLUIJM, B.A. 2000. Clay gouge. *Journal of Structural Geology*, **21**, 1039–1048.

WERNICKE, B. & BURCHFIEL, B.C. 1982. Modes of extensional tectonics. *Journal of Structural Geology*, **4**, 105–115.

WILLS, S. & BUCK, W.R. 1997. Stress-field rotation and rooted detachment faults: a Coulomb failure analysis. *Journal of Geophysical Research*, **102**, 20503–20514.

WINTSCH, R.P., CHRISTOFFERSEN, R. & KRONENBERG, A.K. 1995. Fluid–rock reaction weakening of fault zones. *Journal of Geophysical Research*, **100**, 13021–13032.

YASSIR, N.A. & BELL, J.S. 1994. Relationships between pore pressures, stresses and present-day geodynamics in the Scotian Shelf, Offshore Eastern Canada. *AAPG Bulletin*, **78**, 1863–1880.

YOUNG, A., LOW, P.F. & MCLATCHIE, A.S. 1964. Permeability studies of argillaceous rocks. *Journal of Geophysical Research*, **69**, 4237–4245.

ZHU, W. & WONG, T.F. 1997. The transition from brittle faulting to cataclastic flow: permeability evolution. *Journal of Geophysical Research*, **102**, 3027–3041.

Experimental constraints on the mechanical and hydraulic properties of deformation bands in porous sandstones: a review

IAN MAIN[1], KAREN MAIR[1,2], OHMYOUNG KWON[1,3], STEPHEN ELPHICK[1] & BRYNE NGWENYA[1]

[1]*Department of Geology and Geophysics, University of Edinburgh, Grant Institute, West Mains Road, Edinburgh EH9 3JW, UK (e-mail: ian.main@ed.ac.uk)*

[2]*Present address: Department of Earth Sciences, University of Liverpool, 4 Brownlow Street, Liverpool L69 3GP, UK*

[3]*Present address: Department of Geology and Geophysics, Texas A&M University, College Station, TX 77843-3115, USA*

Abstract: Deformation bands form in porous, clay-poor, sandstones in the top few kilometres of the Earth's surface, involving the sequential growth of a set of discrete fault strands with minimal individual offset, ultimately culminating in the development of a slip surface with a large offset. We review some of our recent experimental results designed to reproduce the early stages of this sequence, obtained at room temperature and low confining pressure ($P < 70$ MPa) in a large-capacity (10 cm core diameter) deformation rig. We examine the physical weakening and strengthening mechanisms at work in the experiments, and discuss the implications for fault sealing. We describe laboratory experiments where deformation occurs by the progressive formation of new bands with a finite small offset and a relatively constant fault gouge grain-size distribution, at a relatively constant stress measured at the sample boundaries. The friction coefficient is 0.6, i.e. within the standard range. No large-offset slip surfaces were observed. Cross-fault permeability is transiently increased during dynamic stress drop, associated with the 'suction pump' provided by rapid near-fault dilatancy under conditions of constant flow rate. As the deformation band develops quasi-statically, permeability is then reduced further by up to two orders of magnitude as a result of shear-enhanced compaction and porosity loss of the poorly sorted gouge fragments. A simple microstructural model successfully predicts the physical sealing rates in the post-failure stage. Finally, we estimate the chemical sealing rates from mass balance calculations based on direct measurement of the pore fluid chemistry during constant flow experiments at temperatures up to 120°C. When extrapolated to longer timescales, these account quantitatively for the differences between permeability reductions measured in the laboratory and in the field.

When initially formed, faults are weak relative to the surrounding country rock. They may then weaken further, or alternatively strengthen, in response to a variety of competing underlying mechanisms. Temporal changes in fault strength may occur physically, as a result of the formation of a fault gouge by comminution and compaction processes; hydraulically, by changes in local pore pressure; or chemically (for example, weakening by stress corrosion microcracking, or strengthening by cementation). In nature all three mechanisms are strongly coupled. For example, changes in pore pressure as a result of compaction or dilatancy of fault gouge under undrained conditions can lead in turn to changes in the effective stress. Faults oriented in such a way that they are near failure tend to be hydraulically conductive (Barton *et al.* 1995; Townend & Zoback 2000). Alternatively, the creation of a fault gouge with a broad bandwidth of poorly sorted fragments leads to more efficient packing and sealing. The newly created fault gouge is also more reactive than the country rock, because of its high surface area to volume ratio, and hence is a preferred location for mineral dissolution and precipitation, leading to local healing (strengthening) and sealing (local reduction of permeability). The deformation style also depends on rock type, pressure and temperature. In the brittle field the localization of deformation is preceded by sample dilatancy as a result of microcracking, associated with an

From: HOLDSWORTH, R.E., STRACHAN, R.A., MAGLOUGHLIN, J.F. & KNIPE, R.J. (eds) 2001. *The Nature and Tectonic Significance of Fault Zone Weakening*. Geological Society, London, Special Publications, **186**, 43–63. 0305-8719/01/$15.00 © The Geological Society of London 2001.

increase in porosity and permeability. Alternatively, grain comminution in porous rocks can lead to subsequent compaction. The deformation style also depends strongly on initial porosity. For example, low-porosity crystalline rocks deform at high pressure by pervasive dilatant microcracking, whereas high-porosity sandstones deform at high pressure by grain crushing and shear-enhanced compaction (Wong & Zhu 1999). At shallow depths in the brittle crust (low temperature and pressure), fault damage zones tend to be wider than at greater depth (Sibson 1977), probably because of a change in rheology from velocity strengthening to velocity weakening friction (Scholz 1990). At even higher temperatures and pressures, deformation becomes more distributed again, in the form of ductile shear zones.

In this paper we review some recent experimental constraints on the mechanisms of formation of deformation bands in porous sedimentary rock. We examine mechanical, structural, hydraulic and chemical changes associated with the formation of a localized gouge-filled fault in porous sandstone samples, at conditions appropriate for the upper 2 km or so in the crust ($T < 120\,°C$, $P < 70\,MPa$). First we briefly describe a model that has been developed for deformation band formation based on field observation (Aydin & Johnson 1983), and then we review some of our recent experiments (Mair *et al.* 2000) to elucidate some of these processes in the laboratory. We confirm experimentally that deformation bands do grow sequentially in the form of a set of discrete strands, and infer a cyclical sequence of positive and negative feedback as a result of weakening and strengthening from mechanical and microstructural measurements (Mair *et al.* 2000).

The changes we observe in porosity and microstructure are used to explain successfully the complete cycle of quasi-static permeability change during their formation. We observe the effects of an active influx of fluid into and along the strike of the deformation bands by a dynamic 'suction pump' effect as a result of a rapid increase in storage by near-fault dilatancy. Finally, we examine the first stages of longer-term chemical healing and sealing processes by observing changes in pore fluid chemistry during the formation of reactive gouge particles. These observations are used to constrain the scale of local chemical precipitation and associated hydraulic sealing as a result of the formation of a reactive fault gouge.

A conceptual model for the formation of deformation bands

Deformation bands occur in porous, quartz-rich, clay-poor, sedimentary rocks at relatively shallow depths in the Earth's crust. They may occur as individual strands of comminuted fragments of the country rock, or in a zone of deformation bands made up of several discrete individual strands, often containing relatively underformed 'pods' of country rock (Aydin 1978; Aydin & Johnson 1978). Their typical structural style is illustrated in Fig. 1. Several workers have discussed the mechanical implications of the observed structure (Jamison & Stearns 1982; Underhill & Woodcock 1987; Antonellini & Aydin 1994), with a consensus supporting the conceptual model suggested first by Aydin (1978). The current generic model for the sequential development of deformation bands is shown in Fig. 2, after Aydin & Johnson (1983). This model is based on structural and microstructural field observation, and a mechanical analysis of materials with strain hardening and strain softening rheology. The critical microstructural observations from field investigation can be summarized as follows: (1) the deformation is highly localized within a narrow band; (2) permanent deformation in a band is by cataclastic fracturing and displacement of grains, by distortion of the matrix; and by reduction of pore volume; (3) there is both volume decrease and shear displacement across the band. The porosity reduction in the band is of the order of 60%; (4) microstructural properties, including density and grain size, change as deformation proceeds; (5) multiple deformation bands develop sequentially, maintaining similar dip and strike; (6) some multiple bands ultimately develop large-offset slip surface.

A complex interplay of local strain hardening and strain softening rheology is required to produce complex zones of deformation bands. Aydin & Johnson (1983) pointed out that initial compaction would result in strain hardening by an increase in total inter-grain contact area, and a resulting work-hardening by an increase in the frictional strength. At some point, however, grains begin to fracture at grain–grain contacts, leading to local weakening at zones of high stress concentration, and eventual localization into an incipient deformation band. Once formed, the newly fractured grains are then fractured further, eventually leading to the cataclastic demolition of the original fabric. In this stage the volume of the deformation band reduces, although some volume dilatancy next to the band continues to occur as a result of the

Fig. 1. Photographs of deformation bands (**a**) in horizontal plan and (**b**) vertical section. The photograph in (**a**) is looking along the strike direction of the fault (viewed from the right-hand side of the diagram as shown here). The anastomosing texture in plan is shown in (**a**). In (**b**) the individual bands are aligned sub-parallel in section (after Muir 1997).

finite shear offset. Once the band is formed, the decrease in grain size leads to a further increase in frictional contact area, resulting in a mechanical re-strengthening. As the fault gouge is formed in this hierarchical way, the load within the band is distributed more evenly between a greater number of grain–grain contacts within the band, resulting in a reduction of the stress concentration within the band. Both of these imply that the relative strength of the deformation band increases with increasing grain comminution. The band then grows in length by crushing of sandstone at its periphery.

A more recent quantitative analysis of local (Hertzian) contact stresses between neighbouring grains during inelastic compaction under hydro-

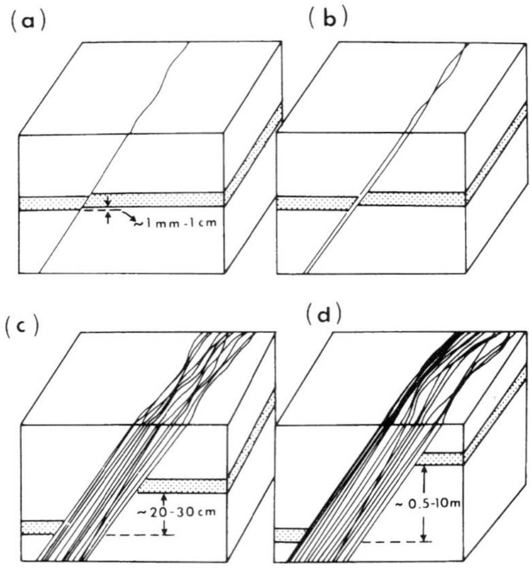

Fig. 2. Conceptual model for the sequential formation of deformation bands and slip surfaces, after Aydin & Johnson (1983), during normal faulting with maximum stress vertical, and minimum stress horizontal from left to right. (Note the more planar habit in section (in the direction of dip) than in plan (the direction of strike), as in Fig. 1.) Reproduced with permission of Elsevier Science.

static pressure component has been given by Zhang *et al.* (1990). This model provides a satisfactory explanation for pervasive grain crushing at high confining pressure. However, deformation bands occur under a finite differential stress at much lower confining pressures. Under these conditions, shear-enhanced compaction occurs at much lower mean stresses than for the hydrostatic loading case. To investigate this, Baud *et al.* (2000) plotted failure strength in porous sandstones as a function of effective mean stress. The results are in good agreement with the 'critical state' theory for failure in unconsolidated granular materials, more commonly associated with soil mechanics. In this theory the differential stress for failure is not a constant, as it would be for an ideal plastic material, but instead reduces as a function of confining pressure until it converges to the isotropic solution for a zero shear stress (Wong & Zhu 1999). This occurs because of the increasingly compactive rheology of the granular material, as observed in measurements of volumetric strain. In particular, Baud *et al.* (2000) have shown a transition from dilatant, localized failure at low confining pressure, to more pervasive shear-enhanced compaction at higher confining pressures. However, even when the volumetric strain of the whole sample (parent rock plus deformation band) is dilatant, the microstructural properties of the deformation band confirm local compaction and porosity loss as a result of the more efficient packing that can be attained with a broad grain-size distribution of cataclastic fragments. This is consistent with the field observations of Aydin & Johnston (1983).

The theories used to explain grain crushing and shear-enhanced compaction essentially ignore the degree of cementation of the grains (Zhang *et al.* 1990). Many deformation bands do form in unconsolidated sands at temperatures below 80°C (for a recent example, see Cashman & Cashman (2000)). Thus their formation is inextricably linked with granular mechanics, although deformation bands also form in better-cemented lithologies. The main effect of cementation is to alter the absolute failure stresses by up to an order of magnitude, without changing the qualitative form of the failure envelopes. Similarly, grain composition also affects the absolute failure stresses, with a 10% weakening associated with feldspar content (Zhu & Wong 1997).

In Fig. 1a, individual strands accommodate dip-slip offset, and anastomose in the horizontal plane, hence forming potential fluid pressure compartments. In Fig. 1b the deformation bands are more tabular in vertical section. This geometrical arrangement would occur for extensional horizontal strain. However, deformation bands with similar properties, but different orientations relative to the surface, are also seen in strike-slip and reverse faulting environments. Depending on small variations in composition and cementation, the zone of deformation bands may be very narrow, or very wide, often with large variations over a small distance (Fig. 1a). This is consistent with the results of Wong & Wu (1995); Baud *et al.* (2000). When significant clay is present, the deformation style shows a single narrow zone of deformation, and multiple deformation bands are absent (Antonellini *et al.* 1994).

Individual strands show very little shear offset, and slip is accumulated by sequentially increasing the number of bands (Aydin 1978; Mair *et al.* 2000). This implies that the newly created strand immediately strengthens on a timescale that is too rapid to be accounted for solely by chemical processes. Physical strengthening mechanisms may include the locking of rotated angular fragments, the reduction in porosity, and an increase in the number of grain–grain contacts in a comminuted material, all increasing the surface area for frictional contact within the gouge (Aydin & Johnson 1983; Nakatani 1998). Thus the first strand to form involves a mechanical bifurcation (Rudnicki & Rice 1975), in the sense that deformation is localized, and the fault is initially weaker than the surrounding rock. The second and later strands show a sequential growth that implies a repeated cycle of local weakening and strengthening (Fig. 2).

Once there has been significant slip, the mechanical behaviour shows a second bifurcation, this time in the form of the creation of a slip surface. Slip surfaces accommodate a large offset caused by repeated dynamic slip on the same fault plane. The fault plane is heavily striated with slickenlines, and often has a glassy texture. Deformation bands are often seen without slip surfaces, but slip surfaces are never observed on their own (Aydin 1978; Aydin & Johnson 1983). This implies that slip surfaces form during a process that initially involves the sequential development of individual strands, when a critical condition for more dramatic weakening is met. In the model of Fig. 2, this occurs at a critical strain. Our laboratory results described below show only the initial phase of the sequential development of deformation bands, and not the formation of large-offset slip surfaces. However, Mair (1997) did observe smaller-offset slickensides on the fault surface in some tests, indicative of abrasive wear of the fault wall even for the small-offset deformation bands.

In summary, the overall properties of natural deformation bands depend on a complex interplay between mechanical, structural, chemical and hydraulic processes. These properties are strongly dependent on rock composition, porosity, cementation, stress, strain, and pore fluid pressure. Therefore, it is difficult to isolate their effects from the study of field outcrop alone. We now summarize the results of some of our recent experiments designed to study some of these processes individually.

Physical and microstructural properties of experimentally generated deformation bands

Here we review the work of Mair (1997); Mair *et al.* (2000), which describes the results of an experimental study of deformation bands. This work was the first to generate realistic deformation bands in the laboratory, by deforming large core samples (10 cm diameter) in a specially designed large-capacity (2 MN) rock press. Previous tests on smaller samples (*c.* 4 cm diameter) have usually produced a single fault strand, followed by stable slip on the new surface at constant stress (e.g. Sammonds *et al.* 1992; Main *et al.* 1994). The larger samples have two main advantages in producing realistic structures. First, they allow a greater number of individual grains per sample, leading to more variety of solutions for the strain accommodation problem. Second, they allow a lower strain rate of $5\times10^{-6}\,s^{-1}$ for a standard test (run in 1 day) compared with *c.* $10^{-5}\,s^{-1}$ for a sample of 4 cm diameter at the same axial displacement rate. Whatever the cause, the larger samples show the same sequential development of discrete strands with increasing strain as inferred from the field observations.

Figure 3 shows a 'plan' view (in radial cross-section) of a sample of Locharbriggs sandstone, deformed to a cumulative axial strain of 6% at a confining pressure of 34.5 MPa. The test was carried out at room temperature under air-dry conditions. The test rock is an initially intact, well-sorted Permian aeolian sandstone with an initial porosity of 22–24%, made up of 83% quartz, 16% feldspar, and 1% hematite–illite.

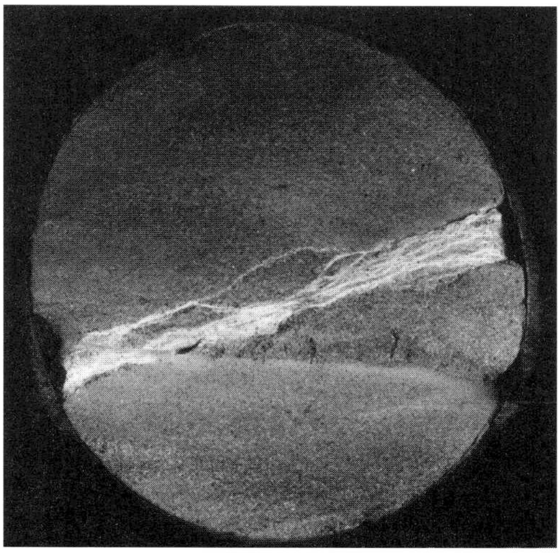

Fig. 3. Photograph in radial cross-section of an experimental deformation band formation in dry Locharbriggs sandstone deformed to a cumulative strain of 6%, at a confining pressure of 34.5 MPa, after Mair *et al.* (2000). The axial stress was applied in a vertical plane (in a direction perpendicular to the paper), and a radial stress applied uniformly around the sample, jacketed with the rubber membrane of 4 mm thickness seen around the sample. The results should be compared with the plan views illustrated pictorially in Fig. 1a, or schematically on the ground surface in Fig. 2.

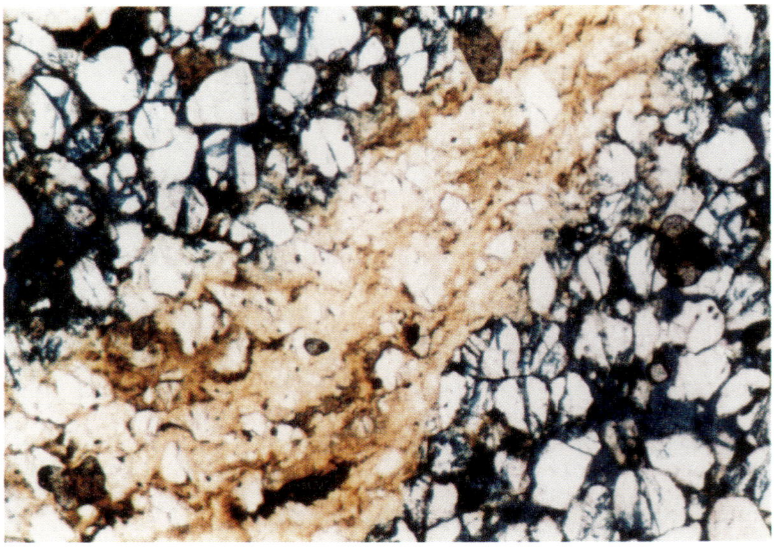

Fig. 4. Photomicrograph of a resin-injected thin section of an individual deformation strand, in cross-section (analogous to the view shown in Fig. 1b), after Mair *et al.* (2000). The resin was dyed blue to highlight the pore space. (Note the presence of a cataclastic zone with poor sorting, angular fragments and very efficient porosity reduction, and a surrounding damage zone of dilatant microcracking.) The photograph was taken with the direction of maximum stress vertical. The alignment of the microfractures is not precisely vertical, implying some stress rotation within the sample. The typical grain size is 200 μm.

The hematite is in the form of coatings on grain surfaces, giving it a characteristic reddish colour, and resulting in a relatively weakly cemented starting material. The individual deformation bands show the same anastomosing texture as in Fig. 1, with isolated relatively undeformed pods. The results confirm the sequential development of a series of individual bands, leading to a widening of the zone of deformation bands with respect to cumulative strain. The individual strands are made of cataclastic zones of reduced grain size and sorting, and show a dramatic loss of porosity as a result of grain crushing and compaction (Fig. 4). Immediately adjacent to the band, there is pervasive microcracking approximately parallel to the direction of maximum compressive stress (Fig. 4), implying dilatant deformation localized near the fault. During deformation the sample overall may compact or dilate, as the pore space first closes elastically, then dilates inelastically because of microcracking, and finally compacts inelastically during cataclasis.

The mechanical processes are intimately tied up with the microstructural characteristics of the resulting deformation. The first band appears with a dramatic mechanical bifurcation, and a large stress drop that produces an audible acoustic emission. However, subsequent bands often occur with no obvious stress drop or slip weakening (Fig. 5). A relatively constant stress in the post-failure stage also occurs in smaller samples, as a result of stable sliding on a single fault plane, but for the large samples this occurs while the number of individual strands is increasing (Fig. 6). This implies a complex interplay between local competing hardening (grain friction and locking) and softening (shear

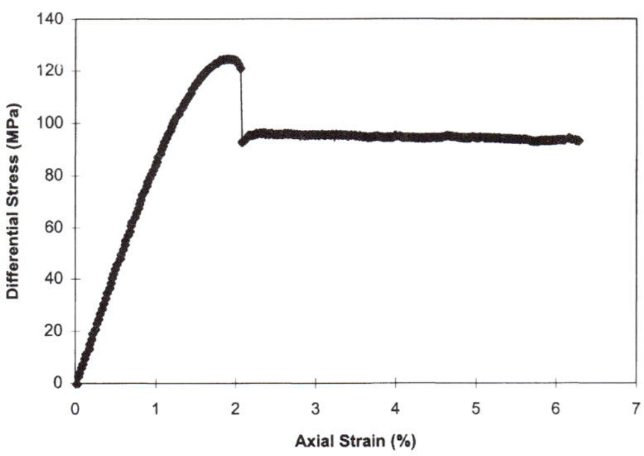

Fig. 5. Stress–strain evolution for the sample shown in Fig. 3, after Mair *et al.* (2000). (Note the phases of quasi-static strain hardening and softening before dynamic failure, and the subsequent relatively constant stress thereafter at increasing strain.)

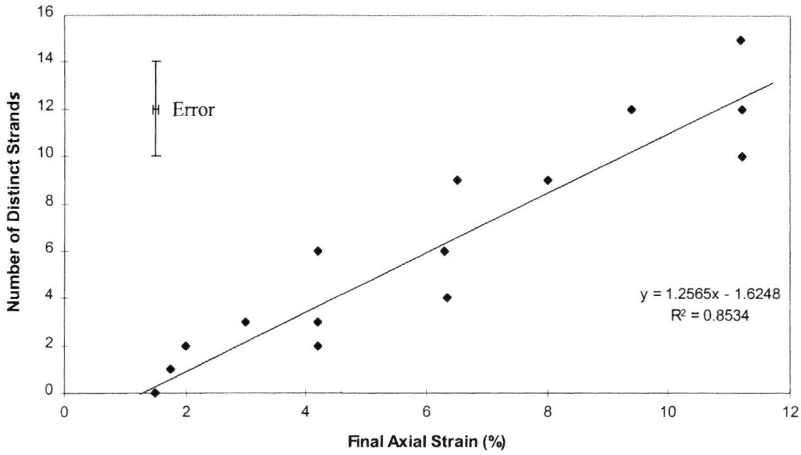

Fig. 6. Plot of the number of discrete strands as a function of cumulative strain, all for experiments conducted at a constant confining pressure of 34.5 MPa. It should be noted that a finite strain is required to produce the first band, and then the number of bands increase approximately linearly with strain at a rate n_o of *c.* 1.2 per unit (percentage) strain. As a result of the difficulty in estimating the number of bands in an anastomosing set, we estimate an error bar or ± 2 in this estimate for all measurements, as shown.

localization) mechanisms during the sequential formation of the individual strands, which cannot easily be distinguished using measurements at the sample boundaries (Mair *et al.* 2000). This implies that the mechanical analysis of Aydin & Johnson (1983) is valid only locally in the immediate vicinity of the band. The sequential growth of the discrete bands is best illustrated by counting the number of strands n at different strains ε (Fig. 6). The results can be summarized in the form $n = n_0(\varepsilon - \varepsilon_c)$, where ε_c is the critical breaking strain for the first band, and n_0 is the number of strands per unit strain. From Fig. 6, n_0 is *c.* 1.2 per unit (percentage) strain. Tests stopped before ε_c showed no cataclastic shear band.

The properties of the fault gouge in individual bands are remarkably uniform (Fig. 7). To examine the detail of the process, we plot results by volumetric fraction, rather than by number of grains, at a certain diameter. The intact rock shows a well-sorted peak at a diameter of 200 μm for the predominantly quartz grains, although there are some smaller peaks owing to detrital fragments. Once cataclasis occurs, the size distribution changes dramatically as a result of grain crushing, being peaked at a particle diameter of 10 μm, with a much wider standard deviation indicating poor sorting. The statistics of the individual bands remain remarkably consistent with increasing strain, although there is a slight tendency to increasing comminution. This implies that, although the microstructural properties of a deformation band may change during formation, as suggested by Aydin & Johnson (1983), the first-order features of the particle-size distribution remain relatively constant. Thus we have a set of discrete bands forming, each with very similar microstructural characteristics. In all other respects, the main points (1)–(5) listed above are reproduced in the laboratory tests.

One of the second-order features of the size distribution is the occurrence of approximately log-periodic peaks in the size distribution, here

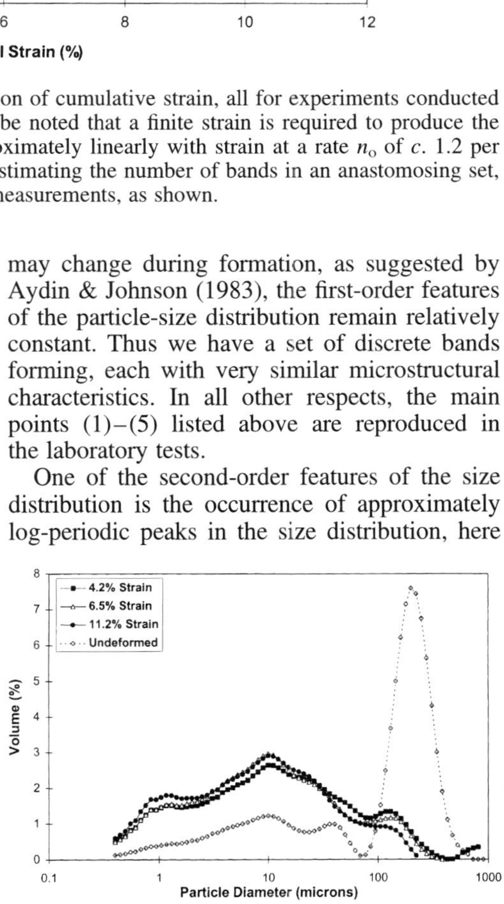

Fig. 7. Particle-size distribution of the undeformed sample, and the cataclastic fault gouge, both plotted by volume fraction in percent for selected samples at different amounts of cumulative strain shown in Fig. 6 for the sample of Locharbriggs sandstone. (Note the presence of a significant proportion of fine particles in the undeformed rock, and relatively constant size distribution once the first strand has been formed. Note also the log-periodic peaks in the fault gouge distribution at around 1, 10 and 100 μm.)

at 1 and 100 μm as well as at the primary peak at 10 μm (Fig. 7). None of this detail would be so obvious on log–log plot of frequency by number against diameter, i.e. the usual method of analysis (e.g. Sammis *et al.* 1987). The significance of log periodic scaling is the more efficient packing that can be achieved with a discrete hierarchy of sizes, increasingly filling in the remaining pore space, rather than the continuous power-law distribution. This allows a more dramatic permeability reduction and is more consistent with a hierarchical development of fractures during comminution as initially proposed by Aydin & Johnson (1983). The position of peaks at 1 and 10 μm is similar to local peaks in the distribution of detrital fragments in the parent rock. It is therefore possible that these exert some control on the scale of comminution. In more recent tests on Clashach sandstone described below we have seen similar log-periodicity at different scales in a rock with a relatively small fraction of detrital fragments, so its presence is not dependent on the pre-existing size distribution of smaller particles, although the location of the peaks might be. Sadovskiy *et al.* (1984) argued for the existence of such a discrete log-periodic hierarchy in rock properties including fracture. Suteanu *et al.* (2000) have confirmed these results on two classes of experiments involving comminution of flat samples (Romceram bricks) by energetic impacts, and fragmentation by desiccation cracking of suspension films. They found that fragments cluster on dominant size intervals that point towards a discrete rather than continuous fractal character of the size distribution, directly analogous to the peaks shown in Fig. 7.

A suite of tests investigating the influence of confining pressures showed a systematic transition to more pervasive fracturing with increasing confining pressure (Mair 1997). This is consistent with the results of Wong *et al.* (1997); Zhu & Wong (1997). The failure and sliding strengths of the first band showed standard frictional coefficients of 0.6 or so, also similar to their results for Darley Dale sandstone. At higher pressures (>100 MPa) porous sandstones fail by pervasive grain crushing in shear-enhanced compaction along a critical state curve (Zhu & Wong 1997; Wong & Zhu 1999). However, our experiments were carried out at confining pressures that were too low to see this transition on the mechanical data, although we do see a tendency towards more pervasive deformation and shear-enhanced compaction at our highest confining pressure (70 MPa) in the microstructural properties. Antonellini *et al.* (1994) concluded from field observation that styles of deformation bands may vary systematically with depth of formation, and hence that there may be several types of deformation bands. Mair (1997) confirmed experimentally that the deformation style is strongly pressure sensitive, and that deformation bands may therefore form within only a narrow depth range within the brittle crust.

The main differences between our results and field observation are that (1) we see no large-offset slip surfaces, and (2) we see many more microfractures than are often apparent at first sight in thin sections from field outcrop (Antonellini *et al.* 1994; Fowles & Burley 1994). The first may be due to the fact that, despite the large samples, we still have a limit on the cumulative axial strain of <16% or so. Perhaps large-offset slip surfaces require a finite minimum amount of critical strain to accumulate before their formation. The formation of large-offset slip surfaces may also be strongly rate dependent, whereas we have examined weakening and strengthening mechanisms only as a function of a state variable (strain). They may therefore require a critical burial depth; for example, a transition to velocity-weakening friction (Scholz 1990). Certainly we see no obvious slip-weakening after the dynamic formation of the first strand.

The difference in the number of microfractures in thin section may not be significant because, when more sophisticated techniques are used, many healed microfractures are observed around faults in field outcrop (Knipe & Lloyd 1994; Mair *et al.* 2000). Calculations based on the reaction kinetics of quartz show that such healing may be rapid on geological timescales (Brantley *et al.* 1990). Data from borehole cores and field outcrop implies that cataclastic deformation in sandstones occurs with greater associated microcracking in the parent rock for thrust zones than in extensional basins (Lloyd, pers. comm.). This could be due to differences in burial depth at the time of formation, or the influence of the gravitational body force that has not yet been reproduced in the laboratory. However, the main outstanding difference between our laboratory tests and field conditions is the large difference in temporal and spatial scales. For example, pre-failure dilatancy is known to reduce systematically with strain rate (Fujii *et al.* 1998), which may lead to fewer microfractures under field conditions than seen in our laboratory tests. The tests here are at mean stresses that are significantly larger than those inferred from the burial depth of many natural deformation bands. This will also tend to produce a greater proportion of distributed deformation in the form of microcracking in the laboratory samples.

Hydraulic properties

Exposed deformation bands generally show a greatly reduced permeability compared with the country rock (Pittman 1981; Antonellini & Aydin 1994; Fowles & Burley 1994), of up to six orders of magnitude. Some of this may occur immediately; for example, as a result of more efficient packing of poorly sorted gouge fragments. Over much longer timescales mineral precipitation may seal the fault further, either from a local source in the reactive gouge, or from more remote pore fluids that travelled along the fault during the dilatant phase of deformation. Fluid transport along faults may occur by seismic pumping (Sibson 1981) or by fault valve action (Sibson 1990). Such models were derived from laboratory experiments based on observations of low-porosity crystalline rock (e.g. Brace & Martin 1968; Teufel 1981) where the fault is essentially under undrained conditions. These results therefore need not necessarily scale simply to more porous and permeable granular media (Wong & Zhu 1999). Under undrained conditions (or rapid dilatancy relative to the fluid diffusion time in a more porous rock), the pore pressure reduction on the fault can induce a local dilatant hardening effect (Brace & Martin 1968; Rudnicki & Chen 1988; Lockner & Byerlee 1994; Raleigh & Marone 1997). Thus there is a strong coupling between the evolution of porosity and fluid permeability and the generation or suppression of dynamic instability. Near-fault dilatancy is also seen in porous rocks, consistent with the observation of microcracks around faults exposed in both of the field areas shown in Fig. 1 (Navajo sandstone, Utah, by Antonellini *et al.* 1994; Hopeman sandstone, Lossiemouth, Scotland, by Edwards *et al.* 1993). It is also consistent with the inferred elevated hydraulic conductivity of faults parallel to their orientation during hydrocarbon migration.

In the experiments described below, we show that near-fault dilatancy can provide a mechanism for the 'suction pump' effect during the dynamic formation of the first deformation band in a sandstone sample. In contrast, the subsequent deformation bands occur with a more quasi-static rheology, allowing the development of a simple 1D geometric model for permeability reduction during the formation of deformation bands. This model is based on a minimum scale

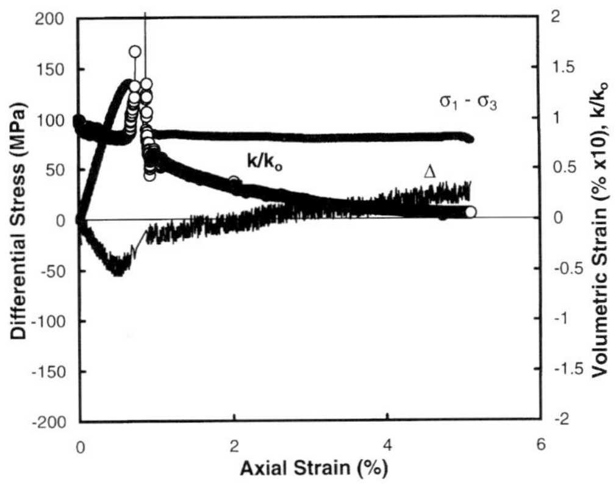

Fig. 8. Plot of differential stress $\sigma_1 - \sigma_3$, percentage volumentric strain Δ and permeability k, normalized to its starting value k_0, all as a function of axial strain, for a sample of Clashach sandstone deformed at a constant confining pressure of 27.6 MPa, after Main *et al.* (2000). It should be noted that the scale for Δ is vertically exaggerated by a factor of 10 to plot in the same diagram as the permeability (right-hand axes).

(one deformation band width) that is larger than those used in more sophisticated 2D or 3D pore network models for permeability evolution such as Zhu & Wong (1996; 1997). It is also analytically much simpler. Nevertheless, it predicts the axial permeability evolution adequately with a minimum number (three) of free model parameters.

The permeability evolution for an experiment involving direct measurement of sample volume and permeability is shown in Fig. 8. Permeability was measured using Darcy's law under constant flow rate conditions against a modest back-pressure of 7 MPa. This test was carried out at room temperature, for Clashach sandstone, a Permo-Triassic aeolian sandstone from the Hopeman quarry, near Lossiemouth, Elgin. This sandstone has slightly larger grains (250–300 μm) than the Locharbriggs sample, consisting of subrounded quartz grains (90%) and K-feldspar (10%). Typical starting porosities are *c.* 20%. The Clashach sample is well cemented by quartz overgrowths, so that it is mechanically stronger than the Locharbriggs samples under the same conditions (consistent with Wong & Wu (1995)). The absence of hematite coatings inhibits to some extent the grain translation mechanisms. This sample is also better sorted than the Locharbriggs sandstone, and contains a much smaller fraction of fine particles in the form of detrital minerals.

Despite the differences in cement type and the proportion of detrital minerals, and the

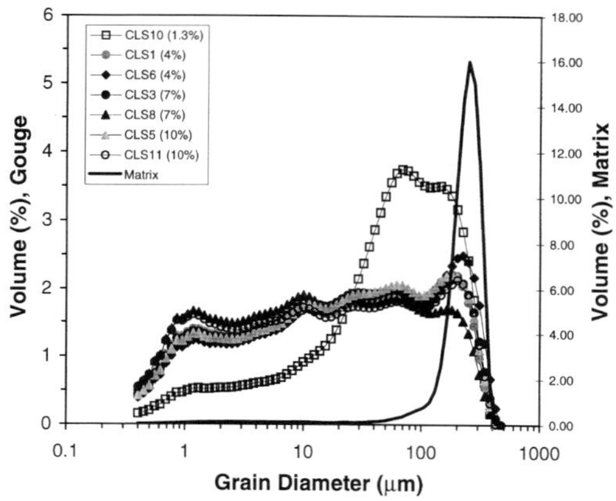

Fig. 9. Particle-size distribution for various end-point strains (in percent) for a suite of Clashach sandstone tests, carried out under constant flow rate conditions using distilled water. The parent rock is very well sorted, so the matrix rock distribution is very peaked at 200 μm. (Note that its distribution is shown with a reduced vertical scale to allow all of the relevant distributions to be shown on the same graph.)

analysis of deformation during fluid flow, the main results for the Clashach sample are very similar to first order to those for the air-dry Locharbriggs tests of Mair (1997); Mair *et al.* (2000). The mechanical results for the Clashach samples also show a dramatic stress drop, with a relatively constant stress thereafter, and again no strong dynamic stress drop associated with the subsequent deformation bands (Fig. 8). Again, the zone of deformation bands grows sequentially, with individual strands having a similar particle-size distribution once the first strand is formed. In this case, the first band this time has a distinct distribution when compared with the subsequent bands (Fig. 9), this time more consistent with the observations of Aydin & Johnson (1983). As a result of the absence of the grain translation mechanism, the resulting gouge distribution overall contains a relatively larger proportion of smaller particles, but a similar minimum size of *c.* 1 μm when compared with Fig. 7, below which the volume fraction decreases. This minimum size is greater than the minimum resolution of the method used (0.4 μm), so it is possible that a smaller milling limit may exist below this size. The peak at 1 μm for both of our rock types is probably due to the similar bulk chemical composition, dominated by quartz. Again, peaks in the particle-size distribution are seen at around 1, 10 and 100 μm. Here there are also two peaks between 10 and 100 μm. Notably, the presence of a pore fluid does not change the first-order microstructural properties of the deformation bands significantly.

The sample volumetric strain Δ was measured from the change in confining fluid volume required to maintain constant confining pressure. It first decreases as a result of pore closure, then increases, first because of quasi-static microfracturing (Scholz 1990) and then by dynamic dilatant slip at the failure time (Rudnicki & Chen 1988). The absolute volumetric strain (<0.1%) is much smaller than observed in samples with a smaller sample-size to grain-size ratio. This may be due to the more efficient strain accommodation in a granular system with a larger number of elements (consider, for example, the limit in which a sample consisting of two particles may deform). Apart from this difference in the absolute scale, our results follow the standard qualitative behaviour in relative volumetric strain seen in smaller core samples. However, the systematic change in the absolute volumetric strain once again illustrates the potential dangers inherent in the linear upscaling of mechanical laboratory data to field conditions.

The permeability shown in Fig. 8 has been inferred from the fluid pressure drop between the inlet and the outlet end of the sample using Darcy's law, assuming a constant flow rate q, corrected for sample shortening. The permeability shows a behaviour that tends to exaggerate the porosity changes. The first phase of permeability decrease can be attributed to elastic pore closure. By analogy with elastic parameters such as the bulk modulus, permeability sensitivity is often assumed to be exponentially dependent on stress (e.g. Rice 1992). However, permeability is essentially a geometrical parameter (units of m^2), which reflects the geometry and connectivity of the pore space. Hence David *et al.* (1994) suggested that permeability is instead more likely to be sensitive to changes in strain. This is consistent with our results in the post-failure stage, where permeability changes despite the stress remaining essentially constant. Therefore we cast our model in terms of strain as a controlling variable. The permeability sensitivity (after Rice (1992), assuming elastic pore closure) can be written

$$k = k_0 \exp(-\gamma_E \varepsilon) \qquad (1)$$

where k is permeability, the subscript zero refers to the initial conditions, and γ_E measures the permeability sensitivity to elastic pore closure with increasing axial strain.

During dilatant microcracking we first see a quasi-static increase, and then a dynamic increase in the permeability inferred from the fluid pressure drop across the sample. The apparent increase in permeability during the dynamic phase of faulting is not likely to be a true increase, but is instead due to dilatancy associated with rapid nucleation and slip on the newly created fault zone. Such dilatancy may occur either by the microcracking we see in the thin sections of Fig. 4, or by the temporary dilatancy that is required for two rough surfaces to slide past one another (Rudnicki & Chen 1988). This implies a sudden increase in storage capacity S, rather than a real increase in permeability k. To explain the large apparent permeability surge during failure in Fig. 8, we modify the theory of Rudnicki & Chen (1988) to account for dilatant slip under conditions of constant flow rate q. If the pore pressure at the inlet sample boundary is p_0, and the pore pressure near the fault is p_F, then the mass flow rate between the inlet and the fault is

$$q = \frac{\rho k}{\eta}\frac{(p_0 - p_F)}{l} = \frac{dm}{dt} \qquad (2)$$

where ρ is fluid density, m is fluid mass, k is the true fluid permeability, η is fluid viscosity and l is sample length. We have used the convention that flow is down the pressure gradient. In our tests we have maintained a constant external flow rate at the pump controlling the inlet pressure q_0, so the total flow rate is $q = q_0 + dS/dt$, where S is the storage capacity (fluid mass). We can neglect the changes in storage if they occur quasi-statically ($dS/dt \ll q_0$). However, during the dynamic phase, this approximation is not valid.

The mass of fluid is the product of its density ρ and its volume ϕ (proportional to porosity), so the total change in storage is

$$\frac{dS}{dt} = \rho\frac{d\phi}{dt} + \phi\frac{d\rho}{dt}. \qquad (3)$$

If the fluid bulk modulus is $K_B = \rho dp/d\rho$, where the average pore pressure in the sample $p \approx p_0$, then

$$\frac{d\rho}{dt} = \rho\frac{dp_0/dt}{K_B} \qquad (4)$$

Thus, under conditions of constant flow rate, we have from equations (2)–(4)

$$q = \frac{\rho k}{\eta}\frac{(p_0 - p_F)}{l} = q_0 + \rho\left(\frac{d\phi}{dt} + \frac{dp_0/dt}{K_B}\right). \qquad (5)$$

The first two terms in equation (5) reduce to Darcy's law for flow between the inlet end and the fault, and the final term is the balancing term for changes in storage capacity. Most pore fluids can be regarded as relatively incompressible, so the main effect is the time-dependent porosity change $d\phi/dt$. At low volumetric strain rates, $d\phi/dt$ is negligible, and hence changes in storage capacity can be neglected in the calculation of permeability. When $d\phi/dI$ becomes significant as a result of near-fault dilatancy in the final stages of deformation, its effect must be balanced by a drop in p_F, resulting in a lower entry pressure p_0 being required to retain the same flow rate. This mechanism acts as a 'suction' pump, pulling fluids into and along the rapidly dilating fault zone, as predicted qualitatively by Sibson (1981) for lower porosity crystalline rock.

In summary, at low volumetric strain rates the suction pump effect is negligible, and Darcy's law holds. At high volumetric strain rates the sample responds by reducing the inlet pressure. This appears as a dramatic increase in apparent permeability near the dynamic failure time (Fig. 8b), but is actually due to a dynamic increase in the rate of change of storage capacity seen in the observed volumetric strain (Fig. 8b). Thus our inferred permeabilities are exaggerated near the dynamic failure time. We are currently modelling this process in more detail, but hereafter, we concentrate on the quasi-static phase of deformation after the first mechanical bifurcation.

The microstructural measurements of Mair *et al.* (2000) can be used to construct a conceptual model for the permeability of deformation bands during their phase of sequential growth (Main *et al.* 2000). This conceptual model is shown in Fig. 10. It assumes an exponential relationship between porosity and permeability (Fig. 10d), which is reasonable for high-porosity rock (Wong & Zhu 1999). For low-porosity rocks the theory may be adjusted easily for a power-law relationship (Wong & Zhu 1999). The model assumes a constant permeability of each strand, k_{strand}, because of the similarity in microstructural characteristics of the individual strands discussed above. It allows for linear dilatancy or compaction in the rock matrix, depending on confining pressure (Fig. 10c). This neglects the non-linear behaviour of the bulk porosity inferred from Fig. 8a, but is a reasonable approximation in the quasi-static phases of deformation seen there. The bulk axial permeability in the model is calculated using a harmonic average of permeabilities along the path shown in Fig. 10a. The model assumes a binary (fault or matrix) local permeability, so that the matrix permeability is itself a global average. For example, this assumption effectively neglects the effects of microcrack-enhanced dilatancy of the pods or

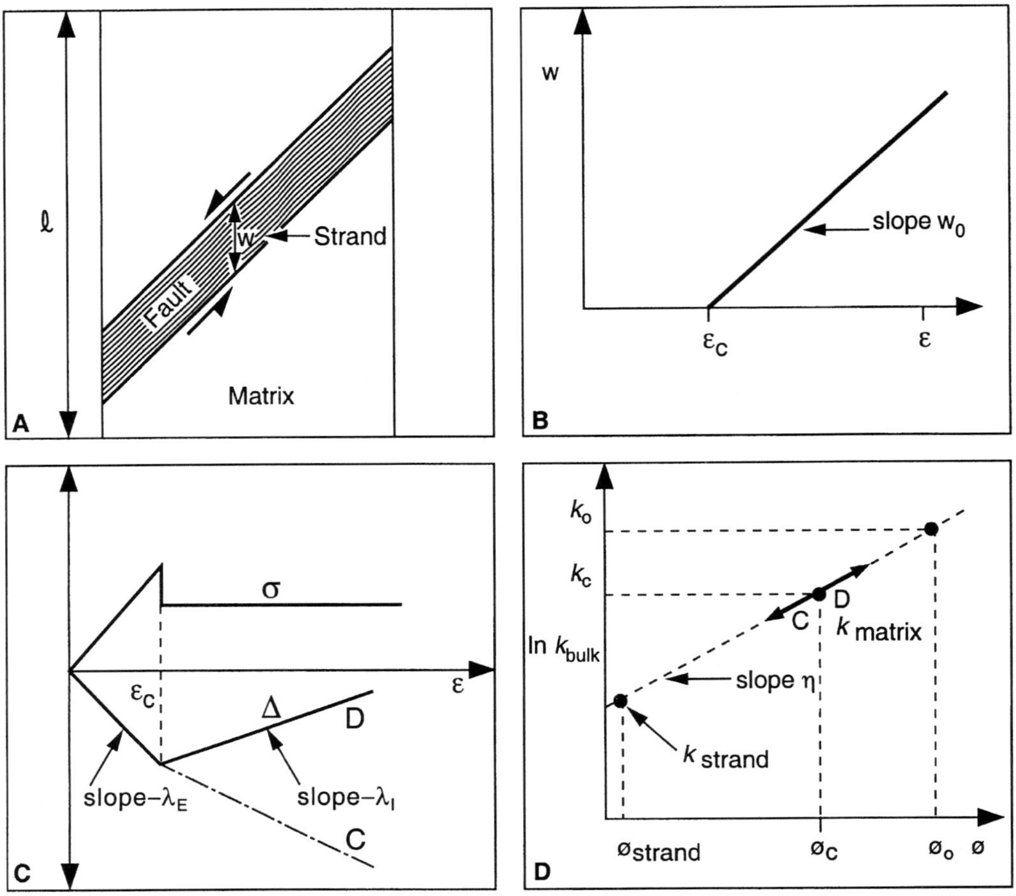

Fig. 10. A conceptual model for the permeability of deformation bands, after Main *et al.* (2000). (**a**) Sample geometry; (**b**) increasing axial band width w with increasing strain (as in Fig. 6); (**c**) linearized version of the results of Fig. 8, showing sample dilatancy (D) (or alternatively compaction (C) at higher confining pressures); (**d**) exponential porosity–permeability relationship for fault strands and country rock. It should be noted that the individual fault strands are assumed to have constant permeability and porosity.

lenses of damaged rock between some of the strands seen in Fig. 3 (Mair *et al.* 2000).

Given these assumptions, the model predicts a decrease in permeability with respect to inelastic strain in the post-failure stage $x = \varepsilon - \varepsilon_c$, of the form

$$y_{\text{bulk}} = y_c + a_1\left(\frac{x}{1-x}\right) + a_2\left(\frac{x^2}{1-x}\right)$$

$$a_1 = -\left[\frac{w_o(y_c - y_{\text{strand}})}{l_c} + \gamma_I\right]$$

$$a_2 = \gamma_I\left(1 + \frac{w_o}{l_c}\right) \quad (6)$$

where $y_i = \ln k_i$, and subscripts bulk, c and strand refer to the bulk measurement, the measurement at the critical strain for the bifurcation, and an individual strand, respectively; w_o is the rate of widening of the band with respect to inelastic strain x; and l_c is the sample length immediately after dynamic failure. In a laboratory test, we measure the bulk permeability and volumetric strain. Equation (6) can then be used to infer model parameters such as the fault zone permeability. The parameter γ_I represents the permeability sensitivity to inelastic volumetric changes, analogous to equation (1), defined positive in compaction. For dilatant deformation ($\gamma_I < 0$) a_1 is negative (the permeability reduces to first order) when the sealing rate s_{fault} on the fault exceeds the matrix permeability enhancement rate, or

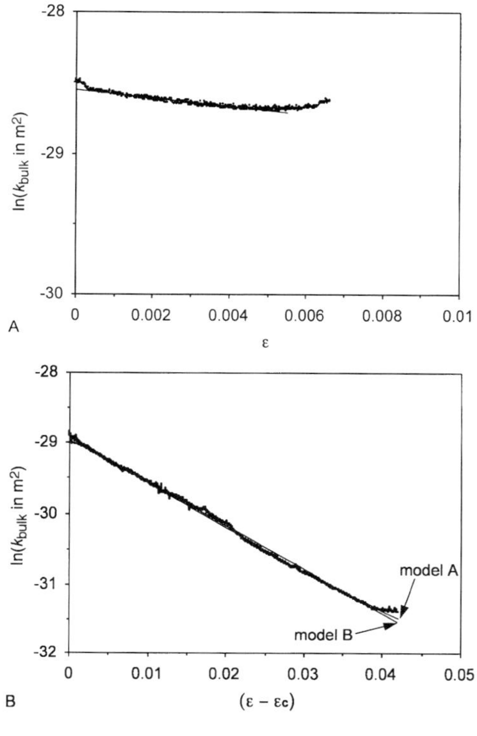

Fig. 11. Permeability in (**a**) the elastic phase of pore closure, and (**b**) during the sequential growth of deformation bands, compared with regression fits to equations (1) and (6) respectively for the results of Fig. 8, after Main *et al.* (2000). Model A includes all of the terms in equation (6), and model B is truncated at the first-order term a_1.

$$s_{\text{fault}} = \frac{w_0(y_c - y_{\text{strand}})}{l_c} > -\gamma_I. \quad (7)$$

Our results are consistent with the criterion given by equation (7), i.e. in the competition between fault sealing and matrix dilatancy, the fault sealing effect dominates. This is consistent with the very efficient porosity reduction in the fault gouge illustrated in Fig. 4.

Equations (1) and (6) provide a very satisfactory fit to the data in the inelastic phase (Fig. 11) with regression coefficients of the order of 0.997. Although a model truncated at the first-order term a_1 also provides a reasonable fit to the data (denoted Model B in Fig. 11), a detailed statistical analysis shows that the second-order term is statistically significant for this dataset (Main *et al.* 2000). Thus fluid permeabilities can be predicted accurately on the basis of the microstructural results presented by Aydin & Johnson (1983); and Mair *et al.* (2000).

Observation of pore fluid chemistry changes during deformation

Chemical sealing by mineral precipitation requires the presence of a supersaturated fluid, i.e. a local chemical disequilibrium. This may be caused by transport from a different *P*, *T* regime, such as the crack-seal mechanism (Brantley *et al.* 1990). Such coupled transport–chemical sealing processes, involving quartz solution and precipitation, may have significant implications for episodic fluid flow and faulting in the Earth (Fisher & Brantley 1992; Matthai & Fisher 1996). When combined with pre-existing permeability heterogeneity, complex dissolution–deposition textures can result (e.g. Bolton *et al.* 1996) including the oscillatory deposition of overgrowths commonly seen in the analysis of thin sections. In systems with low clay content, mineral precipitation in cataclastic fault zones may also be purely local, as a result of pressure solution (Rutter 1983; Revil 2000), or a process known as Ostwald ripening, where small particles dissolve and precipitate on larger ones (Ortoleva 1994). When significant clay is present, quartz dissolution occurs at clay–quartz contacts, and is far more dependent on temperature than on pressure because it is governed by quartz precipitation kinetics rather than dissolution kinetics (Bjorkum 1996; Fisher & Knipe 1998; Fisher *et al.* 2000). Field evidence from direct investigation of borehole cores in sedimentary basins exploited for hydrocarbons has confirmed that the majority of quartz cementation in such rocks can in fact be explained by local rather than regional transport of solutes (Fisher & Knipe 1998). Long-range transport can occur only when the fault is maintained dilatant for long periods of time; for example, in sandstones where deformation occurred under lower effective stress conditions than the rocks had previously experienced (Fisher & Knipe 1998; Fisher *et al.* 2000).

Until recently, it has been difficult to observe such processes in the laboratory at the relevant temperatures and pressures where deformation bands form. In one approach, experimental studies concentrate on the purely chemical properties of crushed powders of minerals such as quartz at the appropriate temperatures, but without the application of a finite confining pressure (e.g. Rimstidt & Barnes 1980). Such 'batch' tests provide basic information primarily on dissolution rates (reviewed for application to real systems by Lasaga (1984)). In another approach, chemical fluid–rock interactions have been investigated under pressure, but at relatively elevated temperature, to observe mineral dissolution and precipitation on a reasonable timescale (Scholz *et al.* 1995; Olsen *et al.* 1998; Kanagawa *et al.*, 2000). Many of these results can be understood in terms of simple reaction-rate theory for the relevant minerals and pore fluids, appropriately scaled to the conditions of the experiment (e.g. Aharanov *et al.* 1998; Tenthorney *et al.* 1998). Such studies are directly appropriate for mid-crustal depths. Together with kinetic chemical reaction rates from batch tests, they elucidate the qualitative processes at work, and provide powerful quantitative constraints, for lower-temperature behaviour. The quantitative scaling may involve significant extrapolation to conditions appropriate for the formation of deformation bands.

The advent of sophisticated measurement techniques has allowed the direct quantification of chemical rate processes at the low temperatures and pressures where deformation bands form. Using high-performance liquid chromatography (HPLC), we have observed mineral dissolution and precipitation reactions both on intact rock (Main *et al.* 1994, 1996; Ngwenya *et al.* 1995) and dissolution reactions only during the cataclastic failure (Ngwenya *et al.* 2000) of porous sandstones at low temperatures and pressures in real time, during deformation experiments. The pore fluids are circulated through the sample, and their chemical composition is measured on-line. Fluid–rock interactions are isolated by using a chemically inert (poly-ethyl-ethyl ketone) tubing and titanium loading platens, thereby avoiding previous problems with contamination from steel tubing. Any iron content in the pore fluid therefore comes from the rock. HPLC also allows the analysis of pore fluids in real time, thereby avoiding the complications involved in 'quenching' the sample. These measurements were carried out before the large-capacity deformation rig became available, and are hence carried out on smaller (3.8 cm diameter) samples. Although they do not produce multiple deformation bands, the results are relevant to the formation of single strands at low strain rates.

Mineral dissolution rates (for quartz and K-feldspar) were measured for different inlet fluid compositions at constant flow rates. Sometimes evidence of dissolution could also be discerned on the rock itself by inference from microstructural analysis of overgrowths (Main *et al.* 1994). More rarely, evidence for precipitation was obtained by chemical fingerprint using ^{18}O-doped pore fluid and ion microprobe analysis (Elphick *et al.* 1996). Although the changes in concentration of dissolved species is very small (of the order of parts per million) in these experiments, they can nevertheless be measured accurately above the noise, to within 2% or so. The reactions themselves are very fast compared with geological timescales, with chemical steady-state concentrations for reaction–advection–diffusion being re-established a few hundred hours after the perturbation to the chemical potential, in turn as a result of under- or oversaturated pore fluids or mechanical disequilibria. Chemical equilibrium was not obtained in these experiments because of the constant circulation of an undersaturated or oversaturated pore fluid at a constant flow rate.

Cataclastic deformation of porous sandstones results in a fault gouge that has a large number of small particles, and a large surface area of fresh grains. Such particles are highly reactive because (a) the cataclastic debris contains a step increase in total reactive surface area for the same volume and (b) the smaller particles produced by cataclasis have a tighter radius, and therefore a larger specific surface energy. These two elements imply a higher steady-state concentration for small particles in the pore fluid, a prerequisite condition for Ostwald ripening (Ortoleva 1994). The large specific surface energy for small particles is also ultimately the physical cause of the minimum particle size or milling limit seen in the particle-size distributions in Figs. 7 and 9.

To examine chemical fluid–rock interactions during deformation we have carried out a series of creep-to-failure tests on Locharbriggs and Clashach sandstones, mainly to examine the effect of differences in diagenetic minerals (silica or hematite) on the resulting fault microstructure and pore fluid chemistry. Figure 12 shows typical results for the resulting axial strain and silica concentration in the pore fluid, measured using HPLC with undersaturated (distilled) water as a pore fluid flowing at a constant rate, at room temperature. The stress field was triaxial, with constant axial and radial stresses (a typical 'creep' test). The differential stress was set to be sufficiently close to the ultimate strength to induce dynamic failure over timescales of a few hundred hours, to observe the chemical changes within a reasonable experimental timescale.

The strain shows the classic three-stage signature, of decelerating, steady-state and accelerating creep, leading to a dynamic strain increase marked by the failure of the sample and the dynamic production of a cataclastic

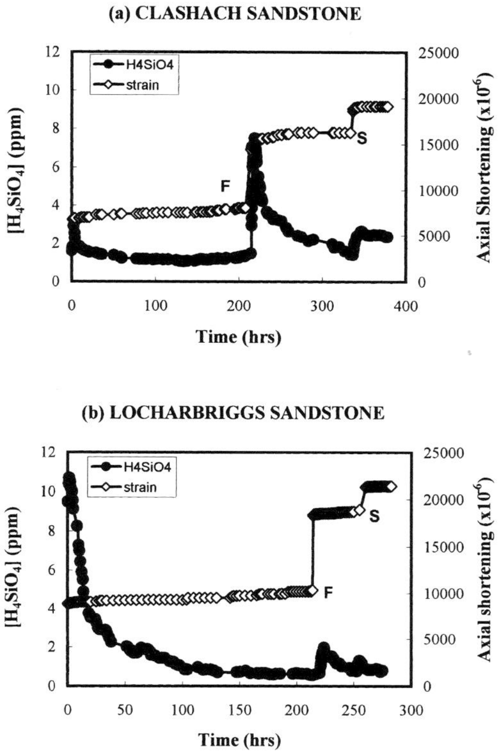

Fig. 12. Evolution of axial strain (in units of microstrain) and fluid silica concentration as a function of time during tests at constant load for Clashach and Locharbriggs sandstones, after Ngwenya *et al.* (2000). The confining pressure for these tests was 6.9 MPa, the flow rate 0.1 μl min^{-1}, and the test was run at room temperature. The time of dynamic faulting at time F is marked by a sudden increase in strain rate to the inertial limit, as the slip on the fault catches up with the elastic strain in the rock. This dynamic event is followed by stable sliding on the fault up to the point marked S in the diagram, where a second dynamic slip event occurs.

shear band at point F in the diagrams, where the strain rate reaches the inertial limit. After dynamic failure, the sample was resheared, producing at first quasi-static, and then a second dynamic shearing event on the same fault plane at time S.

On first application of the stress, both tests show a sharp spike in silica concentration. Before the test the samples were flushed to minimize the number of reactive fine particles in the bulk sample. The initial spike here is therefore more likely to be due to the effect of pressure solution and/or stress corrosion reactions at grain–grain contacts following the application of a differential stress at the start of the test. During the phase of primary or decelerating creep, both samples show a transient spike in silica concentration, with a re-establishment of steady-state concentrations over timescales of *c.* 100 h for Clashach, and *c.* 200 h for Locharbriggs sandstones, respectively. In both cases steady-state chemical concentrations were re-established within the phase of secondary or steady-state creep. The steady-state concentrations were similar for both sample types (1.2 ppm for Clashach; 0.8 ppm for Locharbriggs).

During tertiary creep there is a small but detectable increase in silica concentration immediately before failure at F in the Clashach sample. This precursory increase is not seen in the Locharbriggs sample. However, when the fault forms, both tests show a sudden increase in silica concentration associated with the formation of a cataclastic shear band. This pattern is repeated during the second creep phase, culminating in a dynamic strain rate at point S. The Clashach test shows a large increase in silica concentration at dynamic failure of 7 ppm, compared with a peak of 2 ppm for the Locharbriggs sample. This is consistent with the larger proportion of reactive smaller particles in the size distribution of the fault gouge for the Clashach test.

The microstructure of the two samples after the test is shown in Fig. 13. First, it should be noted that, for these 3.8 cm diameter samples, there is only one strand, implying that deformation after S occurs by reshear on the first strand, perhaps involving further comminution, or a continuous widening of the strand. Second, the Clashach sample shows much more evidence of cataclastic deformation under the same conditions, including microcracking in the country rock, and more effective grain crushing along the fault strand. Consistent with the previous comparison of particle-size distributions in Figs. 7 and 9, the Locharbriggs sample shows much less pervasive microcracking, and also a smaller degree of comminution in direct comparison of these sections (Fig. 13). In fact, the finest gouge fragments in the Clashach could not be preserved in a cohesive thin section because of difficulties in injecting the resin before polishing. Again, the differences in particle-size distributions are likely to be due to the relative enhancement of the grain translation mechanism in the Locharbriggs sample.

Thus two porous aeolian sandstones with very similar chemical composition, but different diagenetic and detrital phases, can show large differences in quartz solubility during different stages of deformation, and large differences in

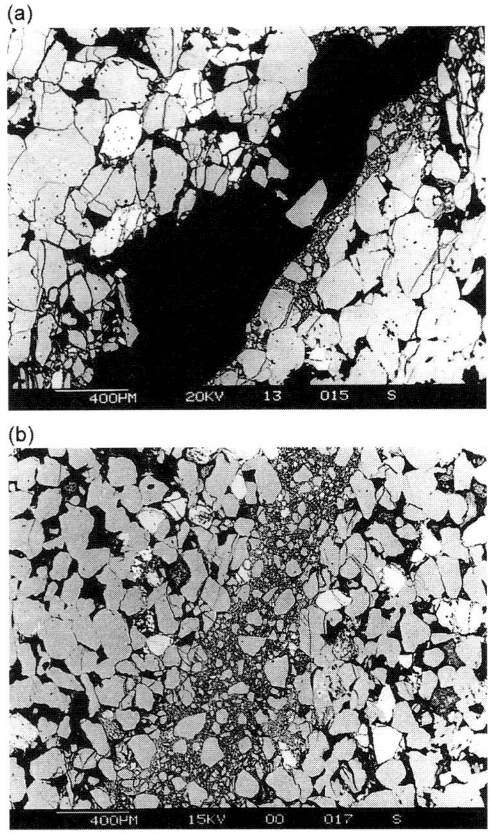

Fig. 13. Thin sections of (**a**) Clashach and (**b**) Locharbriggs sandstones from the end point of the tests shown in Fig. 12, shown with the direction of maximum principal stress in the vertical direction. The more pervasive microcracking of the Clashach sample, and the smaller amount of grain comminution in the Locharbriggs sample, should be noted. The microcracks are again approximately aligned with the direction of maximum principal stress.

associated microstructure. During the tertiary stage the main difference in quartz solubility can be attributed to the greater number of microcracks observed in the Clashach test. Increased silica concentration in the tertiary creep stage may be caused actively by stress corrosion reactions (Dove 1995) or passively by dissolution from the fresh reactive surfaces. Finally, on production of the finer cataclastic deformation band, with a larger surface area, the Clashach test shows a greater silica concentration as a result of the creation of a more reactive fault gouge than the Locharbriggs sample. The differences can be attributed to the remaining differences in detrital fragments, cement type and deformation mechanism. In particular, the soft hematite coatings on the Locharbriggs sample allows a greater proportion of deformation mechanisms that do not create new reactive surfaces, such as grain boundary sliding and/or rotation. In contrast, the Clashach sample has relatively more deformation by microcracking and comminution mechanisms, and hence has a much higher reactivity during the phase of tertiary creep and dynamic failure.

We have seen above that mechanical sealing effects can reduce permeability by up to 2–3 orders of magnitude, but field permeametry can show a larger difference of 4–7 orders of magnitude (Antonellini & Aydin 1994). Figure 13 shows no significant mineral precipitation under flow-through conditions, partly because the dissolved silica is washed out of the sample, so we cannot directly test the hypothesis that the remaining difference is due to chemical sealing. However, using quantitative mass balance calculations, Ngwenya *et al.* (2000) showed that the dissolved silica can indeed provide a large enough source to explain the difference, if mineral precipitation is concentrated within the volume represented by the observed dimensions of the reactive fault gouge. Such local chemical precipitation may lead both to self-sealing and healing of the fault. Experiments on samples composed entirely of synthetic fault gouge are currently under way under undrained conditions to attempt to observe mineral precipitation and quantify its kinetic rate properties.

Discussion

We have considered independently the mechanical, structural, hydraulic and chemical processes associated with the formation of deformation bands in porous sandstone. However, in nature, they are likely to be strongly coupled. For example, during dynamic slip, our calculations show that the fault zone is likely to act as a suction pump, with dilatant slip stabilizing the fault against failure (Rudnicki & Chen 1988), and pore fluids being sucked into the fault zone from the country rock. This is consistent with the dynamic increase in storage we see during the dynamic phase of strain localization. This transport is likely to be transient, as the local flow rate is proportional to the temporal rate of porosity change, and may be an important mechanism of intermittently charging hydrocarbon reservoirs or aquifers in active fault zones.

The behaviour for our large (10 cm diameter) samples is very different from that in more standard tests carried out on smaller samples of the same material (3.8 cm diameter) at the same confining pressure (<70 MPa, i.e. in the brittle field). The small samples produce only a single strand at these confining pressures, whereas the large samples produce sequential deformation bands similar to those seen in field exposure. Although multiple fault zones have been seen previously in laboratory tests on smaller samples (e.g. Jamison & Stearns 1982; Wong *et al.* 1997) these have all been at elevated pressures near the brittle–ductile transition between localized faulting and distributed shear-enhanced compaction and pervasive grain crushing. The difference in response between the two sample sizes calls into question the validity of linear upscaling of results from laboratory tests carried out on smaller samples of porous sedimentary rocks at low confining pressure. We also see much lower dilatant volumetric strain before failure in the large samples. Both observations are likely to be due to the higher sample-volume to grain-volume ratio for large samples.

At different pressures the samples show friction coefficients of the order of 0.6, similar to that for rocks and other natural materials (Scholz 1990). The immediate strengthening implies a local strain hardening rheology during the formation of deformation bands, although this is not always observed in the stresses measured at the sample boundaries.

We see no evidence for a second bifurcation in the form of a large-offset slip surface, although samples at larger strains (11%) did show slickenlines associated with abrasive wear (Mair 1997). Aydin & Johnson (1983) suggested that a critical amount of strain is required to initiate a large-offset slip surface. If this is true, perhaps this critical strain is above that achievable in our tests (2 cm displacement). The model of Aydin & Johnson (1983) describes the first-order features of field outcrops very well, but their model neglects two important elements. First, their model implicitly assumes a homogeneous rheology along the fault's strike and dip. From detailed quantitative field observation of faults both Antonellini *et al.* (1994) and Shipton & Cowie (2001) showed that the microstructural properties of the faults can be locally very heterogeneous, as a consequence of small fluctuations in material type, cementation and particularly clay content. This is consistent with the large local variations in the width of deformation bands seen in Fig. 1a. Second, the model of Aydin & Johnson (1983) assumes the faults grow instantaneously, whereas Shipton & Cowie (2001) demonstrated that the displacement field around deformation bands and slip surfaces can be understood only in terms of the nucleation, growth and coalescence of faults as an evolving process. In their study, Shipton & Cowie (2001) found evidence for localized nucleation of slip patches with identifiable slickenlines at an early stage of deformation, similar to those seen by Mair (1997). They showed that slip surfaces nucleate in small patches, and then propagate to form a connected network that can have very large offsets (tens of metres). The critical difference between their model and that of Fig. 2 is the 3D nature of the fault growth process, so that large offset slip surfaces evolve by propagating through a damage zone consisting of deformation bands ahead of the tip line of the slip surface.

Shipton & Cowie (2001) also investigated the relationship between the total offset and width of the damage zone, defined as the zone containing faults of all types, including slip surfaces as well as deformation bands. They observed a linear relationship between strain and damage zone width, analogous to that shown in Figs 6 and 10b. This implies that at least one of the assumptions in the hydraulic model for permeability evolution is also valid at the field scale. From analysis of thin sections, Shipton & Cowie (2001) found that the porosities of individual and multiple deformation bands are statistically indistinguishable, consistent with the relatively constant grain-size distribution we see in Clashach and Locharbriggs sandstone samples under laboratory conditions. However, they also observed a large further reduction in porosity in the large-offset slip surfaces. The hydraulic model of Fig. 10 could be modified to account for this by adding two extra parameters, i.e. the local strain for slip surface development, and the slip surface porosity or permeability. Antonellini & Aydin (1994) also concluded that slip surfaces induce a further reduction in fault-normal permeability, although possibly increasing fault-parallel permeability. More work would still be needed to include time-dependent chemical sealing effects into the strain-dependent model of equation (6), in particular measuring rates of precipitation as well as dissolution, before quantitative estimates of overall (physical and chemical) fault sealing potential in the subsurface can be made with confidence.

Many outstanding questions remain; for example, the precise mechanism of the nucleation of slip surfaces and their mechanical and hydraulic properties, quantification of longer-term chemical healing and sealing effects, and quantification of the effect of clay on the sample response. For example, clay-rich rocks affect the diagenetic reactions significantly in quartz–clay mixtures (Bjorkum 1996), and tend not to produce the multiple deformation bands illustrated in Fig. 1 (Antonellini *et al.* 1994).

Sealing rartes under true triaxial stress conditions can also be higher than those observed here (Main *et al.* 1996). The range of field conditions remains much wider than those investigated so far in the laboratory. For example, recent direct field evidence from observation of refraction paths of mineral phases deposited diagenetically after the formation of deformation bands actually implies lower permeability reductions under true *in situ* conditions than we have observed here (Taylor & Pollard 2000). This may be due to the bands forming at lower effective confining pressures than investigated here, or due to later mineral dissolution of finer fragments without local precipitation in the fault zone. At this stage we conclude that the formation of deformation bands in porous, sedimentary rocks at low temperature depends on a range of mechanical, structural, chemical, and hydraulic processes. These are likely to be strongly coupled in nature, and many remain to be determined.

Conclusion

We have observed the evolution of mechanical, microstructural, chemical and hydraulic properties of porous sandstones during the formation of deformation bands at low temperature and pressure. Larger samples (10 cm diameter cores) deform by the sequential development of a series of strands, whereas smaller (3.8 cm diameter) samples show only continuing slip on a single strand. The larger samples show a smaller volume dilatancy, but the same qualitative temporal evolution as that seen for smaller samples. The deformation bands show standard frictional coefficients (0.6), and no large-offset slip surfaces are observed. The microstructural properties of the deformation bands predict the hydraulic properties of the sample with a correlation coefficient of 0.997. An apparent permeability spike at the sample failure time can be attributed to a transient 'suction pump' effect caused by rapid local dilatancy on the incipient and moving new fault plane. The fault gouge is highly reactive, even at low temperature and pressure, and chemical steady state for advective flow is re-established over timescales of the order of a few hundred hours. The six orders of magnitude reduction in permeability in deformation bands in field exposure can plausibly be explained by a combination of the physical and chemical sealing rates observed or inferred from our results. Finally, although we have attempted to isolate the different effects, we infer that they are also very strongly coupled in nature.

The 'Big Rig' was donated by the British Geological Survey, Keyworth. We are grateful to D. McCann and S. Horsman and to the University Grants Committee for making this possible, and to M. Rotter of the Department of Civil Engineering, Edinburgh University, for providing space for its recommissioning at Edinburgh. The chemical flow rig was constructed with funding from the Petroleum Science and Technology Institute. We remain grateful to R. Johnson for his personal support. B.N. was funded by the Offshore Supplies Office LINK grant 951/7174/1, with matching funding from BP, Amerada Hess and Statoil. K.M. was funded by NERC studentship GT4/93/157/G. O.K. was funded by an industry consortium including BP–Amoco, Exxon–Mobil, the Japanese National Oil Corporation, LASMO, Shell, and Statoil. We are grateful to J. Constanty, J. Craig, K. Heffer, A. McCann, D. Yale, and M. Takahashi for their constructive advice during its experimental programme. We are grateful to E. Scholz and A. Jackson for their technical support in developing and maintaining the laboratory equipment, to P. Cowie and D. Pollard for constructive reviews of an earlier manuscript, to D. Sornette for correspondence on log-periodicity in fragmentation problems, and to Z. Shipton for providing a preprint from her field study. We appreciate the constructive comments of reviewers G. Lloyd, Q. Fisher, and D. Faulkner, whose comments significantly improved the text.

References

AHARANOV, E., TENTHORNEY, E. & SCHOLZ, C.H. 1998. Precipitation sealing and diagenesis 2. Theoretical analysis. *Journal of Geophysical Research*, **103**, 23969–23981.

ANTONELLINI, M.A. & AYDIN, A. 1994. Effect of faulting on fluid flow in porous sandstones: petrophysical properties. *Bulletin of the American Association of Petroleum Geologists*, **78**, 355–377.

ANTONELLINI, M.A., AYDIN, A. & POLLARD, D.D. 1994. Microstructure of deformation bands in porous sandstone. *Journal of Structural Geology*, **16**, 941–959.

AYDIN, A. 1978. Small faults formed as deformation bands in sandstone. *Pure and Applied Geophysics*, **116**, 913–930.

AYDIN, A. & JOHNSON, A.M. 1978. Development of faults as deformation bands in porous sandstones. *Pure and Applied Geophysics*, **116**, 931–942.

AYDIN, A. & JOHNSON, A.M. 1983. Analysis of faulting in porous sandstones. *Journal of Structural Geology*, **5**, 19–31.

BARTON, C.A., ZOBACK, M.D. & MOOS, D. 1995. Fluid flow along potentially active faults in crystalline rock. *Geology*, **23**, 683–686.

BAUD, P., ZHU, W. & WONG, T.-f. 2000. Failure mode and weakening effect of water on sandstone. *Journal of Geophysical Research*, **105**, 16371–16390.

BJORKUM, P.A. 1996. How important is pressure in causing dissolution of quartz in sandstones? *Journal of Sedimentary Research*, **66**, 147–154.

BOLTON, E.W., LASAGA, A.C. & RYE, D.M. 1996. A model for the kinetic control of quartz dissolution and precipitation in porous media flow with spatially variable permeability: formulation and examples of thermal convection. *Journal of Geophysical Research*, **101**, 22157–22187.

BRACE, W.F. & MARTIN, R.J. 1968. A test of the law of effective stress for crystalline rocks of low porosity. *International Journal of Mining Science and Geomechanical Abstracts*, **5**, 415–426.

BRANTLEY, S.L., EVANS, B., HICKMAN, S.H. & CRERAR, D. 1990. Healing of microcracks in quartz: implications for fluid flow. *Geology*, **18**, 136–139.

CASHMAN, S. & CASHMAN, K. 2000. Cataclasis and deformation-band formation in unconsolidated marine terrace sands. *Humboldt County, California. Geology*, **28**, 111–114.

DAVID, C., WONG, T.-f., ZHU, W. & ZHANG, J. 1994. Laboratory measurements of compaction-induced permeability change in porous rocks: implication for generation and maintenance of pore pressure excess in the crust. *Pure and Applied Geophysics*, **143**, 425–456.

DOVE, P.M. 1995. Geochemical controls on the kinetics of quartz fracture at subcritical tensile stresses. *Journal of Geophysical Research*, **100**, 22349–22359.

EDWARDS, H.E., BECKER, A.D. & HOWELL, J.A. 1993. Compartmentalisation of an aeolian sandstone by structural heterogeneities: Permo-Triassic Hopeman Sandstone, Moray Firth, Scotland. *In*: NORTH, C.P. & PROSSER, D.J. (eds) *Characterisation of Fluvial and Aeolian Sandstones by Structural Heterogeneities*. Geological Society, London, Special Publications, **73**, 339–365.

ELPHICK, S.C., NGWENYA, B.T. & SHIMMIELD, G.B. 1996. Ion microprobe imaging as a tool for elucidating waterflood reactions. *American Association of Petroleum Geologists Bulletin*, **80**, 713–720.

FISHER, D.M. & BRANTLEY, S. 1992. Models of quartz overgrowth and vein formation: deformation and episodic fluid flow in an ancient subduction zone. *Journal of Geophysical Research*, **97**, 20043–20061.

FISHER, Q.J. & KNIPE, R.J. 1998. Fault sealing processes in siliciclastic sediments. *In*: JONES, G., FISHER, Q.J. & KNIPE, R.J. (eds) *Faulting, Fault Sealing and Fluid Flow in Hydrocarbon Reservoirs*. Geological Society, London, Special Publications, **147**, 117–134.

FISHER, Q.J., KNIPE, R.J. & WORDEN, R.H. 2000. Microstructures of deformed and non-deformed sandstones from the North Sea: implications for the origins of quartz cement in sandstones. Special Publications of the International Association of Sedimentologists, **29**, 129–146.

FOWLES, J. & BURLEY, S. 1994. Textural and permeability characteristics of faulted, high porosity sandstones. *Marine and Petroleum Geology*, **11**, 608–623.

FUJII, Y., KIMADA, T. & KODAMA, J. 1998. Examination of rock failure criterion based on circumferential tensile strain. *Pure and Applied Geophysics*, **152**, 551–577.

JAMISON, W.R. & STEARNS, D.W. 1982. Tectonic deformation of the Wingate sandstone, Colorado National Monument. *American Association of Petroleum Geologists Bulletin*, **66**, 2584–2608.

KANAGAWA, K., COX, S.F. & ZHANG, S. 2000. Effects of dissolution–precipitation processes on the strength and mechanical behaviour of quartz gouge at high-temperature hydrothermal conditions. *Journal of Geophysical Research*, **105**, 11115–11126.

KNIPE, R.J. & LLOYD, G.E. 1994. Microstructural analysis of faulting in quartzite, Assynt, NW Scotland—implications for fault zone evolution. *Pure and Applied Geophysics*, **143**, 229–254.

LASAGA, A.C. 1984. Chemical kinetics of water–rock interactions. *Journal of Geophysical Research*, **89**, 4009–4025.

LOCKNER, D. & BYERLEE, J. 1994. Dilatancy in hydraulically isolated faults and the suppression of instability. *Geophysical Research Letters*, **21**, 2353–2356.

MAIN, I., NGWENYA, B., ELPHICK, S., SMART, B., CRAWFORD, B. & POUX, C. 1996. Scale limits on fluid pressure diffusion during rapid self-sealing deformation and fluid flow. *In*: AUBERTIN, M., HASSANI, F. & MITRI, H. (eds) *Rock Mechanics, 2*. Balkema, Rotterdam, 1161–1168.

MAIN, I.G., KWON, O., NGWENYA, B.T. & ELPHICK, S.C. 2000. Fault sealing during deformation band growth in porous sandstone. *Geology*, **28**, 1131–1134.

MAIN, I.G., SMART, B.G.D., SHIMMIELD, G.B., ELPHICK, S.C., CRAWFORD, B.R. & NGWENYA, B.T. 1994. The effects of combined changes in pore-fluid chemistry and stress state on permeability in reservoir rocks: preliminary results from analogue materials. *In*: AASEN, O. (ed.) *North Sea Oil and Gas Reservoirs III*. Kluwer, Dordrecht, 357–370.

MAIR, K. 1997. *Experimental studies of fault zone development in a porous sandstone*. Ph.D. thesis, Edinburgh University.

MAIR, K., MAIN, I.G. & ELPHICK, S.C. 2000. Sequential growth of deformation bands in the laboratory. *Journal of Structural Geology*, **22**, 25–42.

MATTHAI, S.K. & FISHER, G. 1996. Quantitative modelling of fault-fluid-discharge and fault-dilation-induced fluid-pressure variations in the seismogenic zone. *Geology*, **24**, 183–186.

NAKATANI, M. 1998. A new mechanism of slip weakening and strength recovery of friction associated with the mechanical consolidation of gouge. *Journal of Geophysical Research*, **103**, 27239–27256.

NGWENYA, B.T., ELPHICK, S.C., MAIN, I.G. & SHIMMIELD, G.B. 2000. Experimental constraints on the diagenetic self-sealing capacity of natural faults in high-porosity rocks. *Earth and Planetary Science Letters*, **183**, 187–199.

NGWENYA, B.T., ELPHICK, S.C. & SHIMMIELD, G.B. 1995. Reservoir sensitivity to water flooding: an experimental study of sea-water injection in a North Sea analogue reservoir analogue. *American Association of Petroleum Geologists Bulletin*, **79**, 285–304.

OLSEN, P.O., SCHOLZ, C.H. & LEGER, A. 1998. Healing and sealing of a simulated fault gouge under hydrothermal conditions: implications for fault healing. *Journal of Geophysical Research*, **103**, 7421–7430.

ORTOLEVA, P.J. *Geochemical Self-organisation*. Oxford University Press, Oxford.

PITTMAN, E.D. 1981. Effect of fault-related granulation on porosity and permeability of quartz sandstones. Simpson group (Ordovician) Oklahoma. *American Association of Petroleum Geologists Bulletin*, **65**, 2381–2387.

RALEIGH, B. & MARONE, C. 1997. The transition from brittle faulting to cataclastic flow in porous sandstones: mechanical deformation. *Journal of Geophysical Research*, **102**, 3009–3025.

REVIL, A. 2000. Pervasive pressure-solution transfer: a poro-visco-plastic model. *Geophysical Research Letters*, **26**, 255–258.

RICE, J.R. 1992. Fault stress states, pore pressure distribution, and the weakness of the San Andreas Fault. *In*: EVANS, B. & WONG, T.-f. (eds) *Fault Mechanics and Transport Properties of Rocks*. Academic Press, London, 475–504.

RIMSTIDT, J.D. & BARNES, H.L. 1980. The kinetics of silica–water reactions. *Geochimica et Cosmochimica Acta*, **44**, 1683–1699.

RUDNICKI, J. & CHEN, C.-H. 1988. Stabilization of rapid frictional slip on a weakening fault by dilatant hardening. *Journal of Geophysical Research*, **93**, 4745–4757.

RUDNICKI, J. & RICE, J.R. 1975. Conditions for the localization of deformation in pressure-sensitive dilatant materials. *Journal of the Mechanics and Physics of Solids*, **23**, 371–391.

RUTTER, E. 1983. Pressure solution in nature, theory and experiment. *Journal of the Geological Society, London*, **140**, 725–740.

SADOVSKIY, M.A., GOLUBEVA, T.V., PISARENKO, V.F. & SHNIRMAN, M.G. 1984. Characteristic dimensions of rock and hierarchical properties of seismicity. *Izvestiya, Earth Physics*, **20**, 87–96.

SAMMIS, C., KING, G.C.P. & BIEGEL, R. 1987. The kinematics of gouge deformation. *Pure and Applied Geophysics*, **125**, 777–812.

SAMMONDS, P.R., MEREDITH, P.G. & MAIN, I.G. 1992. Role of pore fluids in the generation of seismic precursors to shear fracture. *Nature*, **359**, 228–230.

SCHOLZ, C.H. 1990. *The Mechanics of Earthquakes and Faulting*. Cambridge University Press, New York.

SCHOLZ, C.H., LEGER, A. & KARNER, S.L. 1995. Experimental diagenesis: exploratory results. *Geophysical Research Letters*, **22**, 719–722.

SHIPTON, Z.K. & COWIE, P.A. 2001. Damage zone and slip surface evolution over mm to km scale in high-porosity Navajo sandstone, Utah. *Journal of Structural Geology*, in press.

SIBSON, R. 1977. Fault rocks and fault mechanisms. *Journal of the Geological Society, London*, **133**, 191–213.

SIBSON, R. 1981. Fluid flow accompanying faulting: field evidence and models. *In*: SIMPSON, D.W. & WILLIAMS, P.G. (eds) *Earthquake Prediction—an International Review*. American Geophysical Union, Maurice Ewing Series, **4**, 593–603.

SIBSON, R. 1990. Conditions for fault valve behaviour. *In*: KNIPE, R.J. & RUTTER, E.H. (eds) *Deformation Mechanisms, Rheology and Tectonics*. Geological Society, London, Special Publications, **54**, 15–28.

SUTEANU, C., ZUGRAVESCU, D. & MUNTEANU, F. 2000. Fractal approach of structuring by fragmentation. *Pure and Applied Geophysics*, **157**, 539–557.

TAYLOR, W.L. & POLLARD, D.D. 2000. Estimation of *in situ* permeability of deformation bands in porous sandstone, Valley of Fire, Nevada. *Water Resources Research*, 9, **36**, 2595–2606.

TENTHORNEY, E., SCHOLZ, C.H. & AHARANOV, E. 1998. Precipitation sealing and diagenesis 1. Experimental results. *Journal of Geophysical Research*, **103**, 23951–23967.

TEUFEL, L. 1981. Pore volume changes during frictional sliding of simulated faults. Geophysical Monograph, American Geophysical Union, **24**, 135–145.

TOWNEND, J. & ZOBACK, M.D. 2000. How faulting keeps the crust strong. *Geology*, **28**, 399–402.

UNDERHILL, J.R. & WOODCOCK, N.H. 1987. Faulting mechanisms in high porosity sandstones: New Red Sandstone, Arran, Scotland. *In*: JONES, M.E. & PRESTON, R.M.F. (eds) *Deformation of Sediments and Sedimentary Rocks*. Geological Society, London, Special Publications, **29**, 91–105.

WONG, T.-f. & WU, L.-C. 1995. Tensile stress concentration and compressive failure in cemented granular material. *Geophysical Research Letters*, **22**, 1649–1652.

WONG, T.-f. & ZHU, W. 1999. Brittle faulting and permeability evolution: hydromechanical measurement, microstructural observation, and network modelling. *In*: HANEBERG, W.C., MOZLEY, P.S., MOORE, J.C. & GOODWIN, L.B. (eds) *Faults and Subsurface Fluid Flow in the Shallow Crust*. Geophysical Monographs, American Geophysical Union, **113**, 93–100.

WONG, T.-f., DAVID, C. & ZHU, W. 1997. The transition from brittle faulting to cataclastic flow in porous sandstones: mechanical deformation. *Journal of Geophysical Research*, **102**, 3009–3025.

ZHANG, J., WONG, T.-f. & DAVIS, D.M. 1990. Micromechanics of pressure-induced grain crushing in porous rocks. *Journal of Geophysical Research*, **95**, 341–352.

ZHU, W. & WONG, T.-f. 1996. Permeability reduction in a dilating rock: network modelling of damage and tortuosity. *Geophysical Research Letters*, **23**, 3099–3102.

ZHU, W. & WONG, T.-f. 1997. The transition from brittle faulting to cataclastic flow: permeability evolution. *Journal of Geophysical Research*, **102**, 3027–3041.

Thermal–rheological controls on deformation within oceanic transforms

KEVIN P. FURLONG[1], STEVEN D. SHEAFFER[1,2] & ROCCO MALSERVISI[1]

[1]*Geodynamics Research Group, Department of Geosciences, Pennsylvania State University, University Park, PA 16802, USA (e-mail: kevin@geodyn.psu.edu)*

[2]*Department of Geophysics, Colorado School of Mines, Golden, CO 18401, USA*

Abstract: Transform faults that offset mid-ocean ridge (MOR) segments accommodate plate motion through deformations that involve complex thermal and mechanical feedbacks involving both brittle and temperature-dependent ductile rheologies. Through the implementation of a 3D coupled thermal–mechanical modelling approach, we have developed a more detailed picture of the geometry of plate boundary deformation and its dependence on plate velocity and the age offset of MOR transforms. The modelling results show that cooling of near-ridge lithosphere (lateral heat transfer) has significant effects in the ductile mantle lithosphere for both the location and style of deformation. The region where strain is accommodated in the subjacent mantle lithosphere is systematically offset from the position of the overlying linear transform fault in the brittle crust. This offset causes the boundary to be oblique to plate motions along much of the transform's length, producing extension in regions of significant obliquity modifying the location of the surface fault segments. An implication of this complex plate-boundary geometry is that in the near-ridge region, the older (cooler) lithosphere will extend beneath the ridge tip, restricting the upwelling of mantle to the MOR. The melt to generate the oceanic crust adjacent to the transform must migrate laterally from its offset source, resulting in a reduced volume and thinner crust. This near-ridge plate boundary structure also matches the pattern of core-complex extension observed at inside corners of many slow-spreading ridges. The oblique extensional structure may also explain magmatism that is observed along 'leaky' transforms, which could ultimately result in the generation of new ridge segments that effectively 'split' large transforms.

Mid-ocean ridge (MOR) transform boundaries have complex thermal and rheological structures and therefore potentially complex kinematic behaviour. Even though deformation at the surface appears to occur in narrow linear zones along brittle faults, deformation in the deeper lithosphere below the brittle–ductile transition may be more complicated. Below the depth of faulting, the structure of the shear zone where plate motions are accommodated will be determined by coupled thermal–mechanical effects, and thus can be a 3D regime that does not necessarily mimic the overlying brittle faults, but may perturb their behaviour.

Improvements in tools to image bathymetric features of transform zones allow detailed descriptions of the geometry and localization of shear deformation within transform zones to be determined (see, e.g. Searle 1992). The actual transform fault zone (TFZ) within the transform tectonized zone (TTZ) is often slightly oblique to the TTZ (Fox & Gallo 1984; Searle 1986, 1992; Searle *et al.* 1994). The patterns of deformation observed in the TTZ appear to vary with spreading rate. How these near-surface patterns of plate boundary deformation relate to deeper plate boundary shear is not well resolved and serves as a focus of this research.

An exciting set of recent research results is the documentation of core-complex-like 'mega-mullions' at the inside corners of many ridge–transform intersections (e.g. Cann *et al.* 1997; Blackman *et al.* 1998; Tucholke *et al.* 1998; Escartin & Cannat 1999), particularly associated with low-velocity plate boundaries. These extensional structures expose lower-crustal gabbros and upper-mantle peridotites, and require a significant change in deformational style in the vicinity of the ridge–transform intersection. The role that the lithosphere-scale deformation plays in driving this deformation is an issue requiring investigation.

An idealized model comprising two elastic plates separated by a linear strike-slip fault is

From: HOLDSWORTH, R.E., STRACHAN, R.A., MAGLOUGHLIN, J.F. & KNIPE, R.J. (eds) 2001. *The Nature and Tectonic Significance of Fault Zone Weakening*. Geological Society, London, Special Publications, **186**, 65–83. 0305-8719/01/$15.00

often employed in transform studies. Any plate boundary strength below the brittle–ductile transition is often disregarded. However, this ductile material may have significant shear strength, particularly as one moves from the ridge into the older lithosphere along the transform. Ocean-ridge transforms with large age offset may show patterns of deformation that result from this variation in mechanical strength of the plate boundary region. Furthermore, although the ductile strength tends to decrease with depth (rising temperature), this weaker mantle lithosphere may still play an important role in transform behaviour.

We have developed a 3D coupled thermal–deformational model to investigate the nature of deformation along mid-ocean transform systems. The variation in deformational geometries as a function of plate velocity and age offset along the transform are the primary targets of this modelling.

The modelling approach we are taking is essentially an extension of the oceanic transform model of Forsyth & Wilson (1984), which describes the results of a 3D thermal model of the transform regime using finite difference methods. They assumed a transform that is a vertical strike-slip boundary at all depths and prescribed a uniform velocity field on either side parallel to the transform strike, except under the ridges where upwelling was defined. This model resulted in a 3D temperature field that included the effects of lateral heat conduction across the transform. Limitations of this model approach that we have tried to address in our models are that the kinematics was completely prescribed, and the model did not account for the effects of variations in temperature on rheology. Below the depth of faulting, the 'plate boundary' is presumably a shear zone. The results of our modelling indicate that these factors have significant effects on both the thermal structure and the kinematics of the transform.

In addition to the purely conductive model of Forsyth & Wilson (1984); Phipps-Morgan & Forsyth (1988) coupled the thermal calculation to a sub-lithosphere velocity field based on upper-mantle flow calculations. A semi-analytical solution for 3D linear viscous flow from an assumed transform plate geometry was employed to determine the velocity field in the viscous regions. This field was then used in a finite difference thermal calculation to estimate the large-scale transform temperature field. This model assumed a linear rheology with constant, temperature-independent viscosity, and the calculated velocity field depended only on the plate geometry. In essence, the thermal field depended on the calculated velocity field, but the velocity field was independent of the thermal field. In our study, we have explicitly included the feedbacks between the thermal and deformational fields in modelling the plate boundary deformation zone, to evaluate the influence of these interactions in controlling the pattern of plate boundary deformation. We have focused on the deformational response of the lithospheric part of the MOR–transform system. These results are generally compatible with a suite of modelling studies that have focused more on the linkages between the induced flow field in the asthenospheric mantle and the overlying MOR–transform lithosphere (Blackman & Forsyth 1992; Shen & Forsyth 1992; Blackman 1997).

Here we describe our attempt to obtain an improved picture of transform plate boundary deformation by utilizing a model that fully couples the thermal and mechanical fields, and incorporates elastic–brittle material and brittle faulting at shallow depth, with a non-linear temperature-dependent ductile rheology at depth.

Our modelling shows that the resulting thermal and mechanical structures associated with plate kinematics are significantly different from those seen in previous models. Specifically, the localized shear zone that accommodates plate motions in the ductile mantle lithosphere does not mimic the overlying transform fault. It is offset at the MOR tips because of the effects of lateral cooling on the non-linear ductile rheology of the upper mantle. As a result, much of the boundary at depth is oblique to the transform strike. Such a geometry may tend to drive oblique faulting at shallow levels, and produce extension, upwelling, and thinning along the transform, leading to leaky transform magmatism sometimes observed along transforms. In addition, the plate boundary geometry and pattern of strain at the ridge–transform intersection may serve as the root zone for the formation of oceanic core-complexes.

Conceptual model

We assume that near the surface, MOR transforms are dominated by weak, linear faults that are aligned with plate motions to within a few degrees (Spitzak & DeMets 1996). Although the details of the TTZ are clearly more complex than this (Searle 1992), we have assumed this simplified geometry in our modelling, but address the implications of our models for the near-surface deformation patterns later in the paper. This shallow plate boundary structure is implemented in our model as a single vertical fault of low frictional resistance cutting the elastic layer, which strikes normal to the ridge axes and parallel to plate motions. Seismic analyses suggest that the maximum depth of faulting in these regimes is approximately coincident with the depth of the 600 °C isotherm (e.g. Engeln *et al.* 1986). For our models we use this temperature to define the brittle–elastic layer, beneath which temperature-dependent plastic deformation is assumed.

Lateral heat conduction across the transform will cool young near-ridge material because of its proximity to cooler, older material on the opposite side. Thermal modelling (e.g. Forsyth & Wilson 1984; Phipps-Morgan & Forsyth 1988) shows that there is a zone that can extend for >10 km on either side of the surface transform within which this cooling will lower temperatures by more than 100 °C with respect to plate material of equivalent age and depth far from the transform. Although these temperature variations are usually interpreted as being significant primarily in terms of their density or gravity effects (Escartin & Cannat 1999); they can produce significant contrasts in ductile strength (for temperature-dependent power-law creep in the upper mantle). Temperature variations of 100 °C can produce up to an order of magnitude difference in dislocation–creep flow strength in 'dry' olivine (Karato & Wu 1993).

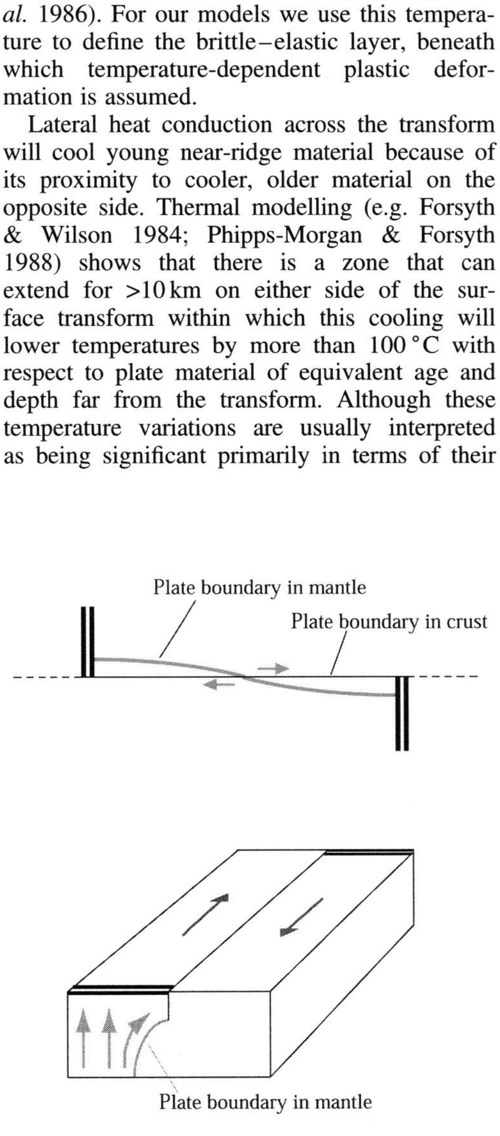

Fig. 1. Conceptual lithospheric and plate boundary structure. Top (map view): location of crustal faulting and the approximate location of the mantle plate boundary at depth. Bottom: cutaway cross-section at the ridge showing the form of the boundary and a schematic interpretation of the plate boundary geometry effect on magma supply.

With the effects of lateral cooling, young near-ridge mantle material in the transform zone will be anomalously strong compared with material of the same age and depth in the rest of the plate. The weaker part of the plate (and thus the potential locus of localized plate boundary strain) will not lie directly below the overlying fault, but will be offset into the young plate in regions where the temperature contrasts between plates are significant. The net result will be a narrow deformation zone in the mantle where plate motions are accommodated, but that is not everywhere aligned with the overlying linear fault. This deformation zone will define the plate boundary in the mantle, and will be referred to as such in the rest of the paper, even though it does refer to a shear zone of finite width.

Figure 1 is a schematic illustration of this concept. The plate boundary structure is complex and 3D in nature. The largest offset between the near-surface plate boundary (transform fault) and the deeper shear zone will be near the ridges, where the temperature contrast across the transform is the greatest. Here the lateral cooling effect requires that the older plate should extend beneath the younger plate (and MOR) at depth. In this region, the details of the magma supply to the ridge will be complex. The magma required to generate new crust in this zone must migrate laterally over the offset and therefore might be reduced in volume (Fig. 1b). This is consistent with studies of the crust along fracture zones that show it to be anomalously thin and have a fundamentally different velocity structure, both of which have been attributed to a reduced magma supply (Detrick *et al.* 1993).

The shape of the boundary itself will result in mismatch between the strike of the near-surface faults and the deeper shear zone. This may lead to a reorientation of surface faults and also result in transform parallel extension along parts of the mantle boundary that are oblique to the transform strike (Fig. 1a). Such extension and the associated upwelling may produce thinning and magmatism along the transform if the upwelling velocity (thinning rate) is large compared with the competing secular cooling rate. Near the ridge, the plate boundary may include a subhorizontal section connecting the base of the brittle fault to the deeper shear zone (Fig. 1b). Such a low-angle shear zone may help nucleate core-complex extension on the inside corners of slow-spreading ridge–transform intersections.

Numerical models

We modelled these structures using two separate numerical codes. The mechanical code is a 3D finite element (FE) code based on TECTON (Melosh & Raefsky 1980, 1981, 1983 which was modified to be fully 3D and to allow temperature-dependent viscous rheologies (Govers & Wortel 1995). Deformation is elasto-viscous, except along faults, which are included via the 'slippery node' technique (Melosh & Williams 1989).

This finite element code is based on a Lagrangian FE formulation, where nodal points are attached to material points and therefore move along with the deforming material. The large amounts of relative displacement across the transform boundary create problems for such a formulation, which is restricted in terms of total displacement. However, we are primarily interested in the final, dynamic steady-state thermal and deformational fields. Through coupling the deformation model with an Eulerian thermal formulation (i.e. the material moves through the mesh according to a prescribed kinematics, and arbitrarily large displacements are allowed) we can overcome this shortcoming.

The separate thermal model, which is employed to calculate the thermal structures outside the TECTON model, is based on a 3D finite difference (Method of Lines) solution to the heat conduction equation where it is decomposed to a system of coupled linear ordinary differential equations (see Furlong *et al.* 1982). The solution to the resulting system is obtained numerically using the Lawrence Livermore ODEPACK solver (Hindmarsh 1983). The effects of heat advection from material kinematics are added by shifting the nodal grid between conduction steps based on a set of externally defined nodal velocity vectors, as in the model of Forsyth & Wilson (1984). However, here the nodal velocities are not arbitrary but are determined by the 3D mechanical model.

As we are interested in studying the final 'dynamic steady-state' regime for a particular plate-velocity and age-offset scenario, these codes are utilized in a two-step process where we iterate between the two models, using the result of one as the input for the other. This process allows the mechanical model to be run for a sufficient deformation to provide the appropriate kinematics for the thermal calculation, which produces a new thermal regime used in the deformation model, etc. The entire process begins with the assumption of an idealized, uniform velocity field to 'seed' the model. This velocity field is used in the thermal code to calculate the steady-state temperature field produced by the assumed plate kinematics (akin to the model approach of Forsyth & Wilson (1984)). These temperatures are then interpolated onto the FE mesh and used as initial temperatures in the deformational model. Appropriate boundary conditions are applied and the deformational model produces a velocity field compatible for the thermal regime (keeping the total strain small so that the thermal structure is not significantly disturbed spatially under the limitations of the Lagrangian approach). This new velocity field is then applied to the next iteration of the thermal model, producing a refined dynamic steady-state thermal field for the transform. These temperatures are then used in the next iteration of the mechanical model, resulting in more refinements to the velocity field. This iterative process is continued to convergence (i.e. the point at which there are no longer

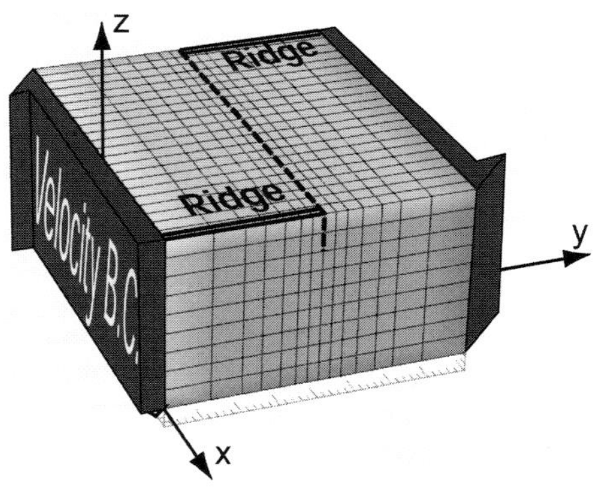

Fig. 2. Schematic diagram of model domain and generalized boundary conditions for mechanical modelling. Plate-velocity boundary conditions are applied along the sides of the model. Thermal and mechanical models covered same domain with slightly different nodal spacing. Width of model (y-direction) is constant (80 km) in all cases, length of model (x-direction) varies with transform offset (100–300 km), thickness of model is 45 km in most cases. It is thicker in the large-offset low-velocity models to accommodate increased plate cooling.

significant changes in either temperature or kinematics between iterations) providing a final estimate of the fully coupled thermal–mechanical field of the transform. In practice, after about 3–5 iterations, the change in either the temperature field or the strain field is small, of the order of 1–2% of the total change from the initial temperature and strain fields.

Model geometry is shown in Fig. 2. Our convention is that the *x*-direction is in the transform-parallel direction, the *y*-direction is ridge parallel, and the *z*-direction is vertical. For the mechanical deformation model, displacement rate (velocity) boundary conditions are applied to the sides of the model in the direction of relative plate motion as shown by the large arrows in Fig. 2. The ridge-parallel ends of the model are free to move as needed, although the location of the ridge itself is specified along the top surface of the model. The top surface of the model is fixed in a vertical direction, but the bottom of the model can move vertically as needed to accommodate any thinning in the model. As there is no gravity acting in this model simulation, the fixing of the top boundary does not in any way affect the patterns of deformation, although the combination of no gravity and the boundary condition precludes directly observing the topographic response. The transform-parallel width of the model is fixed; that is, the model is not allowed to expand or contract in the *y*-direction. The near-surface transform is specified and in the model results shown was allowed to slip freely using 'slippery nodes'. Models were also run (not shown here) with up to 10 MPa of sliding friction along the transform with no significant change in either the final temperature or deformation field. Models with high sliding friction or with locked faults showed distinctly different patterns of strain, with little strain localization, particularly for low plate-velocity models. We believe models with low or no sliding friction best simulate the long-term evolution of transform systems.

In the deformation model the vertical nodal spacing was fixed at 5 km; the *y*-direction (cross-transform) had a variable spacing ranging from 2.5 km near the transform to 12.5 km at the edges of the model; the *x*-direction (transform-strike direction) was constant for each model but changed depending on the total offset of the transform, ranging from 4.2 km for the short offset (100 km) models to 12.6 km for the long offset (300 km) models. In all cases there were 3450 nodes (*x*:*y*:*z* 23:15:10) and 2772 elements in the deformational models. These nodal dimensions are small enough to capture the primary effects of the variable temperature and rheology, although allowing the total model domain to be computationally tractable.

The thermal models covered the same model domain with uniform spacing of nodes in each direction but variable node spacing between directions. The *y*-direction (21 nodes) was spaced at 3.8 km; the *z*-direction (15 nodes) was spaced at 3 km; the *x*-direction (23 nodes) varied from *c.* 4 km to 12 km depending on transform offset. Boundary conditions for the thermal models were 0 °C at the top boundary (ocean floor) and 1300 °C at the bottom boundary (asthenosphere). Boundary nodes at the ends of the models were specified as parts of ridges (1300 °C) or points of material 'flow' where temperatures were set based on the local plate kinematics, to simulate the thermal conditions of material flowing across that boundary. Ridges were assumed to be passive and their length was determined entirely by the specified velocity field (obtained from the previous iteration mechanical model). The remaining sides of the model were set to have zero heat flow (in the direction perpendicular to the boundary).

The model lithosphere rheology for the results presented here is based on a 'dry' olivine composition with the material parameters listed in Table 1. Observations and laboratory experiments suggest that oceanic lithosphere near the mid-ocean ridges is undersaturated by water and 'dry' parameters are appropriate in the upper 100 km of the oceanic lithosphere (Karato & Wu 1993; Hirth & Kohlstedt 1996).

Table 1. *Material parameters in modelling*

Parameter	Value
Thermal properties	
Thermal conductivity (*k*)	3.0 W m^{-1} K^{-1}
Heat generation	0.0 μ W m^{-3}
Density (ρ)	3300 kg m^{-3}
Elastic and plastic properties	
Young's modulus (*E*)	1 × 1011 Pa
Poisson's ratio (υ)	0.25
Mohr–Coulomb angle	35°
Creep properties ('dry' olivine) (Karato & Wu 1993)	
Power-law exponent (*n*)	3.5
Activation energy (*Q*)	544 kJ mol^{-1}
Pre-exponent (*A*)	2.01359 × 10-16 Pa-3.5 s^{-1}
Gas constant (*R*)	8.31441 J mol^{-1} K^{-1}

Modelling results

The results from our coupled thermal–deformational model differ from previous models in both the resulting temperature structure and plate boundary geometry. The thermal field from our coupled model produces very different plate boundary structures and kinematics than implicitly assumed in previous models; that is, where the thermal regime evolved to include effects of lateral conduction, but the kinematics was prescribed (Forsyth & Wilson 1984; Phipps-Morgan & Forsyth 1988). We compare three sets of simulations with the age offset and plate velocity conditions given in Table 2.

Table 2. *Model age and offset conditions*

Model	Age offset (Ma)	Plate velocity (cm a^{-1})
5.04	5	4
5.12	5	12
15.04	15	4

Plate velocity is full spreading rate and age offset is transform length divided by half-spreading rate (i.e. the crustal age on the old side at the ridge).

Thermal structure

The thermal structure of the transform regime is substantially modified by the effects of a 3D kinematic field. Advection of heat within the lithosphere (3D kinematics) couples with the lateral-cooling conductive effects (Forsyth & Wilson 1984) to modify both the kinematics and the resulting thermal field. Phipps-Morgan & Forsyth (1988) showed the effects of advective heat transport from a convective flow field in the mantle; we have focused on the effects of the strain field local to the transform regime in driving a significant component of the heat transfer along the transform.

Figure 3 shows the thermal modelling results for the low plate-velocity and small age-offset model (Model 5.04). The isotherms (at a depth of 20 km) that result from the kinematics determined by our final deformational model are shown in Fig. 3a. A comparison of these results with other models of transform thermal structure and with standard (i.e. 'half-space' in which temperature varies as the error function (erf)) thermal models highlights the importance of the conductive and advective components of heat transport. Figure 3b shows the basic conductive (lateral) cooling effect at a depth of 20 km for the idealized transform (vertical fault–uniform velocity) by plotting the difference between an error function lithosphere (half-space cooling) and the idealized initial-model thermal structure (which includes effects of lateral cooling). This is analogous to the anomalous temperature plotted by Forsyth & Wilson (1984). It can be seen that anomalies of >80 °C exist in regions near the transform. Figure 3c shows the combined conductive (lateral) cooling and the advective effect on the thermal structure from the final deformational model, by plotting the difference between it (Fig. 3a) and an error function lithosphere. It is clear that the combination of heat transfer effects is significant, producing anomalies >100 °C along the transform. Figure 3d shows the difference between the idealized thermal model (conduction alone) and the final thermal model (conductive plus advective effects). Significant differences in the temperature exist, focused near the ridge–transform intersections. These differences are large enough to affect interpretations of crustal structure and/or magma influx derived from gravity analyses (Escartin & Cannat 1999).

Figure 4 shows a similar series of plots for the high plate-velocity transform (Model 5.12). As might be expected, there is a greater difference in final temperature structure between the idealized case and our result. This can be best seen in Fig. 4c, where there is a very large region of >100 °C lateral cooling effect, and Fig. 4d, where the difference between the ideal and final models is more significant than in the low-velocity model (Fig. 3d).

Model 5.04 and Model 5.12 can be normalized according to crustal age. In both models crustal age spans 0–5 Ma, and the pattern of age difference across the transform is the same. In the idealized model, both of these models produce the same thermal structure in the age-normalized reference frame. However, when the feedbacks between thermal structure and deformation are included, as well as the effects of advective and conductive heat transfer, significant thermal differences result. Figure 5 shows the difference between Models 5.04 and 5.12 (normalized for crustal age) at depths of 10 and 20 km. At both depths (10 and 20 km) shown, the low plate-velocity model is hotter than the high plate-velocity model. This is a result of both the relative geometry of the plate boundary zone, which extends further into the 'young' side of the transform in the high-velocity case, and slightly more thinning occurring in the low-velocity model as a consequence of plate boundary curvature (described later).

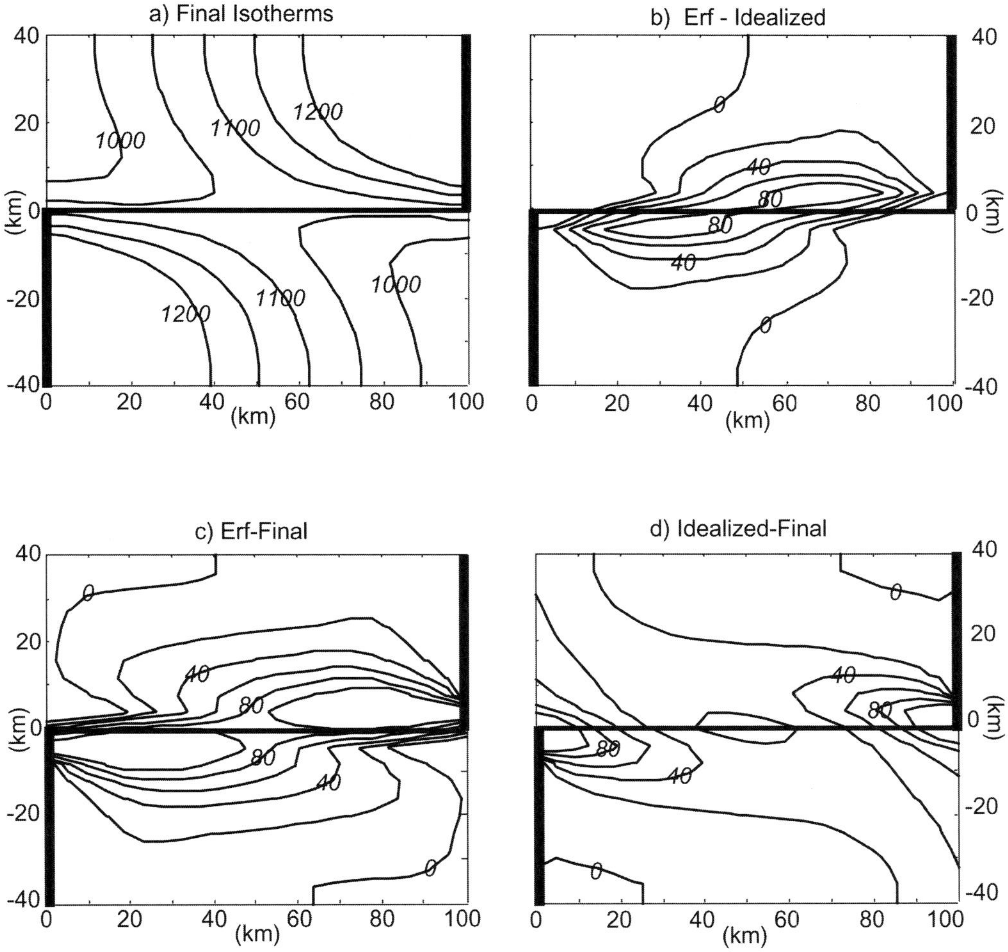

Fig. 3. Temperature model results for Model 5.04 (depth 20 km). (a) Thermal field (isotherms) for our final coupled model, which includes rheological and mechanical effects. (b) Difference between the thermal field of an idealized transform model consisting of a vertical strike-slip fault at all depths with uniform velocity on either side (including 3D heat conduction), and that of an error function lithosphere (infinite half-space) with no lateral heat conduction. (c) Difference between (a), our final transform model, and an error function lithosphere. (d) Difference between thermal fields for the idealized model and our transform model. Contours are labelled in °C.

Kinematics

The influence of the thermal field on the patterns of plate kinematics is shown in Fig. 6, where the velocity distribution across the model transform is shown at a depth of 20 km (below the brittle–ductile transition; at the depth of the temperature results shown in Figs. 3 and 4); a depth where the rheology is very dependent on temperature. The upper plot shows results for Model 5.04; the lower plot is the equivalent result for Model 5.12. Three velocity models are compared in each case. The step-function shows the velocity distribution assumed in the idealized transform model with uniform velocity on either side of a vertical transform. The set of dashed curves shows the velocity distributions at various positions along the transform (from near the ridge to the centre of the transform) as determined by the deformational model using the idealized (step function) thermal model. This is the result one obtains using the Forsyth

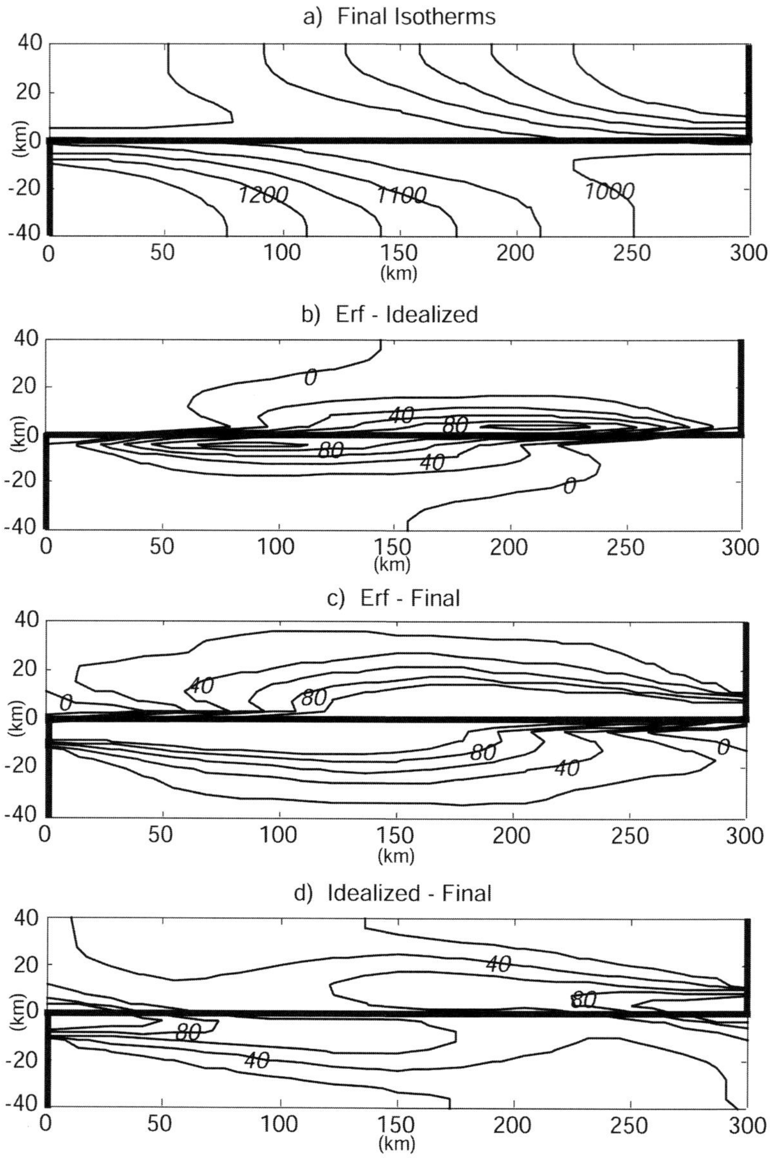

Fig. 4. Temperature model results for Model 5.12 (depth 20 km). (a)–(d) as in Fig. 3.

& Wilson (1984) temperature model to determine plate boundary deformation, and is the result we obtain after the first iteration of our modelling sequence. It is noteworthy that the velocity model produced by the assumed thermal model is inconsistent with the velocity model that produced the thermal field. To correct this inconsistency, we follow the iterative process described earlier to couple the thermal and mechanical fields and produce internally consistent thermal and deformational conditions. The velocity distribution at the same points along the transform of our final model, after model convergence, is shown by the continuous curves, and will be referred to as the final deformational model.

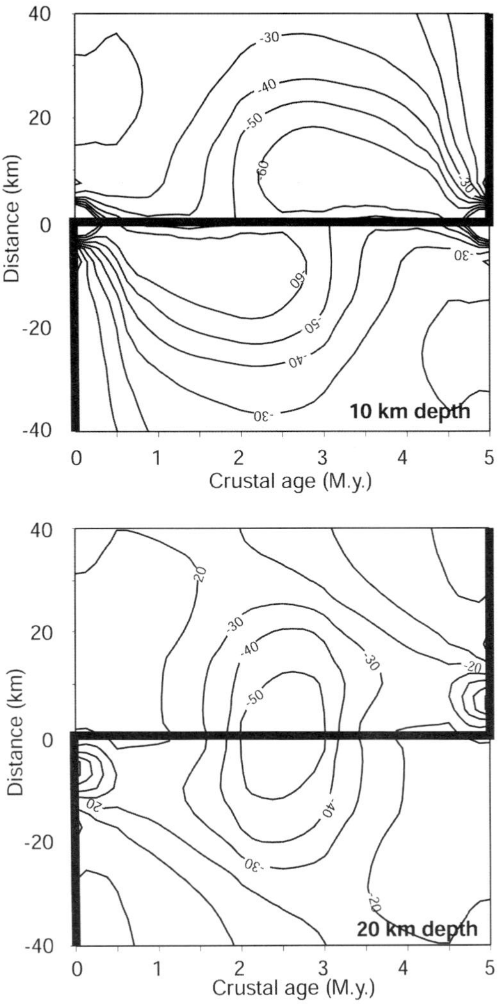

Fig. 5. Difference in temperature results between Models 5.04 and 5.12 normalized for crustal age. Top: temperature difference at 10 km depth within elastic–brittle layer. Bottom: temperature difference results at 20 km depth within ductile region. Contours are labelled in °C. Negative values indicate that Model 5.04 is hotter than Model 5.12.

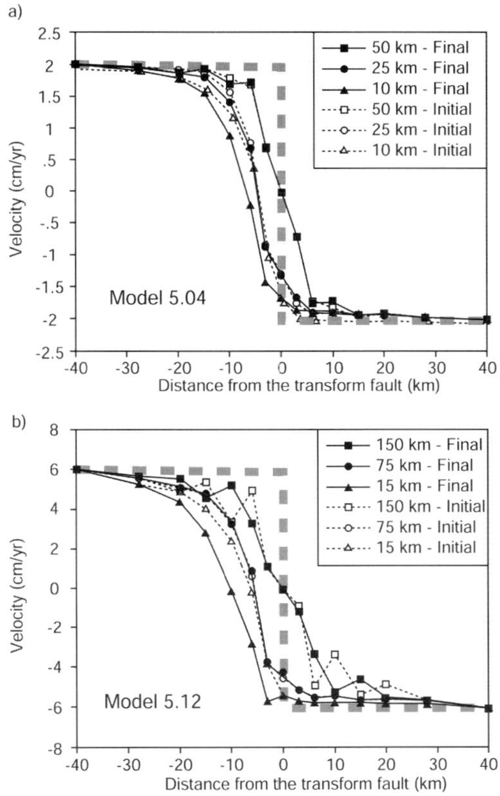

Fig. 6. Velocity profiles across the transform model at a depth of 20 km for (a) the low-velocity model and (b) the high-velocity model. Both have an age offset of 5 Ma. Each graph shows the velocity profiles for three different models. The step function is the velocity distribution for an idealized transform with uniform velocity on either side of a vertical fault. The open symbols show profiles for the initial deformational model, which results from the use of the thermal field from the idealized transform (i.e. the step function) in the deformational model. It is shown at three points along the transform length, 10, 25 and 50 km, the last being at the mid-point. The filled symbols give the velocity profiles for the final deformational model (at the same three locations along the transform), which is the steady-state dynamical result of repeated iterations of coupled thermal–mechanical steps.

For the low plate-velocity transform case, the motion is accommodated in a shear zone roughly 20 km wide (Fig. 6). The middle of the deformation zone (the 'plate boundary') is offset near the ridge onto the 'old side' of the transform by 10 km. Although there are significant differences between idealized (step-function) kinematics and that produced by the coupled deformational model, there is a relatively small difference between the initial and final deformational models, except as we approach the ridge (e.g. the 10 km curve in the upper panel of Fig. 6). This suggests that for low plate-velocity transforms, the idealized thermal structure (Forsyth & Wilson 1984; Phipps-Morgan & Forsyth 1988)) produces a reasonable approximation of the transform kinematics away from the ridges.

In the high plate-velocity model (Fig. 6) the shear zone is wider (*c.* 25 km) and the offset into the young side of the transform is greater. The difference in the kinematics between the initial and final deformation models is again greater near the ridge-ends of the transform. These results suggest that in the near-ridge environment, the idealized thermal structure (derived from simple step-function kinematics) does not produce a good approximation to the deformation field, and a coupling of the thermal and mechanical processes is required to determine the final mechanical configuration.

Three-dimensional deformation along transforms

Our model provides 3D simulations of the strain field occurring along mid-ocean transforms under the thermal conditions that develop in this self-consistent thermal–deformational modelling strategy. Visualizing the 3D deformation patterns is difficult and we have used a series of 2D slices (both horizontal and vertical) through the 3D data blocks to help define patterns of strain and the geometry of the plate boundary. In each cross-section, the colour shading shows strain rates and small arrows represent nodal velocities, which vary with model simulation to accommodate the variations in strain rates and velocity with spreading rate.

Figure 7 shows the results for Model 5.12 (high velocity), which has a transform length of 300 km. Shear strain rate, ε_{xy} (shear strain rate in the transform parallel direction on the vertical plane perpendicular to the ridge (y–z) plane) is shown in map view slices at depths of 5 km, within the brittle–elastic layer (Fig. 7a), and 25 km, in the upper mantle (Fig. 7b). Vertical sections perpendicular to the transform are shown at 5 km from the ridge (Fig. 7c), 60 km from the ridge (Fig. 7d), and 100 km from the ridge (Fig. 7e). The shear zone that acts as the mantle plate boundary can clearly be seen and displays the basic offset form described by the conceptual model in relation to the overlying brittle transform fault (see Fig. 1). The plate boundary, represented by the 'zero' line across which the velocities change direction, is indicated in the plots. The 3D shape of the plate boundary can be inferred from the set of cross-sectional views of the strain rate field in Fig. 7c–e. The location of the brittle fault in the elastic–brittle layer is shown by the grey dashed line, and the orange dashed line traces the zero-velocity fibre through the high-shear zone, which again displays the form shown in the conceptual model; also, the older (cooler) plate can be seen to extend into the younger plate beneath the ridge tips.

For the lower plate-velocity models (Model 5.04 (Fig. 8); Model 15.04 (Fig. 9)), the structure of the shear zone is analogous to Model 5.12. Model 5.04 shows a deformed zone that is less linear than in Model 5.12: the greater curvature (and obliquity) of the strain field as compared with the plate surface-velocity field results from the combination of a shorter transform length with only a slightly smaller offset near the ridges. The increase in the obliquity of the strain rate field for the slower transform may affect the pattern of deformation within the transform domain (Fox & Gallo 1986; Searle 1992).

The results for Model 15.04 (Fig. 9) show some of the characteristics of models of similar strain (i.e. Model 5.04 with same plate velocities) and also some of the characteristics of models of similar fault offset length (i.e. Model 5.12 with the same 300 km offset) in the patterns of transform deformation. The geometrical effects of a longer fault offset allow a more linear plate boundary, reducing the obliquity and curvature in the centre of the transform. The region of greatest obliquity (and possible largest extensional strain) is closer to the ridge at a position about 50 km from the ridge–transform intersection. However, the low velocities and the additional cooling time combine to produce a stronger plate with more lateral cooling. As a consequence, the plate boundary at depth migrates much further into the young side of the transform, as the older plate approaches the ridge. In Model 5.04 the plate boundary extends *c.* 7 km into the young side, whereas in Model 15.04 it extends 15 km or more. Similarly, the region experiencing plate boundary strain has a half-width of *c.* 10 km in Model 5.04 compared with almost 20 km for the larger age-offset case. The equivalence of plate velocities implies the same total integrated strain across both boundaries, but in the larger age-offset case, the impact of cooler lithosphere is to widen the zone of deformation.

Implications of the models for transform tectonics

The models were developed to evaluate the patterns of deformation within the ductile part of the transform plate boundary. As a result, we did not solve for the position of the near-surface faults. However, on the basis of the patterns of strain below the elastic–brittle layer we can investigate the role that subcrustal strain may play in perturbing the deformation at the surface. Four aspects of ridge–transform tectonics are analysed: (a) patterns of strain and faulting within the transform domain; (b) magma supply and asymmetric accretion at ridge tips; (c) development of MOR core complexes; (d) development of leaky transforms and transform splitting ridges.

Strain in transform zones

Using the terminology of Searle (1992) (and also Fox & Gallo (1986)), we can compare components of the transform plate boundary regime with deformation patterns developed in our modelling. The transform domain (TD) contains all of the crust affected by the transform system. The TD is a variable width (<10 km to >50 km), generally wider for larger offsets and lower slip velocities (Searle 1992). The deeper extent of the shear zone in our models shows this behaviour and thus we are tempted to link the broad region of the TD with this zone. This region is also likely to encompass the region of thinner crust as discussed below.

Within the TD is the transform tectonized zone (TTZ), which includes the transform-related faulting. This zone often is oblique to the strike of the plate motion (Roest *et al.* 1984; Searle 1986). This pattern of obliquity for the faults within the TD was well documented for the Romanche Transform (Searle *et al.* 1994). This is a very large offset, low-velocity transform. On the basis of the results shown in Fig. 9, we might expect the cross-over from one side of the transform regime to the other to be located part-way between the ridge and the transform midpoint. The faults within the Romanche TD cross over *c.* 150–200 km from the ridge (of a *c.* 1000 km long transform). The cross-over occurs in the vicinity of the sedimented Vema Deep (Searle *et al.* 1994). Within our model configuration, the Vema Deep would represent the region of extension driven by the local curvature of the subjacent mantle shear zone.

We would expect the patterns of fault obliquity to be more symmetrical in shorter offset transforms, and such is the case for the Atlantis Fracture Zone (Parson & Searle 1986), a low-velocity transform with <100 km of offset. The obliquity of the surface faults relative to the plate motion is as much as 15° for the Atlantis Fracture Zone (Parson & Searle 1986), consistent with the results of Model 5.04 (Fig. 8). Applying the modelling results further we might expect that the lengths of fault segments would be longer along the portions with less curvature, and shorter and/or more oblique to plate motion in the regions of high plate boundary curvature.

Magma supply at ridge ends

Crustal thickness varies along the MOR, with the thinnest crust typically near ridge offsets and adjacent to transforms (Detrick *et al.* 1993), particularly for low plate-velocity systems. A variety of mechanisms have been proposed to explain the variation in crustal structure (e.g. Kuo & Forsyth 1988; Cannat 1996; Cannat *et al.* 1995; Detrick *et al.* 1995) primarily as a consequence of reduction in magma supply near the ridge tips. Our models provide evidence of an additional causative factor for the thinning of crust near the ridge–transform intersection (RTI). The plate boundary geometry near the RTI will mean that the same volume of mantle-derived melt must produce the crust for a longer segment of ridge crust (Figs. 1 and 6–9). This effect will be greatest (i.e. produce the most thinning of the crust) for low-velocity large-offset boundaries (compare geometry of plate boundary structure in Fig. 9 and in Fig. 7). Detrick *et al.* (1995) showed such a pattern of crustal thickness variation with offset length (thinner crust with longer offset), as long as the offset is >50 km. In cases of high plate velocities, the plate boundary offset is somewhat less (compare Fig. 7 and Fig. 9).

Patterns of asymmetric accretion have been observed at the ridge tips of slow-spreading ridges (Tucholke & Lin 1994; Escartin *et al.* 1999; Allerton *et al.* 2000). These appear to be related to regions of anomalous crustal thickness with accretion occurring primarily on the outside corner side of the ridge. The plate boundary geometry from our models in the near-ridge regime would favour that the magma supply that reaches the ridge axis near the RTI preferentially flows to the outside corner side, as that is the direction of material flow in the model (e.g. Fig. 7b).

Core complexes

The recent documentation of extensional core-complex style deformation at the inside corners of many RTIs (Cann *et al.* 1997; Blackman *et al.* 1998; Tucholke *et al.* 1998; Dick *et al.* 1999; Escartin & Cannat 1999), particularly at slow-spreading segments, is an exciting new development in ridge–transform tectonics. The processes that produce these 'mega-mullion'

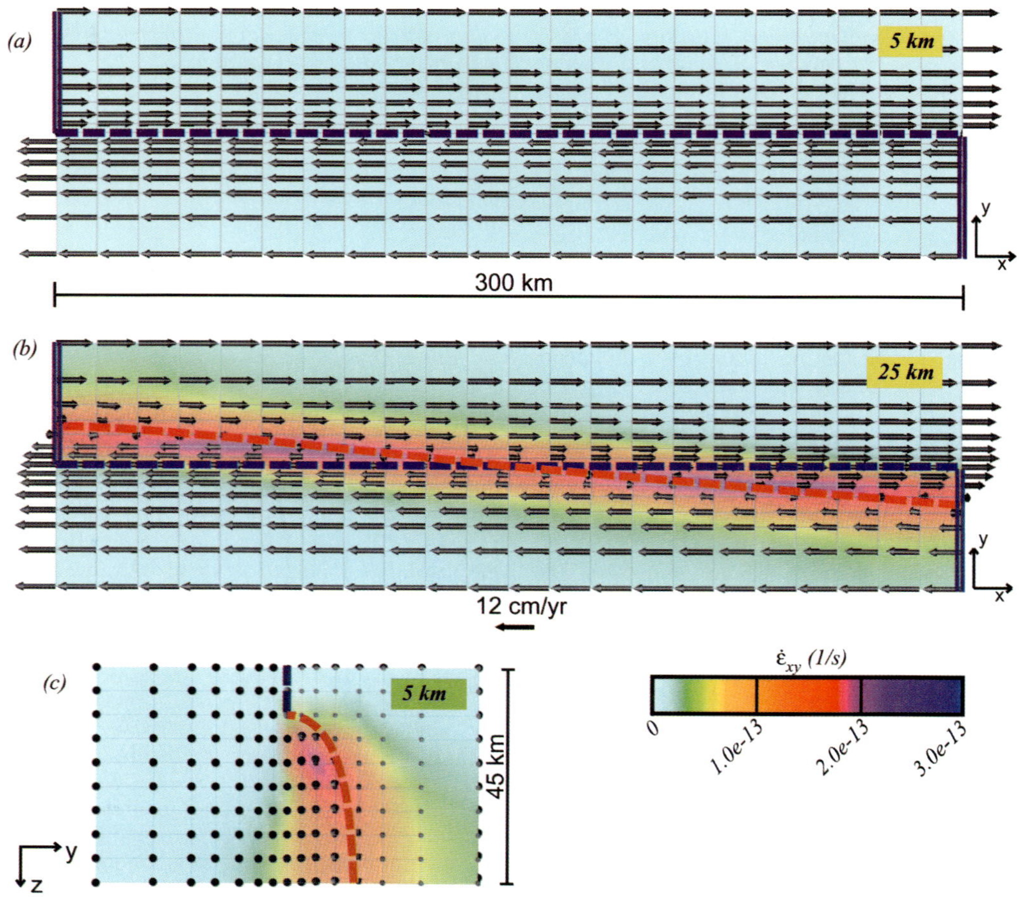

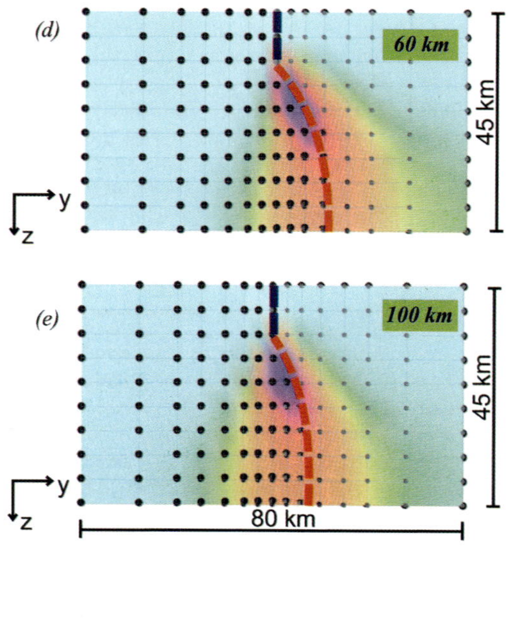

Fig. 7. Results from 3D coupled model for Model 5.12. Strain rate, $\dot{\varepsilon}_{xy}$ (shear strain rate in the transform-parallel direction on the vertical plane parallel to the transform orientation), is shown by colour contours and arrows indicate velocities in plane of each figure. The plate boundary, represented by the ‘zero’ line across which the velocities change direction, is indicated in the plots by the red dashed line. (a) Horizontal slice at depth of 5 km (within elastic–brittle layer) showing near-surface kinematics. (b) Horizontal slice at 25 km depth. Offset in plate boundary is shown by dashed orange line. Width of shear zone varies with position relative to ridge. (c) Vertical cross-sectional slice 5 km from left ridge end of model. (d) Vertical cross-sectional slice 60 km from left ridge end of model in area of maximum plate boundary curvature. (e) Vertical cross-sectional slice 100 km from left ridge end of model near central portion of plate boundary.

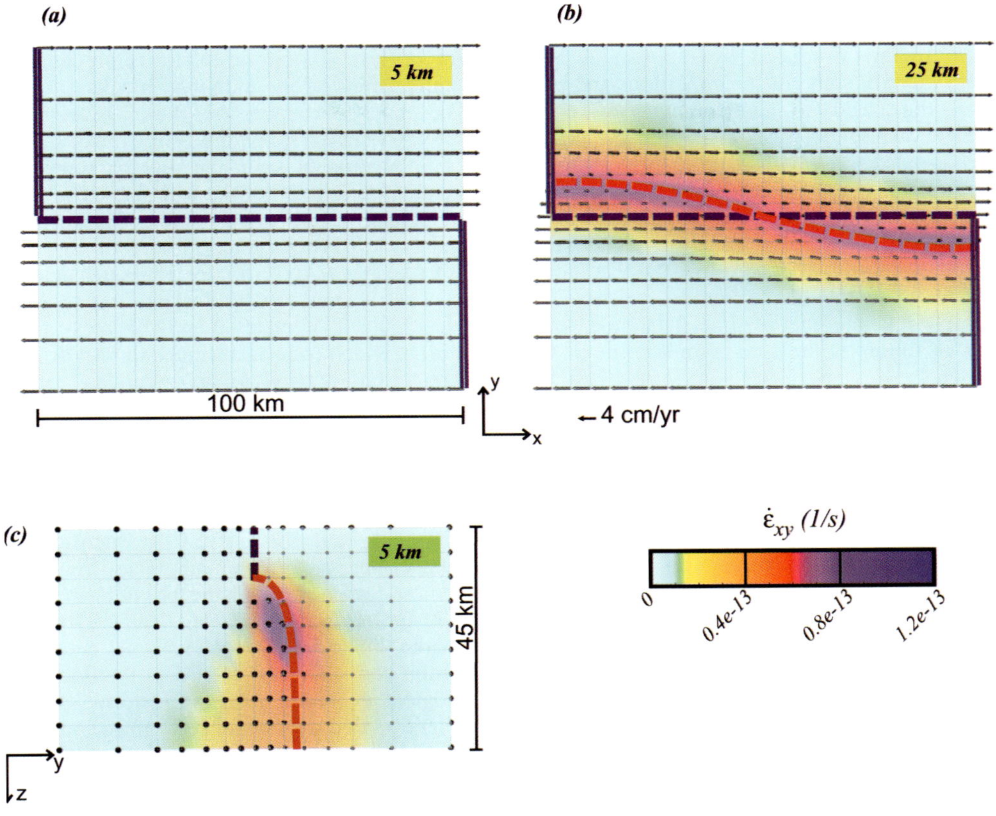

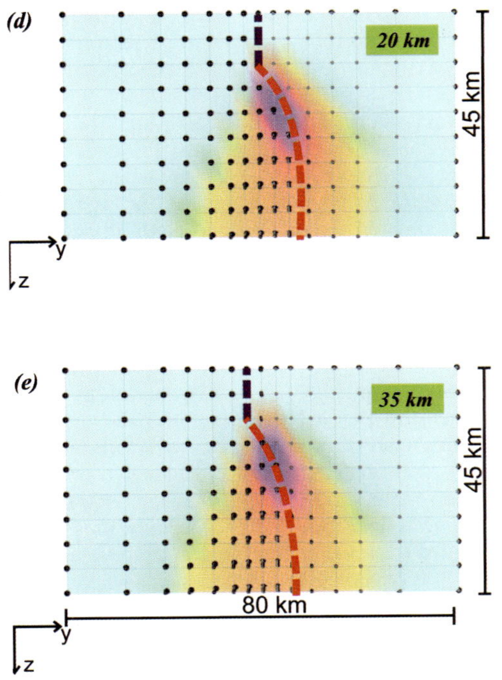

Fig. 8. Results from 3D coupled model for Model 5.04. Strain rate, $\dot{\varepsilon}_{xy}$ (shear strain rate in the transform-parallel direction on the vertical plane parallel to the transform orientation), is shown by colour contours and arrows indicate velocities in plane of each figure. The plate boundary, represented by the 'zero' line across which the velocities change direction, is indicated in the plots by the red dashed line. (a) Horizontal slice at depth of 5 km (within elastic–brittle layer) showing near-surface kinematics. (b) Horizontal slice at 25 km depth. Offset in plate boundary is shown by dashed orange line. Width of shear zone varies with position relative to ridge. (c) Vertical cross-sectional slice 5 km from left ridge end of model. Inferred location of plate boundary shown by dashed line. (d) Vertical cross-sectional slice 20 km from left ridge end of model in area of maximum plate boundary curvature. (e) Vertical cross-sectional slice 35 km from left ridge end of model near central portion of plate boundary. These locations compare in plate age with the cross-sections (c)–(e) in Fig. 7 for the higher plate-velocity case.

structures are not fully understood and continue to drive research. Several aspects of the deformation regime produced by our model may help to identify some of the conditions and processes that lead to development of core-complex tectonics. The characteristic MOR core complex is *c.* 14–42 km (ridge-parallel direction) by 16–35 km (transform-parallel direction) with surface areas between 150 km^2 and 800 km^2 (Tucholke *et al.* 1998). They develop primarily at low-velocity RTIs, form exclusively on the inside corners of the RTI, and represent intervals of time with little magmatic accretion on the core-complex side of the ridge crest.

Because the core complexes develop as shear on planes oriented perpendicular to the transform plane, we have analysed the patterns of $\dot{\varepsilon}_{xz}$ (shear strain rate in the transform-parallel direction on the horizontal plane) for Model 5.04 (Fig. 10). We see that a subhorizontal region of high shear-strain forms with dimensions comparable with observed core-complex size (Fig. 10a). The top of this shear zone (as seen in the transform-parallel vertical cross-section in Fig. 10c), which could act as a detachment structure for core-complex initiation, is at a depth of *c.* 10 km, again compatible with the inferred depth extent of exhumed lower-crust and upper-mantle rocks (Cann *et al.* 1997). In Figure 10 we also show schematically how the model results can be compared with geological models of MOR core complexes (modified from Cann *et al.* (1997)). Although our models are not optimally designed to simulate such tectonics, the general agreement between model behaviour and observation encourages us to pursue the implications of the transform deformation regime for core-complex formation.

The conditions for core-complex initiation implied by our model results lead us to the following speculations. The sporadic nature of core-complex activity as seen in the separate structures observed along several transforms (Blackman *et al.* 1998) may reflect the implications of core-complex extension on the near ridge plate boundary geometry produced in our model. After a large offset on the core-complex detachment fault, much of the lithosphere blocking the influx of mantle-derived magma will have moved downstream, allowing a relative abundance of magma to reach the ridge tip. This would then weaken the MOR axis, making extension more easily accomplished at the axis than along the core-complex detachment fault. This sequence of events would cycle, as plate motions would bring the plate boundary geometry back to our modelled conditions after a few million years or less (based on plate velocities and the dimensions of the extended blocks).

The non-occurrence of these extensional structures at faster-spreading centres may reflect the combination of sufficient magma supply to keep the MOR axis weak, and the more subdued nature of the offset in the deeper plate boundary. For very large offset (but low-velocity) transforms (e.g. Model 15.4) the detachment level is substantially deeper. This leads to a thicker block of material moving away from the ridge, which may again make the MOR axis the preferred location to accommodate extension.

Leaky transforms and transform splitting

Our results indicate that, as a consequence of the curvature and obliquity of the ductile plate boundary shear zones (particularly for low plate velocities), extension and the possible magmatism resulting from associated leaky transform behaviour may be components of transform evolution and behaviour related to the thermal–rheological evolution of the transform system. We believe that several observations could be explained by this process.

There appear to be velocity limits for oceanic transforms, both in an absolute sense (Naar & Hey 1989), and as a function of length (Burr & Solomon 1978; Stoddard 1992). A perhaps better constrained relationship can be seen between the maximum observed transform velocity and age offset; a pattern more consistent with the idea that thermal (rheological) structure is the major control on the transform (Furlong 1992). Figure 11 shows such a dataset where the full plate velocity is plotted against the age offset (approximated as transform length divided by the half-spreading rate). The dashed line is a rough estimate of the stability limit for this relationship. It has been hypothesized that this stability limit bounds the region where transforms are in a mechanical equilibrium. If velocity and/or plate age-offset conditions move a transform beyond the limit into the unstable region, new ridge segments will form, effectively 'splitting' the transform into two or more shorter, more stable offsets (Furlong 1992).

Large transform offsets, especially those near the inferred velocity limit and at high velocities, are often observed to be magmatic, where there is some degree of spreading occurring along the transform ('leaky' transforms). 'Leaky' transforms are often attributed to the effect of changes in plate motions that cause a transform that is originally parallel to spreading to become oblique, causing extension and magmatism (Menard & Atwater 1969; Burr & Solomon 1978; Collette 1986; Searle 1986). Other investigations, however, have concluded that this scenario cannot generally explain the occurrences of these transforms, many of which are not consistent with or cannot be associated with any change in plate motions (Garfunkel 1986; Fornari *et al.* 1989). Therefore, although it is clear that changes in plate rotation poles can enhance magmatism, 'leakiness' in transforms appears to be a more inherent feature of the offsets, as first suggested by Garfunkel (1986), and may be caused by some other upwelling mechanism.

The velocity and age-offset dependences of the observed transform stability limit are probably the combination of several components. Larger age offset generally means larger temperature contrasts, and therefore larger boundary obliquity by our model. It also means increasing length, which will affect the stress distribution in the regime. Finally, there is the issue of brittle failure. In conceptual terms, we could conclude that lower age-offset regimes are hotter and therefore generally more ductile and able to accommodate higher velocities (i.e. higher strain rates and lower shear stress) than larger age offsets (small age-offset transforms are often diffuse shear zones), but the reality is surely more complicated.

Our models are not specifically designed to address these issues. The definition of the stability limit is somewhat subjective, based on observations of surface morphologies and generalized criteria to distinguish a single 'leaky' transform from a set of several small transforms. It is useful to examine a transform that has undergone the inferred process and attempt to correlate it with the stability limit.

Plate velocities are not constant in time, so the location of any transform on a velocity–age-offset plot such as Fig. 11 will not be fixed. Changes in velocity will result in changes on both axes of the plot, and the transform will have some evolutionary path through that parameter space. It is possible that a change in spreading rate could push a previously stable large offset transform across the limit, causing a new segment to form. If this has occurred, we have a reasonable chance of identifying it as such. If the velocity history and path in velocity–age-offset space for the original transform can be ascertained, we can try to correlate the age of the segment with the crossing of the limit.

As an example, we have investigated the Charlie Gibbs Fracture Zone (CGFZ) system on the Mid-Atlantic Ridge. The CGFZ consists of two closely spaced transforms, separated by a small ridge segment (Searle 1981). From the relative lengths of the long and short sets of fracture zones, it can be inferred that the intervening segment is younger than the rest of the system. The short fracture zones associated with the segment allow magnetic anomaly dating of the formation of the segment, which gives an age of roughly 40 Ma.

Using magnetic anomaly data, the velocity history of the transform can be estimated. There has been a significant variation in plate velocity through time (Fig. 11, inset), including a large decrease, which will lead to an increase in age offset (if the length remains constant). Assuming that before the formation of the short median ridge segment the single transform length was approximately equivalent to combined length of the two present-day transforms, then the path of that offset in velocity–age-offset space is as shown in Fig. 11. This path takes the original transform across the stability limit, from the increase in age offset driven by the decrease in plate velocity, at roughly 42 Ma. The new transform-splitting median ridge segment also formed at an off-centre location, consistent with the locations of the zones of maximum extension in the low-velocity and large-offset transform model (Model 15.04; Fig. 9).

Conclusion

Transforms that offset mid-ocean ridges display complex deformational behaviour that cannot be explained by a simple strike-slip plate boundary. Transforms show broad deformation zones encompassing a suite of tectonic structures, often striking obliquely to relative plate motions. Extensional core-complex structures are identified at several ridge–transform intersections, almost exclusively at the inside corners of slow-spreading ridges. Leakiness and the development of transform splitting short ridge segments is also observed at some long offset transforms. The model we have developed, which couples the thermal and deformational regimes, produces a deformation field that provides possible tectonic explanations for many of these observations. The effects of lateral cooling and a migration of the locus of shear deformation along the ductile portion of the plate boundary produce a 3D plate boundary geometry compatible with patterns of deformation along MOR–transform systems. Although

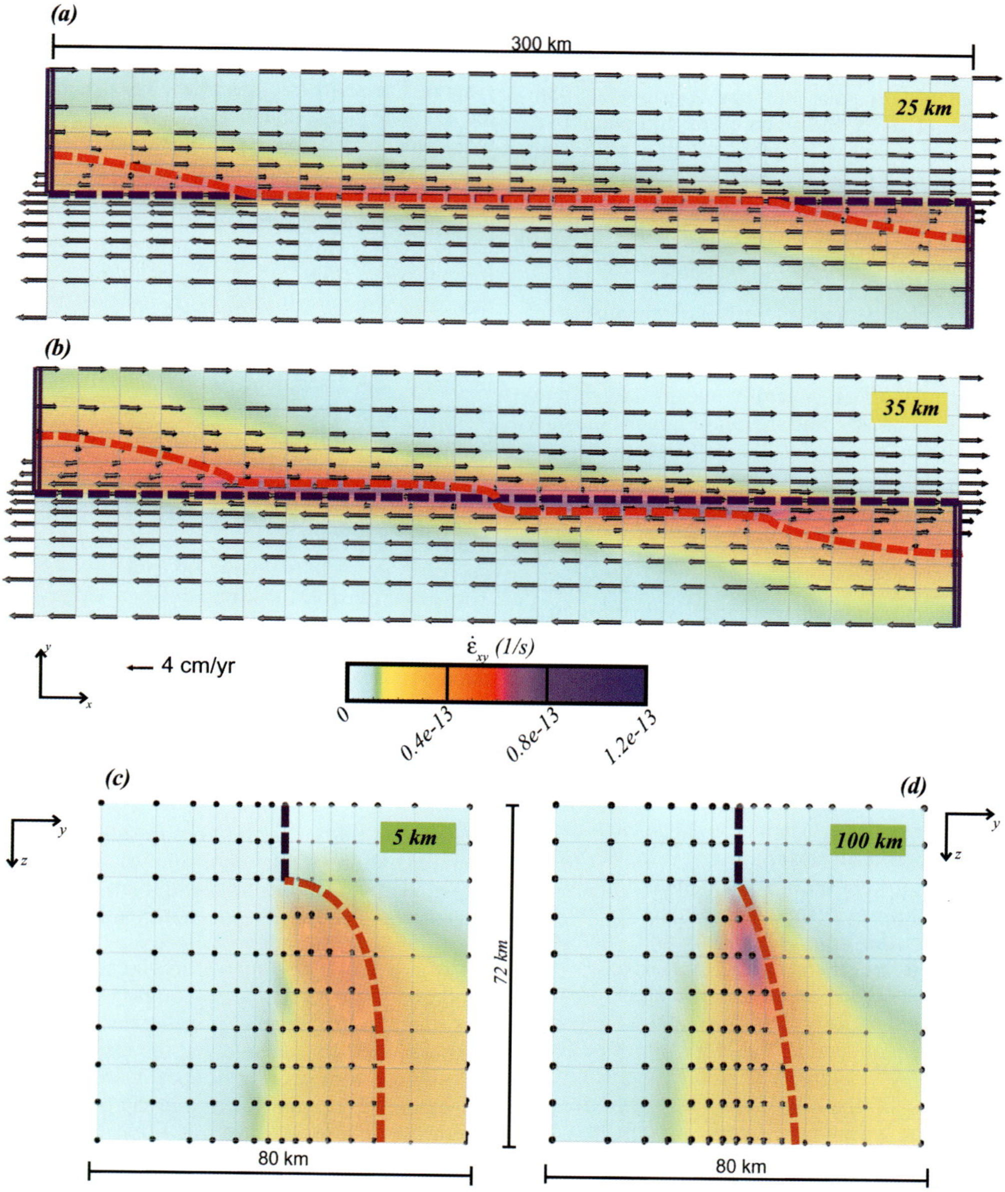

Fig. 9. Results from 3D coupled model for Model 15.04. Strain rate, $\dot{\varepsilon}_{xy}$ (shear strain rate in the transform-parallel direction on the vertical plane parallel to the transform orientation), is shown by colour contours and arrows indicate velocities in plane of each figure. The plate boundary, represented by the 'zero' line across which the velocities change direction, is indicated in the plots by the red dashed line. (a) Horizontal slice at depth of 25 km showing region of high plate boundary curvature and low kinematic obliquity along the interior of the transform. (b) Horizontal slice at 35 km depth. (c) Vertical cross-sectional slice 5 km from left ridge end of model. Inferred location of plate boundary shown by dashed line. (d) Vertical cross-sectional slice 100 km from left ridge end of model.

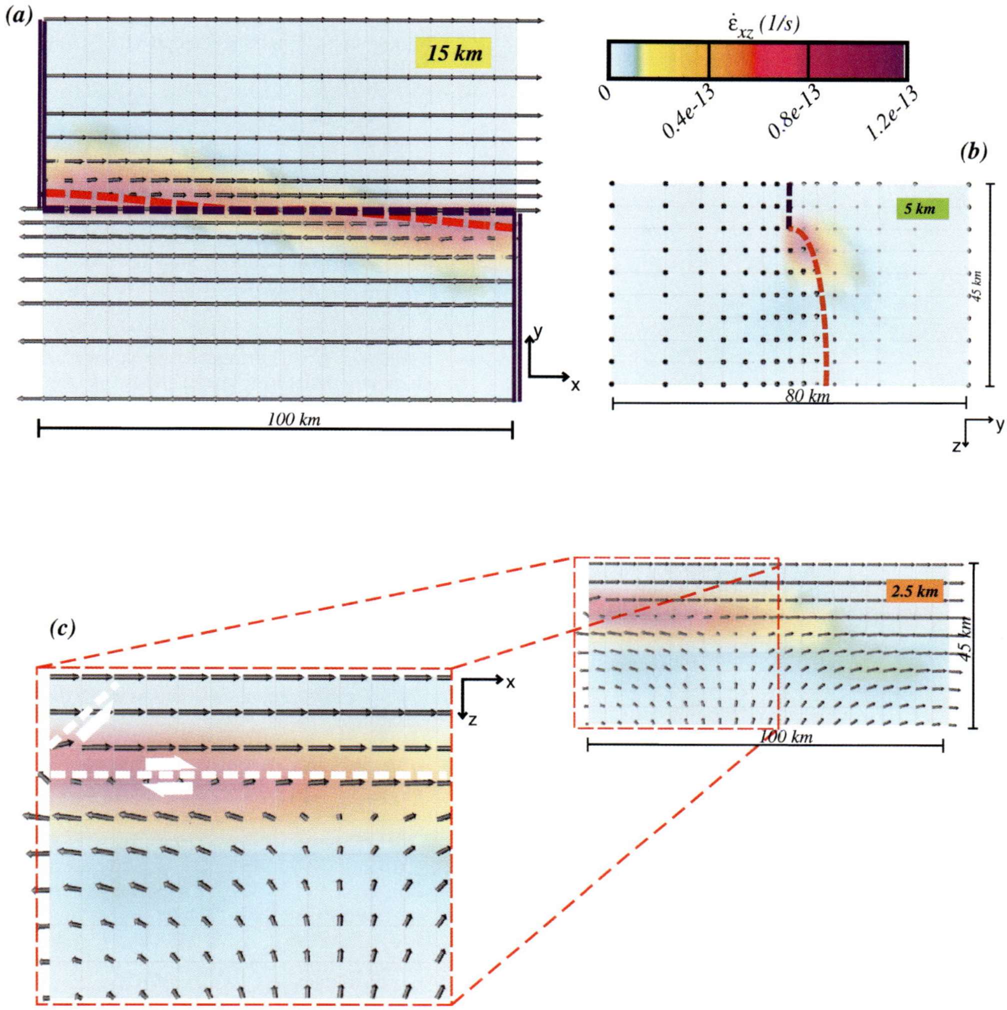

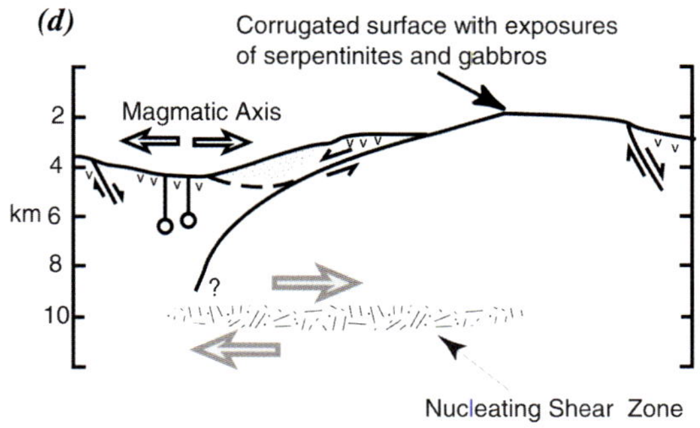

Fig. 10. Results from 3D coupled model for Model 5.04. Strain rate, $\dot{\varepsilon}_{xz}$ (shear strain rate in the transform-parallel direction on the horizontal plane), is shown by colour contours and arrows indicate velocities in plane of each figure. The plate boundary, represented by the 'zero' line across which the velocities change direction, is indicated in the plots by the red dashed line. (a) Horizontal slice at depth of 15 km (below elastic–brittle layer) near inferred initiation depth of MOR core-complex structures, showing spatial extent of potential detachment horizon. (b) Vertical cross-sectional slice 5 km from left ridge end of model. Inferred location of plate boundary shown by dashed line. (c) Vertical slice parallel to and 2.5 km away from transform fault. Possible relationship of detachment surfaces are shown by white dashed lines and arrows. (d) Correlation of our model-derived core-complex geometry with geological model (adapted from Cann *et al.* 1997)

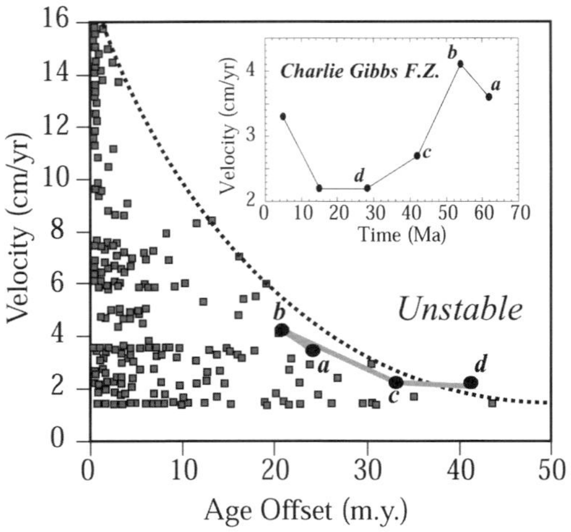

Fig. 11. Global oceanic transform data, plotted with transform velocity v. age offset (approximated as length/half-spreading rate). The inferred stability limit for the velocity–age-offset relationship is shown by the dashed curve. Inset shows the velocity time behaviour for the Charlie Gibbs Fracture Zone, showing a significant decrease in plate velocity at the approximate time of the transform splitting event (*c.* 40 Ma). Path of the transform in velocity–age-offset space is shown by points labeled a–d in main plot. At time of new ridge formation, the plate kinematics had moved the transform into the unstable field.

simple, the coupled 3D thermal–mechanical model provides important insights into deformational processes along transforms. More refined models can be developed to better define the tectonic processes at work along MOR transforms.

This work was supported by a Deike Research Grant (Penn State) to K.P.F. Detailed and very constructive reviews by D. K. Blackman and R. C. Searle are greatly appreciated and helped improve the paper. Insightful discussions with J. Escartin are also acknowledged.

References

ALLERTON, S., ESCARTIN, J. & SEARLE, R.C. 2000. Extremely asymmetric magmatic accretion of oceanic crust at the ends of slow-spreading ridge segments. *Geology*, **28**, 179–182.

BLACKMAN, D.K. 1997. Variation in lithospheric stress along ridge–transform plate boundaries. *Geophysical Research Letters*, **24**, 461–464.

BLACKMAN, D.K. & FORSYTH, D.W. 1992. The effects of plate thickening on three-dimensional passive flow of the mantle beneath mid-ocean ridges. *In*: PHIPPS-MORGAN, J., BLACKMAN, D. & SINTON, J. (eds) *Mantle Flow and Melt Generation at Mid-Ocean Ridges*. Geophysical Monograph, American Geophysical Union, **71**, 311–326.

BLACKMAN, D.K., CANN, J.R., JANSSEN, B. & SMITH, D.K. 1998. Origin of extensional core complexes: evidence from the Mid-Atlantic Ridge at Atlantis Fracture Zone. *Journal of Geophysical Research*, **103**, 21315–21333.

BURR, N.C. & SOLOMON, S.C. 1978. The relationship of source parameters of oceanic transform earthquakes to plate velocity and transform length. *Journal of Geophysical Research*, **83**, 1193–1205.

CANN, J.R., BLACKMAN, D.K., SMITH, D.K. & 6 OTHERS. 1997. Corrugated slip surfaces formed at ridge–transform intersections on the Mid-Atlantic Ridge. *Nature*, **385**, 329–332.

CANNAT, M. 1996. How thick is the magmatic crust at slow spreading oceanic ridges? *Journal of Geophysical Research*, **101**, 2847–2857.

CANNAT, M., MEVEL, C., MAIA, M. & 8 others. 1995. Thin crust, ultramafic exposures, and rugged faulting patterns at the Mid-Atlantic Ridge (22°–24°N). *Geology*, **23**, 49–52.

COLLETTE, B.J. 1986. Fracture zones in the North Atlantic: morphology and a model. *Journal of the Geological Society, London*, **143**, 763–774.

DETRICK, R.S., NEEDHAM, H.D. & RENARD, V. 1995. Gravity anomalies and crustal thickness variations along the Mid-Atlantic Ridge between 33°N and 40°N. *Journal of Geophysical Research*, **100**, 3767–3787.

DETRICK, R.S., WHITE, R.S. & PURDY, G.M. 1993. Crustal structure of North Atlantic Fracture Zones. *Reviews of Geophysics*, **31**, 439–458.

DICK, H.J.B., MACLEOD, C.J., ROBINSON, P.T., ALLERTON, S. & TIVEY, M. A high resolution bathymetric map of Atlantis Bank. *In*: DICK, H.J.B., NATLAND, J. & MILLER, J. (eds) *Proceedings of the Ocean Drilling Program, Initial Reports*, **176**. Ocean Drilling Program, College Station, TX.

ENGELN, J.F., WIENS, D.A. & STEIN, S. 1986. Mechanisms and depths of Atlantic transform earthquakes. *Journal of Geophysical Research*, **91**, 548–577.

ESCARTIN, J. & CANNAT, M. 1999. Ultramafic exposures and the gravity signature of the lithosphere near the Fifteen–Twenty Fracture Zone (Mid-Atlantic Ridge. *14°–16.5°N). Earth and Planetary Science Letters*, **171**, 411–424.

ESCARTIN, J., COWIE, P.A., SEARLE, R.C., ALLERTON, S., MITCHELL, N.C., MACLEOD, C.J. & SLOOTWEG, A.P. 1999. Quantifying tectonic strain and magmatic accretion at a slow spreading ridge segment, Mid-Atlantic Ridge, 29°N. *Journal of Geophysical Research*, **104**, 10421–10437.

FORNARI, D.J., GALLO, D.G., EDWARDS, M.N., MADSEN, J.A., PERFIT, M.R. & SHOR, A.N. 1989. Structure and topography of the Siqueros transform fault system: evidence for the development of intra-transform spreading centers. *Marine Geophysical Research*, **111**, 263–299.

FORSYTH, D.W. & WILSON, B. 1984. Three-dimensional temperature structure of a ridge–transform–ridge system. *Earth and Planetary Science Letters*, **70**, 355–362.

FOX, P.J. & GALLO, D.G. 1984. A tectonic model for ridge–transform–ridge plate boundaries; implications for the structure of oceanic lithosphere. *Tectonophysics*, **104**, 205–242.

FOX, P.J. & GALLO, D.G. 1986. The geology of North Atlantic transform plate boundaries and their aseismic extensions. *In*: VOGT, P.R. & TUCHOLKE, B.E. (eds) *The Geology of North America. Vol. M, The Western North Atlantic Region*. Geological Society of America, Boulder, CO, 157–172.

FURLONG, K.P. 1992. Rheologic complications along large offset transform faults (abstract). *EOS Transactions, American Geophysical Union*, **73**, 550.

FURLONG, K.P., CHAPMAN, D.S. & ALFELD, P.W. 1982. Thermal modeling of the geometry of subduction with implications for the tectonics of the overriding plate. *Journal of Geophysical Research*, **87**, 1786–1802.

GARFUNKEL, Z. 1986. Review of oceanic transform activity and development. *Journal of the Geological Society, London*, **143**, 775–784.

GOVERS, R. & WORTEL, M.J.R. 1995. Extension of stable continental lithosphere and the initiation of lithospheric scale faults. *Tectonics*, **14**, 1041–1055.

HINDMARSH, A.C. 1983. ODEPACK, a systematized collection of ODE solvers. *In*: STEPLEMAN, R.S. (ed.) *Scientific Computing*. North-Holland, Amsterdam, 55–64.

HIRTH, G. & KOHLSTEDT, D.L. 1996. Water in the oceanic upper mantle: implications for rheology, melt extraction and the evolution of the lithosphere. *Earth and Planetary Science Letters*, **144**, 93–108.

KARATO, S. & WU, P. 1993. Rheology of the upper mantle: a synthesis. *Science*, **260**, 771–778.

KUO, B.-Y. & FORSYTH, D.W. 1988. Gravity anomalies of the ridge–transform system in the South Atlantic between 31° and 34.5° S: upwelling centers and variations in crustal thickness. *Marine Geophysical Research*, **10**, 205–232.

MELOSH, H.J. & RAEFSKY, A. 1980. The dynamical origin of subduction zone topography. *Geophysical Journal of the Royal Astronomical Society*, **60**, 333–354.

MELOSH, H. & RAEFSKY, A. 1981. A simple and efficient method for introducing faults into finite element computations. *Bulletin of the Seismological Society of America*, **71**, 1391–1400.

MELOSH, H.J. & RAEFSKY, A. 1983. Anelastic response to dip slip earthquakes. *Journal of Geophysical Research*, **88**, 515–526.

MELOSH, H.J. & WILLIAMS, C.A. 1989. Mechanics of graben formation in crustal rocks: a finite element analysis. *Journal of Geophysical Research*, **94**, 13961–13973.

MENARD, H.W. & ATWATER, T. 1969. Origin of fracture zone topography. *Nature*, **222**, 1037–1040.

NAAR, D.F. & HEY, R.N. 1989. Speed limit for oceanic transform faults. *Geology*, **17**, 420–422.

PARSON, L.M. & SEARLE, R.C. 1986. Strike-slip fault styles in slow-slipping oceanic transform faults—evidence from GLORIA surveys of Atlantis and Romanche Fracture Zones. *Journal of the Geological Society, London*, **143**, 757–761.

PHIPPS-MORGAN, J. & FORSYTH, D.W. 1988. Three-dimensional flow and temperature perturbations due to a transform offset: effects on oceanic crustal and mantle structure. *Journal of Geophysical Research*, **93**, 2955–2966.

ROEST, W.R., SEARLE, R.C. & COLLETTE, B.J. 1984. Fanning of fracture zones and a three-dimensional model of the Mid-Atlantic Ridge. *Nature*, **308**, 527–531.

SEARLE, R. 1986. GLORIA investigations of oceanic fracture zones: comparative study of the transform fault zone. *Journal of the Geological Society, London*, **143**, 743–756.

SEARLE, R. 1992. The volcano-tectonic setting of oceanic lithosphere generation. *In*: PARSON, L.M., MURTON, B.J. & BROWNING, P. (eds) *Ophiolites and their Modern Oceanic Analogues*. Geological Society, London, Special Publications, **60**, 65–79.

SEARLE, R.C. 1981. The active part of the Charlie-Gibbs Fracture Zone: a study using sonar and other geophysical techniques. *Journal of Geophysical Research*, **86**, 243–262.

SEARLE, R.C., THOMAS, M.V. & JONES, E.J.W. 1994. Morphology and tectonics of the Romanche transform and its environs. *Marine Geophysical Research*, **16**, 427–453.

SHEN, Y. & FORSYTH, D.W. 1992. The effects of temperature- and pressure-dependent viscosity on three-dimensional passive flow of the mantle beneath a ridge–transform system. *Journal of Geophysical Research*, 13, **97**, 19717–19728.

SPITZAK, S. & DEMETS, C. 1996. Constraints on present-day plate motions south of 30° S from satellite altimetry. *Tectonophysics*, **253**, 167–208.

STODDARD, P.R. 1992. On the relation between transform fault resistance and plate motion. *Journal of Geophysical Research*, **97**, 17637–17650.

TUCHOLKE, B. & LIN, J. 1994. A geological model for the structure of ridge segments in slow-spreading ocean crust. *Journal of Geophysical Research*, **99**, 11937–11958.

TUCHOLKE, B., LIN, J. & KLEINROCK, M. 1998. Mega-mullions and mullion structure defining oceanic metamorphic core complexes on the Mid-Atlantic Ridge. *Journal of Geophysical Research*, **103**, 9857–9866.

Clay mineral transformations and weakening mechanisms along the Alpine Fault, New Zealand

LAURENCE N. WARR[1] & SIMON COX[2]

[1]*Geologisch-Paläontologisches Institut, Ruprecht-Karls-Universität, Heidelberg, INF 234, 69120, Heidelberg, Germany (e-mail: hq0@ix.urz.uni-heidelberg.de)*

[2]*GNS, Dunedin, New Zealand*

Abstract: The formation of clay minerals within active fault zones, which results from the infiltration of aqueous fluids, often leads to important changes in mechanical behaviour. These hydrous phyllosilicates can (1) enhance anisotropy and reduce shear strength, (2) modify porosity and permeability, (3) store or release significant volumes of water, and (4) increase fluid pressures during shearing. The varying interplay between faulting, fluid migration, and hydrous clay mineral transformations along the central Alpine Fault of New Zealand is suggested to constitute an important weakening mechanism within the upper section of this crustal discontinuity. Well-developed zones of cataclasite and compacted clay gouge show successive stages of hydrothermal alteration, driven by the cyclic, coseismic influx of meteoric fluids into exhumed amphibolite-facies rocks that are relatively Mg rich. Three modes of deformation and alteration are recognized within the mylonite-derived clay gouge, which occurred during various stages of the fault's exhumation history. Following initial strain-hardening and frictional melting during anhydrous cataclastic breakdown of the mylonite fabric, reaction weakening began with formation of Mg-chlorite at sub-greenschist conditions (<320 °C) and continued at lower temperatures (<120 °C) by growth of swelling clays in the matrix. The low permeability and low strength of clay-rich shears are suitable for generating high pore-fluid pressures during faulting. Despite the apparent weakening of the *c.* 6 km upper segment of the Alpine Fault, the upper crust beneath the Southern Alps is known to be actively releasing elastic strain, with small (<M 5) earthquakes occurring to 12 km depth. We predict that larger events will nucleate at *c.* 6–12 km along an anhydrous, strain-hardened portion of the fault.

Clay-rich lithologies have long been demonstrated to be zones of structural weakness. At the Earth's surface, the rapid creep of clayey soil on steeply dipping land surfaces (Mustafayev 1988; Mitchell 1993) and the frequent failure of landslides along clay horizons (Skempton 1985) are good examples of their low resistance to shear. In the upper crust, many active faults are also associated with the occurrence of clay minerals, which have similarly low frictional strengths (Zoback *et al.* 1987; Hickman 1991; Wintsch *et al.* 1995). The clays arise from the infiltration of aqueous fluids during brittle deformation, inducing alteration of fractured rock, typically composed of unstable minerals formed at higher temperatures and pressures.

Faults enriched in these small and hydrous phyllosilicates have been described on a variety of scales ranging from clay-filled joints (Barton 1976) to lithospheric detachments (Moore 1989). In the oceanic environment, hydrothermal alteration of mafic rocks typically produces a wide variety of low-temperature clay mineral phases, notably celadonite mica, chlorite, corrensite and smectite phases (Alt 1999). At convergent margins, clay-rich décollement horizons are commonly located at the base of many accretionary prisms (Moore 1989) and clay mineral transformations have also been suggested to play an important role in the mechanical behaviour of subduction zones (Kagami 1985; Vrolijk 1990). For example, the transition from weak smectite to relatively stronger illite within the subducted pelagic sediments of the Barbados accretionary prism appears to correspond to the onset of interplate seismic activity between 7 and 18 km depth (Vrolijk 1990).

In continental settings, the importance of mechanically weak clay minerals that influence the hydro-mechanical properties of faults has been well documented within extensional, com-

From: HOLDSWORTH, R.E., STRACHAN, R.A., MAGLOUGHLIN, J.F. & KNIPE, R.J. (eds) 2001. *The Nature and Tectonic Significance of Fault Zone Weakening*. Geological Society, London, Special Publications, **186**, 85–101. 0305-8719/01/$15.00

pressional and strike-slip regimes (Knipe 1992; Chester *et al.* 1993; Wibberley 1999). Exploration of the continental shelves in the search for petroleum resources has highlighted the nature of clay mineral diagenesis in extensional fault zones and the role of clay minerals in controlling fluid flow within sedimentary basins (Knipe 1992, 1993). In fold-and-thrust belts, detachments are commonly located in clay-rich horizons (Warr *et al.* 1996; Vrolijk & van der Pluijm 1999), which are also considered to represent sites of important fluid pressure variations and fluid expulsion within deforming rocks (Ruby & Hubbert, 1959; Walden & King 1959; Fitts & Brown 1999). In strike-slip faults, the importance of clay minerals has received particular attention from studies of the San Andreas Fault (Wu *et al.* 1975; Chester *et al.* 1993). The apparent low strength of this fault (Zoback *et al.* 1987) has been related to reaction softening during the formation of smectite- or serpentine-bearing fault rocks formed by alteration of Mg-rich lithologies (Bird 1984; Wintsch *et al.* 1995).

In this paper, we address the dynamic interplay between clay mineral transformations and faulting along another important active continental strike-slip fault and plate boundary: the Alpine Fault in New Zealand. This well-exposed fault presents an ideal opportunity for studying the effect of newly formed clay minerals on the hydro-mechanical behaviour of this seismically active zone. The nature of both non-swelling chlorite and smectitic swelling phases formed along the central Alpine Fault has been characterized in a combined X-ray diffraction and high-resolution transmission electron microscopy study. We suggest that the growth and concentration of these clay minerals has weakened the upper-crustal section of the Alpine Fault over the last c. 2 Ma, as exhumed Mg-rich amphibolite-facies rocks have come in contact with downward-circulating meteoric fluids.

Mechanical and hydrological properties of clays and clay-filled discontinuities

The mechanical and hydrological properties of clay minerals are well studied, particularly in the fields of engineering and structural geology (e.g. Alonso *et al.* 1999; Bird 1984). Clays have a number of rather special properties as a consequence of their thin sheet silicate structure, very small size and charged particle surfaces. They can absorb significant volumes of water, as well as various cations or organic substances onto and/or into their crystal structure, leading to significant variations in material behaviour.

The shear strength of drained pure clays has been extensively investigated by laboratory testing. Such experiments show strong variations dependent on particle anisotropy and layer charge (Olson 1974; Rosenquist 1962; Rosenquist 1984; Müller-Vonmoos & Loken 1989). A rough ordering in terms of decreasing shear strength is: kaolinite > illite > chlorite > illite–smectite > chlorite–smectite > vermiculite > smectite.

The shear strength of clay-filled joints is similarly low, with cohesive strengths of 0–0.18 MPa (Barton 1976). This is in contrast to compacted, phyllosilicate-rich rocks (such as schist or slate), which have cohesive strengths ranging up to 0.38 MPa (Barton 1976; Mesri & Cepeda-Diaz, 1986). Variations in the strength of clays also depend on the type of cation present, the amount of adsorbed water, temperature and pore-fluid pressure, in addition to grain-size effects.

Fewer experimental data are available for the behaviour of water-saturated clays. The presence of a pore fluid significantly reduces shear strength, particularly when the fluid pressure rises above hydrostatic conditions (Wang & Mao 1979; Morrow *et al.* 1992). Smectite (e.g. montmorillonite) is particularly weak as a result of the easy breakage of H_2O–H_2O bonds and slip along basal layers (Bird 1984). At confining pressures of >100 MPa, the differential stresses required to deform montmorillonite are also significantly lower than those required for less hydrated clays such as illite (Wang *et al.* 1980; Shimamoto & Logan 1981; Morrow *et al.* 1984, 1992). Clays typically show a plastic type of behaviour during shearing whereby the dominant mechanism of deformation is by sliding along particle surfaces (Maltman 1987). During progressive shear, a strong anisotropy develops by particle reorientation as critical states are attained for laminar flow. Experimental studies also indicate that phyllosilicates deform by dislocation glide mechanisms at temperatures characteristically lower than for other common rock-forming silicates (Shea & Kronenberg 1992).

In addition to the low shear strengths of clay-rich discontinuities, the development of matrix and pore-filling clay minerals significantly influences the migration and pressure of fluids, particularly in deforming rocks (Fitts & Brown 1999). The rate of fluid migration is clearly reduced as the porosity and permeability are lowered by clay growth. Simultaneously,

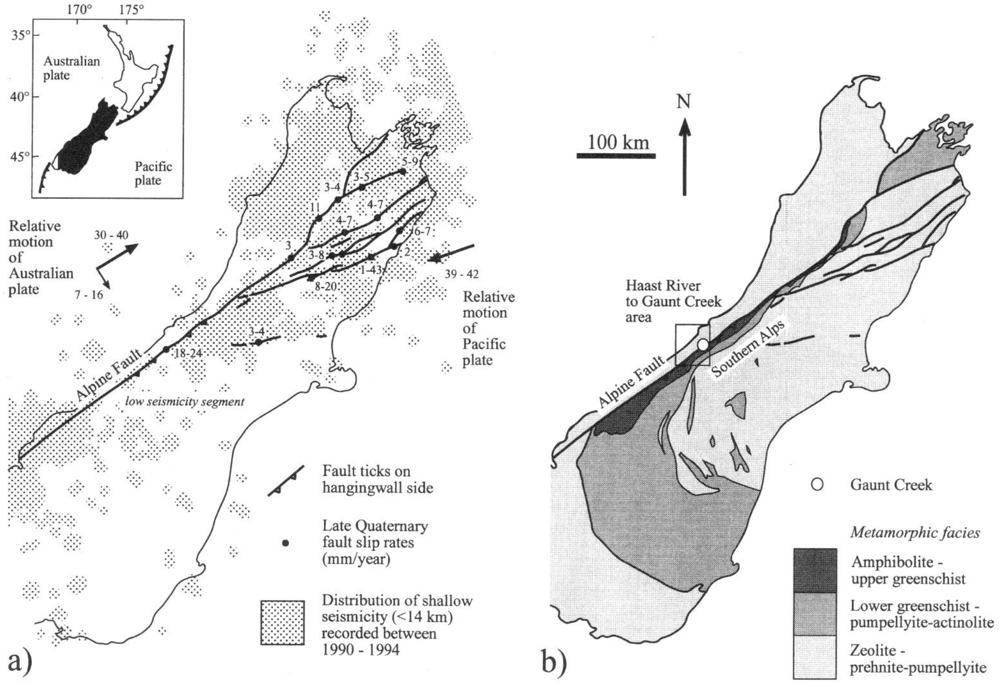

Fig. 1. (**a**) Fault map of the South Island, New Zealand, showing structures associated with the Alpine Fault. The apparent low seismicity segment is after Evison (1971) and is also evident in the data of Anderson & Webb (1994). The late Quaternary fault slip rates (mm a^{-1}) are from Knuepfer (1992); Cooper & Norris (1994). (**b**) Metamorphic map of the South Island, New Zealand, compiled after Grindley (1978); Coombs *et al.* (1996). (Note exhumation of amphibolite- and upper greenschist-facies rocks along the central region of the Alpine Fault.)

because of their hydrous nature, clay minerals have the ability to induce fluid pressure variations by storing or releasing water by hydration–dehydration reactions (Bruce 1985; Vrolijk 1990). This process is particularly important in the presence of smectite, which can generate significant fluid pressures at low confining pressures by particle swelling. Whereas the addition of a single water layer to a pure air-dried smectite (e.g. Na-montmorillonite) can produce *c.* 1.1 MPa of fluid pressure in a confined system, the initial growth and full hydration of smectite may generate as much as 8 MPa (Kahr *et al.* 1990; Tessier *et al.* 1998). Variations in the hydration state of smectite, and hence fluid pressure fluctuations, are strongly influenced by deformation and changes in effective pressure (Morrow *et al.* 1992; Fitts & Brown 1999).

The Alpine Fault, New Zealand

The Alpine Fault forms an important segment of the active Australian–Pacific plate boundary that runs for *c.* 650 km through the continental crust of the South Island, New Zealand (Fig. 1). Contemporary and predicted plate motions indicate a lateral movement of 30–42 mm a^{-1} and a compressional movement of 7–16 mm a^{-1} between the two plates (Fig. 1a, Beavan *et al.* 1999). Some 70–90 km of convergence has occurred since *c.* 6.4 Ma, resulting in uplift of the Southern Alps. Motion along the central part of the Alpine Fault has involved oblique-slip displacement, with minimum late Quaternary slip rates over the past 14 ka of 18–24 mm a^{-1} and hanging-wall uplift of *c.* 10 mm a^{-1} (Cooper & Norris, 1994; Norris & Cooper 1995).

In the NE of the South Island, the main Alpine Fault diverges into a splay where late Quaternary displacements have been partitioned across a number of NE–SW-trending faults (Fig. 1a). Here, the rate of movement along the Alpine Fault appears to decrease, with slip rates of <11 mm a^{-1} recorded at its northeastern end (Knuepfer 1992). Whereas much of the plate boundary is accompanied by shallow seismicity, the central region is considered to represent a segment of relatively low seismic activity (Evison 1971; Reyners 1989; Anderson & Webb 1994; Leitner *et al.* 2001). Small-scale geodetic networks crossing the fault indicate no significant near-surface aseismic fault slip on the central Alpine Fault over the past 25 years (Beavan *et al.* 1999). There is, however, ample evidence that shallow crust in the vicinity of the fault is actively releasing elastic strain in the form of small earthquakes (Eberhart-Phillips 1995; Leitner *et al.* 2001) and major earthquakes that are not regular but occur on average every *c.* 200 years ($\pm$100 years) (Wells *et al.* 1999).

With an apparent dextral displacement of at least 480 km, movement along the Alpine Fault has been accompanied by a component of thrusting, which has increased over the past 6.4 Ma and exhumed metamorphic rocks along the Pacific plate margin (Fig. 1b). A narrow zone, of 5–12 km width, of amphibolite-facies rocks has been uplifted some 20 km to the surface in the last 2 Ma (Cooper 1980; Koons 1987; Grapes 1995). The ductile and brittle deformation that occurred at various depths during exhumation along the Alpine Fault is exposed across a zone of *c.* 1 km width composed of fault-related mylonite, cataclasite and gouge material (Sibson *et al.* 1979). The coherent mylonite is derived from well-foliated quartzofeldspathic (Alpine) schist and interleaved amphibole-bearing metabasite lithologies that are relatively Mg rich. The occurrence of pseudotachylyte, reported to cut all fault rocks apart from the gouge, provides clear evidence for frictional melting by shear heating (Sibson *et al.* 1979; Johnston & White 1983; Bossiere 1991). In contrast, the fault zone is also characterized by a wide range of fluid-related deformation and transformation mechanisms that are responsible for a broad zone of retrogressive hydrothermal-type alteration found within the fault zone (White & White 1983). Significant quantities of low-temperature phyllosilicates developed to produce altered mylonites, clay-rich cataclasite and a clay gouge, which becomes sticky mud when wet.

The abundance of deformed veins and mineralized fissures, located both within the fault zone and in the adjacent Alpine Schist, also provides evidence for intense fluid activity. Fluid inclusions and stable isotopes of calcite and quartz veins collected within 5 km of the fault are largely of meteoric origin (Holm *et al.* 1989; Craw & Norris 1993; Jenkin *et al.* 1994). Immiscible fluid inclusions trapped at temperatures of 200–350 °C and pressures of 10–180 MPa suggest that fluids circulated to depths of at least 4–5 km and although largely hydrostatic, locally approached lithostatic conditions (Jenkin *et al.* 1994). Meteoric fluid may also have been driven by deformation as deep as ductile rocks of the biotite zone (Upton *et al.* 1995). Today, a series of warm (40–56 °C) springs are located along the hanging wall of the Alpine Fault. The water is of meteoric composition and contains Na as the major cation (Barnes *et al.* 1978), although Ca- and Sr-rich carbonate sinter are deposited on cooling at discharge zones. Geothermometry of the warm spring waters indicates equilibration with rock at 90–150 °C, corresponding to 2–3 km depth (Allis & Shi 1995). The hot springs are thought to reflect a thermal anomaly and elevated geothermal gradient in the shallow crust about the Alpine Fault, developed in response to the rapid exhumation of the hanging wall.

Estimates of upper-crustal geothermal gradients based on modelling and palaeothermal evidence vary greatly, from *c.* 40 to 90 °C km^{-1} depending on whether purely rock- or fluid-related temperatures are considered (e.g. Koons 1987; Holm *et al.* 1989; Allis & Shi 1995; Grapes 1995). A shallow 220 m drillhole, 4 km east of the Alpine Fault in the hanging wall yielded c. 60 $\pm$ 15 °C km^{-1}, similar to models that match the present-day uplift and erosion rate of 8–10 mm a^{-1} (Allis & Shi 1995). In contrast to the elevated geothermal gradients characterizing the upper crust, mid- and lower-crustal levels are considered to be significantly cooler as a result of conductive heat transfer and the effects of crustal thickening (Holm *et al.* 1989; Allis & Shi 1995). Where calculations in this paper have required a geothermal gradient, we have estimated 60 °C km^{-1} at depths <5 km (reflecting convective fluid circulation) and 10 °C km^{-1} at depths >5 km. This gradient approximates to the thermal profile and uplift path defined by Holm *et al.* (1989).

Sample material and field relationships

A total of 53 samples consisting of clay-bearing mylonite, cataclasite and gouge were investigated from three well-known localities along the Alpine Fault: Havelock Creek, Hare Mare Creek and Gaunt Creek, situated NE of the Haast River (Fig. 1b). This central part of the Alpine Fault forms a regional structure that dips *c.* 50° toward the SE, but in detail is divided near the surface into strike-slip and thrust segments of 1–3 km length (Cooper & Norris 1994). The following lithologies were sampled across a thrust segment of the fault at Gaunt Creek

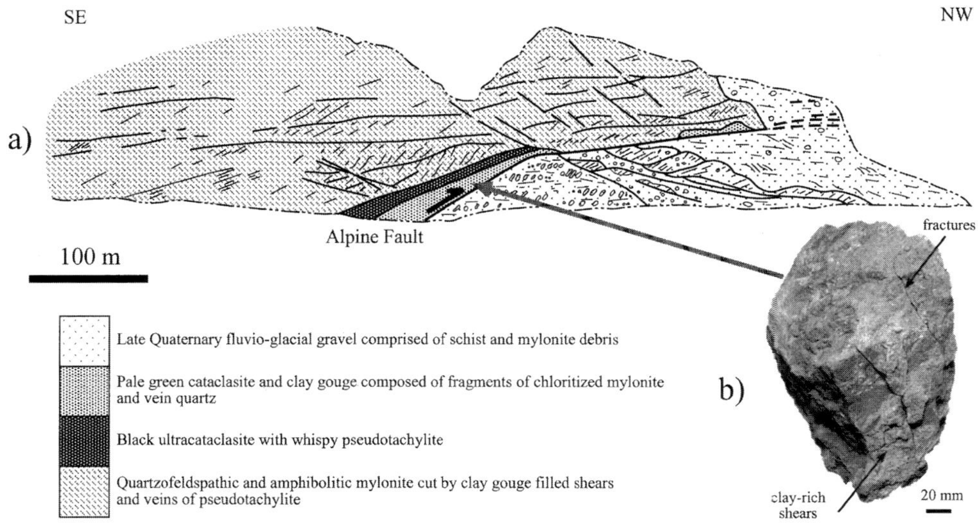

Fig. 2. (**a**) Geological cross-section of the Alpine Fault at Gaunt Creek, showing the main rock types sampled in this study (after Cooper & Norris 1994). (**b**) Photograph of clay gouge collected from a *c.* 70 cm thick layer at the base of the Alpine Fault. (Note the clay-rich shears and intersecting fracture sets.)

(Fig. 2a): (1) quartzofeldpathic and amphibolitic mylonite cut by clay gouge filled shears and veins of pseudotachylyte; (2) black ultracataclasite with wispy pseudotachylyte; (3) pale green clay-rich cataclasite composed of fragments of chloritized mylonite and vein quartz; (4) moderately indurated grey–green clay gouge from the main fault plane. The clay gouge at the base of the Alpine Fault ranges up to 70 cm thick and is rather variable in nature, ranging from homogeneous in appearance to visibly layered. Gouge samples (Fig. 2b) typically contain numerous clay-rich shears, fault surfaces bearing slickensides and intersecting fracture sets.

Analytical methods

All clay-bearing fault rocks were prepared for X-ray diffraction (XRD) study following the analytical methods described by Moore & Reynolds (1997). Samples were gently disaggregated using an ultrasonic bath, the clay fraction was separated by centrifugation (applying Stokes' Law), and orientated mounts were prepared by pipetting 45 mg of clay suspended in 1 ml of water onto 3 cm × 3 cm glass slides. Identification of the clay mineralogy was aided by saturation with exchangeable cations (Na, K, Ca, Sr and Mg), ethyl glycol treatment, and selective heat treatment to 300 °C for 1 h. Although no quantitative analysis of mineral abundance was undertaken in this study, a relative assessment was made by comparing the intensity of XRD reflections. In addition to routine mineral identification, representative samples of mylonite, cataclasite and gouge were selected and the weight percent of clay-sized particles was determined. To recover all of the <2 μm fraction, well-disaggregated samples were repeatedly centrifuged until all fine particles were removed.

The clay gouge was investigated by transmission electron microscopy (TEM) using a CM20 Philips STEM microscope (University of Granada) following the analytical procedure of Warr & Nieto (1998). Samples were vacuum impregnated with a low-viscosity Spurr resin and whole-rock ultra-thin ion-milled sections prepared. The small (1 mm diameter) sections were picked from clay-rich portions of the heterogeneous gouge. TEM study of the material combined lattice-fringe imaging, selected-area electron diffraction (SAED) patterns and microchemical analyses of selected sites of interest. Routine microchemical data were obtainable by energy dispersive spectroscopy (EDS) and scanning transmission electron microscopy (STEM).

Results

Clay mineralogy of Alpine Fault rocks

In agreement with field observations, the largest amount of clay-sized fraction was recorded from cataclasite and gouge samples collected along the base of the Alpine Fault. Some 45–57 wt % of the gouge and 20–50 wt % of the cataclasite has a grain size <2 μm. In contrast, <5 wt % of altered mylonite samples consists of clay-sized particles.

XRD analysis reveals that the majority of the clay-sized fraction is composed of phyllosilicates, namely (in general order of decreasing abundance), chlorite, illite–muscovite, biotite, smectite and vermiculite. The non-sheet silicate accessory minerals recognized are amphibole, quartz, feldspar (albite and K-feldspar), epidote and calcite. The alteration mineralogy of sampled fault rocks is remarkably similar and no diagnostic assemblages could be distinguished.

The quartzofeldpathic and amphibole-bearing mylonites contain highly variable amounts of chlorite, illite–muscovite and biotite as the principal phyllosilicates. As expected, the relatively mafic-rich (green) mylonites show higher concentrations of chlorite, and scattered occurrences of smectite and vermiculite. Cutting the mylonites, clay-rich shears and hydrothermal alterations along quartz vein contacts were found to be highly enriched in chlorite. Black ultracataclasite and interleaved wispy pseudotachylyte lenses have a comparable mineralogy to the mylonite rocks. In contrast, the pale green cataclasite is more strongly altered to chlorite and contains correspondingly lower amounts of mica (illite–muscovite and biotite) and amphibole. The clay gouge displays similar abundance in chlorite and sporadic occurrence of smectite and vermiculite.

In the natural air-dried state (with no cation saturation), the XRD pattern of a typical clay gouge (Fig. 3) show 001 basal peaks of smectite (15 Å), chlorite (14.3 Å) and vermiculite (*c.* 10.5 Å). The 1.5 nm lattice spacing of smectite reflects a two-water-layer structure, suggesting Ca as the dominant exchangeable cation, rather than Na or K (Fig. 3a). Saturation with ethyl glycol enhances smectite layers to 16.5 Å, which upon heat treatment collapse to *c.* 10 Å. The nature of the smectite was confirmed by saturation with K and Ca cations, producing both one-water-layered (*c.* 13 Å) and two-water-layered smectites (15 Å), respectively (Fig. 3b and c). The vermiculite in the gouge sample collapses to 10.5 Å in its natural and K-saturated condition, and at first glance could be mistaken for a mixed-layer illite–smectite phase (Fig. 3a and b). However, upon saturation with Ca (or Mg), its vermiculite nature is revealed as lattice layers swell to *c.* 14.4 Å and then collapse to 10 Å after heat treatment (Fig. 3c). In other gouge samples, the air-dried vermiculite is characteristically variable and produces broad 001 reflections ranging between 10.5 and 12.7 Å.

Microchemical analyses of the clay gouge

In addition to the typical mineral assemblages described above, minor quantities of rutile, pyrite, pyroxene and a Ca-amphibole were

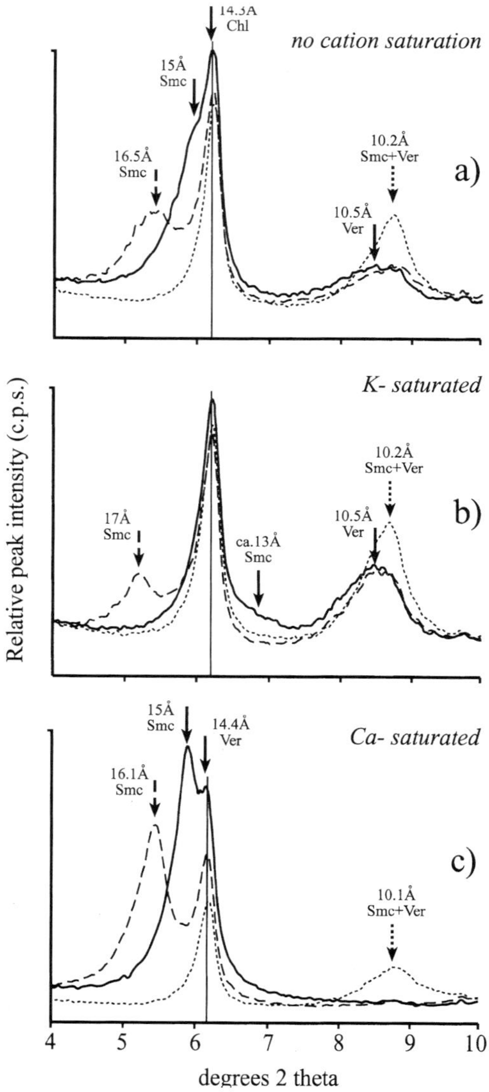

Fig. 3. XRD characteristics of clay gouge from Hare Mare Creek. Clay mineral 001 reflections, occurring between 4 and 10 °2θ are shown for the following preparation treatments: (**a**) no cation saturation; (**b**) K saturated; (**c**) Ca saturated. Continuous lines and arrows are for air-dried samples. Dashed lines and arrows are for ethyl glycol treated samples. Dotted lines and arrows are for heat-treated samples.

Table 1. *Microchemical data for smectite, chlorite and biotite in Alpine Fault clay gouge, Gaunt Creek*

Normalized to $O_{10}(OH)_2$	Si	Al^{iv}	Al^{vi}	Mg	Fe	$\sum$oct	K	Na	Ca	$\sum$int
Smectite	3.92	0.08	1.23	0.33	0.34	1.90	0.21	0.14	0.15	0.65
	3.38	0.62	1.37	0.32	0.38	2.07	0.30	0.23	0.02	0.57
	3.92	0.08	1.22	0.40	0.33	1.95	0.20	0.14	0.23	0.80
Chlorite	2.77	1.23	1.37	2.06	2.47	5.90	–	–	–	–
	2.92	1.08	1.34	2.15	2.36	5.85	–	–	–	–
	2.91	1.09	1.30	2.70	1.68	5.68	0.24	0.01	0.02	–
	2.83	1.17	1.33	2.69	1.74	5.76	0.13	0.01	0.03	–
Biotite	2.69	1.31	0.30	1.00	1.37	2.67	1.03	–	–	1.03
	2.80	1.20	0.24	1.41	1.06	2.71	0.91	0.01	0.06	0.98
	2.77	1.23	0.31	1.29	1.01	2.61	0.94	0.1	0.01	0.96
	2.80	1.20	0.40	1.23	1.01	2.64	0.92	0.02	–	0.94

int, interlayer charge.

identified by microchemical analyses of the whole-rock clay gouge (by EDS and STEM). Although good quality analyses of the clay mineral phases were difficult to determine because of the very small crystallite size, some compositions for smectite, chlorite and biotite were obtainable (Table 1). The smectite is a dioctahedral variety and contains a mixture of K, Na and Ca interlayered cations, with a layer charge of 0.57–0.65. Occasional high layer charges (e.g. 0.80) probably reflect the presence of vermiculite phases.

The chlorite is notably Mg rich and has a relatively low Al^{iv} content ranging between 1.08 and 1.28, with Fe/(Fe + Mg) ratios of <0.6. Such compositions are considered typical of chlorite formation at sub-greenschist-facies temperature conditions (Kranidiotis & MacLean, 1987; Cathelineau 1988), although the compositional variation is more likely to reflect varying amounts of interlayered smectite, rather than variations in Al^{iv} within tetrahedra sites (Shau *et al.* 1990). This is indicated by the excess of Al^{vi} over Al^{iv}, the apparent deficiency in the octahedral site, and small K, Na and Ca content present in the analytical results (Table 1).

The composition of biotite is also Mg rich, with the main interlayer cation K reflecting a layer charge close to unity. The biotite in the gouge is probably derived from the adjacent amphibolite-grade schist-derived mylonites and has compositions comparable with the petrology of adjacent mylonite units (Bossière 1991). On the basis of thermobarometry, these rocks formed under metamorphic temperatures of 490–540 °C and pressures of 5.4–6.6 kbar (Johnston & White 1983; Grapes 1995).

Microscopic characteristics of the clay gouge

Mineralogical and microstructural observations from TEM images of the clay gouge are outlined as follows (Fig. 4a–e). A general view at low magnification shows numerous clasts of randomly orientated metamorphic minerals derived from the Alpine Schist mylonite. Subangular grains of amphibole are set in a broken mass of fine mica fragments composed of both biotite and muscovite (Fig. 4a). These mica particles are commonly 20–30 nm in thickness and have particularly small aspect (length-to-thickness) ratios in comparison with their parent metamorphic crystals. They are also relatively low in intracrystalline defects such as zones of lattice distortion and appear to have formed by breakage along fractures running either perpendicular or parallel to basal lattice layers. The haphazard orientation of fractured grains within the gouge indicates subsequent rotation, presumably by granular flow.

Some biotite fragments show various degrees of layer-by-layer, solid-state transformation to chlorite (Fig. 4b). Whereas weakly altered biotite includes the occasional extra brucite layer, with little disruption of the crystallite packet, strongly chloritized biotite crystals show uneven addition of brucite layers that disrupts the regularity of lattice spacing and causes grains to thicken. According to Veblen & Ferry (1983), this mechanism involves substantial incorporation of octahedral cations (Mg, Fe, Al), the release of K and a significant increase in volume (up to 39.8%).

Thick sheets of chlorite are observable throughout the gouge. They are typically composed of numerous subparallel stacked packets of variable crystallite size, and in contrast to the

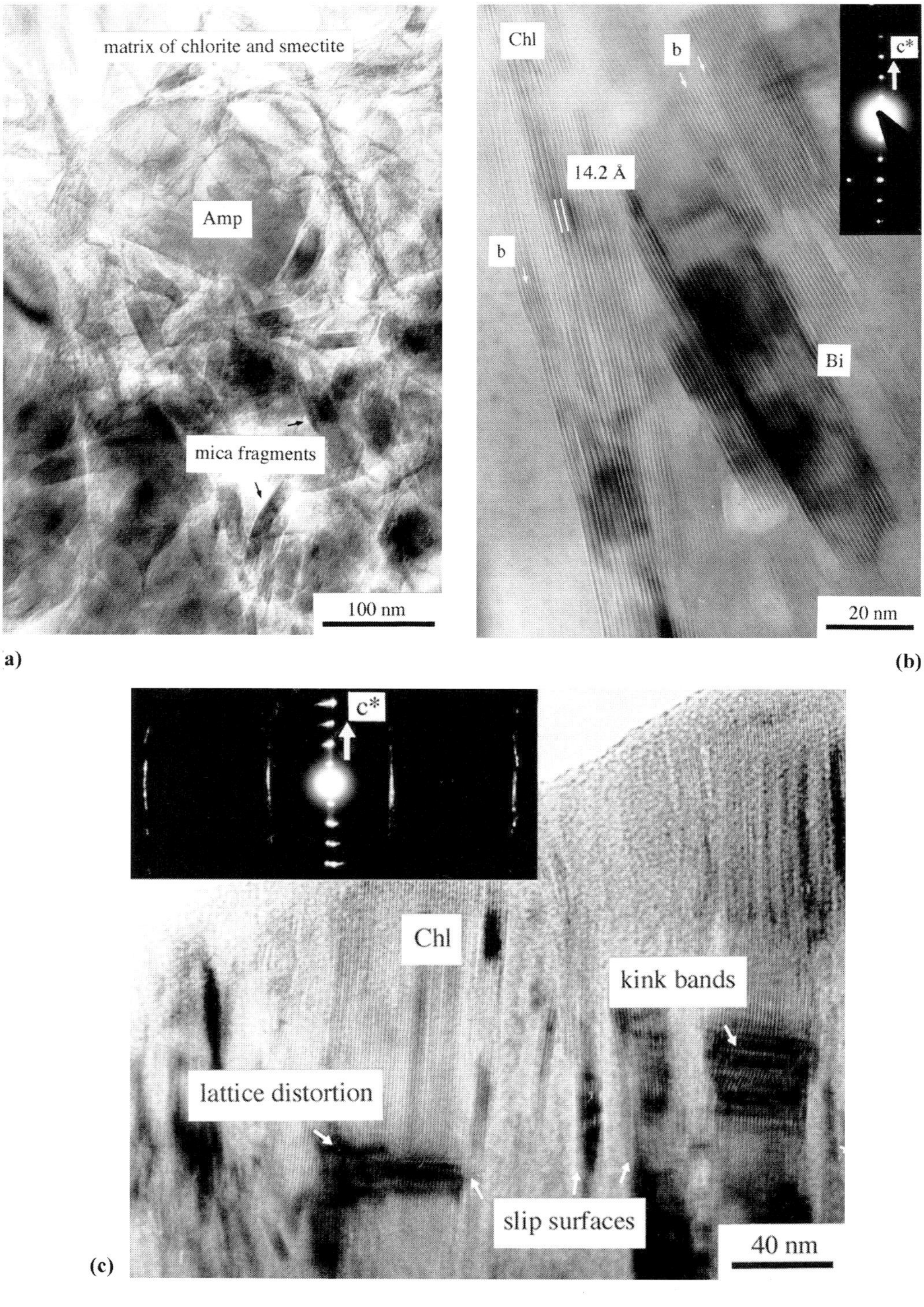
matrix of chlorite and smectite
Amp
mica fragments
100 nm
Chl
b
c*
14.2 Å
b
Bi
20 nm
c*
Chl
kink bands
lattice distortion
slip surfaces
40 nm

(a) (b) (c)

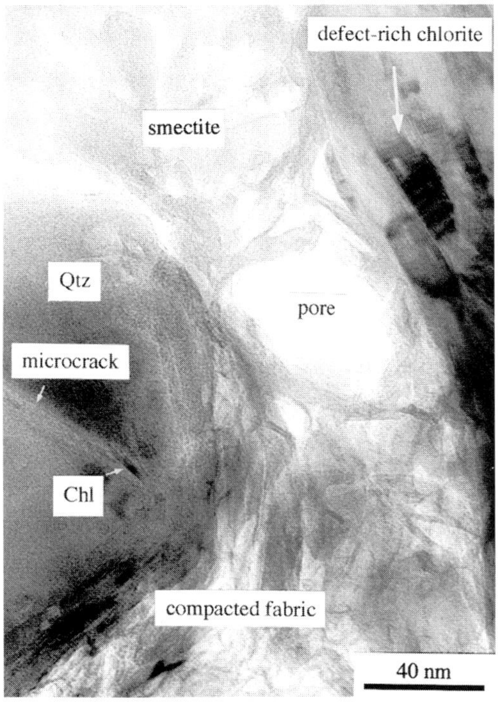

(d)

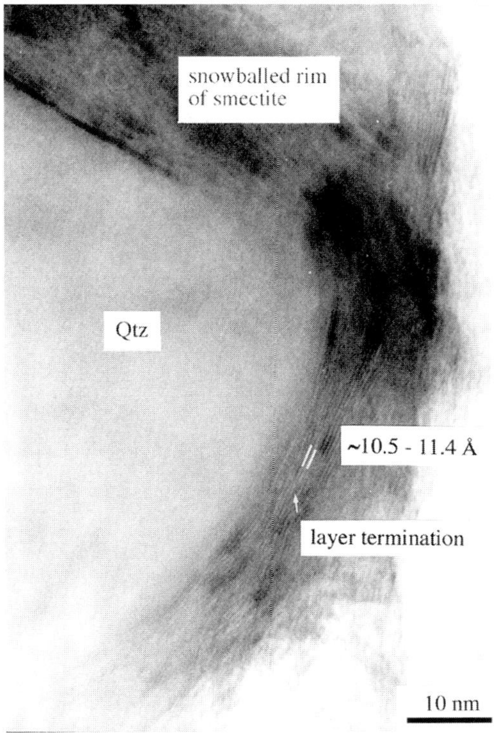

(e)

mica fragments are characterized by a high degree of particle orientation (Fig. 4c). Crystallites show notable signs of intracrystalline deformation, particularly kink bands marked by distinct changes in lattice orientation and zones of crystal distortion. These zones of deformation are typically developed at high angles to packet boundaries and provide clear evidence of slip along both lattice layers and low-angle crystallite boundaries. SAED patterns of the chlorite show streaking parallel to c^* in non-00l reflections, indicating disordering in *hkl* directions of crystals. This type of randomly stacked structure is presumably growth related and comparable with the structure of low-temperature chlorites formed under sub-greenschist-facies conditions (Schmidt & Livi 1999). Some areas of the chlorite are characterized by the occurrence of sub-domains, which are commonly 20–30 nm thick (Fig. 4d). These microstructures presumably reflect the migration of dislocations to form low-angle sub-grain boundaries.

Fig. 4. TEM lattice images of clay gouge from Gaunt Creek (Fig. 2b). (**a**) Patch of gouge containing numerous clasts of metamorphic minerals derived from the Alpine Schist mylonite. The mica fragments are composed of both biotite and muscovite. The large clast in the central upper part of the image is a Ca-amphibole. The matrix consists of a fine network of chlorite and smectite crystallites. (**b**) Biotite grains showing various stages of chloritization by the addition of brucite layers. The weakly altered biotite packet (top right) has two extra brucite layers, added without major disruption of the crystal structure. The more strongly chloritized grain (left side) shows more irregular addition of brucite layers, causing the phyllosilicate packet to thicken upwards. The insert is an SAED pattern of biotite. (**c**) Sheared Mg-chlorite sheets, containing kink bands and zones of lattice distortion. Slip surfaces are parallel to both lattice layers and packet boundaries. The insert is a SAED pattern of Mg-chlorite showing strong disordering in the *hkl* direction of crystals. (**d**) This area of the gouge shows a thick porous matrix of smectite lying between a sheet of defect-rich chlorite and a sub-rounded quartz clast. The microcrack in the quartz is sealed by a small chlorite crystallite (Chl). (Note the domainal structure of the defect-rich chlorite (top right) suggestive of dislocation glide.) (**e**) A sub-rounded quartz clast surrounded by a thick 'snowballed' rim of smectite. The smectite contains numerous layer terminations and has a lattice layer thickness of 10.5–11.4 Å in the vacuum of the microscope.

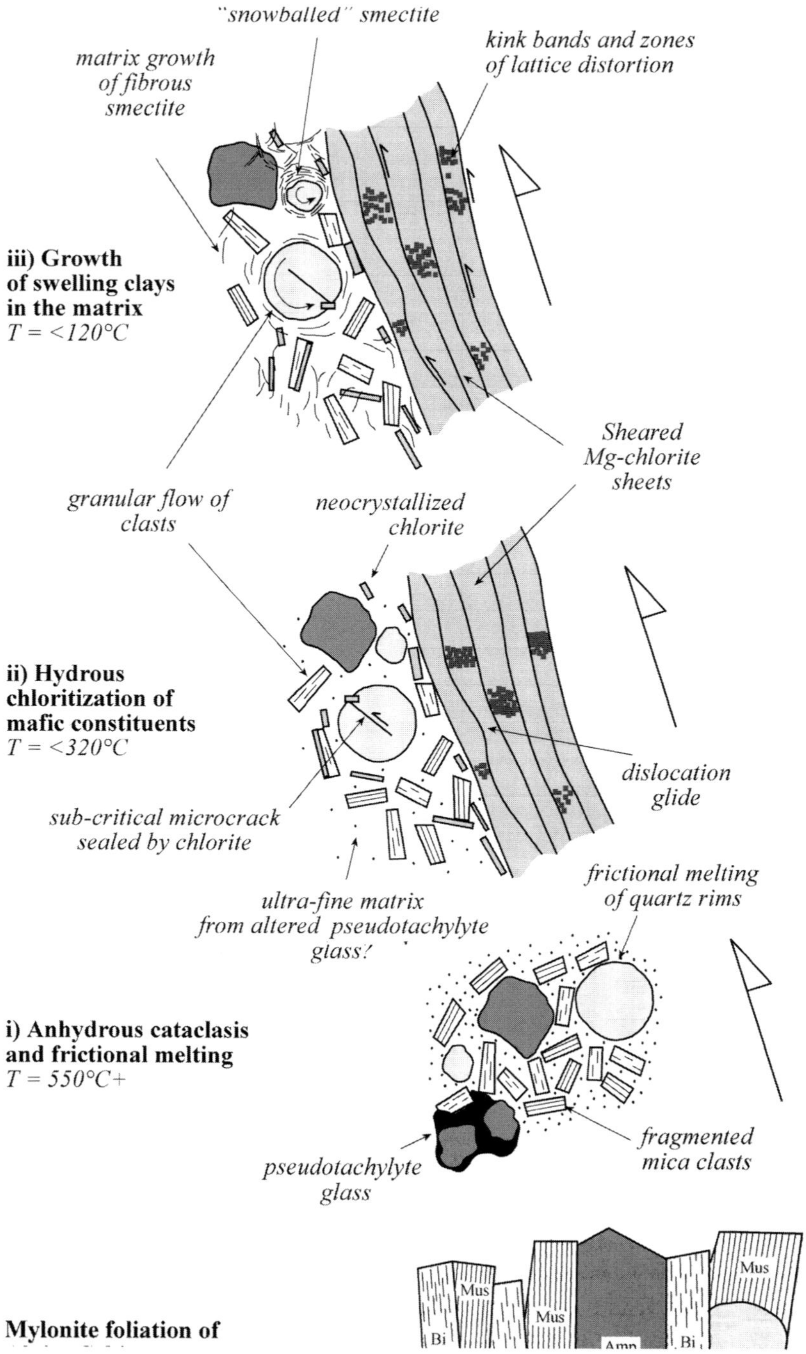

Fig. 5. Schematic representation of the development of mylonite-derived clay gouge, based on TEM observations.

Other features of the gouge are the occurrence of sub-rounded quartz clasts, commonly sandwiched between thick sheets of defect-rich chlorite (Fig. 4d). Such clasts contain microcracks that do not always propagate entirely through grains and are partly sealed by neocrystallized chlorite crystallites. Overgrowing the fragmented grains and forming the matrix between clasts is a fine, interwoven network of chlorite and smectite crystallites (Fig. 4a and d). The network appears to form an open porous structure and shows no signs of particle fragmentation. Weak compaction fabrics around clasts and the porous nature of the matrix in part reflect the high water content that has been lost during sample dehydration (Fig. 4d). A thick coating of dioctahedral smectite with a lattice-layer spacing between 10.5 and 11.4 Å surrounds some sub-rounded quartz grains (Fig. 4e). This 'snowballed' smectite contains thin wavy packets and numerous layer terminations. The geometry and thickness of smectite packets within the grain coating suggests a primary origin for the rimmed texture, although some collapse of expandable layers has occurred during sample preparation and/or within the vacuum of the microscope. The smectite appears to be growing in an ultra-fine matrix, which in part may be amorphous.

Interpretation and discussion

The central Alpine Fault represents a highly active crustal-scale discontinuity and plate boundary with large volumes of clay-sized minerals within its upper segment. A major fraction (*c.* 50%) of the gouge and cataclasite cropping out along the surface trace of the fault is <2 μm in grain size and comprises mostly phyllosilicate (clay) minerals. Our combined XRD and TEM study of the clay fraction reveals that notable quantities of Mg-chlorite and some swelling clay (dioctahedral smectite and vermiculite) have formed within the fault zone, as the hanging-wall mylonites were progressively deformed and exhumed through downward-circulating meteoric fluids.

On the basis of the mineralogical and microstructural characteristics of clay-rich shears within the mylonite-derived clay gouge, collected from the base of the Alpine Fault at Gaunt Creek, three successive modes of deformation and mineral alteration are defined (Fig. 5): (i) anhydrous cataclasis and frictional melting; (ii) hydrous chloritization of mafic constituents; (iii) growth of swelling clays in the matrix.

Anhydrous cataclasis and frictional melting

The first recognizable stage of development in the clay gouge is the cataclastic fragmentation of metamorphic grains during brittle deformation of the mylonite units. This fragmentation process significantly reduced the size of biotite and muscovite particles both in thickness and in length, but without accumulating much lattice distortion. The low level of intracrystalline strain is probably a consequence of the small size and elastic nature of the mica grains.

As the parent mylonite rocks are characterized by a relatively good foliation formed during amphibolite-facies metamorphism (White & White 1983), initial cataclasis led to loss of fabric and strain-hardening in the direction of fault slip. As neither partly fractured grains nor associated clay mineral transformations are related to this grain-size breakdown, there is no evidence to demonstrate the existence of hydrous fluid phases during the fragmentation process. Dry cataclasis provided ideal conditions for generating the high frictional shear stresses and localized melting that have been well documented along the Alpine Fault (Sibson *et al.* 1979; Johnston & White 1983; Bossière 1991). According to Lin (1999), the roundness of quartz and feldspathic clasts in pseudotachylyte and cataclasitic rocks are directly attributable to frictional melting, rather than fracturing or chipping of grains. We interpret the sub-rounded nature of the metamorphic quartz clasts (Fig. 4d and e) as evidence for such frictional melting. A genetic link between cataclasis and frictional melting is also consistent with the petrology of the vesicle-free pseudotachylytes described by Bossière (1991). These formed in an initially anhydrous environment by selective melting of hydrous mafic minerals (mainly biotite and amphibole), at temperatures higher than the amphibolitic host rocks. Such conditions clearly occurred at some depth, away from infiltrating meteoric fluids. If hydrous fluid phases were present, such as those released from the melting of mafic minerals, they would presumably be lithostatically pressured.

Hydrous chloritization of mafic constituents

Following dry cataclastic breakdown and pseudotachylyte generation, hydrous alteration occurred by intense chloritization of mafic constituents. The dominant reaction mechanism of chlorite formation was that of solid-state transformation of biotite crystals by the additional of brucite layers (Veblen & Ferry 1983). However, the large amount of Mg-rich chlorite is unlikely to have formed by chloritization of biotite alone, but would also have involved alteration of other mafic minerals, such as amphibole and pyroxene, and garnet, feldspar and muscovite, which are common constituents of the Alpine Schist. Dissolution and neocrystallization processes were also operative in the matrix, as is

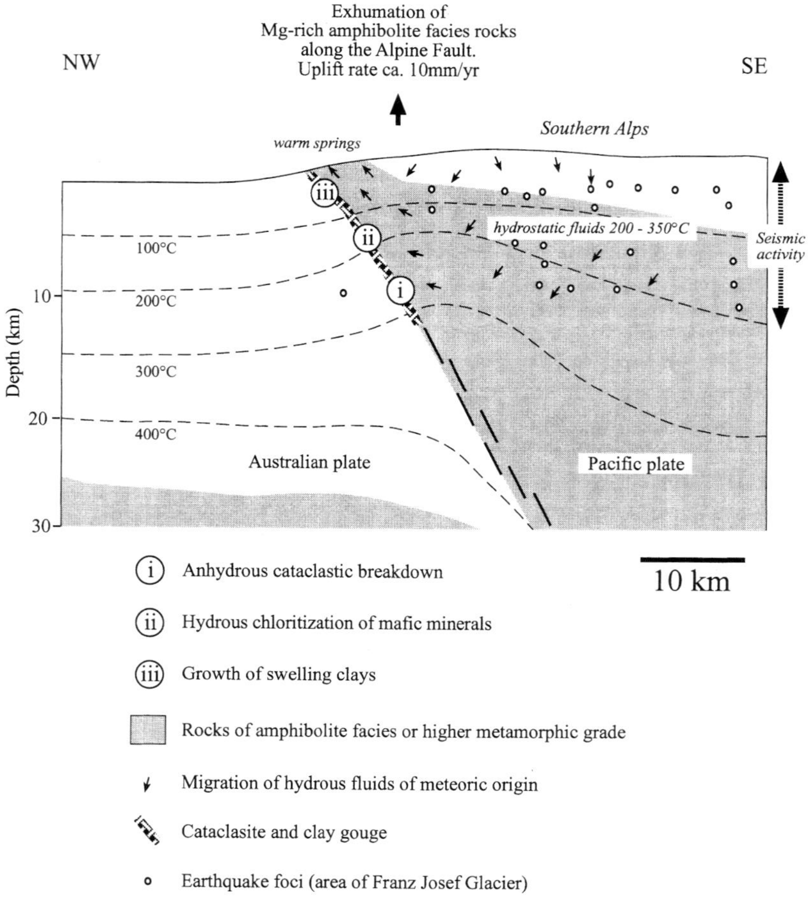

Fig. 6. Hydro-mechanical reconstruction of the Alpine Fault based on the regional thermal modelling of Allis & Shi (1995) and the temperature of infiltrating hydrothermal fluids (Holm *et al.* 1989; Jenkin *et al.* 1994). The seismogenic zone is defined by small earthquakes <M 5 and extends to 13 km depth (Leitner *et al.* 2001).

evident from the growth of chlorite crystallites (Fig. 4a and d).

In contrast to mica minerals within the gouge, Mg-chlorite shows no sign of particle fragmentation along the length of grains, but largely forms well-orientated, sub-parallel phyllosilicate packets that define a good planar fabric. Reaction weakening is indicated by the strong particle anisotropy of chlorite sheets, pervasive slip along basal lattice layers and packet boundaries (producing kink bands and zones of lattice distortion), and a partly domainal fabric suggestive of dislocation glide (Fig. 4d). Mineral composition, the lack of recrystallization and disordered stacking in *hkl* directions of chlorite crystals indicate that this stage of reaction weakening occurred under sub-greenschist-facies conditions (<320 °C).

Additional evidence for the infiltration of hydrous fluids during reaction softening is the partial microcracking of sub-rounded quartz clasts (Fig. 4d). Similar microstructures have been described in adjacent mylonite units by White & White (1983) and attributed to fluid-induced sub-critical cracking during the earthquake cycle at around the brittle–ductile transition zone. Fracturing and dilatation presumably occurred during high rates of coseismic slip along clay-rich shears. Whereas chlorite packets were sheared parallel and sub-parallel to lattice layers, competent quartz grains underwent partial fracturing and subsequent sealing by the growth of fine chlorite crystallites during post-seismic periods. As both the solid-state biotite to chlorite transformation and neocrystallization of chlorite in the matrix involved a significant increase in mineral volume, these hydrous reactions presumably led to notable reductions in permeability and porosity. Such conditions were highly suitable for short-lived enhancement of pore-fluid pressures during slip events, similar to those envisaged to occur along the San Andreas Fault (Chester *et al.* 1993). The resulting low effective confining pressures were also appropriate for granular flow of grains within the fault rock matrix (Fig. 4a).

Growth of swelling clays in the matrix

Further alteration reactions continued at lower temperatures, with the growth of swelling smectite and vermiculite in the ultra-fine matrix of the gouge. Despite the low abundance of these minerals detected by XRD analyses, higher concentrations along thin clay shears suggest an influence on the mechanical weakening and hydrological properties of the Alpine Fault at shallow crustal levels. Reaction weakening and ductile flow of the fault rock matrix is evident from compactional fabrics and 'snowballed' smectite rims observed around rotated, subrounded quartz clasts (Fig. 4e).

The high water content and low permeability of the smectitic matrix is highly suitable for maintaining high pore-fluid pressures during slip events. Additional variations in pore-fluid pressure may arise by swelling pressure changes during hydration and dehydration cycles. At low confining pressures, the growth of 1% Ca-smectite, with a two-water-layer structure, may induce as much as 0.8 MPa pressure (Kahr *et al.* 1990). However, as the number of water layers within smectite decreases under effective pressure, this mechanism of fluid pressure generation will be significant only at shallow depths (Morrow *et al.* 1992; Fitts & Brown 1999). As the formation of discrete smectite particles is reported mostly at hydrothermal conditions <120 °C (e.g. Steiner 1968), a maximum depth estimate of 2–4 km can be predicted for the occurrence of swelling phases. This corresponds to the estimated depth at which the fault is divided into strike-slip and thrust segments (Norris & Cooper 1995), and is considered to be the weakest level of the fault zone.

Clay mineral transformations and the hydromechanical behaviour of the Alpine Fault

The distinct modes of deformation and mineral transformations leading to formation of the clay gouge provide insight into the physical–chemical conditions at various stages of the fault's exhumation history. A reconstruction of fault rock evolution is presented in a temperature-depth profile (Fig. 6), based on the tectono-thermal modelling of Allis & Shi (1995) for the lithosphere beneath the Southern Alps. Two sets of temperature data are shown: calculated isotherms that represent the distribution of present-day regional rock temperatures and hydrostatically pressured, convective fluid temperatures trapped within mineralized fissures (Holm *et al.* 1989; Jenkin *et al.* 1994). The influence of frictional heating is not considered to be significant at this scale, as models including this effect produced near-surface temperature gradients higher than observed (Allis & Shi 1995).

The depth of anhydrous cataclasis and strain-hardening is suggested to occur between 6 and 12 km. The upper limit of this zone is defined by the onset of hydrous chloritization at <320 °C (applying a gradient of 60 °C km^{-1} at depths <5 km for convective fluid circulation). The lower limit corresponds to the base of the seismogenic zone at 12 km (Leitner *et al.* 2001). The anomalously high-temperature minerals reported within pseudotachylyte (Bossière 1991), which form within the seismogenic zone, reflect local frictional heating along discrete slip surfaces and not regional rock or fluid temperatures. As few deeper earthquake events have been recorded at mid- to lower-crustal levels along the Alpine Fault (Reyners 1987; Leitner *et al.* 2001), deformation at these depths appears to be largely accommodated aseismically.

The upper-crustal section of the Alpine Fault is shown to be a zone undergoing heterogeneous reaction weakening, governed by the periodic infiltration of hydrous fluids of meteoric origin. This infiltration caused main-stage chloritization at depths <6 km, apparently under low regional rock temperatures (<200 °C) and short-term, higher fluid temperatures (200–350 °C). The evidently higher temperatures of upward-circulating fluid (Holm *et al.* 1989; Jenkin *et al.* 1994) and higher geothermal gradients (perhaps 90 °C km^{-1}) induce mineral alteration and reaction weakening at much shallower levels within the fault zone than envisaged from the regional rock temperatures alone. Important variables governing the alteration process are the rate of fluid cooling and the rate of mineral transformation that occurs between seismic events. The distribution of clay mineral alterations within the upper crust is probably more complex than the simple depth relationship shown (Fig. 6). Small-scale complexities may also result from localized shear heating within segments of the fault characterized by partial infiltration of hydrous fluids.

The weakest upper 3–4 km level of the fault is similarly controlled by downward-circulating meteoric waters, which are pumped back up to the surface within the hanging wall, and extruded as warm springs (Barnes *et al.* 1978; Allis & Shi 1995). In this scenario, the up-flow of hydrous fluids can be viewed as a consequence of the low-permeability barrier presented by the clay-rich fault zone. It could alternatively reflect the shape and position of the thermal anomaly, rather than spatial variations in permeability as a result of an impervious clay-rich fault.

In relation to the four distinct phases (α, β, γ, δ) of the seismic cycle (Sibson 1986, Sibson, 1989), the long α phase of secular, mainly elastic strain accumulation represents the principal period of clay mineral formation. Progressive sealing by dissolution and growth processes, and volume increase associated with diffusion-controlled solid-state replacement transformations, resulted in lowering of rock permeability and the gradual build-up of fluid pressure. These pressures reached their peak during the β phase of anelastic deformation and accelerating slip, before been released during the γ phase of main-shock rupture. During coseismic slip, the majority of fracture porosity and permeability should be generated near the principal slip surface (Chester *et al.* 1993), and this effect facilitates short-term infiltration of fluids into the fault. Such pressure drops continue into the δ phase of decelerating after-slip, allowing local infiltration of hotter, hydrous fluids and renewed clay mineral transformations.

Despite the reaction weakening inferred from microstructures for the upper levels of the fault zone, it is apparent that clay mineral transformations have not weakened the fault as a whole. No continuous creep has been recorded along the central Alpine Fault (Beavan *et al.* 1999) and calculated stress tensors provide no evidence that the fault is weak, or has weak sections (Leitner *et al.* 2001). The episodic and violent nature of movements along the fault therefore appears to be controlled by sticking of the fault at depth. Such a sticking point could well correspond to the zone of anhydrous cataclasis and frictional melting described, and forms the focus for larger earthquake movements. Similar violent slip behaviour is observable in simulated clay gouge experiments consisting of 85% anhydrite and 15% chlorite at confining pressures of 100 MPa (Shimamoto & Logan 1981).

As the main volume of clay results from alteration of amphibolite-grade metabasites and metapelites, the rate of reaction softening in the upper reaches of the fault has probably increased since coming into contact with the hydrous zone. The relatively Mg-rich amphibolite-facies rocks are particularly important, as they alter easily to weaker chlorite and smectite when in contact with meteoric fluids, in contrast to felsic lithologies, which produce assemblages of illite and kaolinite (Wintsch *et al.* 1995). Assuming a constant uplift rate of 10 mm a^{-1} (Cooper 1980; Koons 1987), such an increase in clay production is likely to have occurred at least within the last 1 Ma, or possibly earlier, when amphibolite rocks first reached *c.* 320 °C conditions at *c.* 6 km depth.

Conclusions

(1) Mylonite-derived cataclasite and clay gouge along the central Alpine Fault, between the Haast River and Gaunt Creek, contain significant volumes of clay minerals. Mineralogical and microstructural observations indicate that three stages of development occurred during the fault's exhumation history: (a) anhydrous cataclasis and frictional melting; (b) hydrous chloritization of mafic constituents; (c) growth of swelling clays in the matrix.

(2) Mg-chlorite formed by solid-state transformation of mafic minerals (notably biotite) and by dissolution and neocrystallization at sub-greenschist-facies conditions (<320 °C). Hydrous alteration resulted from the coseismic infiltration of meteoric fluids, which probably extends to depths of at least 6 km. Minor quantities of swelling smectite and vermiculites are present. These phases are considered to extend down to *c.* 2 km (possibly 4 km) depth, corresponding to the estimated level at which the fault is divided into strike-slip and thrust segments.

(3) Clay mineral transformations within the upper-crustal section of the Alpine Fault led to reaction weakening, and provided ideal conditions for generating short-lived high fluid pressures. Despite this weakening, the fault appears to be sticking at depth, presumably releasing larger quantities of elastic strain in large earthquakes from a zone of anhydrous cataclasis and frictional melting at *c.* 6–12 km.

(4) The significant quantities of clay in the upper section of the Alpine Fault are related to the exhumation of relatively Mg-rich metabasite-rich amphibolite-facies rock over the last 1 Ma. Exhumation of varying metamorphic units and mineral compositions along the

Australian–Pacific plate boundary is an important factor in the production of weaker hydrous minerals along the fault.

We thank F. Nieto (Granada) for his support in obtaining the TEM results. R. Norris, D. Craw, D. Eberhart-Phillips and R. Sibson are thanked for discussions and ideas that helped clarify our understanding of the geology and tectonics of the Alpine Fault. A. Bolton, D. Peacor and B. van der Pluijm provided constructive reviews. L.N.W. acknowledges financial assistance from the DFG (Wa 803/3-2 and GRK 273) that made this study possible.

References

ALLIS, R.G. & SHI, Y. 1995. New insights to temperature and pressure beneath the central Southern Alps, New Zealand. *New Zealand Journal of Geology and Geophysics*, **38**, 585–592.

ALONSO, E.E., VAUNAT, J. & GENS, A. 1999. Modelling the mechanical behaviour of expansive clays. *Engineering Geology*, **54**, 173–183.

ALT, J.C. 1999. Very low-grade hydrothermal metamorphism of basic igneous rock. *In*: FREY, M. & ROBINSON, D. (eds) *Low-Grade Metamorphism*. 169–201. Blackwell Science, Oxford.

ANDERSON, H. & WEBB, T. 1994. New Zealand seismicity: patterns revealed by the upgraded National Seismograph Network. *New Zealand Journal of Geology and Geophysics*, **37**, 477–493.

BARNES, I., DOWNES, C.J. & HULSTON, J.R. 1978. Warm springs, South Island, New Zealand and their potential to yield laumontite. *American Journal of Science*, **278**, 1412–1427.

BARTON, N. 1976. The shear strength of rock and rock joints. *International Journal of Rock Mechanics and Mining Sciences*, 9, **13**, 255–279.

BEAVAN, J., MOORE, M., PEARSON, C.& 8 OTHERS 1999. Crustal deformation during 1994–1998 due to oblique continental collision in the central Southern Alps, New Zealand, and implications for seismic potential of the Alpine Fault. *Journal of Geophysical Research*, **104**(B11), 25233–25255.

BIRD, P. 1984. Hydration-phase diagrams and friction of montmorillonite under laboratory and geologic conditions, with implications for shale compaction, slope stability, and the strength of fault gouge. *Tectonophysics*, **107**, 235–260.

BOSSIÈRE, G. 1991. Petrology of pseudotachylytes from the Alpine Fault of New Zealand. *Tectonophysics*, **196**, 173–193.

BRUCE, C.H. 1985. Smectite dehydration—its relation to structural development and hydrocarbon accumulation in Northern Gulf of Mexico basin. *AAPG Bulletin*, **68**, 673–683.

CATHELINEAU, M. 1988. Cation site occupancy in chlorites and illites as a function of temperature. *Clay Minerals*, **23**, 471–485.

CHESTER, F.M., EVANS, J.P. & BIEGEL, R.L. 1993. Internal structure and weakening mechanisms of the San Andreas Fault. *Journal of Geophysical Research*, **98**, 771–786.

COOMBS, D.S., KAWACHI, Y. & FORD, P.B. 1996. Porphyroblastic manganaxinite metapelites with incipient garnet in prehnite–pumpellyite facies, near Meyers Pass, Torlesse Terrane, New Zealand. *Journal of Metamorphic Geology*, **14**, 125–142.

COOPER, A.F. 1980. Retrograde alteration of chromian kyanite in metachert and amphibole whiteschist from the Southern Alps, New Zealand, with implications for uplift on the Alpine fault. *Contributions to Mineralogy and Petrology*, **75**, 153–164.

COOPER, A.F. & NORRIS, R.J. 1994. Anatomy, structural evolution, and slip rate of a plate-boundary thrust: the Alpine fault at Gaunt Creek, Westland, New Zealand. *Geological Society of America Bulletin*, **106**, 627–633.

CRAW, D. & NORRIS, R.J. 1993. Grain boundary migration of water and carbon dioxide during uplift of garnet-zone Alpine Schist, New Zealand. *Journal of Metamorphic Geology*, **11**, 371–378.

EBERHART-PHILLIPS, D. 1995. Examination of seismicity in the central Alpine Fault region, South Island, New Zealand. *New Zealand Journal of Geology and Geophysics*, **38**, 571–578.

EVISON, F.F. 1971. Seismicity of the Alpine Fault, New Zealand. *In*: COLLINS, B.W. & FRASER, R. (eds) *Recent Crustal Movements*. Royal Society of New Zealand, Bulletin **9**, 161–165.

FITTS, T.G. & BROWN, K.M. 1999. Stress-induced smectite dehydration: ramifications for patterns of freshening and fluid expulsion in the N. Barbados accretionary wedge. *Earth and Planetary Science Letters*, **172**, 179–197.

GRAPES, R.H. 1995. Uplift and exhumation of Alpine Schist, Southern Alps, New Zealand: thermobarometric constraints. *New Zealand Journal of Geology and Geophysics*, **38**, 525–533.

GRINDLEY, G.W. 1978. The Rangitata Orogeny. *In*: SUGGATE, R.P., STEVENS, G.R. & TE PUNGA, M.T. (eds) *The Geology of New Zealand, Vol. 1*. Government Printer, Wellington, 318–343.

HICKMAN, S.H. 1991. Stress in the lithosphere and the strength of active faults. *Reviews of Geophysics*, **29**, 759–775.

HOLM, D.K., NORRIS, R.J. & CRAW, D. 1989. Brittle and ductile deformation in a zone of rapid uplift: Central Southern Alps, New Zealand. *Tectonics*, **8**, 153–168.

JENKIN, G.R.T., CRAW, D. & FALLICK, E. 1994. Stable isotopic and fluid inclusion evidence for meteoric fluid penetration into an active mountain belt; Alpine Schist, New Zealand. *Journal of Metamorphic Geology*, **12**, 429–444.

JOHNSTON, D.C. & WHITE, S. 1983. Shear heating associated with movement along the Alpine Fault, New Zealand. *Tectonophysics*, **92**, 241–252.

KAGAMI, H. 1985. Internal structures of the accretionary wedge in the Nankai Trough off Shikoku, southwestern Japan. *In*: NASU, N., KOBAYASHI, K., UYEDA, S., KUSHIRO, I. & KAGAMI, H. (eds) *Formation of Active Ocean Margins*. Terra, Tokyo, 193–219.

KAHR, G., KRAEHENBUEHL, F., MÜLLER-VONMOOS, M. & STOECKLI, H. 1990. Study of the water–bentonite system by vapour adsorption, immersion calorimetry and X-ray techniques: II. Heats of immersion, swelling pressures and thermodynamic properties. *Clay Minerals*, **25**, 499–506.

KNIPE, R. J. 1992. Fault processes and fault seal. *In*: LARSEN, R.M., BREKKE, H., LARSEN, B.T. & TALLERAAS, E. (eds) *Structural and Tectonic Modelling and its Application to Petroleum Geology*. Norwegian Petroleum Society Special Publications, **1**, 325–342.

KNIPE, R. J. 1993. The influence of fault zone processes and diagenesis on fluid flow. *In*: HORBURY, A. D. & ROBINSON, A. G. (eds) *Diagenesis and Basin Development*. American Association of Petroleum Geologists, Studies in Geology, **36**, 135–148.

KNUEPFER, P.L. 1992. Temporal variations in latest Quaternary slip across the Australian–Pacific Plate Boundary, northeastern South Island, New Zealand. *Tectonics*, **2**, 449–464.

KOONS, P.O. 1987. Some thermal and mechanical consequences of rapid uplift: an example from the Southern Alps. *Earth and Planetary Science Letters*, **86**, 307–319.

KRANIDIOTIS, P. & MACLEAN, W.H. 1987. Systematics of chlorite alteration at the Phelps Dodge sulfide deposit, Matagami, Quebec. *Economic Geology*, **82**, 1898–1911.

LEITNER, B., EBERHART-PHILLIPS, D., ANDERSON, H. & NABELEK, J.L. 2001. A focused look at the Alpine Fault, New Zealand: seismicity, focal mechanisms and stress observations. *Journal of Geophysical Research*, **106(B2)**, 2193–2220.

LIN, A. 1999. Roundness of clasts in pseudotachylytes and cataclasitic rocks as an indicator of frictional melting. *Journal of Structural Geology*, **21**, 473–478.

MALTMAN, A. 1987. Shear zones in argillaceous sediments—an experimental study. *In*: JONES, M.E. & PRESTON, R.M.F. (eds) *Deformation of Sediments and Sedimentary Rocks*. Geological Society, London, Special Publication, **29**, 77–87.

MESRI, G. & CEPEDA-DIAZ, A.F. 1986. Residual shear strength of clays and shales. *Géotechnique*, **36**, 269–274.

MITCHELL, J.K. 1993. *Fundamentals of Soil Behavior*. Wiley, New York.

MOORE, J.C. 1989. Tectonics and hydrogeology of accretionary prisms: role of the décollement zone. *Journal of Structural Geology*, **11**, 95–106.

MOORE, D.M. & REYNOLDS, R.C. 1997. *X-Ray Diffraction and the Identification and Analysis of Clay Minerals* (2nd edition). Oxford University Press, Oxford.

MORROW, C.A., RADNEY, B. & BYERLEE, J. 1992. Frictional strength and the effective pressure law of montmorillonite and illite clays. *In*: EVANS, B. & WONG, T.-F. (eds) *Fault Mechanics and Transport Properties of Rocks*. Academic Press, New York, 69–88.

MORROW, C.A., SHI, L.Q. & BYERLEE, J.D. 1984. Permeability of fault gouge under confining pressure and shear stress. *Journal of Geophysical Research B*, **89**, 3193–3200.

MÜLLER-VONMOOS, M. & LOKEN, T. 1989. The shearing behaviour of clays. *Applied Clay Science*, **4**, 125–141.

MUSTAFAYEV, A.A. 1988. Rheological peculiarity of clayey soil deformation under swelling. *In*: KEEDWELL, M.J. (ed.) *Proceedings of the International Conference on Rheology and Soil Mechanics*. Elsevier Applied Science, London, 139–146.

NORRIS, R.J. & COOPER, A.F. 1995. Origin of small-scale segmentation and transpressional thrusting along the Alpine fault, New Zealand. *Geological Society of America Bulletin*, **107**, 231–240.

OLSON, R. 1974. Shearing strength of kaolinite, illite and montmorillonite. *Journal of Soil Mechanics and Foundations Division, ASCE*, **100**, 1215–1229.

REYNERS, M. 1987. Subcrustal earthquakes in the central South Island, New Zealand, and the root of the Southern Alps. *Geology*, **15**, 1168–1171.

REYNERS, M. 1989. New Zealand seismicity 1964–87: an interpretation. *New Zealand Journal of Geology and Geophysics*, **32**, 307–315.

ROSENQUIST, I. 1962. The influence of physico-chemical factors upon the mechanical properties of clays. *Clays and Clay Minerals*, **9**, 12–23.

ROSENQUIST, I. 1984. The importance of pore water chemistry on mechanical and engineering properties of clay soils. *Philosophical Transactions of the Royal Society, Series A*, **311**, 369–383.

RUBY, W.R. & HUBBERT, M.K. 1959. Role of fluid pressure in mechanics of overthrust faulting. *Geological Society of America Bulletin*, **70**, 167–206.

SCHMIDT, D. & LIVI, K.J.T. 1999. HRTEM and SAED investigations of polytypism, stacking disorder, crystal growth, and vacancies in chlorites from subgreenschist facies outcrops. *American Mineralogist*, **84**, 160–170.

SHAU, Y.H., ESSENE, E.J. & PEACOR, D.R. 1990. Corrensite and mixed-layer chlorite/corrensite in metabasalt from northern Taiwan: TEM/AEM, EMPA, XRD and optical studies. *Contributions to Mineralogy and Petrology*, **105**, 123–142.

SHEA, W.T. & KRONENBERG, A.K. 1992. Rheology and deformation mechanisms of an isotopic mica schist. *Journal of Geophysical Research*, **97**, 201–237.

SHIMAMOTO, T. & LOGAN, J.M. 1981. Effects of simulated clay gouges on the sliding behaviour of Tennessee sandstone. *Tectonophysics*, **75**, 243–255.

SIBSON, R.H. 1986. Brecciation processes in fault zone: inferences from earthquake rupturing. *Pure and Applied Geophysics*, **124**, 159–175.

SIBSON, R.H. 1989. Earthquake faulting as a structural process. *Journal of Structural Geology*, **11**, 1–14.

SIBSON, R.H., WHITE, S.H. & ATKINSON, B.K. 1979. Fault rock distribution and structure within the Alpine fault zone: a preliminary account. *In*: WALCOTT, R.I. & CRESSWELL, M.M. (eds) *The Origin of the Southern Alps*. Royal Society of New Zealand, Bulletin **18**, 55–65.

SKEMPTON, A.W. 1985. Residual strength of clays in landslides, folded strata and the laboratory. *Géotechnique*, **35**, 3–18.

STEINER, A. 1968. Clay minerals in hydrothermally altered rocks at Waireki, New Zealand. *Clays and Clay Minerals*, **16**, 3081–3096.

TESSIER, D., DARDAINE, M., BEAUMONT, A. & JAUNET, A.M. 1998. Swelling pressure and microstructure of activated swelling clay with temperature. *Clay Minerals*, **33**, 255–267.

UPTON, P., KOONS, P.O. & CHAMBERLAIN, C.P. 1995. Penetration of deformation-driven meteoric water into ductile rocks: isotopic and model observations from the Southern Alps, New Zealand. *New Zealand Journal of Geology and Geophysics*, **38**, 525–533.

VEBLEN, D.R. & FERRY, J.M. 1983. A TEM study of the biotite–chlorite reaction and comparison with petrologic observations. *American Mineralogist*, **68**, 1160–1168.

VROLIJK, P. 1990. On the mechanical role of smectite in subduction zones. *Geology*, **18**, 703–707.

VROLIJK, P. & van der Pluijm, B.A. 1999. Clay gouge. *Journal of Structural Geology*, **21**, 1039–1048.

WALDEN, R.W. & KING, H.M. 1959. Overthrust belt in geosynclinal area of western Wyoming in light of fluid-pressure hypothesis, [Part] 2 of Role of fluid pressure in mechanics of overthrust faulting. *Geological Society of America Bulletin*, **70**, 2, 167–205.

WANG, C.Y. & MAO, N.H. 1979. Shearing of saturated clays in rock joints at high confining pressures. *Geophysical Research Letters*, **6**, 825–828.

WANG, C.Y., MAO, N.H. & WU, F.T. 1980. Mechanical properties of clays at high pressures. *Journal of Geophysical Research*, **85**, 1462–1468.

WARR, L.N. & NIETO, F. 1998. Crystallite thickness and defect density of phyllosilicates in low-temperature metamorphic pelites: a TEM and XRD study of clay-mineral Crystallinity-Index Standards. *Canadian Mineralogist*, **36**, 1453–1474.

WARR, L.N., GREILING, R.O. & ZACHRISSON, E. 1996. Thrust-related very low-grade metamorphism in the marginal part of an orogenic wedge, Scandinavian Caledonides. *Tectonics*, **15**, 1213–1229.

WELLS, A., YETTON, M.D., DUNCAN, R.P. & STEWART, G.H. 1999. Prehistoric dates of the most recent Alpine fault earthquakes, New Zealand. *Geology*, **27**, 995–998.

WIBBERLEY, C. 1999. Are feldspar-to-mica reactions necessarily reaction-softening processes in fault zones? *Journal of Structural Geology*, **21**, 1219–1227.

WHITE, J.C. & WHITE, S.H. 1983. Semi-brittle deformation within the Alpine fault zone, New Zealand. *Journal of Structural Geology*, **5**, 579–589.

WINTSCH, R.P., CHRISTOFFERSEN, R. & KRONENBERG, A.K. 1995. Fluid–rock reaction weakening of fault zones. *Journal of Geophysical Research*, **100**, 13021–13032.

WU, F.T., BLATTER, L. & ROBERSON, H. 1975. Clay gouges in the San Andreas Fault System and their possible implications. *Pure and Applied Geophysics*, **113**, 87–95.

ZOBACK, M.D., ZOBACK, M.L., MOUNT, V. & 9 OTHERS 1987. New evidence on the state of stress of the San Andreas fault system. *Science*, **238**, 1105–1111.

Deformation microfabrics of clay gouge, Lewis Thrust, Canada: a case for fault weakening from clay transformation

YONGHONG YAN, BEN A. VAN DER PLUIJM & DONALD R. PEACOR

Department of Geological Sciences, University of Michigan, 2534 C. C. Little Building, Ann Arbor, MI 48109, USA (e-mail: vdpluijm@umich.edu)

Abstract: A sequence of bentonite and shale samples in a gouge zone of the Lewis Thrust (Alberta, Canada) that display increasing degree of transformation of clay minerals toward the hanging wall of the thrust has been studied by X-ray diffraction (XRD), X-ray texture goniometry (XTG), scanning electron microscopy (SEM), and transmission and analytical electron microscopy (TEM–AEM), to examine the relations among mineral transformations, microfabrics and fault zone properties. TEM images of authigenic clays show abundant smectite in shale away from the hanging wall, characterized by anastomosing layers with an average orientation that parallels bedding, coexisting with uncommon *R*1 illite–smectite (I–S). In the sample nearest the hanging wall, by contrast, the dominant clay is mixed-layered, illite-rich illite–smectite (*R*1 I–S), coexisting with discrete illite, occurring in individual packets of relatively straight layers with well-defined boundaries. Deformed clay packets are common. Pore space, where packets intersect at high angles to one another and to bedding, is abundant (*c.* 25%). The microfabric and proportion of illite of intermediate samples are transitional to these end-members. Interlayered bentonite samples show properties that are similar to those of shale. TEM observations are supported by quantification of the fabrics using XTG, which shows that the intensity of clay preferred orientation decreases significantly with increasing illitization. These relations imply that faulting was the cause of mineral transformations and formation of secondary pore space. The illitization reaction rate was enhanced both by stress-induced defects in clays, and by increased water/rock ratio resulting from deformation-related pore space, resulting in lowering of the effective stress. The deformation-enhanced reaction thus created a positive feedback for further faulting in clay gouge, leading to enhanced weakening of the fault zone.

Many upper-crustal faults contain clay-rich fault gouge, which is typically light- to dark-coloured, fine-grained material that becomes sticky when wet. Whereas it was traditionally assumed that gouge is primarily formed by mechanical processes (e.g. Sibson 1977; Chester & Logan 1986; Rutter *et al.* 1986), more recent work increasingly indicates that mineral transformations may be a significant process in clay-rich fault rocks. For example, Vrolijk & van der Pluijm (1999) described clay transformation in fault gouge from the Lewis Thrust in Canada, based on X-ray diffraction (XRD) of various grain-size fractions. Determinations of per cent illite in mixed-layer illite–smectite (I–S) showed that the per cent illite in I–S increases from <30% to >80% toward the hanging wall. In this paper the detailed microstructure of selected samples from this fault zone is analysed using X-ray texture goniometry (XTG) and through direct observations using electron microscopy (EM). The goal of our study is to examine the fabric characteristics and role of clay transformations in fault gouge evolution, and explore the consequences for faulting.

X-ray texture goniometry has been used to constrain the development of preferred orientation of fine-grained phyllosilicates in both depositional and tectonic environments (e.g. Oertel 1985; Sintubin 1994; Ho *et al.* 1995; 1996; 1999). Of particular relevance to the present study, Ho *et al.* (1999) showed that the onset of the smectite-to-illite transition is associated with an increase in the preferred orientation of phyllosilicates in the classic Gulf Coast section. The anastomosing smectite layers that are found in samples before the transition zone have a weak fabric that parallels bedding, whereas well-defined post-transition illite packets display a strong fabric that is parallel to bedding (Ho *et al.* 1999). Thus in the Gulf Coast sequence, the smectite-to-illite transformation is characterized by a significant strengthening of the microfabric.

From: HOLDSWORTH, R.E., STRACHAN, R.A., MAGLOUGHLIN, J.F. & KNIPE, R.J. (eds) 2001. *The Nature and Tectonic Significance of Fault Zone Weakening*. Geological Society, London, Special Publications, **186**, 103–112. 0305-8719/01/$15.00 © The Geological Society of London 2001.

Transmission electron microscopy (TEM) is a powerful means to study the detailed properties of individual clays. Many workers have studied the transformation of smectite to illite in prograde sequences, from diagenesis to low-grade metamorphism (e.g. Hower *et al.* 1976; Ahn & Peacor 1986; Freed & Peacor 1987, Freed & Peacor 1992; Jiang *et al.* 1990*a*; 1997; Dong *et al.* 1997). These transformations probably follow the Ostwald step rule, from metastable heterogeneous mineral assemblages, compositions and textures to a system with stable, homogeneous and relatively large crystals (Essene & Peacor 1995). The sequence of clay minerals involved in the transformation from smectite to illite has been modelled in two ways: (1) continuously increasing proportions of illite layers, from smectite through mixed-layered I–S, to illite; the sequence includes I–S with R (Reichweite) values increasing from $R = 0$ to values as large as $R = 3$; (b) a sequence of three phases, smectite, $R1$ I–S, and illite, with the relative proportions of the individual phases changing continuously with increasing degree of diagenesis (Dong *et al.* 1997), the latter being based on a unique structure for $R1$ I–S. TEM observations suggest that smectite and I–S phases coexist rather than occurring as random mixing of illite- and smectite-like layers (Yan *et al.* 2001). Moreover, it is reported that the dominant polytypes of prograde sediment sequences are $1M_d$ to $2M_1$ with no $1M$ (e.g. Dong & Peacor 1996; Jiang *et al.* 1990*a*; Masuda *et al.* 1996; Li *et al.* 1997).

Clay minerals are generally thermodynamically metastable, with transformations controlled by reaction kinetics (Velde *et al.* 1986; Morse & Casey 1988; Peacor 1992; Essene & Peacor 1995). Temperature is a well-known factor that controls transformation, as illustrated by Gulf Coast mudstones (Perry & Hower 1970; Hower *et al.* 1976; Bruce 1984; Freed & Peacor 1987; Freed & Peacor 1989, 1992). However, any factor that affects reaction rates can promote transformations among clay minerals, among which water/rock ratio and stress-induced defects, both of which are related to faulting, are significant. This study was undertaken in part to determine the role played by these factors during faulting where the affected rocks are dominated by clay minerals.

Sample descriptions and methods

Five representative samples of dark- and light-coloured clay gouge from the Lewis Thrust were analysed in detail. The samples were collected at Gould Dome, north of Crowsnest Pass, Alberta, Canada, where the Lewis Thrust emplaces the Mississippian Rundle Formation over the Upper Cretaceous Belly River Formation. Details of the site and fault gouge samples have been given by Vrolijk & van der Pluijm (1999). Figure 1 shows two representative samples of gouge. The sample shown in Fig. 1a is from *c.* 3 m below the contact with the Rundle Formation, which is the hanging wall. The large, angular clast in this sample is derived from this limestone unit. The sample shown in Fig. 1b shows a well-layered section of gouge *c.* 10 cm from the contact. The light layer is bentonite and the dark layer is shale, and the layers parallel the base of the Lewis Thrust as marked by the overlying limestone unit.

X-ray diffraction

Both bentonite and shale bulk powder samples were prepared by air-dried and ethylene glycol treatments. XRD data were obtained using an automated diffractometer equipped with a graphite monochromator, scintillation counter and CuK_α radiation (35 kV; 15 mA). The scanning angle ranged from 2° to 32° (2θ), with step size of 0.020° 2θ and counting time of 2 s per step.

X-ray texture goniometry

Clay samples were impregnated in epoxy to prevent sample rupture, then cut normal to bedding or foliation. Sections were polished to slices of about 0.2 mm thickness and mounted on a square aluminium holder. The crystallographic preferred orientation was determined using an X-ray pole figure device on an Enraf–Nonius CAD4 single-crystal diffractometer with Mo source and scintillation detector. A distinct feature of this instrument is its ability to measure small, relatively thick regions (of the order of *c.* 1 mm diameter and 0.2 mm thickness) with minimal absorption effects. The transmission mode was used, characterized by low diffraction angles for phyllosilicates. Pole figures are presented as relative intensities in multiples of random distribution (m.r.d.). Detailed descriptions of corrections and other procedures have been given by van der Pluijm *et al.* (1994).

Electron microscopy

To avoid collapse of smectite, all samples used for TEM were prepared with L. R. White resin using the method described by Kim *et al.* (1995). Well-polished thin sections were made

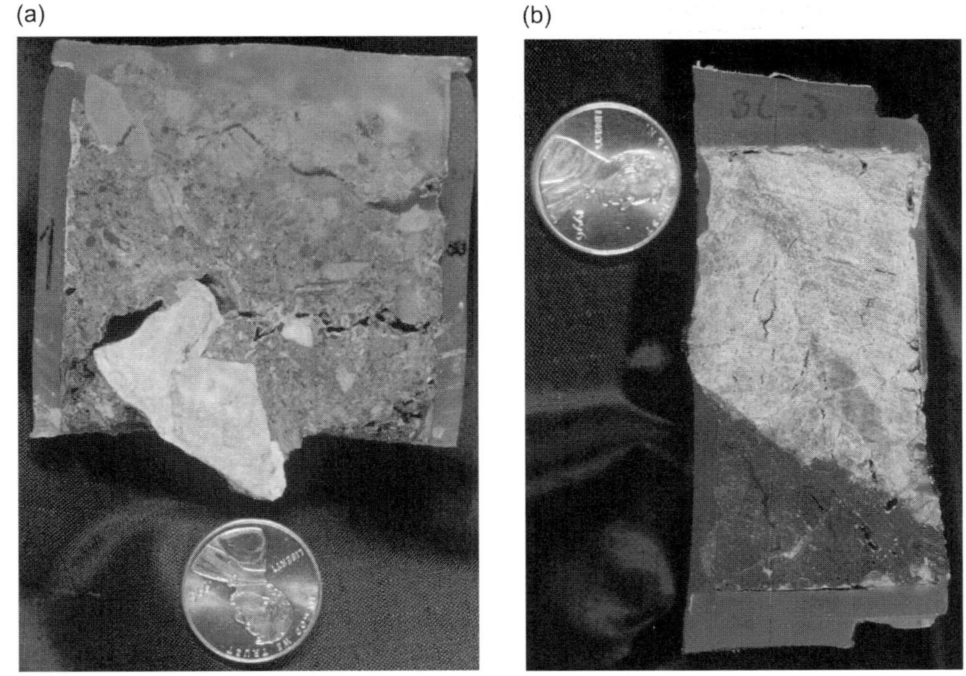

Fig. 1. Fault gouge samples from the Lewis Thrust, Alberta (Canada). (**a**) Sample BC8 is located *c*. 10 m from the overlying limestone unit that marks the top of the fault zone. (Note the coarse, angular limestone fragments that are clasts of the overlying limestone unit.) (**b**) Sample BC3 shows layered bentonite (light grey) and shale (black). The contact parallels the orientation of the fault zone. Coin (2 cm) for scale; plastic liner used for sampling is preserved along the edges.

from surfaces perpendicular to bedding, to optimize the alignment of (001) planes of phyllosilicates for obtaining lattice-fringe images. SEM observations were carried out at an operating voltage of 20 kV. The scanning transmission electron microscope was operated at 120 kV with a beam current of 20 μA. Most fringe images were obtained at 100 000 magnification, with an objective aperture of 0.030 mm. Through-focus sequences of images were obtained from 100 nm underfocus to 100 nm overfocus to obtain optimum contrast as a function of composition and structure. Selected-area electron-diffraction (SAED) patterns were obtained with a camera length of 770 mm and aperture of 0.010 mm. Intensity ratios of energy-dispersive spectra for determination of chemical compositions were processed with procedures described by Jiang *et al.* (1990*a*).

Results

Identification of clay minerals based on XRD data for each of the five bulk samples shows that mixed-layered I–S dominates all samples, with per cent illite in I–S ranging from *c*. 30% to >80% with decreasing distance from the hanging wall, consistent with results of Vrolijk & van der Pluijm (1999). Bulk XRD patterns of material identified as bentonite and shale primarily on the basis of colour in hand specimen were surprisingly similar, with major peaks ascribed to I–S, chlorite, kaolinite and quartz. This implies that the black shale differs little from the bentonite units; that is, the shales apparently were originally largely composed of volcanic ash, with only a small component of terrestrially derived material. The results of XRD (and XTG; see below) are summarized in Table 1. The Reichweite (R) values, which refer to the probability of finding a layer of illite (I) followed by a layer of smectite (S) (e.g. $R = 1$ for the ordered sequence…ISISIS…) and per cent illite were determined according to the method of Moore & Reynolds (1997). Compositions of I–S as obtained by analytical electron microcopy (AEM) are listed in Table 2.

Table 1 lists the peak d values (D_{mix}) from 2θ scans of each of the five samples used in this study. X-ray goniometry results for these peaks are presented as the maximum intensity

Table 1. *Summary results of Lewis Thust samples, ordered by lithology and distance from hanging wall*

Sample	Distance (m)	Lithology	% illite in I–S	Reichweite (*R*) value	D_{mix} (nm)	Maximum intensity (m.r.d.)
BC2S	0.12	Shale	70–80	3	1.08	2.83
BC9S	3.7	Shale	60–70	2	1.20	2.96
BC13S	12	Shale	30	0	1.06	3.83
BC2B	0.12	Bentonite	70–80	3	1.20	2.50
BC10B	6	Bentonite	50	1	1.27	1.99

Illite percentages in I–S are obtained from XRD. 'Distance' is distance to top of fault zone. D_{mix} indicates the *d* value of selected 2θ peaks in XTG scans. Maximum intensity values and contours are obtained from pole figure analysis and are shown in Fig. 2. m.r.d., multiples of random distribution.

of diffracted X-rays, contoured in multiples of random distribution. They are plotted in Fig. 2 in contoured equal-area, lower-hemisphere projections, with the peak orientation rotated to the centre. Shale (S) samples show significant decreasing preferred orientation toward the base of the overlying limestone unit, which marks the top of the fault zone. The results from bentonite appear to show a slight increase in fabric intensity toward the top of the fault zone, but the sampling range is too small and the contour patterns are not sufficiently well defined to attach great significance to this difference.

TEM images verified that smectite and *R*1 I–S are the dominant minerals. Chlorite and kaolinite were observed in two samples. The descriptions below, which are ordered by rock type (shale or bentonite) and decreasing distance from the top of the fault zone (hanging wall), summarize the results from hundreds of TEM images.

Shale BC13S (20–30% I, 1200 cm)

Figure 3a shows that the dominant clay mineral is smectite, occurring as wavy, ill-defined and anastomosing smectite layers with 1.2–1.3 nm (001)-interplanar spacings, and variable orientations whose average is that of bedding. Pore space is sparingly visible. Illite-like layers occur rarely in separate packets of *R*1 I–S (2.1–2.2 nm periodicity). The SAED pattern inset to Fig. 3a shows only 00*l* reflections that are weak and diffuse, showing streaking parallel and normal to c^*. The streaking parallel to c^* is the result of variable layer space of 00*l* fringes, whereas streaking normal to c^* is the result of variable layer orientation.

Table 2. *Representative AEM-determined chemical formulae of I–S*

	BC2S	BC9S	BC13S	BC2B	BC10B
Si	7.07	7.70	7.84	7.13	7.29
$Al^{[4]}$	1.05	0.34	0.20	0.87	0.71
$Al^{[6]}$	3.36	3.35	3.03	3.50	2.98
Fe	0.37	0.37	0.22	0.30	0.83
Mg	0.35	0.21	0.80	0.21	0.30
Ca	n.d.	n.d.	0.18	n.d.	0.25
Na	n.d.	n.d.	0.15	0.14	n.d.
K	1.07	0.99	0.41	1.10	1.00

Formulae are normalized to a total of 22 oxygen atoms, assuming Fe to be ferrous. They are an average for each sample. n.d., not detected.

Shale BC9S (60–70% I, 370 cm)

Figure 3b shows the microfabric intermediate to the extremes shown by other samples. The lattice fringes show periodicity of 2.1–2.2 nm, as defined by alternating dark and light contrast, which is typical of images of *R*1 I–S obtained at proper overfocus conditions (Veblen *et al.* 1990; Guthrie & Reynolds 1998). However, the texture is one in which packets are bent and commonly intersect at high angles, in sharp contrast to the fabric described for the preceding sample. Pore space is common, especially where individual packets intersect at high angles. Chlorite (lattice fringes with $d = 1.4$ nm) occurs rarely as separate packets parallel to I–S. SAED patterns show 0*kl* reflections ($k/=3$ N) that are weak, diffuse, and non-periodic, corresponding to $1M_d$ polytypism.

Shale BC2S (70–80% I, 12 cm)

Figure 3c shows that the dominant clay mineral is *R*1 I–S with a high proportion of illite-like layers which occur locally as *R*2-like and *R*3-like units. Although quantitative values of the relative proportions of illite- and smectite-like layers could not be determined, TEM images were consistent with the estimate of *c*. 80% illite-like layers as determined by XRD data. Individual packets are very different from those

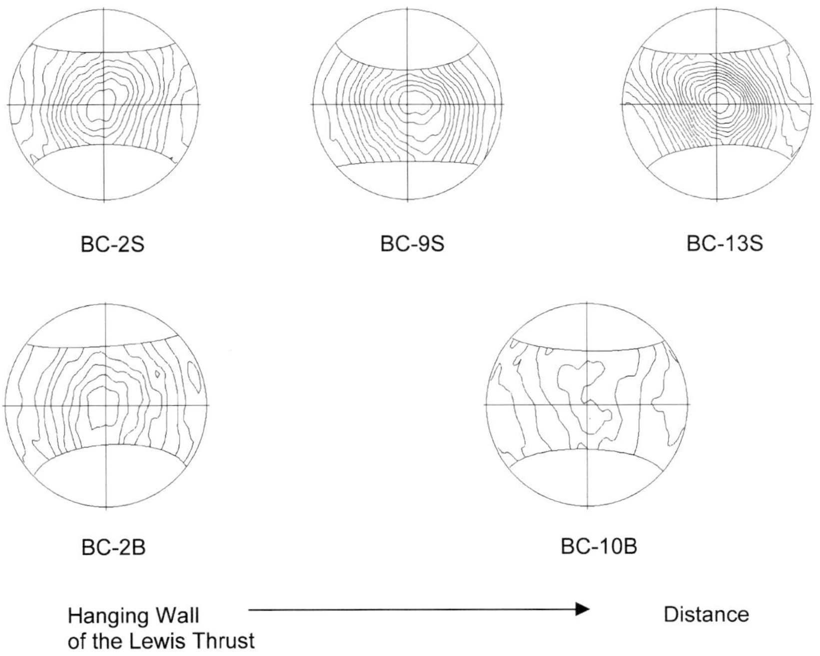

Fig. 2. Pole figures showing preferred orientation for shale (S) and bentonite (B) samples. Samples are grouped by lithology and ordered by distance from the top of the fault zone (hanging wall). Pole figures are contoured plots of basal-plane poles. The maximum of each pole figure is parallel to the dominant foliation measured in the field, and is rotated to the centre of the plot for comparison. Each contour represents 0.2 m.r.d.

dominated by smectite-like layers in other samples, as they are composed of well-defined and relatively straight lattice fringes of 1 nm periodicity. The average thickness of packets is *c*. 30.0 nm. The relative orientation of packets is variable, especially as they commonly intersect at high angles. Kinking and bending of packets are common, and pore space is abundant at the junctions of packets where they intersect at high angles. The SAED patterns are typical of $1M_d$ polytypism.

Bentonite BC10B (50% I, 605 cm)

Figure 4a shows smectite-rich *R*1 I–S lattice-fringe images with 2.1–2.2 nm periodicity. Boundaries of I–S packets are poorly defined, layers occurring as a somewhat continuous array. Lens-shaped layer separations are common, but larger units of pore space are rare. The inset SAED pattern has diffuse and weak, low-order 00*l*reflections.

Bentonite BC2B (70–80% I, 12 cm)

Figure 4b shows the bentonite microfabric nearest the contact and adjacent to shale sample BC2S (see above). The image shows typical illite-like lattice fringes with 1 nm periodicity in packets. Illite-rich I–S occurs as separate packets of layers in other images. Packet boundaries are well defined and have a thickness of *c*. 30.0 nm. Bent packets, intersecting relationships and pore space are less common than in the neighbouring shale, as reflected in the stronger fabric measured by XTG.

Table 2 lists representative chemical formulae of each sample obtained by AEM, using concentration ratios that are normalized to 22 oxygen atoms. The AEM data were collected in the same areas as shown in the TEM images above; that is, they were obtained from areas that were first characterized by lattice-fringe images and SAED patterns. Each composition is an average of at least three analyses. Both shale and bentonite sequences show a trend of increasing Si/$Al^{([6]+[4])}$ ratio and decreasing K content with increasing proportion of illite-like layers. Fe and Mg contents are slightly larger than expected for such I–S, largely as a result of the occurrence of minor chlorite within analysed areas, as observed in lattice-fringe images.

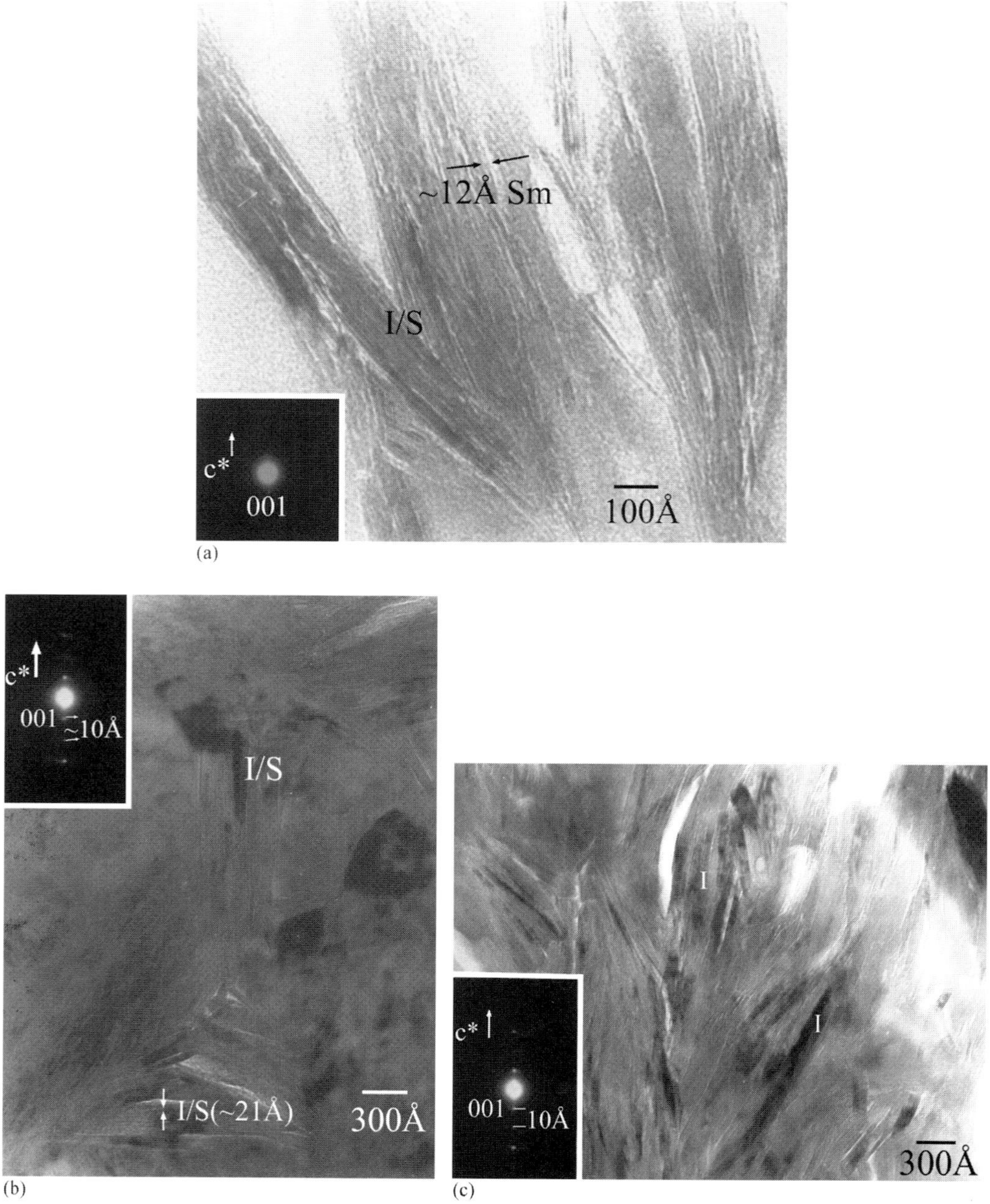

Fig. 3. Electron micrographs of shale samples. (**a**) Lattice-fringe image of sample BC13S, *c.* 12 m from the hanging wall. Smectite displays typical wavy and anastomosing fringes with 1.2–1.3 nm interlayer spacing that approximately parallels bedding. *R*1 I–S is present in addition to smectite, but smectite is the dominant clay mineral. Pore space is rare. The inset SAED pattern shows typical weak and diffuse 00*l* reflections. (**b**) Lattice-fringe images of sample BC9S, 3.7 m from the hanging wall. Relatively straight *R*1 I–S fringes of 2.1–2.2 nm periodicity occur in packets 10.0–20.0 nm thick that show bending and intersecting relationship. Pore spaces are more common than in sample BC13S, but less common than in sample BC2S (see below). (**c**) Lattice-fringe image of sample BC2S, 0.12 m from the hanging wall. Well-defined and straight illite-like packets (I) of 1.0 nm periodicity show extensive bending and intersecting relationships. Pore space is abundant.

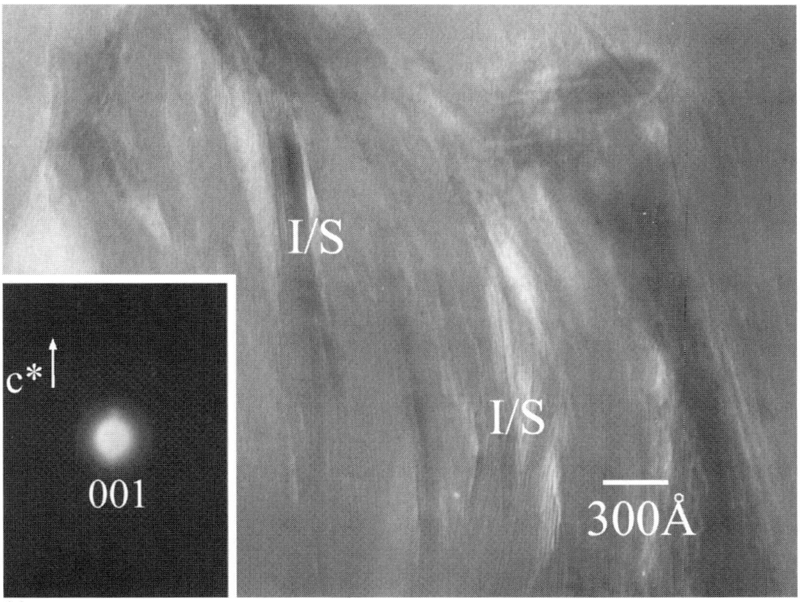

(a)

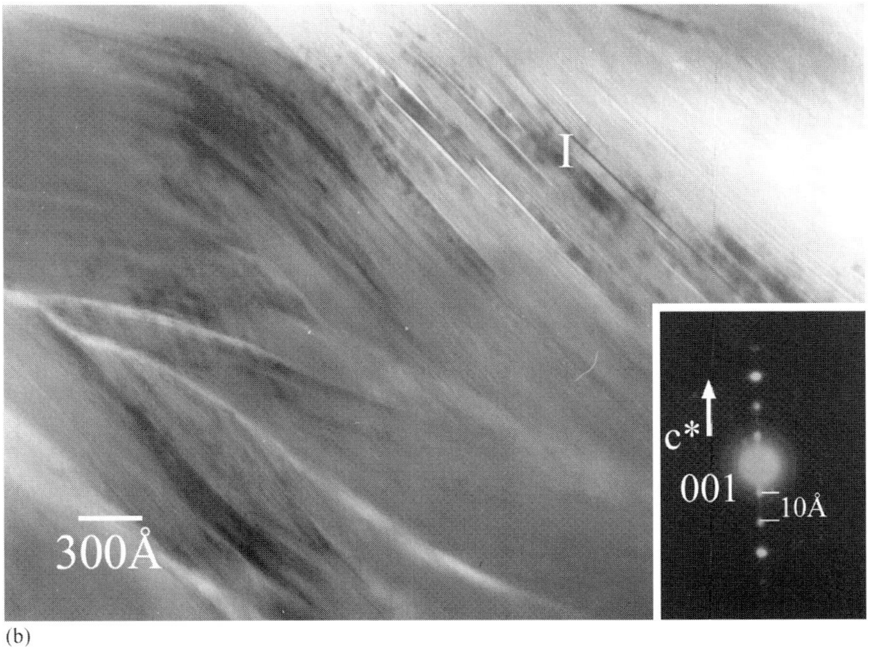

(b)

Fig. 4. Electron micrographs of bentonite samples. (**a**) Lattice-fringe image of sample BC9B, 6.5 cm from the hanging wall. *R*1 I–S has relatively straight fringes with 2.1–2.2 nm periodicity. Lens-shaped pore spaces reflect curvature of smectite layers. (**b**) Lattice-fringe image of sample BC2B, 0.12 m from the hanging wall. Well-defined and straight illite-like fringes are observed with a periodicity of 1.0 nm in discrete packets. Bending and intersecting features, and pore space are less common than in the neighbouring shale.

Discussion and conclusions

Among many models describing the smectite-to-illite transformation, most can be described as either solid-state transformation (SST) or dissolution–neocrystallization (DN), as reviewed by Altaner & Ylagan (1997). The SST model emphasizes layer-by-layer replacement of smectite by illite, with continuous changes in composition and structure. The DN model, on the other hand, considers new growth of authigenic matrix material by means of dissolution of smectite-rich clays and neocrystallization of illite-rich clays, accompanied by transport of chemical components in a fluid phase. Our observations show that the smectite-to-illite sequence is incomplete, with pure smectite coexisting with *R*1 I–S at low degrees of transformation, and illite-rich *R*1 I–S coexisting with illite at higher degrees of transformation. Such discrete and separate phases as products and reactants can only occur via dissolution and neocrystallization. The well-studied transformation in the Gulf Coast section is similar in this respect, in that Dong *et al.* (1997) demonstrated that the prograde diagenetic transition progressed from smectite to smectite-rich *R*1 I–S to illite.

Although alike in their clay mineralogical sequences, the Gulf Coast and Lewis Thrust systems differ in important respects. Clays in Gulf Coast shales transformed in a passive burial-metamorphic, compaction-strain environment over a broad range of sediment depth (*c.* 500 m) and temperature (e.g. Hower *et al.* 1976; Morton 1985; Ohr *et al.* 1991; Dong *et al.* 1997; Ho *et al.* 1999). Pore space is so sparse in samples studied by TEM that it cannot be detected. Stress-induced defects cannot be detected, and although the transition is marked by an increase in preferred orientation, almost all clay packets are at least approximately oriented parallel to bedding (Ho *et al.* 1999). It is generally agreed that rising temperature with increasing depth of burial is the driving force for the smectite-to-illite transformation.

The smectite-to-illite transformation in Lewis Thrust gouge occurs over a distance of <10 m, by contrast. The microfabric is also distinctly different from that of Gulf Coast shales. With increasing illitization, occurring toward the top of the gouge zone, the grain fabric becomes more poorly oriented. The fabric degradation recorded by X-ray goniometry is supported by direct observations of clays by electron microscopy. The decreasing fabric intensity is the result of irregularly oriented packets of illite that are otherwise well ordered internally. These packets are separated by pore space geometries that are not found in host rock, smectite-rich, samples. This implies that deformation was primarily responsible for the observed microfabric of bending and intersecting illite packets. Because illite was not originally present in these samples and shear heating was of limited importance (see Vrolijk & van der Pluijm 1999), strain must have largely reduced the kinetic barriers to promote clay transformation from smectite to illite through local dissolution and neocrystallization.

The transformation of smectite to illite releases a significant volume of fluid, which is accompanied by volume reduction of solid phases (Colten-Bradley 1987; Jiang *et al.* 1990b; Elliot & Matisoff 1996; Mello & Karner 1996). We surmise that the H_2O-rich fluid produced during this reaction occupied the volume that was generated from illite formation in these fault rocks. The presence of this free fluid would produce a positive feedback to faulting, as fluid pressure lowers the effective stress and thus promotes sliding along the fault zone.

In conclusion, clay gouge of the Lewis Thrust preserves major mineralogical change in response to faulting. Illite-rich clays are favoured at conditions of high strain energy, which can be considered analogous to illitization in response to elevated temperatures (thermal energy) in Gulf Coast shales (see also van der Pluijm *et al.* (1998)). The formation of fluid that accompanies this transformation produced dynamic pore space, as recorded by fabric degradation toward the contact of the gouge zone with competent limestone. The associated volume change created permeability–porosity, which produced fluid pathways that promoted further dissolution and neocrystallization.

Deformation-induced smectite-to-illite transformation may significantly affect the strength of clay-rich fault zones. The mechanical strength of clay gouge cannot be estimated by the strength of the individual mineral phases, as the fluid-assisted transformation process is progressive and cannot be represented by solid–solid reactions. Moreover, the production of fluid-filled pore space lowers the effective stress in the fault zone. Collectively, the positive feedback of fluid-assisted mineral transformation, volume reduction and fluid production will significantly weaken clay-rich fault zones, allowing for weak, easily sliding faults.

We are grateful to Nei-Che Ho for XTG instruction, and thank Li-Shun Kao, Weiming Zhou and B. Bauluz for help with STEM analysis. This research was supported by NSF grant EAR-9614407. The single-crystal diffractometer and the scanning transmission electron microscope were acquired under NSF grants EAR-8917350 and EAR-9628196, respectively. We thank L. Warr and Q. Fisher for stimulating reviews, although they do not necessarily agree with our interpretations.

References

AHN, J.H. & PEACOR, D.R. 1986. Transmission electron microscopic study of diagenetic chlorite in Gulf Coast argillaceous sediments. *Clays and Clay Minerals*, **33**, 228–236.

ALTANER, S.P. & YLAGAN, R.F. 1997. Comparison of structure modes of mixed-layer illite/smectite and reaction mechanisms of smectite illitization. *Clays and Clay Minerals*, **45**, 517–533.

BRUCE, C.H. 1984. Smectite dehydration—its relation to structural development and hydrocarbon accumulation in northern Gulf of Mexico basin. *AAPG Bulletin*, **68**, 673–683.

CHESTER, F.M. & LOGAN, J.M. 1986. Implications for mechanical properties of brittle faults from observations of the Punchbowl Fault, California. *Pure and Applied Geophysics*, **124**, 79–106.

COLTEN-BRADLEY, V.A. 1987. Role of pressure in smectite dehydration—effects on geopressure and smectite-to-illite transformation. *AAPG Bulletin*, **71**, 1414–1427.

DONG, H. & PEACOR, D.R. 1996. TEM observations of coherent stacking relations in smectite, I/S and illite of shales: evidence for MacEwan crystallites and dominance of $2M_1$ polytypism. *Clays and Clay Minerals*, **44**, 257–275.

DONG, H., PEACOR, D.R. & FREED, R.L. 1997. Phase relations among smectite, *R*1 illite–smectite, and illite. *American Mineralogist*, **82**, 379–391.

ELLIOT, W.C. & MATISOFF, G. 1996. Evaluation of kinetic models for the smectite to illite transformation. *Clays and Clay Minerals*, **44**, 77–87.

ESSENE, E.J. & PEACOR, D.R. 1995. Clay mineral thermometry—a critical perspective. *Clays and Clay Minerals*, **43**, 540–553.

FREED, R.L. & PEACOR, D.R. 1987. New insights on diagenesis and I/S reactions in Texas Gulf Coast sediments. *Clay Minerals*, **24**, 667–668.

FREED, R.L. & PEACOR, D.R. 1989. Geopressured shale and sealing effect of smectite to illite transition. *AAPG Bulletin*, **73**, 1223–1232.

FREED, R.L. & PEACOR, D.R. 1992. Diagenesis and the formation of authigenic illite-rich I/S crystals in Gulf Coast shales: TEM study of clay separates. *Journal of Sedimentary Petrology*, **62**, 220–234.

GUTHRIE, G.D. JR & REYNOLDS, R.C. JR. 1998. A coherent TEM- and XRD-description of mixed-layer illite/smectite. *Canadian Mineralogist*, **36**, 1421–1434.

HO, N.-C., PEACOR, D.R. & VAN DER PLUIJM, B.A. 1995. Reorientation mechanisms of phyllosilicates in the mudstone-to-slate transition at Lehigh Gap, Pennsylvania. *Journal of Structural Geology*, **17**, 345–356.

HO, N.-C., PEACOR, D.R. & VAN DER PLUIJM, B.A. 1996. Contrasting roles of detrital and authigenic phyllosilicates during slaty cleavage development. *Journal of Structural Geology*, **18**, 615–623.

HO, N.-C., PEACOR, D.R. & VAN DER PLUIJM, B.A. 1999. Preferred orientation of phyllosilicates in Gulf Coast mudstones and relation to the smectite–illite transition. *Clays and Clay Minerals*, **47**, 495–504.

HOWER, J., ESLINGER, E.V., HOWER, M.E. & PERRY, E.A. 1976. Mechanism of burial metamorphism of argillaceous sediments: 1. Mineralogical and chemical evidence. *Geological Society of America Bulletin*, **87**, 725–737.

JIANG, W.-T., PEACOR, D.R. & ESSENE, E.J. 1990*a*. Transmission electron microscopic study of coexisting pyrophyllite and muscovite: direct evidence for the metastability of illite. *Clays and Clay Minerals*, **38**, 225–240.

JIANG, W.-T., PEACOR, D.R., MERRIMAN, R.J. & ROBERTS, B. 1990*b*. Transmission and analytical electron microscopic study of mixed-layer illite–smectite formed as an apparent replacement product of diagenetic illite. *Clays and Clay Minerals*, **38**, 449–468.

JIANG, W.-T., PEACOR, D.R., ARKAI, P., TOTH, M. & KIM, J.W. 1997. TEM and XRD determination of crystallite size and lattice strain as a function of illite crystallinity in pelitic rocks. *Journal of Metamorphic Geology*, **15**, 267–281.

KIM, J.W., PEACOR, D.R., TESSIER, D. & ELLASS, F. 1995. A technique for maintaining texture and permanent expansion of smectite interlayers for TEM observations. *Clays and Clay Minerals*, **43**, 51–57.

LI, G., PEACOR, D.R. & COOMBS, D.S. 1997. Transformation of smectite to illite in bentonite and associated sediments from Kaka Point, New Zealand: contrast in rate and mechanism. *Clays and Clay Minerals*, **45**, 54–67.

MASUDA, H., O'NEIL, J.R., JIANG, W.-T. & PEACOR, D.R. 1996. Relation between interlayer composition of authigenic smectite, mineral assemblages, I/S reaction rate and fluid composition in silicic ash of the Nankai Trough. *Clays and Clay Minerals*, **44**, 443–459.

MELLO, U.T. & KARNER, G.D. 1996. Development of sediment overpressures and its effect on thermal maturation: application to the Gulf of Mexico Basin. *AAPG Bulletin*, **80**, 1367–1396.

MOORE, D.M. & REYNOLDS, R.C. 1997. *X-Ray Diffraction and the Identification and Analysis of Clay Minerals* (2nd edition). Oxford University Press, New York.

MORSE, J.W. & CASEY, W.H. 1988. Ostwald processes and mineral paragenesis in sediments. *American Journal of Science*, **288**, 537–560.

MORTON, J.P. 1985. Rb–Sr evidence for punctuated illite/smectite diagenesis in the Oligocene Frio Formation, Texas Gulf Coast. *Geological Society of America Bulletin*, **96**, 67–76.

OERTEL, G. 1985. The relationship of strain and preferred orientation of phyllosilicate grains in rocks—a review. *Tectonophysics*, **100**, 413–447.

OHR, M., HALLIDAY, A.N. & PEACOR, D.R. 1991. Sr and Nd isotopic evidence for punctuated clay diagenesis, Texas Gulf Coast. *Earth and Planetary Science Letters*, **105**, 110–126.

PEACOR, D.R. 1992. Diagenesis and low-grade metamorphism of shale and slates. *In*: BUSECK, P.R (ed.) *Minerals and Reactions in Atomic Scale: Transmission Electron Microscopy*. Mineralogical Society of America, Reviews in Mineralogy **27**, 335–380.

PERRY, E. & HOWER, J. 1970. Burial diagenesis in Gulf Coast pelitic sediments. *Clays and Clay Minerals*, **18**, 165–177.

RUTTER, E.H., MADDOCK, R.H., HALL, S.H. & WHITE, S.H. 1986. Comparative microstructure of natural and experimentally produced clay-bearing fault gouges. *Pageoph*, **124**, 3–30.

SIBSON, R.H. 1977. Fault rocks and fault mechanisms. *Journal of the Geological Society, London*, **133**, 190–213.

SINTUBIN, M. 1994. Clay fabric in relation to the burial history of shales. *Sedimentology*, **41**, 1161–1169.

VAN DER PLUIJM, B.A., HO, N.-C., MERRIMAN, R.J. & PEACOR, D.R. 1998. Contradictions of slate formation resolved? *Nature*, **392**, 348.

VAN DER PLUIJM, B.A., HO, N.-C. & PEACOR, D.R. 1994. High-resolution X-ray texture goniometry. *Journal of Structural Geology*, **16**, 1029–1032.

VEBLEN, D.R., GUTHRIE, G.D. JR, LIVI, K.J.T. & REYNOLDS, R.C. JR. 1990. High-resolution transmission electron microscopy and electron diffraction of mixed-layer illite/smectite: experimental results. *Clays and Clay Minerals*, **38**, 1–13.

VELDE, B., SUZUKI, T. & NICOR, E. 1986. Pressure–temperature–composition of illite/smectite mixed-layer minerals: Niger Delta mudstones and other examples. *Clays and Clay Minerals*, **34**, 435–441.

VROLIJK, P. & VAN DER PLUIJM, B.A. 1999. Clay gouge. *Journal of Structural Geology*, **21**, 1039–1048.

YAN, Y., TILLICK, D.A., PEACOR, D.R. & SIMMONS, S.F. 2001. Genesis of dioctahedral phyllosilicates during hydrothermal alteration of volcanic rocks: II. The Broadlands–Ohaaki hydrothermal system, New Zealand. *Clays and Clay Minerals*, **49**, 141–155.

Microfracturing associated with reactivated fault zones and shear zones: what can it tell us about deformation history?

GAUTAM MITRA & ZESHAN ISMAT

Department of Earth and Environmental Sciences, University of Rochester, Rochester, NY 14627, USA (e-mail: mitr@troi.cc.rochester.edu)

Abstract: Deformation in fault zones is commonly characterized by grain-scale microfracturing, with microcrack density typically increasing toward the middle of the zone. The cracks can form under a wide variety of conditions and need to be used with great caution in making tectonic interpretations, particularly in areas with a complex history of fault reactivation. Microcracks may be intragranular (contained within single grains) or intergranular (with a length of several grain diameters). Intragranular cracks formed under dominantly plastic deformation conditions are crystallographically controlled and may not be directly related to regional stresses. Intragranular cracks formed during initial fracturing under cataclastic conditions develop only in grains that are optimally oriented to the deforming stresses. Intergranular cracks form during progressive cataclasis as intragranular cracks grow to join one another: they may develop as transgranular cracks that cut across several grains or as grain-boundary cracks. Once formed, microcracks may be preserved in a variety of ways (e.g. sintering, healing, cementation) depending on post-deformation conditions, and may be distinguished from one another on the basis of microstructural characteristics. Distinguishing between successive generations of microcracks in areas of fault reactivation is particularly important in determining the deformation history and obtaining deformation conditions. For example, Proterozoic quartzites collected from the central Utah Sevier belt have undergone multiple episodes of contractional deformation followed by Basin-and-Range extension. The use of polarized and dark-field optical microscopy and scanning electron microscopy allows microcracks related to the separate episodes of deformation to be distinguished on the basis of morphology, mode of preservation and consistent cross-cutting relationships. Variations in microcrack density and volume of cataclasized rock for the different generations of microcracks are used to establish the patterns of overprinting during fault reactivation. Anastomosing zones of intense deformation formed during successive episodes of faulting may not coincide with one another, as grain-size reduction and cementing during each episode hardens the zones, causing deformation to shift to adjoining weaker rock. However, the fault zone as a whole is a sufficiently large inhomogeneity that it is reactivated during successive faulting events.

Large-scale brittle and brittle–ductile faults at shallow to intermediate crustal levels and ductile shear zones at deeper levels are all characterized by localized intense deformation within a relatively thin zone. The general term ‘fault zone’ is typically used to describe such a zone (Sibson 1977). A fault zone formed early in the history of an orogenic belt may be reactivated during later parts of its history. For example, basin-margin normal faults, originally formed in a passive margin setting where miogeoclinal sediments are deposited, may subsequently be reactivated as thrust faults during an orogeny that results in the development of a fold–thrust belt (FTB) by inversion tectonics (e.g. Boyer & Mitra 1988; Cooper & Williams 1989; and references therein). Similarly, extension during the late stages of deformation in an FTB is often accommodated along listric normal faults that are localized at thrust ramps (Royse *et al.* 1975). Thus, we should expect many fault zones to have complex structural histories, and this should be reflected in their internal structure at all scales.

In this paper we address two important questions regarding the distribution of deformation within a fault zone based on detailed studies of fault zones with prolonged and complex tectonic histories. First, is there a uniform decrease in deformation intensity away from a fault? Many fault zone models based on detailed field studies of fault zones (e.g. Brock & Engelder 1977;

From: HOLDSWORTH, R.E., STRACHAN, R.A., MAGLOUGHLIN, J.F. & KNIPE, R.J. (eds) 2001. *The Nature and Tectonic Significance of Fault Zone Weakening*. Geological Society, London, Special Publications, **186**, 113–140. 0305-8719/01/$15.00

Aydin & Johnson 1978; Mitra 1984, 1992, 1993; Chester & Logan 1986; Evans 1993; Newman & Mitra 1993; Yonkee & Mitra 1993; Knipe & Lloyd 1994) suggest that the zones typically have a pattern, at many different scales, of relatively undeformed pods of rock bounded by anastomosing zones of strongly deformed rocks. We should, therefore, anticipate a non-uniform distribution of deformation intensity, although there may still be an overall progressive decrease in deformation away from a fault. Second, when an old fault is reactivated, does the new fault zone always follow the old zone or can the new zone develop in rocks adjoining the old zone? Also, what effect does such reactivation have on the distribution of deformation within the composite zone?

We focus our attention on fault zones associated with major thrust faults in the internal portions of FTBs. Such faults typically originate at intermediate to deep crustal levels (Boyer & Elliott 1982; Yonkee 1992) and are brought progressively closer to the surface during synorogenic erosion (Schonborn 1992; DeCelles & Mitra 1995; Mitra 1997). They are sometimes reactivated during late-stage normal faulting in FTBs (Royse *et al.* 1975; Yonkee 1992). We start with a typical fault zone model, based on many published studies, describing the distribution of mesoscopic and microscopic deformation features. We then suggest models for reactivated fault zones that address a number of possible scenarios that might arise during reactivation. We then look in detail at one reactivated fault zone in the Sevier FTB of the western USA, and interpret the observed distribution of deformation on the basis of our models. Finally, we discuss the implications of our results for interpretation of other fault zones.

Fault zone models

Much of the deformation in an FTB takes place at shallow crustal levels, above the elasticofrictional to quasiplastic (EF–QP) transition (Sibson 1977) for crustal rocks. Thus FTB fault zones typically develop within the EF regime or the transitional zone between the EF and QP regimes (Sibson 1977). Under such conditions fault rocks such as coherent cataclasites and breccias develop along the fault: the terminology used to describe these rocks follows the classification by Sibson (1977), or slightly modified versions thereof (Marshak & Mitra 1988; Scholz 1990). The faults commonly initiate by forming a process zone of distributed cracking with the cracks eventually linking up to form a throughgoing fault (Blenkinsop & Rutter 1986; Twiss & Moores 1992; Knipe & Lloyd 1994). During continued movement on the fault, the fault zone generally grows in thickness (Brock & Engelder 1977; Hull 1988; Wojtal & Mitra 1988; Means 1995) and is commonly characterized by intense fracturing and comminution together with rolling and sliding of fragments. This eventually results in the formation of a thick deformation zone of brecciated rocks, which contains within it thinner zones of intense deformation (Mitra 1984; Wojtal & Mitra 1986; Newman & Mitra 1994) characterized by cataclasites (Sibson 1977; Snoke *et al.* 1998).

The fracturing takes place at all scales (Lloyd & Knipe 1992; Evans 1993; Mitra 1993; Anders & Wiltschko 1994), and quantitative measurements of fracture density vary with the scale at which observations are made (Mitra 1993). At any scale, the fracture density generally decreases away from the fault in both the hanging wall and the footwall (e.g. Brock & Engelder 1974; Mitra 1993; Anders & Wiltschko 1994) although the structural patterns on the two sides of the fault may not be symmetric. In detail, the variation of fracture density can be irregular, with the density increasing in zones of intense deformation that surround relatively undeformed pods of rock (Brock & Engelder 1977; Mitra 1984; Chester & Logan 1986; Wojtal & Mitra 1988; Yonkee & Mitra 1993). It is, therefore, worthwhile to study the architecture of fault zones at both the mesoscopic and the microscopic scales. In addition to the geometry, the deformation regimes reflected in the fault rocks in the hanging wall and footwall of the fault may be the same or different (Schmid & Handy 1991) and this determines how the rocks can be used to interpret the deformation conditions.

Mesoscopic structure

At the outcrop scale, a large fault with significant displacement typically has a zone of intense deformation in both its hanging wall and its footwall, with the deformation intensity progressively decreasing away from the fault (Brock & Engelder 1974; Wojtal 1986). We refer to the entire zone of more intense deformation as the fault-related deformation zone (FRDZ) (Newman & Mitra 1993, 1994). The variation in deformation intensity within the FRDZ may be asymmetrically distributed on either side of the fault and the asymmetry may vary along the length of the fault. The total thickness of the FRDZ may vary by as much as two orders of magnitude along the length of a fault, and such variation can take place both in the transport direction and perpendicular to it (Newman & Mitra 1993, fig. 9). The zone of deformation shows evidence for progressive increase in thickness with increasing displacement along the fault (Robertson 1983; Mitra 1984, 1992; Wojtal & Mitra 1988; Scholz 1990); at the same time some older parts of the zone continue to accumulate deformation, giving rise to the most intensely deformed rocks there. Thus, on the basis of regional patterns, the FRDZs of most large faults generally show an internal zonation, with the most strongly deformed rocks (generally cataclasites) closest to the fault, and less deformed rocks (e.g. breccias and fractured rock) farther away from the fault (Brock & Engelder 1974; Wojtal & Mitra 1986) (Fig. 1). Although such a simplified model provides a useful starting point for interpreting reactivation patterns along faults, in detail both the internal structure of an FRDZ and its movement history are spatially and temporally complex (e.g. Wojtal & Mitra 1988; Schmid & Handy 1991; Chester & Chester 1998), so that locally the FRDZ may show significant variations from the simple zonal structure. The boundaries between the internal zones of an FRDZ are typically irregular, showing embayments and protrusions, and are sometimes offset by late minor faults (Newman & Mitra 1993) that may juxtapose intensely deformed rocks directly against 'undeformed' wallrock.

The zone of cataclasites and ultracataclasites closest to the fault, which we refer to as the 'fault zone' (Mitra 1984; Wojtal & Mitra 1986, 1988; Evans 1993; Yonkee 1992; Newman & Mitra 1993, 1994; Yonkee & Mitra 1993), is generally relatively thin (1–10 m, e.g. Chester & Logan 1986) but may be as much as 100 m thick on some large faults (Mitra 1984). Most of the fault-parallel shear associated with faulting is concentrated in this zone and is reflected in a strongly concave-up asymptotic deformation profile associated with thrust sheets (Elliott 1977; Wojtal 1982, 1986; Mitra 1994). In addition, there may be significant fault-normal shortening by pressure solution in fine-grained ultracataclasites (e.g. Wojtal & Mitra 1986) or fault-normal extension by vein growth (e.g. Zadins 1989). The portion of the fault zone closest to the fault is characterized by ultracataclasites composed of very fine grained, comminuted material with a penetrative foliation (Mitra 1984; Chester & Logan 1986; Wojtal & Mitra 1988) (Fig. 1); this subzone has also been referred to as the 'gouge zone' (Brock & Engelder 1977) or the 'core zone' (e.g. Hadizadeh 1994; Caine *et al.* 1996; Chester & Chester 1998; Heynekamp *et al.* 1999). Farther away from the fault, the fault zone is typically made up of fine-grained cataclasites that show no penetrative foliation; the rocks may, however, show a crude outcrop-scale foliation defined by elongated pods of less deformed material and subtle colour bands as a result of compositional variations (Mitra 1984; Wojtal & Mitra 1988) (Fig. 1). These deformation characteristics have also been described as belonging to the 'mixed zone', by Heynekamp *et al.* (1999). Both the ultracataclasite and cataclasite subzones show pinch-and-swell structure along the length of the fault and may, therefore, not be present everywhere along a fault. In detail, the structure within the subzones is characterized by thin anastomosing zones of strongly deformed rocks surrounding lenses of less deformed material (Fig. 1); this pattern is observable at all scales down to the microscopic scale, although the degree of development may vary at different scales (e.g. Knipe & Lloyd 1994).

Outside the fault zone, deformation associated with faulting may extend for hundreds of metres away from the fault (Wojtal 1986) in the form of fine- to coarse-grained brecciation (Wojtal & Mitra 1988) and deformation by populations of mesoscopic faults and fractures (Wojtal 1986). Although the rocks in this zone may be highly broken up, the total fault-parallel displacement associated with this zone is relatively small, as indicated by displacement profiles (Wojtal 1982, 1986; Mitra 1994). This deformed region has been referred to as a 'transition zone' to undeformed wallrock (Yonkee 1992), or as a 'damage zone' (Caine *et al.* 1996). Once again, the deformation within this zone is also inhomogeneous, with pockets of less deformed rock surrounded by diffuse areas of more deformed rocks (Fig. 1). The contact of the damage zone with the adjoining 'undeformed wallrock' is typically an irregular boundary, and must be carefully defined for each fault on the basis of mesoscopic and microscopic criteria.

Microscopic structure

At the microscopic scale the deformation in a fault zone typically takes place by grain-scale microcracking, with the density of microcracks showing an overall decreasing trend away from the fault zone into the wallrocks, in both the hanging wall and the footwall (e.g. Engelder 1974; Mitra 1984, 1993; Chester & Logan 1986; Anders & Wiltschko 1994; Hadizadeh 1994; Newman & Mitra 1994). Microcracks may develop under a wide variety of conditions,

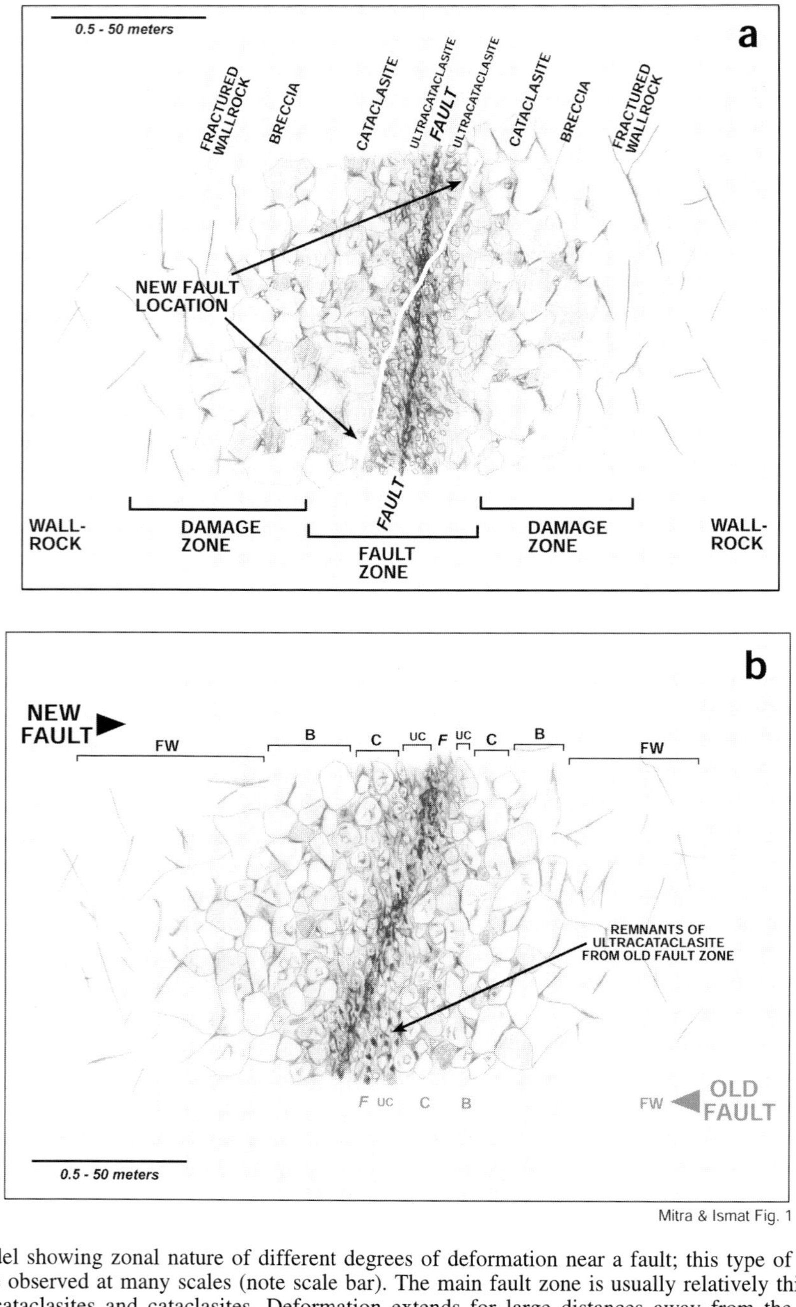

Fig. 1. Model showing zonal nature of different degrees of deformation near a fault; this type of zonal structure may be observed at many scales (note scale bar). The main fault zone is usually relatively thin and made up of ultracataclasites and cataclasites. Deformation extends for large distances away from the fault in the damage zone, made up of breccias and fractured wallrock. Pods of less deformed material are found in all the zones. The asymmetry of deformation extending on either side of the fault proper should be noted. (**a**) New fault location influenced by an existing fault zone and its associated deformation. (**b**) Reactivation of an old fault zone along a new, subparallel fault zone (shown in (**a**)) where the centres of the two zones do not coincide, resulting in complex overprinting patterns. The sense of shear on the new fault is hanging wall up; remnants of the older fault zone in the footwall of the new fault have remained in place whereas remnants in the new hanging wall have moved up and out of the frame of the figure. FW, fractured wallrock; B, breccia; C, cataclasite; UC, ultracataclasite; F, fault.

ranging from EF to QP (Mitra 1978; Knipe 1989; Lloyd & Knipe 1992), and hence measurement of microcrack morphology and density offers a readily available way of comparing the deformation patterns of different fault zones developed under a wide variety of conditions. Increase in microfracture density has therefore been used as a measure of the intensity of deformation along a fault zone (e.g. Chester & Logan 1986) and as a tool for identifying fault zones that are 'blind' (i.e. not exposed at the surface) (e.g. Anders & Christie-Blick 1994). However, because of the inhomogeneous nature of fault zones, microcrack distribution can be rather inhomogeneous, and hence, measures of microcrack density need to be used with great caution in making interpretations about fault zone deformation.

For fault zones with the simplest geological history, involving one episode of deformation during which the fault zone grew in thickness with progressive displacement along the fault, the density of microcracks decreases with distance away from the fault (Engelder 1974) and may more specifically decrease in a logarithmic manner (Anders & Wiltschko 1994). Microcrack density is generally high within the main fault zone (zone of cataclasites), although it may be very low in fine-grained ultracataclasites where grain size is sufficiently reduced so that cracks do not nucleate easily (Chester & Logan 1986). The microcrack density decreases dramatically in the damage zone, whose spatial extent can be defined as the region near the fault where microcrack densities are higher than background levels present in the wallrock. However, in areas with a complex deformation history, where a fault has been reactivated, the patterns of microcrack density variation can be expected to be much more complex.

Reactivated zones

During any single episode of fault movement in the EF regime, grain-scale fracturing results in reduction of grain size within the fault zone (e.g. Mitra 1984, 1994; Chester & Logan 1986; Knipe & Lloyd 1994; Schulz & Evans 1998), and also increases dilatancy. Fluids drawn into the zone by dilatancy pumping (Mitra 1978; Sibson 1990) may precipitate new material to cement the zone before the next episode of fracturing occurs (Mitra 1993); the fluids may also aid in fracturing by enhancing pore-fluid pressure or promoting subcritical crack growth (Atkinson 1982). As cycles of fracturing and cementation are repeated (e.g. Knipe & Lloyd 1994) the grain size of the fault zone rocks is progressively reduced and their overall fracture strength increases (Lawn 1993). This mechanism of strain hardening causes new fracturing to start in adjoining coarser-grained wallrock so that the fault zone grows in thickness with progressive displacement on the fault (Newman & Mitra 1994); such zones are referred to as Type 1 zones (Means 1984, 1995; Hull 1988; Mitra 1992).

Some fault zones may soften with progressive deformation if the grain size is reduced sufficiently for the dominant deformation mechanism to switch from cataclasis to grain-size sensitive mechanism such as pressure solution creep (Mitra 1984; Schmid & Handy 1991). In polymineralic rocks, softening may occur by chemical alteration to weaker phases (e.g. Kerrich *et al.* 1980; Gilotti 1989; Yonkee & Mitra 1993), or by geometric softening by reorienting grains so that their slip systems are aligned parallel to each other (e.g. Schmid & Casey 1986). Softening leads to localization of the deformation in a thin zone within the larger fault zone, and the fault zone ceases to grow in thickness even though displacement along the fault may continue. Such zones are referred to as Type 2 zones (Means 1984, 1995; Hull 1988; Mitra 1992).

Reactivation of a fault zone during a new orogeny suggests that the new deformation utilizes the existing fault zone architecture. As Type 2 zones are weaker than the surrounding wallrock it is likely that a new generation fault may follow the older weak zone. Type 1 zones, on the other hand, are strain-hardened and have a higher fracture strength than the wallrock; they are, therefore, unlikely candidates for reactivation within the EF regime. New faulting may, however, localize at the competency contrast between the strain-hardened fault zone and the relatively weaker wallrock. In this case a new fault zone may develop that is centred on the boundary of an old zone.

The two generations of fault zones may be parallel to one another, or a new fault may cut obliquely across an older fault at low angles. In either case, because both zones have irregular boundaries there will be complex overprinting of fault zone and damage zone structures (Fig. 1) from two generations of faulting and offset of the zonal architecture of the older zone. Traverses across such a composite zone might encounter panels of strongly deformed rocks that have undergone two episodes of cataclasis juxtaposed against panels of little deformed rocks that have undergone only one event of brecciation. Repeated reactivation along the same fault may give rise to a thick zone of cataclastic rock, which is easily weathered and within which it would be difficult to find well-preserved evidence for distinguishing successive generations of structures at the outcrop scale. Microscopic evidence for multiple deformation events may, however, be well preserved and is worth studying.

At the microscopic scale, grain-size reduction is a common feature of fault rocks and has been used by many workers to quantify the variation in deformation intensity across fault zones (e.g. Mitra 1984, 1994; Chester & Logan 1986; Knipe & Lloyd 1994; Schulz & Evans 1998). For reactivated zones the grain-size distribution would be a resultant of the multiple deformations, but it would generally not be possible to separate the effects of successive grain-size reduction events in the EF field. Thus, although grain-size distribution is a good measure of the total deformation in a fault zone, it is not particularly useful for understanding reactivation events, and we have not used it in the present study.

Microcrack density variation has also been widely used to study deformation patterns in fault zones (e.g.Chester & Logan 1986; Anders & Wiltschko 1994), and may be useful for distinguishing successive deformation events in the EF regime. Composite zones may show overall microcrack distribution patterns similar to those of simple (i.e. unreactivated) fault zones. There would be an overall decrease in grain-scale microfracturing from the 'fault zone' through the 'damage zone' into the wallrock, with the usual variations caused by inhomogeneities in the deformation. However, the observed microcrack populations in the FRDZ would actually have been produced by more than one episode of faulting. Separation of the microcrack populations into different generations (associated with different deformation episodes) on the basis of microstructural morphological differences (i.e. styles of cracking and modes of preservation) and consistent cross-cutting relationships should allow us to decipher the fault zone architectural complications caused by superimposed deformation. To illustrate the usefulness of this approach we have chosen an area where it is possible to distinguish between successive generations of microcracks on the basis of differences in fracture characteristics and mode of preservation (Sussman 1995); these are summarized later in the paper. Specifically, we focus on a fault zone that shows map-scale evidence of reactivation and use successive generations of microcracks in deciphering fault zone reactivation and evolution.

Our study is based on a normal fault zone formed during Basin-and-Range (Tertiary age) extension in the Sevier belt. The normal faulting is localized along a major thrust ramp formed during earlier Sevier (Cretaceous) thrusting at the miogeocline–shelf transition and also involves a smaller imbricate fault formed during late Sevier thrusting. The fault zone is exposed along the western margin of the Canyon Mountains in west–central Utah, and involves Proterozoic–Eocambrian quartzites that preserve evidence for successive episodes of deformation (Sussman 1995; Mitra & Sussman 1997).

Regional geology

Structural geology and evolution of the Canyon Mountains

The Sevier FTB defines the eastern margin of thin-skinned deformation in the Cordilleran orogen of western North America (Fig. 2) (Armstrong 1968; Burchfiel & Davis 1975; Allmendinger 1992; Miller *et al.* 1992). The Central Utah segment of the belt (Fig. 2) is the type area of the Cretaceous Sevier orogen (Armstrong 1968). The major thrusts in this segment, from west to east, are the Canyon Range (CR), Pavant, (PVT) Paxton (PXT) and Gunnison (GUN) thrusts (DeCelles *et al.* 1995; Lawton *et al.* 1997; Mitra 1997). The Sevier age folds and faults are broken up by Tertiary Basin-and-Range normal faulting, with both types of structures best exposed in ranges such as the Canyon Mountains (Fig. 2). The rocks involved in the deformation are Proterozoic, Palaeozoic and Mesozoic miogeoclinal rocks (Fig. 2).

Restoration of a regional balanced cross-section (Coogan *et al.* 1995) suggests that initial deformation of the Proterozoic quartzites of the CR thrust sheet occurred under deep conditions at the time of its initial emplacement (Fig. 3; Mitra 1997). The sheet was then folded as it moved over a ramp in the Pavant thrust (PVT), forming a ramp anticline with a syncline in front of it (Fig. 3). The Canyon Range thrust (CRT) was continually reactivated by motion on younger thrusts (Paxton, Gunnison) below and in front of it (Fig. 3): slip was transferred to the CRT by means of an antiformal stack connecting splay duplex (Mitra & Sussman 1997) between the lower PVT and the upper CRT (Fig. 4). The duplex constitutes the core of the CR anticline, and its growth was directly responsible for amplification of the anticline and tightening of the adjoining syncline from an open fold to a tight, overturned fold (Mitra & Sussman 1997). Deformation of the complete

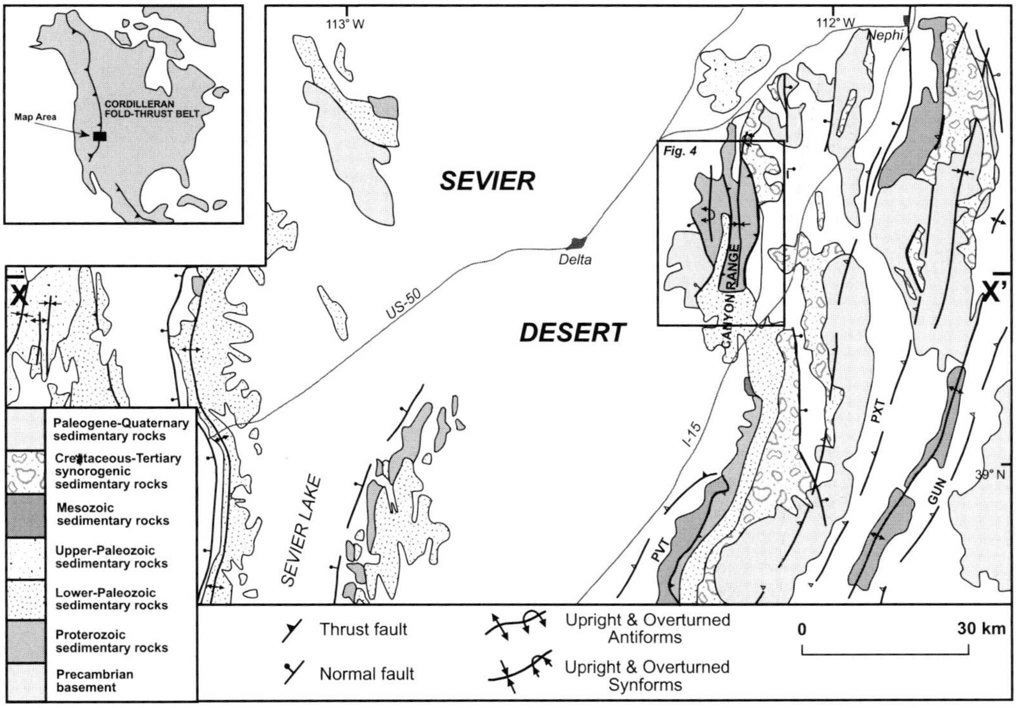

Fig. 2. Generalized geological map of central Utah showing the major Sevier-age structures exposed in the ranges of the Basin-and-Range province (after Hintze 1988). Line of regional cross-section (XX′) used in Fig. 3 is shown. Enlarged map view of boxed area is shown in Fig. 4.

section of the Proterozoic quartzites during the late stages of fold tightening occurred under a shallow overburden (<5 km), within the EF regime for quartzites, and was mainly accommodated by the growth of fracture networks (Ismat & Mitra 2001). CR footwall rocks (belonging to the underlying Pavant sheet) are exposed both along the CR thrust trace on the eastern flank of the range and in an erosional half-window through the anticline on the western flank of the range (Fig. 4).

The western half of the window is truncated by Tertiary normal faults associated with Basin-and-Range deformation (Figs 4–6; Otton 1995). Quartzites in the truncated west limb of the antiform have thus undergone at least three major phases of faulting-associated deformation: CR thrusting during the earliest part of the Cretaceous Sevier orogeny, connecting splay thrust imbrication during late stages of the Sevier orogeny, and Tertiary Basin-and-Range normal faulting. The quartzites of the Canyon Mountains contain mesoscopic (outcrop scale) fracture networks associated with all three of these major deformation events. They also preserve distinctive microstructural evidence for all three deformation episodes (Table 1), together with evidence of inherited cracks from earlier (Precambrian) orogenic events in the source terrane from which the quartz grains were derived (Sussman 1995).

Canyon Mountains normal faults

Large-scale structure

Along the western margin of the Canyon Mountains, the Sevier age structures are truncated by a system of north–south-trending normal faults (Otton 1995). The normal faulting is segmented and the geometry varies along the length of the range (Fig. 4). The middle part of the Canyon Mountains (from Fool Creek to Oak Creek) is characterized by en echelon segments of steeply dipping normal faults along its western margin (Figs 4 and 5a). These faults may have reactivated ramps in the Canyon Range and Pavant thrusts that were stacked atop one another during the growth of the Canyon Range connecting splay duplex (Fig. 7). These faults have modified earlier Sevier thrust structures, placing older (Canyon Range hanging-wall) rocks on top of younger (Canyon Range footwall) rocks (Fig. 6). The faults have downdropped the

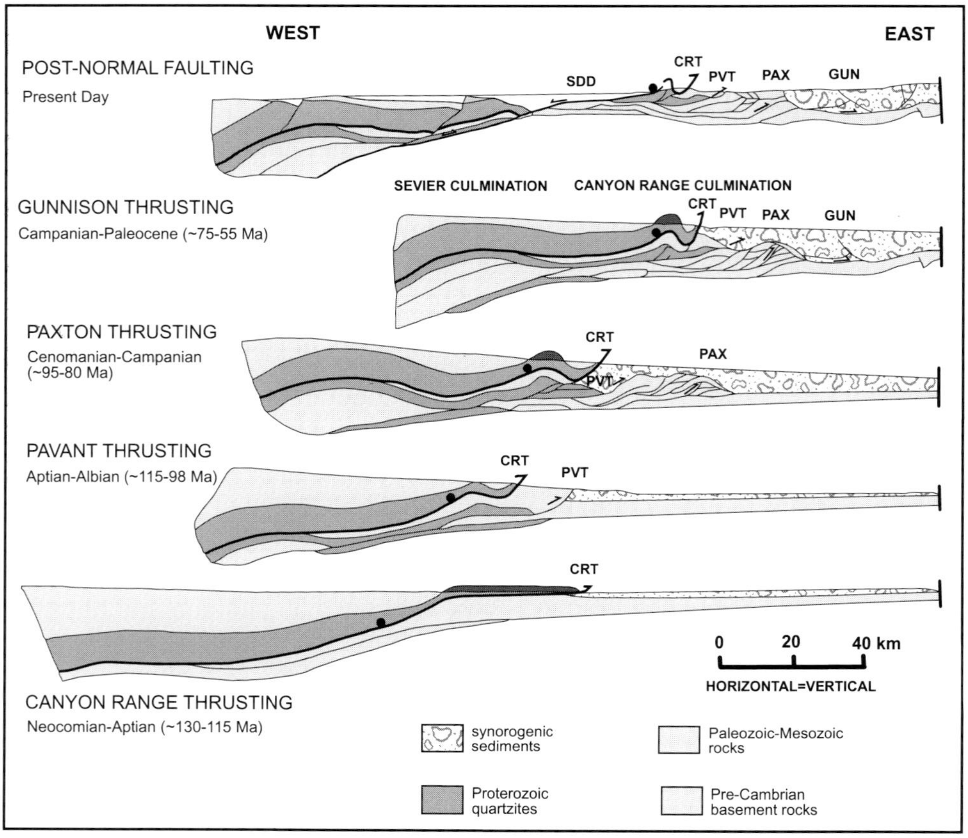

Fig. 3. Incrementally restored regional balanced cross-section (along line XX′ in Fig. 2) of the Central Utah segment of the Sevier FTB showing the approximate shapes of the orogenic wedge at the time of emplacement of successive major thrusts (based on Coogan *et al.* 1995). Black dots indicate location of the rocks studied through time. Thrusts shown are Canyon Range (CRT), Pavant (PVT), Paxton (PAX) and Gunnison (GUN); also shown is the Sevier Desert Detachment (SDD), a low-angle normal fault.

Canyon Range thrust so that it is not exposed anywhere along the western margin of the range; gently dipping CR hanging-wall rocks exposed there are presumably underlain by the thrust at depth. Curvature ('drag') of beds near the steeply dipping faults, in both the hanging wall and the footwall, indicate normal fault motion (Fig. 5b). In places, these steep normal faults are well exposed and can be studied in detail; the variation of mesoscopic and microscopic structures across one such fault zone is described here.

Table 1. *Microcracks preserved in fault-related deformation zone of Canyon Mountains normal fault*

Deformation event	Age	Morphological characteristics
Canyon Range thrusting	Neocomian (120–130 Ma)	Intragranular cracks coeval with plastic deformation; sintered cataclasite zones and associated cracks
Fold tightening	Turonian–Maastrichtian (65–90 Ma)	Healed transgranular cracks (with bubble trails) and cemented cataclasites
Basin-and-Range faulting	Miocene (10–20 Ma)	Transgranular cracks and coherent cataclasite zones (with iron oxide)

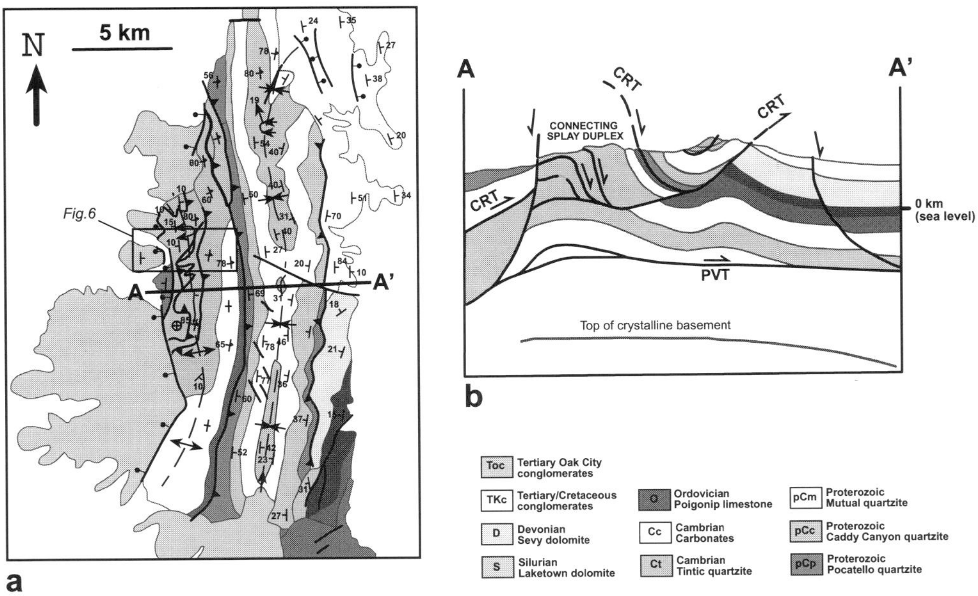

Fig. 4. (**a**) Geological map of the Canyon Mountains showing the major structural elements and representative bedding dips (from boxed area in Fig. 2). The Canyon Range thrust (CRT) is folded into an anticline–syncline pair. The CRT carries Proterozoic–Cambrian rocks in its hanging wall and overrides its own synorogenic deposits on the east limb of the syncline. The thrust is east dipping to overturned (west dipping) on the western limb of the syncline. Cambrian footwall rocks are exposed in the core of the adjoining anticline, truncated by Tertiary normal faults on the west side of the range. Enlarged map view of boxed area is shown in Fig. 6. (**b**) Cross-section (AA') showing the folded CRT and its down-dropped portion along the normal fault on the west side of the range. The subthrust connecting splay duplex, and the underlying Pavant thrust are also shown.

Mesoscopic structure

The best exposed fault zone lies some 4.8 km due east of Oak City, Utah, where the fault dips 60° toward the west. The hanging wall is made up of gently dipping (10–25°W) beds of medium-grained quartzites and metasiltstones of the Proterozoic Pocatello Formation (Figs. 5 and 6), which are part of the Canyon Range thrust sheet that has been downdropped along the normal fault. The beds show outcrop-scale eastward-verging asymmetric folds and small wedge faults related to emplacement of the Canyon Range thrust sheet (Fig. 5c and d). The beds show steeper dips close to the fault as a result of fault 'drag', and this suggests a normal sense of motion on the fault (Figs 5b and 6). The exposed hanging-wall stratigraphy (Proterozoic Upper Pocatello quartzites) suggests that the Canyon Range thrust is at a depth of *c.* 500 m. The FRDZ associated with this major thrust is possibly fairly thick and normal fault 'drag' might have caused these rocks to be exposed at the surface in the study area (Fig. 6).

The footwall is made up of more steeply dipping (25–40°W) and folded beds of coarse-grained Cambrian Tintic quartzites (Fig. 6) that belong to the Pavant thrust sheet. Beds close to the normal fault are part of the west limb of the antiformal stack connecting splay duplex between the Canyon Range and Pavant thrusts (Figs. 5b and 6); along the length of the normal fault, the Mahogany Hollow–Cascade Canyon and Devils Den slices (Sussman 1995) are in contact with the fault in its footwall (Fig. 6). At the location studied, the Devils Den thrust (of the connecting splay duplex) curves into the normal fault (Figs. 5a and 6). As for the hanging wall, bedding in footwall rocks also show steeper dips close to the fault, suggesting a normal sense of motion on the fault (Figs. 5b and 6).

Both the hanging-wall and footwall rocks show populations of older mesoscopic fractures (Fig. 8) that are the cumulative effect of deformation (Ismat & Mitra 2001) during Canyon Range and Pavant thrusting, and folding of the Canyon Range sheet as a result of the growth of

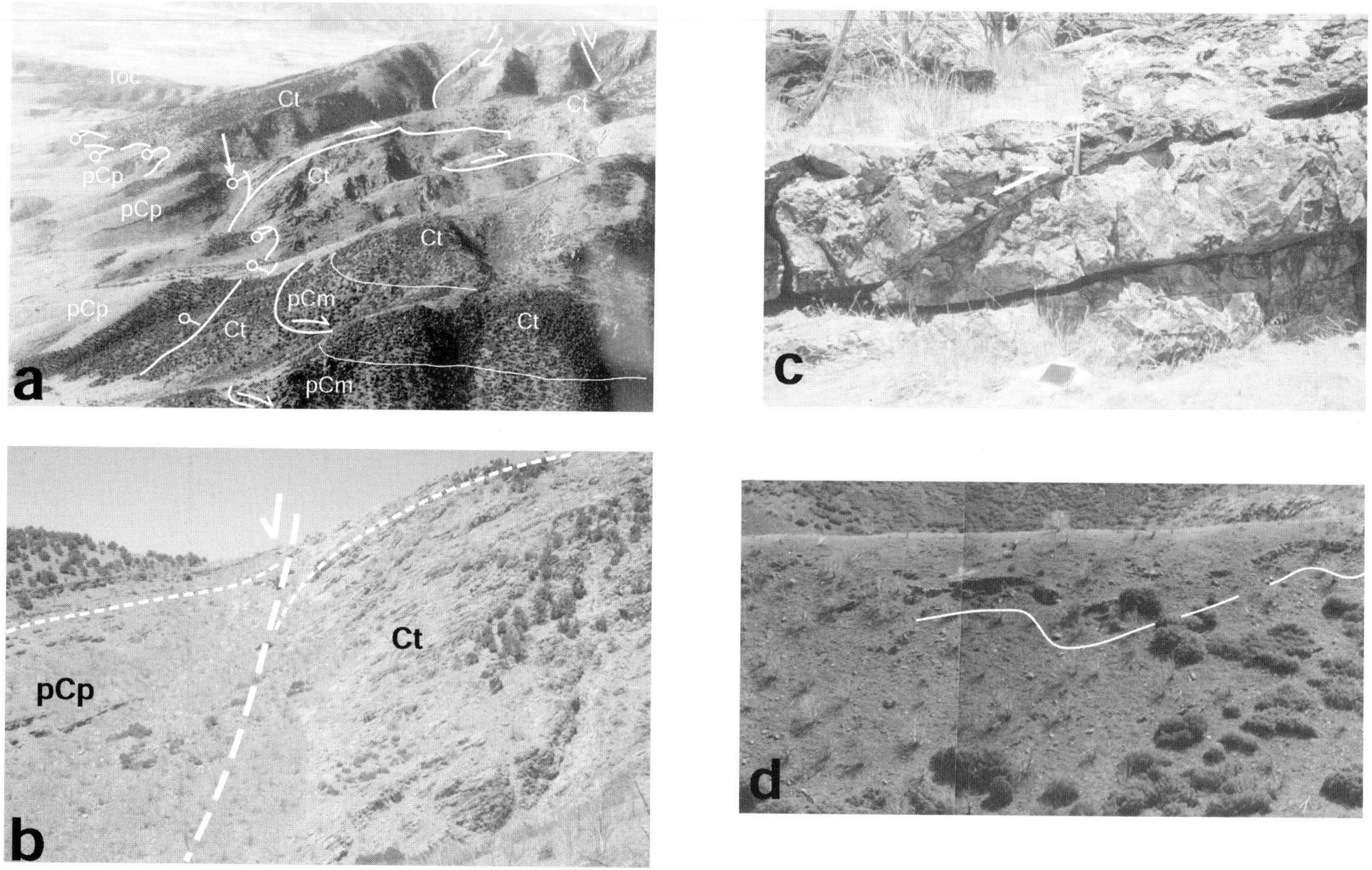

Fig. 5. Photographs of study area (all viewed looking north) showing large-scale and outcrop-scale structural features. (**a**) The west flank of the Canyon Mountains, showing gently dipping Proterozoic Pocatello quartzites (pCp) in the hanging wall and strongly deformed Proterozoic Mutual (pCm) and Eocambrian Tintic quartzites (Ct) in the footwall of the range-bounding normal fault system. The ridges on the right expose folded imbricate thrust faults of the Canyon Range footwall duplex. Toc, Tertiary Oak City Formation. (**b**) Closer view of the normal fault in the detailed study area. The gently dipping pCp in the hanging wall and the steeper Ct beds in the footwall both curve into the steep fault. (**c**) Wedge fault in pCp showing top to the east motion associated with CR thrusting. (**d**) Z-folds in pCp formed by top to the east shearing related to CR thrusting.

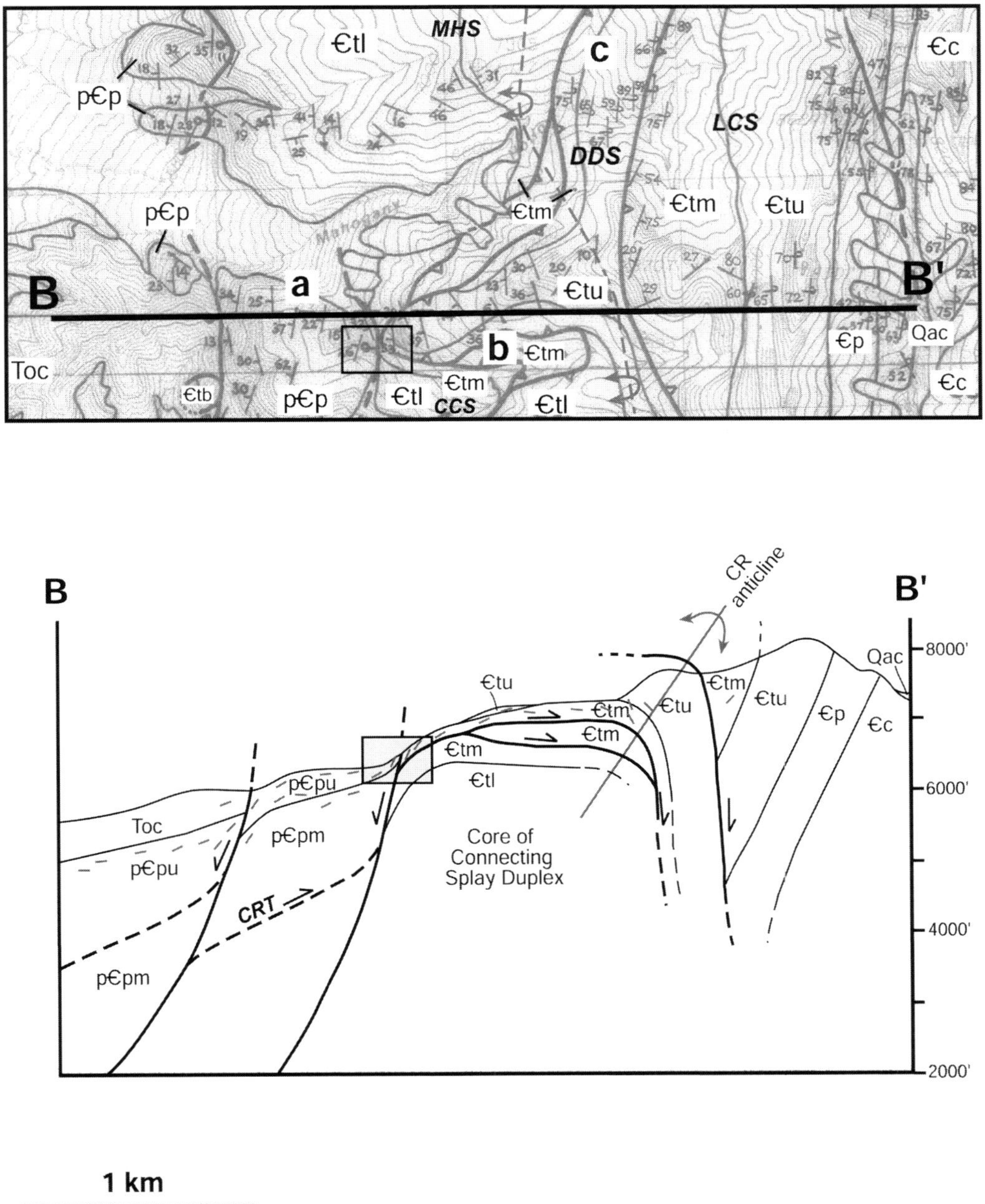

Fig. 6. (**a**) Detailed geological map, on a topographic base with 20' contour intervals, from boxed area in Fig. 4. Connecting splay duplex slices shown are: LCS, Limekiln Canyon slice; DDS, Devils Den slice; CCS, Cascade Canyon slice; MHS, Mahogany Hollow slice. (**b**) Cross-section BB' showing structural relationships along the west flank of the Canyon Mountains. Bedding dip rakes are shown as small dashes. Normal faults truncate the west limb of the duplex antiformal stack. (Note fault 'drag' in both the hanging wall and footwall of the normal faults.) Stratigraphic units: pCp, Proterozoic Pocatello Formation, pCpm, pCpu, middle and upper Pocatello Formation; Ctl, Ctm, Ctu, Cambrian Tintic Formation (lower, middle and upper); Ctb, brecciated Tintic Formation; Cp, Cambrian Pioche Formation; Cc, Cambrian carbonate rocks; Toc, Tertiary Oak City Formation; Qac, Quaternary alluvial deposits. Boxed areas indicate area of detailed study.

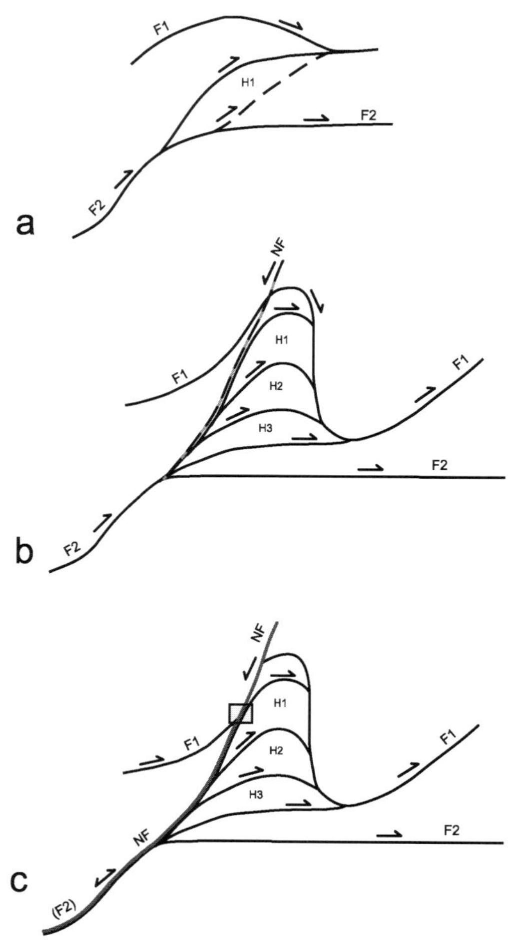

Fig. 7. Simplified kinematic model showing successive deformation events preserved within the study area (boxed area in (**c**)). F1, older fault; F2, younger fault; H1, H2, H3, oldest to youngest horses in connecting splay duplex (see Mitra & Sussman 1997, fig. 15); NF, normal fault. (**a**) Initial folding and reactivation of the older fault (F1) as the first horse (H1) of a subthrust connecting splay duplex forms and transfers slip from the younger fault (F2) to the older fault (F1). (**b**) Slip is continuously transferred to and reactivates F1 during later stages of duplex growth and amplification (H2 and H3), resulting in folding of F1 and its sheet into a tight antiform–synform pair. Emplacement of duplex slices H2 and H3 also reactivates the faults bounding the older slice, H1. Location of future normal fault (NF) is shown. (**c**) Normal fault is located along pre-existing thrusts (F1, F2) and duplex slices (H1, H2 and H3). NF reactivates parts of older thrusts, and may weave in and out of older fault zones (see Fig. 1). The boxed area contains rocks that preserve the deformation associated with all three of the events ((**a**)–(**c**)).

the connecting splay duplex (Mitra & Sussman 1997). These episodes are essentially coaxial and should have similar motion planes. Motion planes derived from the fracture populations suggest general eastward transport, but somewhat different emplacement directions for the Canyon Range sheet and for the various slices of the duplex (Fig. 8). The older mesoscopic fracture populations are made up of healed fractures and cemented cataclasite zones, and are morphologically rather different from the open fractures and unhealed breccia and cataclasite zones associated with normal faulting (Ismat & Mitra 2001). Except for a few preserved cross-fractures, most of the older mesoscopic fracture populations are obliterated within the normal fault zone as they are overprinted by close-spaced fault-parallel fractures and cataclasite zones.

The fault zone is *c.* 10 m thick. Small (10–30 cm) lenses of strongly deformed quartzite cataclasite are preserved within the zone whereas the surrounding intensely deformed material (ultracataclasite) has generally been eroded away. Within the cataclasite lenses are anastomosing seams of ultracataclasite that surround smaller lenses of less deformed cataclasite. Beyond the fault zone the rocks show close-spaced fracturing extending *c.* 10 m into both the hanging wall and the footwall. Although the older mesoscopic fracture populations are overprinted and not easily recognizable within the FRDZ of the normal fault, the older deformation fabrics can be distinguished at the microscopic scale. This provides us with a tool for recognizing reactivation along a fault zone even if overprinting has obliterated most outcrop-scale evidence. The FRDZ was, therefore, studied in detail for its microstructural characteristics.

Microstructures

Methods of study

A total of 41 samples were collected in three separate transects across the fault zone and reaching 30 m into the wallrock in the hanging wall and footwall. Samples were impregnated in epoxy and thin sections were cut parallel to the transport plane, which trends approximately east–west and is subvertical. Thus the plane of section cross-cuts most prominent deformation zones at high angles and is approximately parallel to the slip lineation. The microstructures were studied using polarized and dark-field optical microscopy and scanning electron microscopy; these techniques helped us in separating

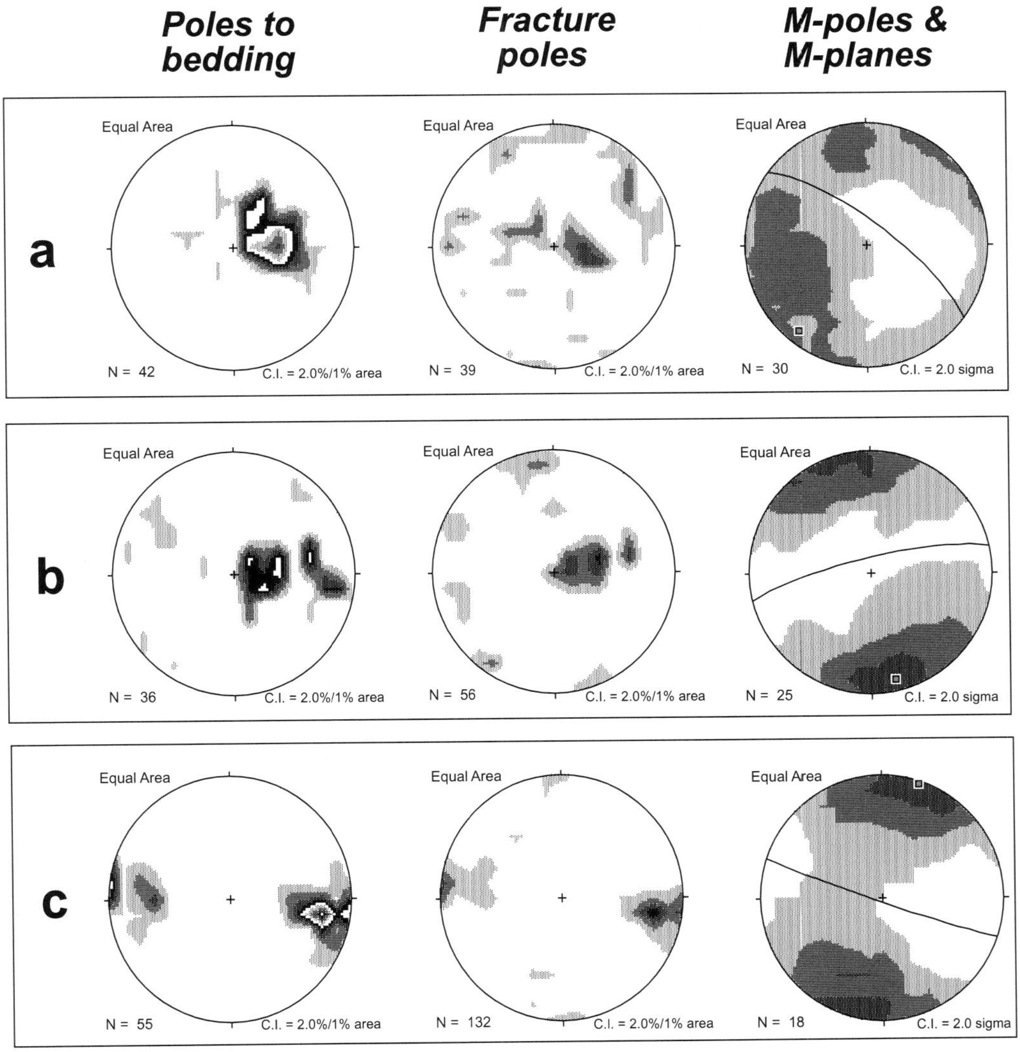

Fig. 8. Equal-area plots from the hanging wall and footwall of the normal fault in areas a, b, c indicated in Fig. 6 of poles to bedding, fracture poles, and M-poles (with associated motion plane). (**a**) Bedding poles show the gently dipping beds in the hanging wall. (**b**) Bedding poles show variable dips (as a result of pre-existing structure) in the footwall of the normal fault. (**c**) Bedding poles show the steepening of bedding in the east limb of the CR anticline. The variations from the overall east–west subvertical motion plane, as a result of variations from the hanging wall to the footwall, and of variable local motion of duplex slices, should be noted.

several generations of microcracks on the basis of their morphological characteristics and cross-cutting relationships. On the basis of the initial microstructural study samples were selected for quantitative analysis of microcrack density.

The microcracks may be unhealed fractures (which are easily recognizable optically) or they may be healed cracks. Depending on the amount of impurities preserved within healed crack systems these features may or may not be easily visible optically. In clean quartzites healed cracks are completely transparent in polarized light and failure to recognize them may result in significant undercounts of crack densities (Fig. 9a); they do, however, show up clearly under cathodoluminescence (Lloyd & Knipe 1992) or under dark-field illumination (Fig. 9b), and we have relied on the latter in our crack density counts. Even where healed cracks are optically visible in polarized light, cracks dipping at low angles are generally difficult to recognize, and there is greater likelihood of counting high-angle cracks (Anders & Wiltschko 1994); this is less of a problem under dark-field illumination. In addition, the dark-field view highlights a number of other features including old grain boundaries, the extent of overgrowths, inclusions such as rutile needles in some quartz grains, and certain deformation features such as deformation lamellae (Fig. 9b and c).

The morphologies of successive generations of cracks and deformation zones were confirmed with preliminary SEM studies using simple secondary electron emission topographic images. Although we plan further detailed SEM work on these rocks (e.g. Lloyd & Knipe 1992; Knipe & Lloyd 1994) in the future, this was not considered necessary for the present study, as successive generations of fractures and deformation zones could be easily distinguished optically on the basis of morphological characteristics and consistent cross-cutting relationships.

Polarized and dark-field optical microscopy were used to quantify microcrack density on several traverses across each thin section, keeping track of three generations of microcracks. The densities were calculated using the principles of lineal analysis (Underwood 1970; Mitra 1978). This method of measurement also allowed us to determine the volume percentage of rock directly affected by cataclasis during each generation of deformation in each sample. On a linear transect the chord length across a cataclasized zone compared with total transect length is directly proportional to the volume percentage of cataclasite in the whole rock (Underwood 1970). This is a particularly useful measure of deformation intensity in rocks that are intensely cataclasized and comminuted so that microcracks within grains are rarely preserved. The variation of deformation intensity in the FRDZ could then be studied in terms of both microcrack density variations and the volume of rock affected by cataclasis at each site.

Geometry of microcracks

The rocks involved in the deformation in the Canyon Mountains are dominantly framework-

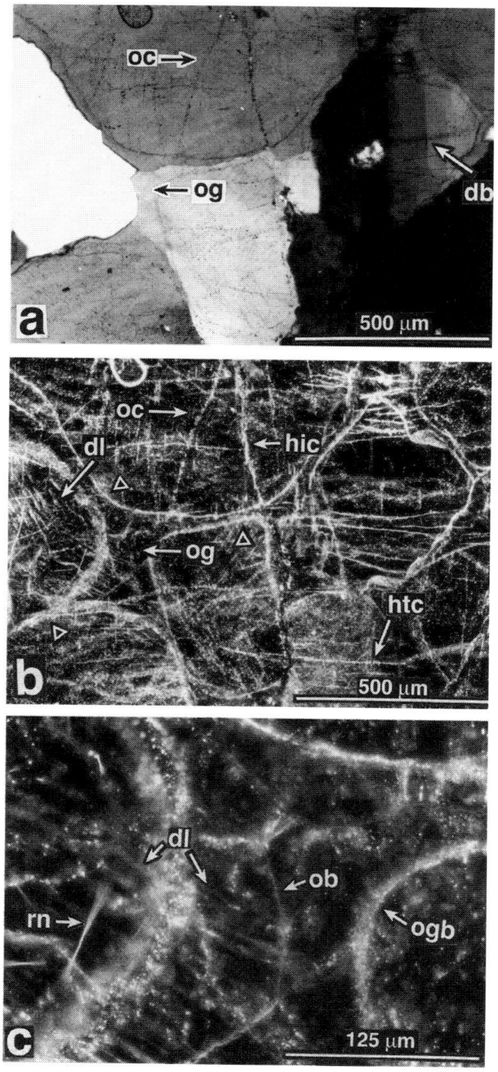

Fig. 9. Photomicrographs showing typical grain-scale fracturing in quartzites. (**a**) Cross-polarized view showing diagenetic overgrowths on grains (og), undulose extinction, deformation bands (db) and healed cracks (subtle features defined by inclusion trails). (**b**) Dark-field view highlights old grain boundaries and healed cracks. Grains show prominent overgrowths (og) (note that old (inherited) cracks (oc) do not extend into overgrowth areas). Healed transgranular cracks (htc) and healed intergranular cracks (hic) cut across grains and overgrowths, and developed after the overgrowths had formed. Overgrowths increase grain-boundary contact areas, reducing the possibility of stress concentrations at grain–grain impingement points (Δ), so that very few cracks originate at these contacts. Faint deformation lamellae (dl) are seen in grain on left. (**c**) Close-up dark-field view of overgrowth area showing boundaries between adjoining overgrowths (ob). Grain on left has rutile needles (rn), and deformation lamellae (dl) extend from the grain into its overgrowth. It should be noted that no cracks originate at overgrowth boundaries (ogb), as a result of large surface area and no impingement points.

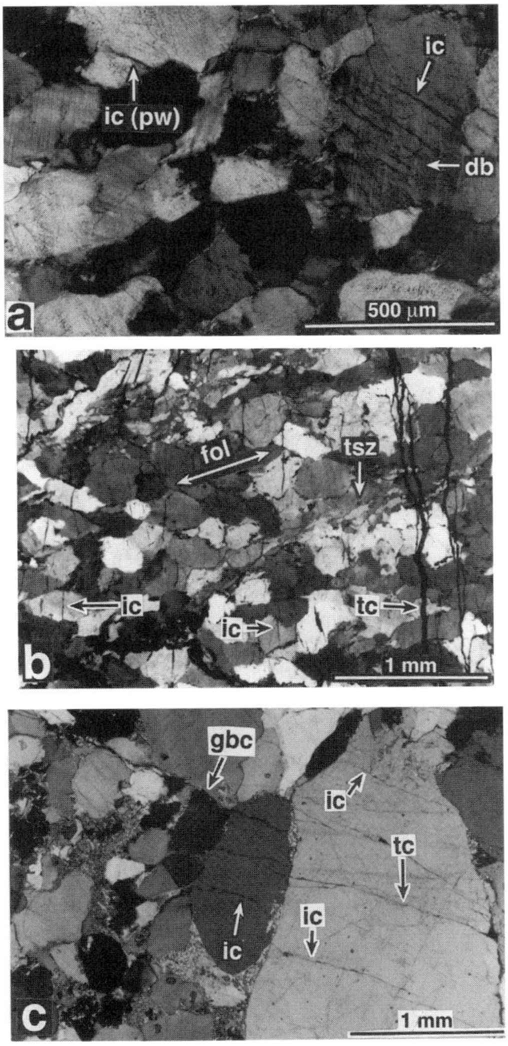

Fig. 10. Photomicrographs in cross-polarized light showing different microcrack geometries. (**a**) Intragranular cracks (ic) associated with considerable plastic deformation (undulose extinction and deformation bands, db) within grains. Cracks also show undulose extinction adjacent to them, formed as a plastic wake during crack propagation (ic (pw)). Crack orientation is crystallographically controlled and varies from grain to grain. (**b**) Intragranular cracks (ic) in the same orientation in many grains, suggesting that they are brittle and result from far-field stresses that also produced the large transgranular crack (tc). It should be noted that the cracks cross-cut foliation (fol) and a thin sintered zone (tsz). (**c**) Intergranular cracks of two kinds: grain boundary crack (gbc) that is parallel to and continuous with transgranular cracks (tc) that cut across grain boundaries. Also visible are some healed, intragranular cracks (ic).

supported quartzites with unimodal grain-size distribution. A variety of crack geometries, morphologies and orientations are developed in the quartzites. They reflect the variations in deformation conditions throughout a complex deformation history. They are also strongly dependent on inhomogeneities in material behaviour, particularly the relative strengths of framework grains and the surrounding 'material matrix' (which may be overgrowths or cement or sedimentary matrix). The cracks may be broadly classified as *intragranular cracks* or *intergranular cracks* (Kranz 1983).

Intragranular cracks are confined within single grains and die out either within the grain or at a grain boundary. Under plastic or EF–QP transitional deformation conditions, quartzite grains undergo some amount of plastic deformation by dislocation glide before fractures nucleate at stress concentrations that develop at dislocation pile-ups (Mitra 1978). These fractures can also be described as Cleavage II or Cleavage III cracks (Ashby *et al.* 1979; Gandhi & Ashby 1979; Atkinson 1980), and similar features have been described from quartzites in other areas (e.g. Lloyd & Knipe 1992). The cracks are associated with evidence of plastic deformation within grains, such as undulose extinction and deformation bands (Fig. 10a). The cracks are typically blunted at grain boundaries, either because they run into a more ductile matrix or because they impinge on a grain with a suitably oriented glide system that allows the stress concentration to be dissipated (Mitra 1978). This type of deformation results in grain-size reduction in a thin deformation zone in the rock.

In some metamorphosed quartzites we observe unhealed intragranular cracks in the same orientation in many different grains and these cracks are parallel to transgranular cracks with the same morphology (Fig. 10b). Clearly, these cracks did not grow from point-to-point contact of framework grains (e.g. Gallagher *et al.* 1974). Overgrowths on grains may increase contact areas between framework grains reducing the stress concentrations at grain contacts, or, in matrix-supported quartzites, the matrix may shield the framework grains from coming into contact, lessening the likelihood of stress concentrations at grain-to-grain contacts (Fig. 9). Under EF deformation conditions, the finer-grained matrix is stronger than the framework, and microcracks probably grow from pre-existing flaws within the framework grains (Cleavage I cracks of Gandhi & Ashby (1979)), and only in those grains that contain flaws that are in the correct orientation to grow under the externally applied stresses (Sussman 1995). As the cracks are blunted at grain boundaries, this gives rise to many grains with intragranular cracks in the same orientation (Fig. 10b). With increasing deformation (perhaps under higher local stresses) more grains may develop cracks in the same orientation, and some eventually coalesce to form a single throughgoing fracture (Fig. 10b).

Intergranular cracks range in length from a few grain diameters to throughgoing fractures in the rock. They generally form under EF deformation conditions, under which cracks, once nucleated, can coalesce with other cracks or propagate unstably as a result of stress concentrations at crack tips. These types of cracks have been described as Brittle Intergranular Fracture type 1 (BIF1) (Ashby *et al.* 1979; Gandhi & Ashby 1979; Atkinson 1980) and have been previously described from quartzites (Lloyd & Knipe 1992). Crack propagation is blocked by high pressures (i.e. is pressure dependent) and is also controlled by material inhomogeneities. If the 'matrix' has a lower fracture strength than framework grains cracks tend to nucleate at grain–grain or grain–matrix interfaces and to propagate along grain boundaries (Fig. 10c); these intergranular cracks are *grain boundary cracks*. If there is no strength difference between framework and 'matrix', cracks may nucleate within framework or matrix, and can propagate across grain boundaries and through successive grains; these intergranular cracks are referred to as *transgranular cracks* (Fig. 10c).

Microcracks preserved in fault-related deformation zone

Proterozoic–Cambrian quartzites exposed in the Canyon Mountains show evidence for five generations of fracturing, including inherited cracks in quartz grains from older source terranes from which the sediments were derived (Sussman 1995). Quartzites of the Pocatello and Tintic Formations exposed within the FRDZ of the normal fault show evidence for a sequence of at least three brittle deformation events that gave rise to intragranular and intergranular cracks as well as cataclasite zones of various sizes. Successive generations of fractures can be recognized on the basis of the geometry of the fractures, consistent cross-cutting relationships, and types of cementation and healing of the fractures (Table 1). The features are described in the order in which they were developed.

Initial emplacement of the Canyon Range thrust sheet took place in Neocomian time (130–115 Ma) (DeCelles *et al.* 1995) under deep crustal conditions (⩾10 km) (Fig. 3, Table 1) and a penetrative foliation was developed in the hanging-wall Pocatello quartzites by plastic deformation processes (Fig. 11a). In some rocks, stylolites parallel to the foliation indicate that pressure solution was also active (Sussman 1995). Cataclasis also appears to have occurred generally coeval with the plastic deformation. In zones of strong deformation, smaller framework grains show particularly strong undulose extinction whereas larger grains are fractured, presumably by the formation of Cleavage 2 or Cleavage 3 cracks (Fig. 10a). These cracks typically form at dislocation pile-ups and are oriented at high angles to glide planes. They are intragranular cracks and lead to reduction of grain size in the rock. The cracks also show localized plastic deformation at the boundaries, left behind as the wake of a propagating crack tip. Transgranular healed cracks associated with thin sintered zones are observed in a number of different mutually cross-cutting relationships with plastic deformation features; the zones are sometimes recognizable by pockets of misoriented cataclastic fragments (Fig. 11b).

Wider transgranular zones of fine-grained cataclasite that have been completely re-formed show microstructures resembling sintered material (Anderson *et al.* 1974) with fine-grained material showing interlocking grain boundaries (Fig. 11c). Because of their fine grain size and 3D interlocking nature the boundaries often appear a little fuzzy under the optical microscope. Sintering in metals and ceramics is generally described as reduction of porosity involving grain growth assisted by diffusion (Cottrell 1964) and typically occurs at high temperatures ($\geqslant 0.7T_m$). Sintered microstructures in ceramics closely resemble granular microquartz formed from chert by a time-dependent annealing mechanism involving fluid-assisted diffusion at low temperatures (Knauth 1994). A similar low-temperature (*c.* $0.2T_m$) mechanism has also been suggested for densification of powdered compacts (Raj 1982) where diffusion-assisted grain growth eliminates porosity along grain boundaries. This last mechanism best explains the fine-grained interlocking (sintered) textures that are developed in the oldest cataclasites (Fig. 11c). With SEM these zones show an interlocking network of extremely fine ($\leqslant 1\,\mu m$) grains (Fig. 11d); newer cataclasis would localize in adjoining wallrock because of the higher fracture strength of the sintered zones.

During the late stages of Sevier thrusting in Turonian to Maastrichtian time (65–90 Ma)

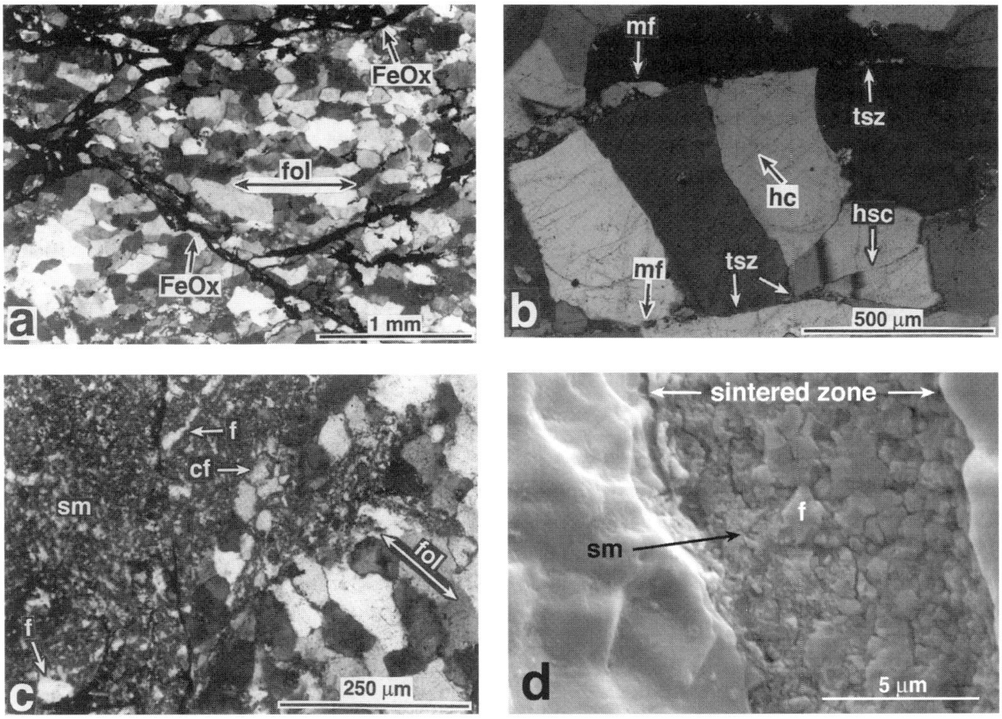

Fig. 11. Photomicrographs showing features developed during initial Canyon Range thrusting. Photos (**a**)–(**c**) between crossed polarizers. (**a**) Penetrative foliation (fol) in the Pocatello quartzite with the grains showing plastic deformation (undulose extinction). The foliation is cross-cut by cataclasite zones with iron oxide cement (FeOx). (**b**) Transgranular thin sintered zones (tsz) with a few misoriented fragments (mf) recognizable along them. There are associated healed shear cracks (hsc) with grain boundaries and deformation bands offset along them. Also evident are subtle bubble trails indicating transgranular healed cracks (hc). (**c**) Thick sintered zone cross-cutting foliation (fol). The zone contains cataclasite fragments (cf) and grain fragments (f) in a fine-grained matrix that shows interlocking sintered texture (sm). (**d**) Scanning electron micrograph of thin sintered zone with small (1–2 μm) fragments (f) in extremely fine-grained sintered matrix (sm).

(DeCelles *et al.* 1995; Mitra & Sussman 1997) the Canyon Range thrust sheet underwent folding and fold tightening under a shallow overburden (<5 km) (Fig. 3, Table 1) within the EF regime (Pequera *et al.* 1994; Mitra & Sussman 1997; Ismat & Mitra 2001). The fold tightening was accomplished mainly by rotation and steepening of the common limb between the antiform and the synform, and vertical stretching of this limb as an antiformal stack connecting splay duplex grew in the core of the anticline (Mitra & Sussman 1997). Most of this deformation was accomplished by large-scale cataclastic flow on a network of mesoscopic fractures (Ismat & Mitra 2001) that can be correlated with distinctive grain-scale cataclastic features that are different from older and younger features. In the area studied in detail one of the imbricate faults of the duplex and its associated FRDZ is intersected by the normal fault and is affected by it (Fig. 6). The rocks of this older FRDZ have been redeformed during movement on the normal fault and contain multiple generations of fractures. At the grain scale, the late Sevier age deformation is represented by transgranular cracks that generally appear to be clear and healed (Fig. 9a). Within any particular grain the crystal structure across a healed crack boundary is re-formed, resulting in optical continuity in the grain, although bubble trails (with possible fluid inclusions) or tiny particles of insoluble material are left behind along the boundary (Lloyd & Knipe 1992; Anders & Wiltschko 1994) (Fig. 9a and b). These are clearly visible in dark-field microscopy, and can be seen as parallel transgranular features cutting across grains and grain boundaries (Fig. 9b). Some of the healed cracks were shear fractures and show offset of grain boundaries along them (Fig. 12a and b). Healed transgranular cracks are also found at the submicroscopic scale, cutting through fine-grained

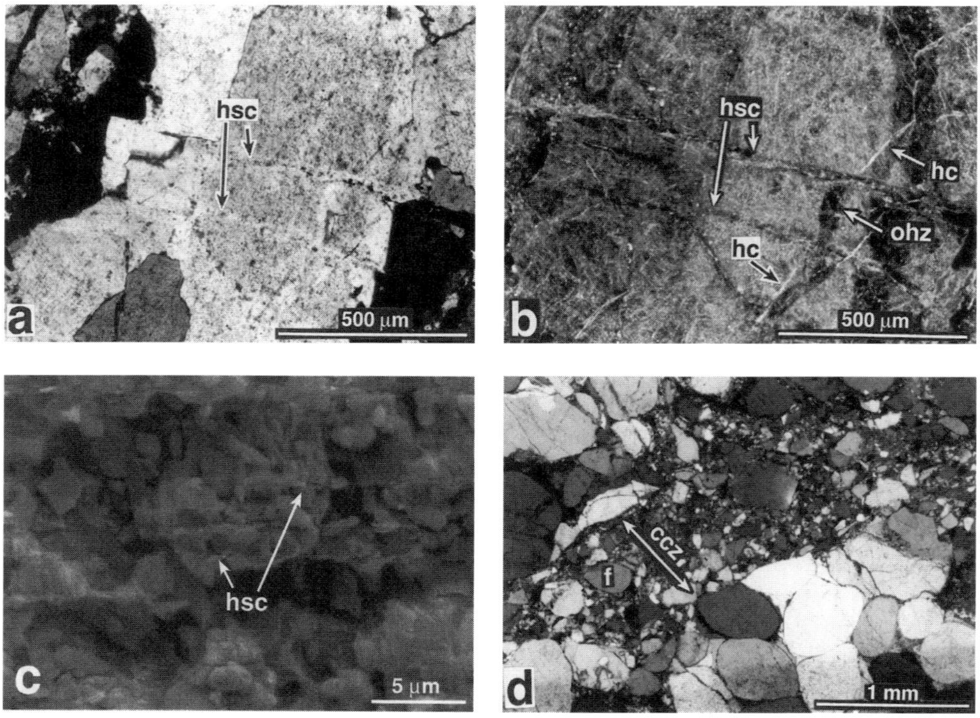

Fig. 12. Photomicrographs showing features developed during the late stages of Sevier thrusting and Canyon Range sheet folding. (**a**) Healed shear cracks (hsc) offsetting grain boundaries. (Crossed polarizers.) (**b**) Dark-field view showing finely comminuted material along one of the healed shear cracks (hsc). Also seen are healed cracks (hc) oblique to the shear cracks, and an old healed zone (ohz) that has been offset by the shear cracks. (**c**) Scanning electron micrograph showing closely spaced, thin, healed shear cracks (hsc) with fine rotated fragments along them. (**d**) Thick cataclasite zone (ccz) with sharp boundaries truncating grains. The cataclasite has a silica cement holding together subangular to subrounded fragments (f) of various sizes showing undulose extinction. (Crossed polarizers.)

foliated cataclasites formed during earlier deformation (Fig. 12c). In addition, there are very thin (*c.* 0.1 mm) zones of healed, cemented cataclasite that cross-cut the early foliation. There are also thick (*c.* 1 mm) zones of cemented cataclasite with angular, randomly oriented fragments that probably grew from thinner zones by incorporating angular fragments from the wallrock into the zones (Fig. 12d).

The last formed brittle deformation features are fractures or zones that are consistently oriented north–south and are generally steeply dipping everywhere along the western flank of the range. The zones are characterized by iron oxide staining in many cases (Otton 1995). At the grain scale, thin open cracks and thin zones of coherent cataclasite, with the same north–south trends, cross-cut all previous features (Figs. 10b and c, and 13a and b). The cracks are generally unhealed and may be stained with iron oxide (Fig. 13a); in some cases they are cemented. Extension fractures are preserved as thin, hairline cracks (Fig. 10b and c); shear fractures are also seen as thin, dark lines, but they have grains offset along them (Fig. 13c). Thin cataclasite zones (<1 mm) contain small subangular to subrounded fragments cemented with iron oxide cement (Fig. 13d). The zones also contain fragments of older generation cataclasites, where they cross-cut or reactivate older zones (Fig. 13d). With SEM the cataclasite zones generally show a wide size range of fragments. The fragments may be quartz grains that show conchoidal fracture faces or fragments of earlier cataclasite (Fig. 13e and f). In cross-sectional view, the zones show fragments held together by finer material, and a network of transgranular cracks. Thicker zones (*c.* 5 mm) of coherent cataclasite also developed during this phase of deformation, and contain a wide size

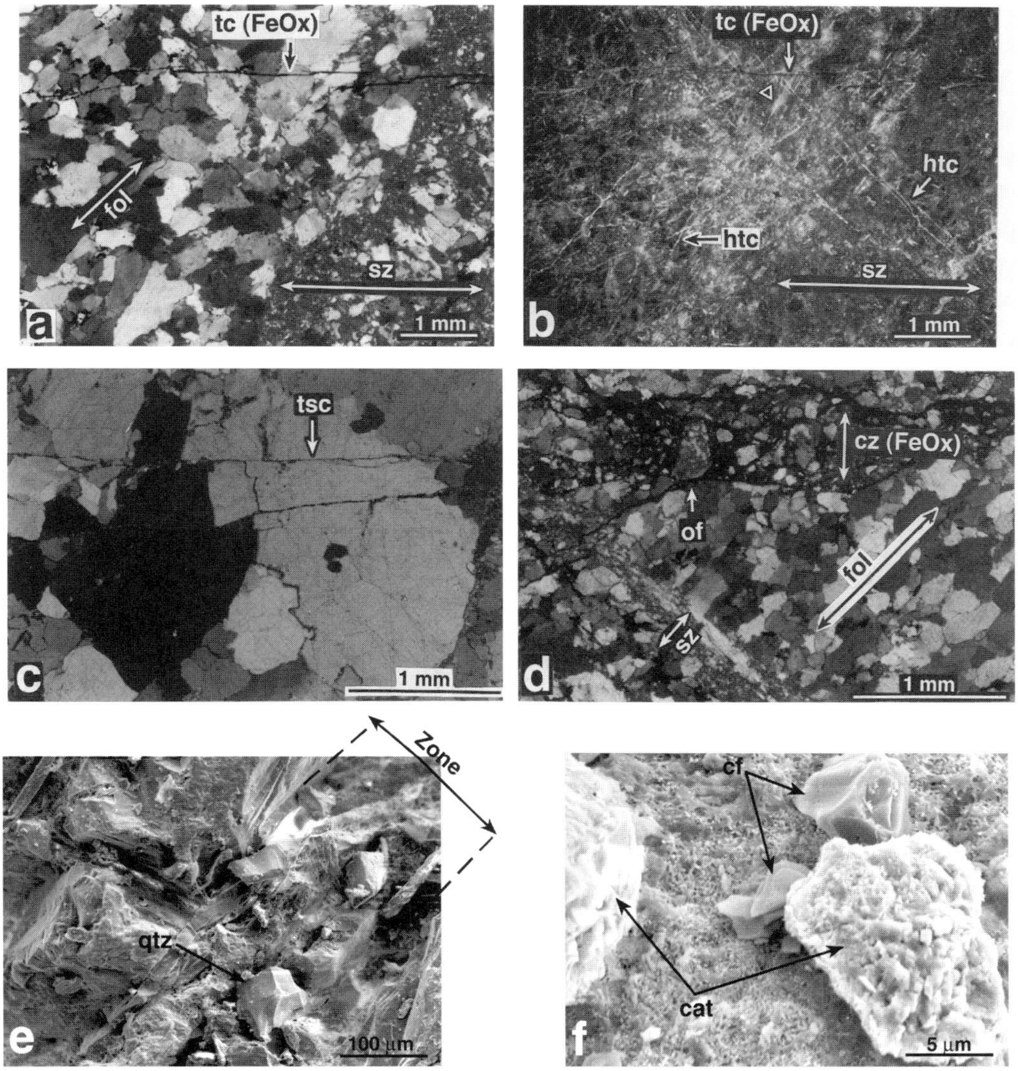

Fig. 13. Photomicrographs showing grain-scale cataclastic features developed during Basin-and-Range extensional faulting. (**a**) Unhealed transgranular crack with iron oxide staining (tc (FeOx)) cutting across early foliation (fol) and a sintered zone (sz). (Crossed polarizers.) (**b**) Dark-field view (of same frame as (**a**)) showing the unhealed transgranular crack (tc (FeOx)) cross-cutting earlier-formed healed transgranular cracks (htc) in many orientations; some of these healed cracks are offset (at Δ). The healed transgranular cracks (htc) clearly cut across the sintered cataclasite zone (sz) indicating that they post-date the zone. (**c**) Cross-polarized view showing iron oxide stained transgranular shear cracks (tsc) with grain boundaries offset along them. Subtle healed cracks (bubble trails) are also visible. (**d**) Cross-polarized view of cataclasite zone with iron oxide cement (cz (FeOx)) cutting across a prominent early foliation (fol) and a thin sintered cataclasite zone (sz) with its own foliation. The iron oxide cemented cataclasite zone contains fragments of older cemented cataclasite (of) and other rotated angular fragments. (**e**) Scanning electron micrograph showing cross-section through a cataclasite zone with quartz fragments (qtz), showing conchoidal fracture, surrounded by very fine-grained matrix. (**f**) Scanning electron micrograph showing fragment of cataclasite (cat) and small fragments of quartz with conchoidal fracture (cf) within a younger fine-grained cataclasite matrix.

range of subangular to subrounded fragments held together by fine-grained non-cemented matrix; portions of these fine-grained zones may show iron oxide staining. We interpret all these features (at different scales) to have formed at low pressures and temperatures at shallow crustal levels and associate them with Miocene (*c.* 10–20 Ma) normal faulting along the western margin of the range (Otton 1995; Sussman 1995; Coogan & DeCelles 1996) (Fig. 3, Table 1).

Microfracture populations

Field evidence suggests that the quartzites along the western flank of the Canyon Mountains have undergone a complex deformation history, including repeated episodes of contractional deformation during the Cretaceous Sevier orogeny followed by extensional deformation during Tertiary Basin-and-Range normal faulting. In these rocks it is obviously important to separate the generations of microcracks, so as to avoid overcounts or undercounts of crack densities associated with each deformation event. Healed and unhealed cracks and zones that formed during one deformation event generally show some form of continuity from grain to grain within a rock, indicating that the cracking took place after lithification of the rock, and reflect the orientation of the far-field stresses that gave rise to the fracturing. Cracks formed during successive deformation events can be distinguished from one another on the basis of crack morphologies and consistent intersection relationships.

Microcrack populations were determined from 26 sites on several overlapping transects across the fault zone. Crack densities were quantified by crack counts on multiple transects across thin sections of samples, with 100–200 cracks counted per sample and a total of over 2500 cracks counted. In addition, the thicknesses of cracks and zones were measured during the crack counts to obtain the volume percentage of rock affected by cataclasis in each sample, using the principles of lineal analysis (described above).

Microcrack density

The Pocatello quartzites in the hanging wall of the normal fault show a general decrease in total microfracture density away from the fault (Fig. 14a). We attempted fitting linear, polynomial, exponential and logarithmic curves to the data (using Cricket Graph). The best fit overall shows that the decrease follows a logarithmic trend, although simple interpolation between data points indicates that there are wide variations from the general trend (Fig. 14). We believe that these variations indicate real trends reflecting the inhomogeneity of deformation in fault zones, with pods of undeformed rock (valleys) surrounded by anastomosing zones of intense deformation (peaks). Such variations are observed at many different scales and may be a reflection of the fractal nature of distribution of high-intensity deformation associated with fault zones that has been suggested by a number of other workers (e.g. King 1983; Turcotte 1989; Knipe & Lloyd 1994). Recognition of the variations in microfracture density is clearly dependent on the choice of sampling interval near a fault, and even with the proper scale-representative interval it is possible to find wide variations in density if some of the samples are from the undeformed pods within the fault zone.

The overall decreasing trend in total microfracture density may be intuitively pleasing, and it certainly fits earlier models (e.g. Anders & Wiltschko 1994), but it may be misleading. When only the healed fractures (formed during

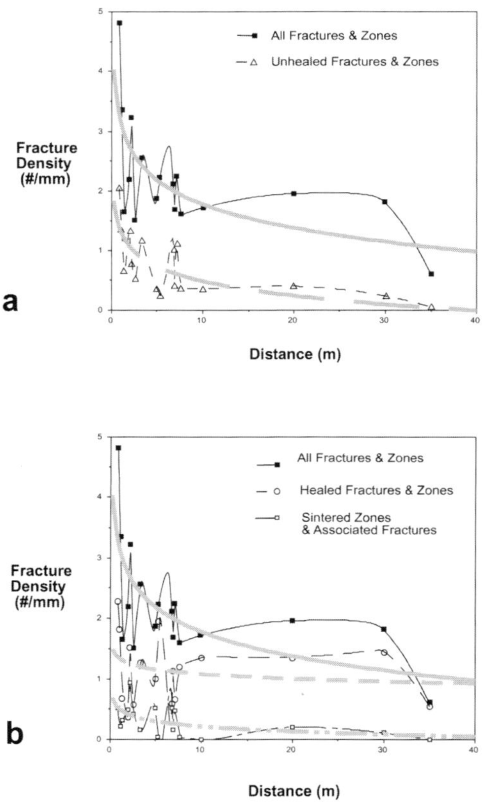

Fig. 14. (**a**) Variation of total microcrack density and unhealed crack density in the hanging wall of the fault. Thick grey lines show the overall trends, whereas the thin dark lines show the local variations. (**b**) Variations of total microcrack density, healed crack density and sintered crack density in the hanging wall of the fault; the two latter measures show approximately uniform distribution with distance away from the fault.

late-stage Sevier deformation) or only the sintered deformation zones (formed during Canyon Range thrusting) are considered, the microfracture densities remain approximately constant across the entire transect (Fig. 14b); once again, there are local variations from this trend with significant increases or decreases from the constant density. One peak in healed fracture density (at 6 m from the fault) is the principal contributor to the total fracture density at that location. The iron-oxide cemented deformation zones and open fractures (related to the normal faulting), on the other hand, show a logarithmic decrease in density away from the fault, mimicking the trend of the total crack density; there are, however, wide variations from this general trend with significant peaks that are the dominant contributors to peaks in the total trend (Fig. 14a).

The Tintic quartzites in the footwall of the normal fault also show a general decrease in total microfracture density away from the fault (Fig. 15a), following an overall logarithmic trend; there are, however, significant variations from this trend. The fracture density in sintered zones remains constant across this transect (Fig. 15b). The density of unhealed fractures shows the same general logarithmic decrease as the total count, but its peaks fail to explain one of the major peaks in density in the total trend (at 6 m from the fault) (Fig. 15a). The density of healed fractures appears to show a slight increase away from the fault zone, and its major peak (at *c.* 6 m from the fault) is clearly the major contributor to the total count peak at that distance (Fig. 15b). As the density counts give equal weight to single fractures and to cataclasite zones (where the rocks have undergone intense fracturing and granulation), they may not be the best measure of the role played by different types of zones in the total deformation. The cataclasite zones affect a certain volume of rock, and within each zone the deformation is sufficiently strong that most microcracks are obliterated.

Cataclasized volume fraction

Measurement of the volume fraction of cataclasized rock at each site may give a better estimate of the relative importance of the different generations of deformation zones to the total deformation. Figure 16 shows the variation of volume fraction of cataclasized rock at different sites in the Tintic quartzite in the footwall. It should be noted that the relative importance of the total count peaks is dramatically different from the pattern seen with the fracture density

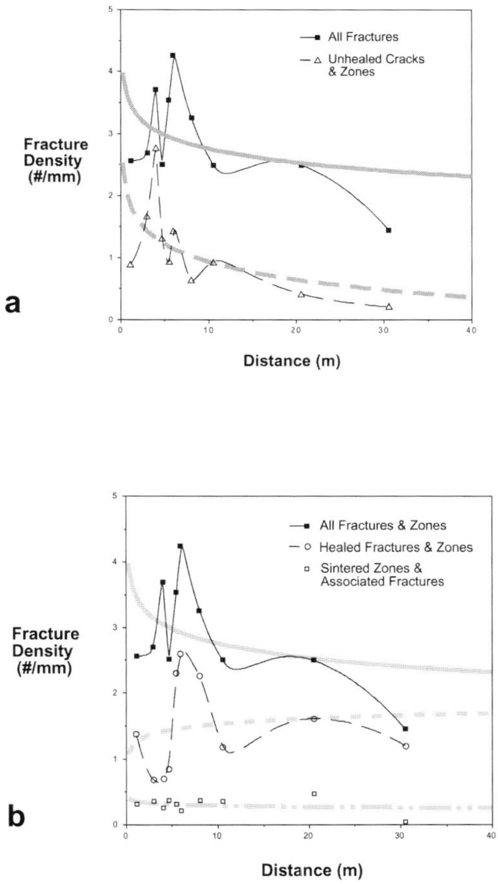

Fig. 15. (**a**) Variation of total microcrack density and unhealed crack density in the footwall of the fault. Thick grey lines show the overall trends, whereas the thin dark lines show the local variations. (**b**) Variations of total microcrack density, healed crack density and sintered crack density in the footwall of the fault; sintered crack density remains constant whereas healed crack density increases slightly with distance away from the fault.

counts (Fig. 15a and b), with the peaks at *c.* 4 and *c.* 11 m being the most important. The unhealed fractures and zones are clearly the major contributors to these peaks. Interestingly, the peak at *c.* 6 m also sees its major contribution from unhealed fractures and zones even though the density of these features is lower than the healed fractures and zones (Fig. 15b).

Comparison of the total cataclasized volume fraction data from the footwall with that from the hanging wall (Fig. 17) shows that for this FRDZ a much larger volume fraction of rock in the footwall is strongly cataclasized even though the density of fractures and zones close to the

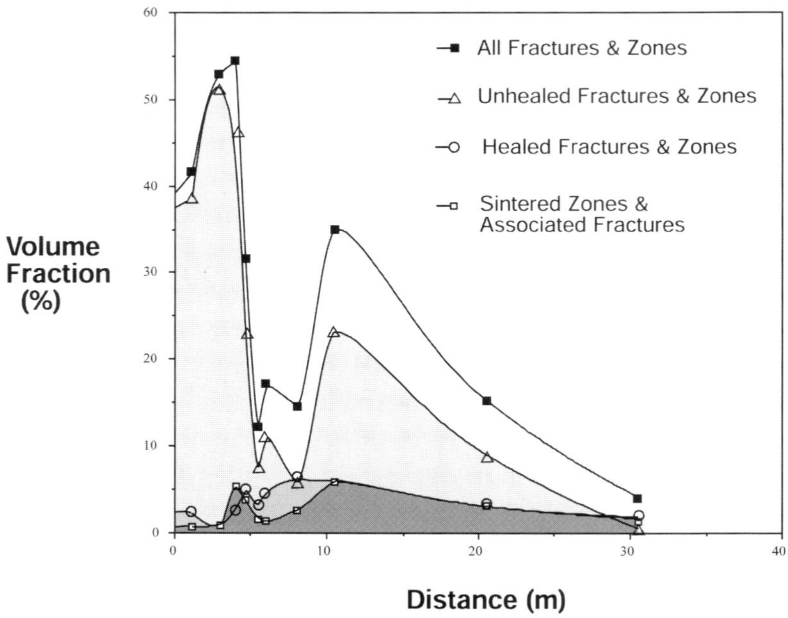

Fig. 16. Variation of volume fraction of deformed rock with distance from the fault produced by the different events of cataclasis in the footwall of the fault. (Compare this with fracture density variations shown in Fig. 15.)

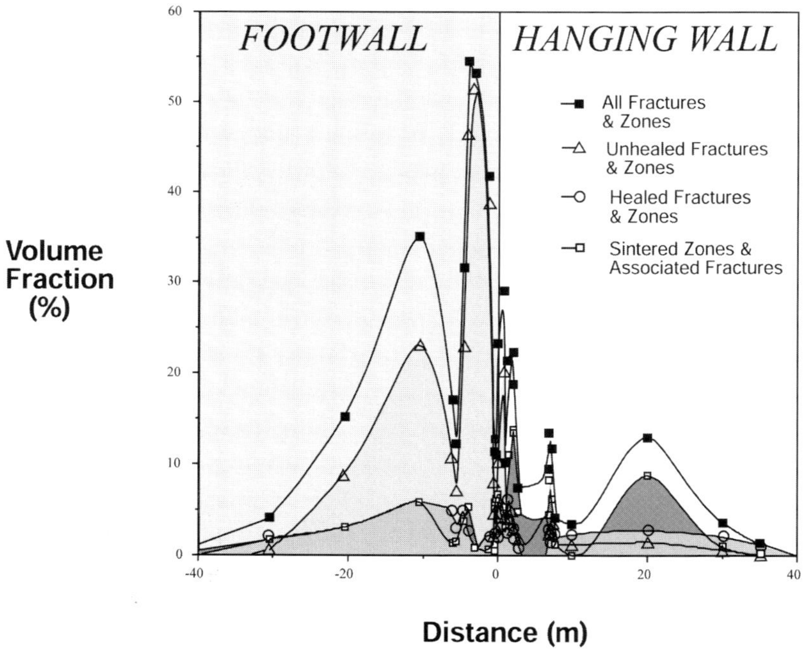

Fig. 17. Variation of volume fraction of deformed rock with distance from the fault produced by the different events of cataclasis in the footwall and hanging wall of the fault and the adjoining damaged zones. (Note that the footwall is more strongly deformed than the hanging wall.)

fault is somewhat higher in the hanging wall (compare Figs. 14 and 15). Much of the footwall deformation is accomplished on unhealed fractures and zones, which are associated with normal faulting on the basis of mesoscopic and microscopic criteria (described above). Thus, the latest deformation in the fault zone as a whole is very asymmetric, with much of the deformation associated with the normal faulting concentrated in the footwall. The distribution of total volume fraction of cataclasized deformed rock in the hanging wall shows some interesting trends. The main peaks in the FRDZ (at 7 and 20 m) mainly arise from contributions of early Sevier sintered zones.

A closer look at the volume fraction of cataclasized rock in the fault zone proper (Fig. 18) shows three main peaks in the total count. The peaks at 0.75 and 3.5 m are clearly related to unhealed fractures, but the major contribution to the peak at 2 m is from sintered zones. In both the fault zone and the FRDZ, where the contribution from sintered zones (the oldest zones) is high, the contribution from unhealed fractures and zones (the youngest cataclastic features) is very low. This suggests that the grain-size reduction in certain zones during the older deformation and their subsequent healing increases the fracture strength of the rock in those areas, making it unlikely that later EF deformation will occur within the same zone (Fig. 15). Instead, during later faulting, deformation is concentrated in adjoining rock that is coarser grained and has lower fracture strength.

Discussion

In areas with a complex and prolonged deformation history fault zones may be reactivated during successive deformation episodes, either because they are weaker than surrounding rocks or because they are distinct inhomogeneities in the rock. The old fault zone need not be reactivated along its entire length; rather, new deformation takes place along segments of the old fault zone that are appropriately oriented to be active in a new stress field. Microcrack density generally decreases away from the fault in some manner, although there may be significant variations from the general decreasing trend resulting from inhomogeneities in the deformation, with pods of relatively undeformed material surrounded by strongly deformed zones. In reactivated zones, the variations in microcrack density may be partly a result of superimposition of multiple microcracking events in an irregular manner.

It is clear from the microcrack density variation and cataclasized volume variation graphs that the patterns of deformation in a reactivated

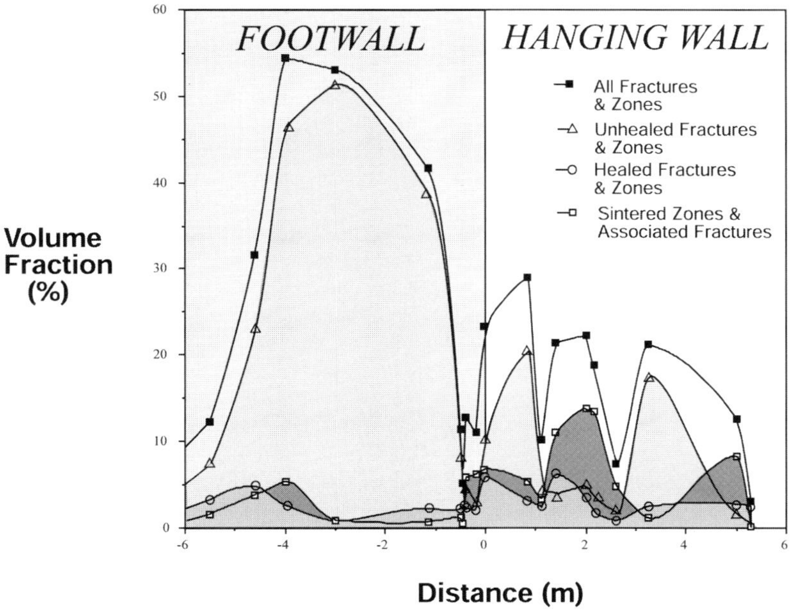

Fig. 18. Variation of volume fraction of deformed rock with distance from the fault produced within the fault zone proper by the different events of cataclasis. (Note that the footwall is more strongly deformed than the hanging wall.)

fault zone are complex. Inhomogeneous deformation patterns from successive deformation events are superimposed to produce a final pattern. Although it is possible to recognize the reactivation that has taken place along this fault zone from macroscopic (map-scale) and mesoscopic (outcrop-scale) data, it is not possible to separate out the effects of successive faulting events from this type of information. At the outcrop scale, rotation of blocks of all sizes within the fault zone makes it impossible to recognize all but the last fracture population from orientation data. Detailed sampling of microstructures provides a tool for understanding the superimposition of successive deformation events.

At the microscopic scale, grain-size variation is a commonly used measure for tracking the variations in intensity of deformation across a fault zone. In an area of fault reactivation, grain-size variation provides only a measure of the total deformation, but does not help in separating the effects of successive deformation events. Microcrack density is another useful measure that can be used not only to measure the total intensity of deformation, but also to separate out the effects of successive deformation events. Correlation of microcracks (2D data from thin sections) with their mesoscopic counterparts (3D fracture network data from outcrops) on the basis of orientation is difficult under the best of circumstances. Such correlation is fruitless for all but the youngest microcracks in reactivated zones because of the rotation of earlier features during the latest faulting.

It is possible, however, as we have shown, to separate successive generations of microcracks on the basis of differences in morphology, mode of preservation and consistent cross-cutting relationships. Being able to distinguish between microcracks formed at different stages of deformation is critical to successfully carrying out this kind of analysis. Because cracks can form in a variety of orientations during a single deformation episode, a conservative approach is warranted in distinguishing successive generations of cracks, with distinctive features and consistent cross-cutting relationships being used to separate generations. A variety of tools can be used to track diagnostic features of different generations of cracks; we have found dark-field optical microscopy to be a particularly useful and affordable method for distinguishing between crack types and rapidly collecting quantitative information on them.

Correlating the different generations of microcracks with large-scale tectonic events is more difficult. In the Canyon Mountains, the different generations of cracks can be recognized and distinguished everywhere within the range, not just close to the range-bounding normal fault. This makes it possible to correlate the older microcracks with early tectonic features (e.g. thrust-related foliation, fold tightening) on the basis of orientation and mutually cross-cutting relationships in areas where they have not been reoriented or otherwise affected by later deformation. The youngest microcracks are clearly related to normal faulting on the basis of their style, orientation and frequency of occurrence near normal faults.

The intensity of deformation is quantified not only by the density of microcracks but also by the volume of rock affected by cataclasis. Because both extension and shear fractures have finite thicknesses, as do zones of microbreccia or gouge or cataclasite, it is possible to determine the volume fraction of rock affected by cataclasis. This is a particularly useful measure in strongly comminuted rock where the fragments are too small and the matrix too weak to show the development of microcracks. Overall, the fault zone displays symmetric fault rock distribution (Schmid & Handy 1991) indicating that thermal equilibrium was maintained across the zone during faulting. Because the fault zone is thermally relatively simple, it offers opportunities for understanding the complexities caused by reactivation.

The distribution of deformation features suggests that the earliest cracks and sintered zones (associated with Canyon Range thrusting) are dominantly preserved in the hanging wall of the fault zone, which is also the hanging wall of the Canyon Range thrust. The most recent cracks and cataclasite zones (associated with Basin-and-Range normal faulting) are dominant in the footwall of the normal fault zone. We suggest that these patterns may be related to the localization of movement in the hanging wall or footwall depending on the type of faulting. The second generation cracks (associated with fold tightening) are generally thin and therefore do not involve large volumes of rock; however, the crack density data show that these cracks are present in large numbers everywhere and were probably related to the penetrative cataclastic flow that occurred during that phase of the deformation (Ismat & Mitra 2001).

The volume fraction of rock affected by deformation zones and associated cracks during any single phase of deformation shows inhomogeneous distribution (Figs. 17 and 18), with zones of intense deformation (peaks) separating areas with little deformation (valleys). These patterns suggest anastomosing deformation zones that are developed at scales of 5–10 m (Fig. 17) and 1–2 m (Fig. 18). They are also observed at the hand-sample and thin-section scales. Although these geometries may suggest a fractal distribution of deformation zones, the dominant deformation mechanisms may be very different at different scales (Ismat & Mitra 2001).

It should be clear from the above discussion that it is possible to document the variation in density of microcracks associated with faulting across any recognizable fault zone (either simple or reactivated), and use this as a measure of the intensity of deformation. However, the reverse is not necessarily true; that is, microcrack density cannot always be used as a diagnostic tool to judge the presence or absence of faulting, particularly for blind faults. Because of variations of thickness of a fault zone and inhomogeneities in deformation within a fault zone, microcracking may be unevenly distributed or even completely absent along some segments of a fault; such absences do not necessarily imply that no fault is present.

Conclusions

(1) The Canyon Mountains in central Utah expose the Cretaceous Canyon Range thrust sheet, which is folded as a result of the growth of a footwall connecting splay duplex. Proterozoic to Cambrian quartzites are the dominant lithologies in both the hanging wall and the footwall. Along the western margin of the range Tertiary normal faults truncate the older structure; these normal faults have reactivated the old thrust ramp and imbricate faults of the duplex.

(2) Within the reactivated fault zone three generations of microcracks can be recognized corresponding to the three faulting events. The generations of microcracks are distinguished on the basis of distinctive crack morphologies and modes of preservation observable under polarized light and dark-field optical microscopy, and preliminary SEM study. Microcrack orientations have been used as an additional check to identify the last set of cracks only, as the earlier cracks have presumably been rotated during later deformation.

(3) Variations of total microcrack density and densities of the three generations of cracks, as well as the volume fraction of rock affected by cataclasis during the three separate events, have been documented using stereological techniques. The data indicate that deformation in the fault zone as a whole is very inhomogeneous, with anastomosing zones of intense deformation separating less deformed pods of rock, but with an overall decrease in deformation intensity away from the fault. This pattern is a result of inhomogeneous deformation, with similar patterns, during each of the three faulting events.

(4) The anastomosing zones of intense deformation for successive faulting events do not coincide with one another. Grain-size reduction by cataclasis and recementing during one faulting event hardens the zones sufficiently to make them unlikely candidates for deformation during the next stage. However, the fault zone as a whole is a sufficient inhomogeneity that it is reactivated during successive faulting events.

Acknowledgement is made to the donors of the Petroleum Research Fund, administered by the American Chemical Society, for partial support of this research under grant 33387-AC2. The work was also partly supported by NSF grant EAR-0001030 to G.M. We thank A. Sussman for her help at the start of this project, and George Davis and G. Lloyd, whose thorough reviews and many suggestions helped us in significantly improving the paper. Finally, we acknowledge the special editors, B. Holdsworth and J. Magloughlin, for their patience and their help in finalizing this paper.

References

ALLMENDINGER, R.W. 1992. Fold and thrust tectonics of the western United States exclusive of the accreted terranes. *In*: BURCHFIEL, B.C., LIPMAN, P.W. & ZOBACK, M.L. (eds) *The Cordilleran Orogen, the Conterminous US*. Geological Society of America, The Geology of North America **G-3**, 583–608.

ANDERS, M.H. & CHRISTIE-BLICK, N. 1994. Is the Sevier desert reflection of west–central Utah a normal fault? *Geology*, **22**, 771–774.

ANDERS, M.H. & WILTSCHKO, D.V. 1994. Microfracturing, paleostress and the growth of faults. *Journal of Structural Geology*, **16**, 795–815.

ANDERSON, J.C., LEAVER, K.D., ALEXENDER, J.M. & RAWLINGS, R.D. 1974. *Materials Science*. Van Nostrand Reinhold, Wokingham.

ARMSTRONG, R.L. 1968. Sevier orogenic belt in Nevada and Utah. *Geological Society of America Bulletin*, **79**, 429–458.

ASHBY, M.F., GANDHI, C. & TAPLIN, D.M.R. 1979. Fracture mechanisms maps and their construction for FCC metals and alloys. *Acta Metallurgica*, **27**, 699–729.

ATKINSON, B.K. 1980. Stress corrosion and the rate-dependent tensile failure of a fine grained quartz rock. *Tectonophysics*, **65**, 281–290.

ATKINSON, B.K. 1982. Subcritical crack propagation in rocks: theory, experimental results and applications. *Journal of Structural Geology*, **4**, 41–56.

AYDIN, A. & JOHNSON, A.M. 1978. Development of faults as zones of deformation bands and as slip surfaces in sandstone. *Pure and Applied Geophysics*, **116**, 931–942.

BLENKINSOP, T.G. & RUTTER, E.H. 1986. Cataclastic deformation of quartzite in the Moine thrust zone. *Journal of Structural Geology*, **8**, 669–681.

BOYER, S.E. & ELLIOTT, D. 1982. Thrust systems. *AAPG Bulletin*, **66**, 1196–1230.

BOYER, S.E. & MITRA, G. 1988. Deformation of the basement–cover transition zone of the Appalachian Blue Ridge Province. *In*: MITRA, G. & WOJTAL, S. (eds) *Geometry and Mechanisms of Thrusting, with Special Reference to the Appalachians*. Geological Society of America, Special Papers, **222**, 119–136.

BROCK, W.G. & ENGELDER, T. 1977. Deformation associated with the movement of the Muddy Mountain overthrust in the Buffington window, southeastern Nevada. *Geological Society of America Bulletin*, **88**, 1667–1677.

BURCHFIEL, B.C. & DAVIS, G.A. 1975. Nature and controls of Cordilleran orogenesis, western United States—extensions of an earlier synthesis. *American Journal of Science*, **275A**, 363–396.

CAINE, J.S., EVANS, J.P. & FORSTER, C.B. 1996. Fault zone architecture and permeability structure. *Geology*, **24**, 1025–1028.

CHESTER, F.M. & CHESTER, J.S. 1998. Ultracataclasite structure and friction processes of the Punchbowl fault, San Andreas system, California. *Tectonophysics*, **295**, 199–221.

CHESTER, F.M. & LOGAN, J.M. 1986. Implications for mechanical properties of brittle faulting observations of the Punchbowl fault zone, California. *Pure and Applied Geophysics*, **124**, 79–106.

COOGAN, J.C. & DECELLES, P.G. 1996. Extensional collapse along the Sevier Desert reflection, northern Sevier Desert basin, western United States. *Geology*, **24**, 933–936.

COOGAN, J.C., DECELLES, P.G., MITRA, G. & SUSSMAN, A.J. 1995. New regional balanced cross section across the Sevier desert region and the Central Utah thrust belt. *Geological Society of America, Rocky Mountain Meeting, Abstracts*, **27**, 7.

COOPER, M.A. & WILLIAMS, G.D. (eds) 1989. *Inversion Tectonics*. Geological Society, London, Special Publications **44**.

COTTRELL, A.H. 1964. *The Mechanical Properties of Matter. University of Cambridge*. Wiley, Chichester.

DECELLES, P.G. & MITRA, G. 1995. History of the Sevier orogenic wedge in terms of critical taper models, northeast Utah and southwest Wyoming. *Geological Society of America Bulletin*, **107**, 454–462.

DECELLES, P.G., LAWTON, T.F. & MITRA, G. 1995. Timing of Sevier thrusting, central Utah. *Geology*, **23**, 699–702.

ELLIOTT, D. 1977. Some aspects of the geometry and mechanics of thrust belts. *8th Annual Canadian Society of Petrology and Geology Seminar*, **1**, 17.

ENGELDER, T. 1974. Cataclasis and the generation of fault gouge. *Geological Society of America Bulletin*, **85**, 1515–1522.

EVANS, J.P. 1993. Deformation mechanisms and kinematics of a crystalline-cored thrust sheet: the EA thrust system, Wyoming. *In*: SCHMIDT, C., CHASE, R. & ERSLEV, E. (eds) *Basement Behavior in Rocky Mountain Foreland Structure*. Geological Society of America, Special Papers **280**, 147–161.

GALLAGHER, J.J., FRIEDMAN, M., HANDIN, J. & SOWERS, G.M. 1974. Experimental studies relating to microfractures in sandstone. *Tectonophysics*, **21**, 203–247.

GANDHI, C. & ASHBY, M.F. 1979. Fracture mechanism maps for materials which cleave: FCC, BCC and HCP metals and ceramics. *Acta Metallurgica*, **27**, 1565–1602.

GILOTTI, J.A. 1989. Reaction progress during mylonitization of basaltic dikes along the Särv thrust, Swedish Caledonides. *Contributions to Mineralogy and Petrology*, **101**, 30–45.

HADIZADEH, J. 1994. Interaction of cataclasis and pressure solution in a low temperature carbonate shear zone. *Pure and Applied Geophysics*, **143**, 255–280.

HEYNEKAMP, M.R., GOODWIN, L.B., MOZLEY, P.S. & HANEBERG, W.C. 1999. Controls on fault zone architecture in poorly lithified sediments, Rio Grande rift, New Mexico: implications for fault zone permeability and fluid flow. *In*: HANEBERG, W.C., MOZLEY, P.S., MOORE, J.C. & GOODWIN, L.B. (eds) *Faults and Subsurface Flow in the Shallow Crust*. Geophysical Monograph, American Geophysical Union **113**, 27–49.

HINTZE, L.F. 1988. *Geologic History of Utah*. Brigham Young University Geology Studies Special Publication **7**.

HULL, J. 1988. Short note: Thickness–displacement relationships for deformation zones. *Journal of Structural Geology*, **10**, 431–435.

ISMAT, Z. & MITRA, G. 2001. Folding by cataclastic flow at shallow crustal levels in the Canyon Range, Sevier orogenic belt, west–central Utah. *Journal of Structural Geology*, **23**, 355–378.

KERRICH, R., ALLISON, I., BARNETT, R.L., MOSS, S. & STARKEY, J. 1980. Microstructural and chemical transformations accompanying deformation of granite in a shear zone at Mieville, Switzerland; with implications for stress corrosion cracking and superplastic flow. *Contributions to Mineralogy and Petrology*, **73**, 221–242.

KING, G. 1983. The accommodation of large strains in the upper lithosphere of the earth and other solids by self-similar fault systems: the geometrical origin of *b*-value. *Pure and Applied Geophysics*, **121**, 761–815.

KNAUTH, P.L. 1994. Petrogenesis of chert. *In*: HEANY, P.J., PREWITT, C.T. & GIBBS, G.V. (eds) *Silica; Physical Behavior, Geochemistry and Material Applications*. Mineralogical

Society of America, Reviews in Mineralogy **29**, 233–258.

KNIPE, R.J. 1989. Deformation mechanism: Recognition from natural tectonites. *Journal of Structural Geology*, **11**, 127–146.

KNIPE, R.J. & LLOYD, G.E. 1994. Microstructural analysis of faulting in quartzite, Assynt, NW Scotland; implications for fault zone evolution. *Pure and Applied Geophysics*, **143**, 229–254.

KRANZ, R.L. 1983. Microcracks in rock: a review. *Tectonophysics*, **100**, 449–480.

LAWN, B. 1993. *Fracture of Brittle Solids* (2nd edition). Cambridge University Press, Cambridge.

LAWTON, T.F., SPRINKEL, D.F., DECELLES, P.G., MITRA, G., SUSSMAN, A.J. & WEISS, M.P. 1997. Sevier thrust belt central Utah: Sevier Desert to Wasatch Plateau. *In*: LINK, K.P. & KOWALLIS, B.J. (eds) *Brigham Young University, Geology Studies Field Trip Guide Book*. Geological Society of America, Annual Meeting **II**, 33–68.

LLOYD, G.E. & KNIPE, R.J. 1992. Deformation mechanisms accommodating faulting of quartzite under upper crustal conditions. *Journal of Structural Geology*, **14**, 127–143.

MARSHAK, S. & MITRA, G. 1998. *Basic Methods of Structural Geology*. Prentice Hall, Englewood Cliffs, NJ.

MEANS, W.D. 1984. Shear zones of Type I and II and their significance for reconstruction of rock history. *Geological Society of America, NE Section Meeting, Abstracts*, **16**, 50.

MEANS, W.D. 1995. Shear zones and rock history. *Tectonophysics*, **247**, 157–160.

MILLER, D.M., NILSEN, T.H. & BILODEAU, W.L. 1992. Late Cretaceous to early Eocene geologic evolution of the U.S. Cordillera. *In*: BURCHFIEL, B.C., LIPMAN, P.W. & ZOBACK, M.L. (eds) *The Cordilleran Orogen, the Conterminous US*. Geological Society of America, The Geology of North America **G-3**, 205–260.

MITRA, G. 1978. Ductile deformation zones and mylonites: the mechanical processes involved in the deformation of crystalline basement rocks. *American Journal of Science*, **278**, 1057–1084.

MITRA, G. 1984. Brittle to ductile transition due to large strains along the White Rock thrust, Wind River Mountains, Wyoming. *Journal of Structural Geology*, **6**, 51–61.

MITRA, G. 1992. Deformation of granitic basement rocks along fault zones at shallow to intermediate crustal levels. *In*: MITRA, S. & FISHER, G.W. (eds) *Structural Geology of Fold and Thrust Belts*. Johns Hopkins University Press, Baltimore, MD, 123–144.

MITRA, G. 1993. Deformation processes in brittle deformation zones in granitic basement rocks: a case study from the Torrey Creek area, Wind River Mountains. *In*: SCHMIDT, C., CHASE, R. & ERSLEV, E. (eds) *Basement Behavior in Rocky Mountain Foreland Structure*. Geological Society of America, Special Papers **280**, 177–195.

MITRA, G. 1994. Strain variation in thrust sheets and across the Sevier fold-and-thrust belt (Idaho–Utah–Wyoming): implications for section restoration and wedge taper evolution. *Journal of Structural Geology*, **16**, 585–602.

MITRA, G. 1997. Evolution of salients in a fold-and-thrust belt: the effects of sedimentary basin geometry, strain distribution and critical taper. *In*: SENGUPTA, S. (ed.) *Evolution of Geological Structures in Micro- to Macro-scales*. Chapman and Hall, London, 59–90.

MITRA, G. & SUSSMAN, A.J. 1997. Structural evolution of connecting splay duplexes and their implications for critical taper: an example based on geometry and kinematics of the Canyon Range culmination, Sevier Belt, central Utah. *Journal of Structural Geology*, **19**, 503–521.

NEWMAN, J. & MITRA, G. 1993. Lateral variations in fault zone thickness as influenced by fluid–rock interactions, Linville Falls fault, North Carolina. *Journal of Structural Geology*, **15**, 849–863.

NEWMAN, J. & MITRA, G. 1994. Fluid-influenced deformation and recrystallization of dolomite at low temperatures along a natural fault zone, Mountain City Window, Tennessee. *Geological Society of America Bulletin*, **106**, 1267–1280.

OTTON, J.K. 1995. Western frontal fault of the Canyon Range: is it the breakaway zone of the Sevier Desert detachment? *Geology*, **23**, 547–550.

PEQUERA, N., MITRA, G. & SUSSMAN, A.J. 1994. The Canyon Range thrust sheet in the Sevier fold-and-thrust belt of central Utah: deformation history based on structural analysis. *Geological Society of America, Rocky Mountain Section Meeting, Abstracts*, **26**, 58.

RAJ, R. 1982. Creep in polycrystalline aggregates by matter transport through a liquid phase. *Journal of Geophysical Research*, **87**, 4731–4739.

ROBERTSON, E.C. 1983. Relationship of fault displacement to gouge and breccia thickness. *Mining Engineering*, **35**, 1426–1432.

ROYSE, F., WARNER, M.A. & REESE, D.L. 1975. Thrust belt structural geometry and related stratigraphic problems, Wyoming–Idaho–northern Utah. *Rocky Mountain Association of Geologists 1975 Symposium*, 41–54.

SCHMID, S.M. & CASEY, M. 1986. Complete fabric analysis of some commonly observed quartz *c*-axis patterns. *In*: HOBBS, B.E. & HEARD, H.C. (eds) *Mineral and Rock Deformation: Laboratory Studies. The Paterson Volume*. Geophysical Monograph, American Geophysical Union **36**, 263–286.

SCHMID, S.M. & HANDY, M.R. 1991. Towards a genetic classification of fault rocks: geological usage and tectonophysical implications. *In*: MULLER, D.W., MCKENZIE, J.A. & WEISSERT, H. (eds) *Controversies in Modern Geology: Evolution of Geological Theories in Sedimentology, Earth History and Tectonics*. Academic Press, New York, 339–361.

SCHOLZ, C.H. 1990. *The Mechanics of Earthquakes and Faulting*. Cambridge University Press, Cambridge.

SCHONBORN, G. 1992. Alpine tectonics and kinematic models of the central Southern Alps. *Memorie degli Istituti di Geologia e Mineralogia dell'Universita di Padova*, **44**, 229–393.

SCHULZ, S.E. & EVANS, J.P. 1998. Spatial variability in microscopic deformation and composition of the Punchbowl fault, southern California: implications for mechanisms, fluid–rock interaction and fault morphology. *Tectonophysics*, **295**, 223–244.

SIBSON, R.H. 1977. Fault rocks and fault mechanics. *Journal of the Geological Society, London*, **133**, 191–213.

SIBSON, R.H. 1990. Conditions for fault-valve behavior. *In*: KNIPE, R.J. & RUTTER, E.H. (eds) *Deformation Mechanisms, Rheology and Tectonics*. Geological Society, London, Special Publications **54**, 15–28.

SNOKE, A.W., TULLIS, J. & TODD, V.R. (eds) *Fault Related Rocks: a Photographic Atlas*. Princeton University Press, Princeton, NJ.

SUSSMAN, A. J., 1995. Geometry, deformation history and kinematics in the footwall of the Canyon Range thrust, central Utah. M.S. thesis, University of Rochester, NY.

TURCOTTE, D. 1989. Fractals in geology and geophysics. *In*: SCHOLZ, C.H. & MANDELBROT, B.B. (eds) *Fractals in Geophysics*. Birkhauser, Boston, MA, 171–196.

TWISS, R.J. & MOORES, E.M. 1992. *Structural Geology*. W.H. Freeman, San Francisco, CA.

UNDERWOOD, E.E. 1970. *Quantitative Stereology*. Addison Wesley, Reading, MA.

WOJTAL, S., 1982. *Finite deformation of thrust sheets and their material properties*. Ph.D. thesis, Johns Hopkins University, Baltimore, MD.

WOJTAL, S. 1986. Deformation within foreland thrust sheets by populations of minor faults. *Journal of Structural Geology*, **8**, 341–360.

WOJTAL, S. & MITRA, G. 1986. Strain hardening and strain softening in fault zones from foreland thrusts. *Geological Society of America Bulletin*, **97**, 674–687.

WOJTAL, S. & MITRA, G. 1988. Nature of deformation in some fault rocks from Appalachian thrusting. *In*: MITRA, G. & WOJTAL, S. (eds) *Geometries and Mechanisms of Thrusting with Special Reference to the Appalachians*. Geological Society of America, Special Papers **222**, 17–33.

YONKEE, W.A. 1992. Basement–cover relations, Sevier orogenic belt, northern Utah. *Geological Society of America Bulletin*, **104**, 280–322.

YONKEE, W.A. & MITRA, G. 1993. Comparison of basement deformation styles in the Rocky Mountain Foreland and Sevier Orogenic Belt. *In*: SCHMIDT, C., CHASE, R. & ERSLEV, E. (eds) *Basement Behavior in Rocky Mountain Foreland Structure*. Geological Society of America, Special Papers **280**, 197–228.

ZADINS, Z.Z., 1989. *Structural and thermal evolution of the Hudson Valley Little Mountains thrust belt, eastern New York*. Ph.D. thesis, University of Rochester, NY.

Episodic weakening and strengthening during synmetamorphic deformation in a deep-crustal shear zone in the Alps

KURT STEFFEN, JANE SELVERSTONE & ADRIAN BREARLEY

Department of Earth and Planetary Sciences, University of New Mexico, Albuquerque, NM 87131-1116, USA (e-mail: ksteffen@unm.edu)

Abstract: The Greiner shear zone (western Tauern Window) deformed a variety of metasedimentary, metavolcanic, and plutonic lithologies at conditions of *c*. 525–575 °C and 30–40 km depth. Microstructural relationships point to a succession of weakening and strengthening episodes. Stage I involved softening via a change in deformation mechanism. Grain-size reduction in plagioclase-rich horizons locally produced rocks with an average grain size of <30 μm and microstructural features consistent with deformation via grain boundary diffusion creep (GBDC; a fluid-assisted deformation mechanism similar to pressure solution, which may result in superplastic behaviour). At constant stress, GBDC will result in a significant increase in strain rate relative to neighbouring layers. Stage II involved reaction-induced strengthening. Rapid bulk diffusion rates associated with GBDC allowed rapid growth of large (up to 20 cm) hornblende cystals. This growth of large cross-cutting crystals shut down grain-size-sensitive flow mechanisms in the plagioclase matrix and locally 'locked' the shear zone, shifting ductile deformation to weaker horizons. Stage III involved reaction-induced softening. Local variations in bulk composition and/or fluid availability caused large hornblende grains in some horizons to be partially replaced by biotite. These biotite-rich layers localized subsequent deformation, whereas adjacent layers with intact hornblende record minimal strain. Deformation and metamorphism together exert control on fluid availability, diffusion rates, and reaction kinetics, and these factors collectively control fabric development and rheology. The effects of interaction between small-scale deformational and metamorphic processes are difficult to predict, but can have an important influence on shear zone behaviour at depth. The resulting complexities need to be accounted for in models of crustal strength–depth relationships and shear zone rheologies.

Strain localization leading to shear zone formation in the middle and lower crust is generally thought to result from strain softening owing to cataclasis, dynamic recrystallization, and/or the local development of fine-grained, often hydrous metamorphic reaction products (White & Knipe 1978; Lister & Williams 1983; Brodie & Rutter 1985; Tullis & Yund 1987; Rutter 1999). The resulting decrease in grain size allows grain-size-sensitive deformation–recovery mechanisms to be activated (Handy 1989; Tullis *et al.* 1990). These mechanisms are generally diffusion-rate controlled and allow for higher strain rates than their power-law counterparts at constant stress conditions, leading to strain softening. The possible driving forces for dynamic recrystallization include chemical, surficial, and elastic energies in addition to deformational excess energies (Urai *et al.* 1986; Stünitz 1998). The fact that chemical excess energy (disequilibrium) may provide a control on 'dynamic recrystallization' indicates a strong interrelationship between deformation and metamorphism. In addition, certain recovery mechanisms can enhance diffusion rates and thus reaction rates, especially in the presence of a metamorphic fluid (Wintsch & Knipe 1983; Dipple *et al.* 1990; Yund & Tullis 1991). Because both deformation and metamorphism can affect the thermodynamics and kinetics of dynamic and heterogeneous recrystallization, both processes must be considered in the study of mid- and lower-crustal shear zones. Interactions between metamorphism and deformation, which can lead to localized or bulk strengthening or weakening of shear zones for variable time intervals, will significantly complicate efforts to model the strength of the crust.

In this study of the Greiner shear zone (western Tauern Window, Eastern Alps) we show how metamorphism and deformation interacted to affect the rheology of a deep-crustal shear zone through the following

From: HOLDSWORTH, R.E., STRACHAN, R.A., MAGLOUGHLIN, J.F. & KNIPE, R.J. (eds) 2001. *The Nature and Tectonic Significance of Fault Zone Weakening*. Geological Society, London, Special Publications, **186**, 141–156. 0305-8719/01/$15.00

processes: (1) softening via fluid-assisted, grain boundary diffusion creep (a grain-size- and fluid-dependent deformation–recovery mechanism that involves both grain boundary sliding and diffusional processes) in the plagioclase-rich matrix. Although this mechanism has previously been identified only in laboratory experiments, optical, compositional and structural data allow correlation between natural samples and experimental results. (2) Reaction-induced mechanical strengthening via the growth of large, radiating bundles of hornblende that provide a rigid framework. Development of this unique hornblende morphology is probably due to rapid diffusion rates resulting from deformation-enhanced fluid distribution (Tullis *et al.* 1996) in the plagioclase matrix. (3) Reaction-induced softening as a result of the partial replacement of hornblende by biotite. These reactions, which probably were controlled by local variations in fluid availability, caused strain weakening by destroying the rigid hornblende framework and by production of phyllosilicates that subsequently deformed by slip along cleavage planes and grain boundaries.

These three metamorphic–deformational processes led to both spatial and temporal variations of rheology within the Greiner shear zone. Detailed documentation of the interaction of deformation and metamorphism in exhumed lower-crustal shear zones, such as the Greiner shear zone, is essential for the development of a quantitative model of stress–strain relationships in the deep crust.

Geological setting and earlier work

The Greiner shear zone separates two distinct lobes of basement gneiss in the western half of the Tauern Window in the Eastern Alps (Fig. 1). The principal metamorphism and deformation in the region resulted from Late Cretaceous–Tertiary collision between the Eurasian plate and the Adriatic plate, with closure of the intervening Neo-Tethys ocean basin. Convergence initially resulted in the west- and NNW-directed emplacement of nappes derived from the Adriatic plate over the European and oceanic rocks now exposed in the window (e.g. Dewey *et al.* 1973; Dietrich 1976). Subsequent transpressional motion of the Adriatic plate led to east–west extension of the overthickened crust in the vicinity of the Tauern Window (Selverstone 1988; Behrmann & Frisch 1990; Ratschbacher *et al.* 1991; Lammerer & Weger 1998).

The rocks within the window are subdivided into three lithotectonic packages (Morteani 1974). The structurally lowest unit, the Zentralgneis (ZG), is composed of pre-Alpine plutonic rocks. In contact with the ZG is the autochthonous to parautochthonous Lower Schieferhülle (LSH), which is composed of Palaeozoic amphibolite and graphitic schist and an overlying Mesozoic cover sequence of conglomerate, quartzite, marble, and pelitic and calcareous schist. The Upper Schieferhülle (USH), an allochthonous sequence of greenstone, marble, and graphitic schist derived from the floor of the neo-Tethys ocean basin, is the structurally highest unit. In the area of this study, all of the rocks experienced amphibolite-facies metamorphism during Alpine orogenesis (Selverstone *et al.* 1984; Selverstone 1993).

The Greiner shear zone deforms rocks from all structural levels within the southwestern portion of the Tauern Window and must therefore be a structure of post-nappe Alpine age, although earlier motion related to the Hercynian orogeny and/or Early Mesozoic rifting may also have occurred. The shear zone has a subvertical dip, and can be traced along strike for a distance of >35 km (Fig. 1). In the east, where the shear zone interacts with the Zentralgneis, it is composed of anastomizing bands that are typically between 1 and 10 m thick. In the west, in areas where the shear zone interacts with the Lower Schieferhülle, it widens to >1 km, but preserves strain heterogeneities within this more diffuse zone. De Vecchi & Baggio (1982); Behrmann & Frisch (1990) documented sinistral motion on this shear zone, Selverstone *et al.* (1991) argued for a significant component of pressure solution and volume loss, and Lammerer & Weger (1998) showed large flattening strains within and adjacent to the zone. Selverstone *et al.* (1984, 1991) documented deformation in the shear zone during decompression and heating from *c.* 525 °C and 10–11 kbar to *c.* 550–575 °C and 7–8 kbar, with 'peak' temperature increasing eastward. Direct dating (Rb/Sr) of garnet cores and rims from a sample located at the southern margin of the Greiner shear zone yielded ages of 35 ± 1 Ma for the cores and 30 ± 1 Ma for the rims (Christensen *et al.* 1994). Selverstone *et al.* (1991) argued for the involvement of large amounts of externally derived fluid during deformation at the deepest levels of the shear zone, but at shallower levels fluids appear to have had incomplete access to the shear zone (Selverstone & Munoz 1987).

Where metavolcanic rocks of the Lower Schieferhülle were affected by the Greiner shear zone, they were typically transformed into rocks with spectacular radiating bundles of coarse hornblende (up to 20 cm long) set in a finer-grained matrix (Fig. 2). These rocks are referred

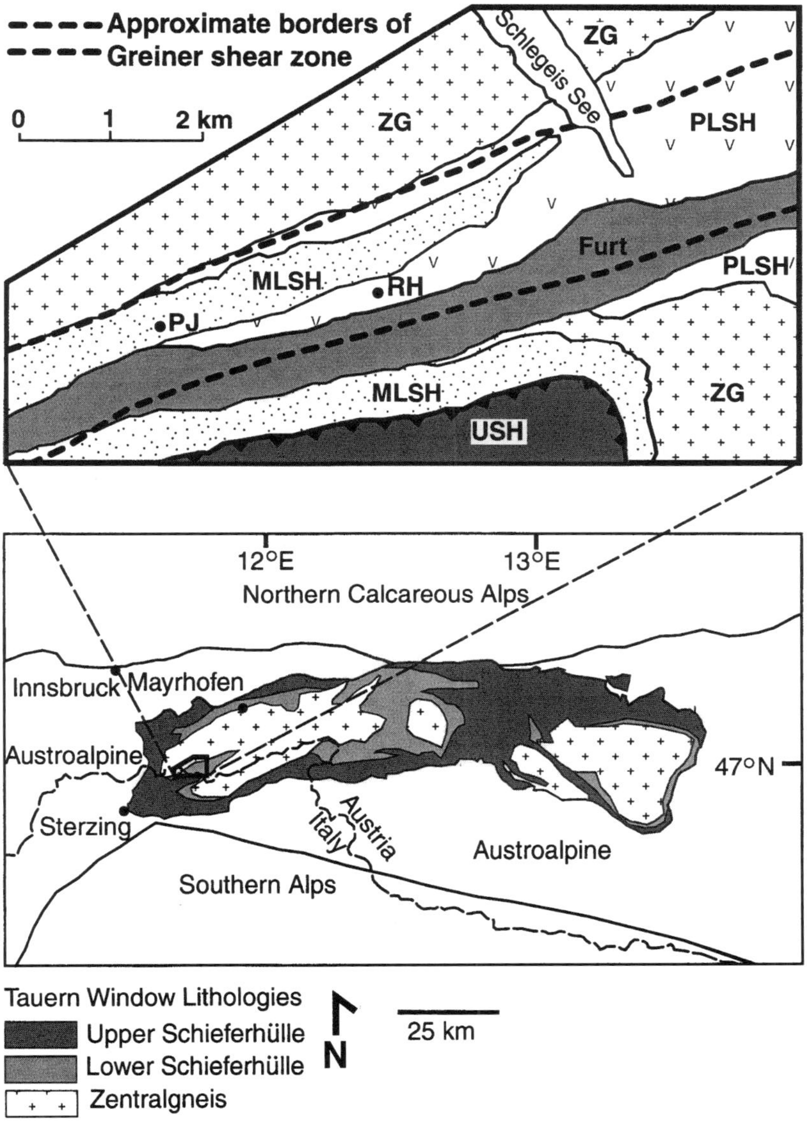

Fig. 1. Below: generalized geological map of the Tauern Window, showing location of field area. Above: geological map of field area showing the location of the Greiner shear zone as it crosses units of the Lower Schieferhülle. ZG, Zentralgneis; PLSH, Palaeozoic Lower Schieferhülle; MLSH, Mesozoic Lower Schieferhülle; Furt, Furtschaglschiefer; USH, Upper Schieferhülle; PJ, Pfitscher Joch Haus; RH, Rotbachlspitze.

to as Garbenschiefer (from the German 'Garbe' (sheaf), and 'Schiefer' (schist)). In some samples, the hornblende sprays lie within the plane of foliation, but in many cases hornblende crosscuts, and thus clearly postdates, the foliation. Garbenschiefer horizons are intimately intermingled with biotite, white mica and chlorite schists. The Furtschaglschiefer, a graphitic schist finely intercalated with the Garbenschiefer along the Garbenschiefer's southern margin, records significant variations in fluid composition over the scale of millimetres to centimetres (Selverstone & Munoz 1987), implying either limited fluid mobility or fluid movement only parallel to the foliation of the shear zone. Local intermingling of graphitic schist within the Garbenschiefer suggests that fluid composition and/or availability may have been heterogeneous within the Garbenschiefer as well.

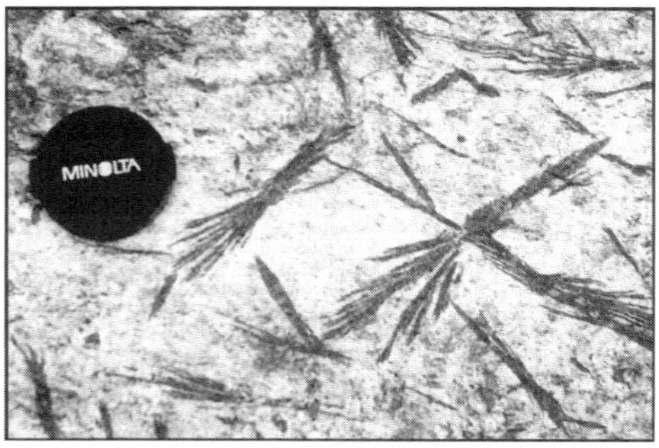

Fig. 2. Field photograph of typical Garbenschiefer framework texture dominated by large hornblende porphyroblasts in a matrix of fine-grained plagioclase. Lens cap for scale.

Minerals identified within the Garbenschiefer include hornblende (pargasite to edenite), plagioclase, quartz, biotite, chlorite, garnet, white mica (phengite ± paragonite ± margarite), epidote, ankerite, staurolite, kyanite, and ilmenite (Selverstone *et al.* 1984). Chlorite and white mica do not coexist within individual samples, and staurolite and kyanite are typically associated only with white mica horizons. Epidote and ankerite are locally abundant. Garnet appears with both chlorite and white mica but is more abundant in white mica horizons. The variation in mineralogy is apparently controlled by variations in bulk chemistry and reaction kinetics, not by variations in $P-T$ conditions, as demonstrated by the intermingling of different assemblages on a centimetre to metre scale throughout the Garbenschiefer.

We observed several distinct domains within the Garbenschiefer that appear to record successive changes in strain localization during deformation. Layers with randomly oriented sprays of hornblende and only a weak foliation alternate on a centimetre to metre scale with phyllosilicate-dominated layers characterized by a well-developed foliation. In thin section, it is apparent that many of the more prominently foliated horizons were derived from unfoliated precursors during replacement of hornblende by biotite ± chlorite. A large variety of plagioclase microstructures is also evident in thin section. Some samples contain an extremely fine-grained plagioclase matrix with no obvious shape-preferred or lattice-preferred orientation, whereas other samples contain coarse-grained plagioclase with high grain aspect ratios and an obvious lattice-preferred orientation.

Textural analysis of samples displaying different microstructural features indicates that the strain localization within the Greiner zone shifted repeatedly in space and time as the interaction of deformational and metamorphic processes weakened and strengthened different domains within the shear zone. Domains that underwent strain weakening through grain-size reduction and the activation of diffusional processes were subsequently strengthened as a result of the formation of large hornblende porphyroblasts (Garbenschiefer texture), probably related to rapid intergranular diffusion and decreased distances for intragranular diffusion. This hardening of originally strain-softened domains presumably caused strain to be transferred to other areas within the shear zone. In some domains, the Garbenschiefer texture was destroyed by metamorphic reactions that produced biotite ± chlorite at the expense of hornblende; these reactions caused strain weakening by destroying the rigid hornblende framework. Still other domains were dominated by dislocation creep processes that are inferred to represent an intermediate strength regime within the Greiner shear zone. The variations in these domains and the complex processes that control their interactions lead us to infer a complex stress–strain relationship involving both temporal and spatial variation throughout the shearing history.

Review of relevant deformation mechanisms

Dislocation creep

Dislocation creep is probably the dominant deformation mechanism affecting the strength of large domains of the crust (e.g. Tullis & Yund 1991). Strain is initially produced by glide of dislocations, leading to production of dislocation tangles and strain hardening; recovery processes (dynamic recrystallization and dislocation climb) act to offset strain hardening by reducing internal strain energy. Processes associated with dislocation creep in quartz- and feldspar-dominated rocks have been documented both experimentally and in natural samples (Simpson 1983; Wintsch & Knipe 1983; Rutter & Brodie 1988; Hacker & Christie 1990; Tullis *et al.* 1990; Tullis & Yund 1991; Hirth & Tullis 1992).

Deformation via dislocation creep leaves a signature in natural quartzofeldspathic rocks that can be recognized petrographically (assuming subsequent annealing has been minor). Grains may show undulatory extinction, subgrain development, sutured grain boundaries, core-and-mantle structure, and development of a pronounced shape-preferred and/or lattice-preferred orientation (Schmid & Handy 1991).

Fluid-assisted grain-boundary diffusion creep: grain-size-sensitive flow

Some workers have described evidence for grain-size-sensitive flow in rocks deformed experimentally and in nature (Boullier & Gueguen 1975; White & Knipe 1978; Schmid 1982; Rubie 1983; Behrmann & Mainprice 1987; Tullis & Yund 1991; Stockhert *et al.* 1999). In some cases, grain-size-sensitive flow is induced by the transient production of fine-grained reaction products during metamorphism (reaction-enhanced ductility). In other cases, it can result from migration of dislocations, leading to a reduction in grain size (dynamic recrystallization). Tullis & Yund (1991) experimentally identified a grain-size-sensitive deformation–recovery mechanism similar to pressure solution with a component of grain boundary sliding, which they named grain-boundary diffusion creep (GBDC). They found that GBDC was active in plagioclase, but that no corresponding mechanism could be identified in quartz at experimental conditions of 900 °C, strain rates of $10^{-5}\,s^{-1}$ in samples containing *c.* 0.9 wt % water, and grain sizes <10 μm. Dry and/or coarse-grained samples of plagioclase deformed under similar temperatures and strain rates displayed microstructural features consistent with dislocation creep.

GBDC is facilitated by grain-size reduction events, such as cataclasis, crossing of a discontinuous reaction boundary, or dynamic recrystallization (e.g. Handy 1989), and requires the presence of a grain-boundary fluid to facilitate grain-scale mass transfer. Deformation via this mechanism will result in a very fine-grained matrix in which mineral grains show lower aspect ratios even after large strains, have low dislocation densities, and exhibit no lattice-preferred orientation (Knipe 1989; Schmid & Handy 1991). Isotopic and chemical homogenization is likely to occur (Tullis & Yund 1991), but solid solution minerals may develop complex zoning patterns with two or more compositions that mutually truncate one another (we expect that this will be particularly likely in minerals that have miscibility gaps, such as plagioclase, amphibole, or epidote).

GBDC is a grain-size-sensitive process requiring rapid intercrystalline diffusion and short distances for intracrystalline diffusion, to be rate competitive with dislocation creep. These high diffusion rates may result from deformation-enhanced fluid distribution (Tullis *et al.* 1996). In laboratory experiments in which GBDC was identified, samples quenched immediately after deformation showed evidence of nearly complete grain boundary wetting (dihedral angle *c.* 0°). Samples that were held hydrostatically after deformation showed completely closed grain boundaries that were produced by fluid migration back to grain triple junctions after deformation had ceased (dihedral angle >60°). Tullis *et al.* (1996) hypothesized that complete grain boundary wetting should increase creep rates by several orders of magnitude. Although they did not formulate a flow law for plagioclase undergoing GBDC, they stated that GBDC probably represents a weakening process in relation to dislocation creep mechanisms. Similarly, Cooper & Kohlstedt (1984) documented a threefold increase in creep rate in the presence of localized melt channels relative to melt-absent experiments. The presence of fully wetted grain boundaries greatly increases the surface area available for fluid-mineral interactions, and will increase bulk transport rates where GBDC is active. If fluid distribution patterns (and therefore fluid diffusion rates) in plagioclase-rich samples are related to deformation, GBDC is likely to operate and could greatly increase rates of transport-limited processes such as heterogeneous nucleation and grain growth (Carlson 1989; Chernoff & Carlson 1997). If complete grain wetting is indeed a transient process related to deformation, however, it will be nearly impossible for natural samples to preserve open grain boundaries during subsequent periods of hydrostatic stress at elevated temperatures.

Rocks deforming via GBDC may accommodate large strains, yet still preserve textures that appear to be indicative of fairly low cumulative strain. Inferring the occurrence of this deformation mechanism in natural samples thus requires detailed petrographic and compositional data. Perhaps as a result of the 'undeformed' appearance of GBDC textures, relatively few published studies have argued for GBDC in natural samples. The importance of this mechanism in nature is thus not well known. If GBDC proves to be a common mechanism, the implications for crustal strength in regions undergoing deformation are enormous (e.g. Handy 1990).

Deformation of amphibole and phyllosilicates

Little is known about deformation mechanisms in amphibole, but it is clear that amphibole is one of the strongest common silicate minerals at depth in the crust (Brodie & Rutter 1985; Cumbest *et al.* 1989; Hacker & Christie 1990). Several slip systems have been identified at high temperatures (Biermann & Van Roermund 1983; Cumbest *et al.* 1989), but they do not appear to operate at the temperature conditions of interest in this study (500–600 °C).

Micas deform primarily by slip along their cleavage planes, or by kinking, folding, and fracturing. Grain-boundary migration recrystallization may be important at moderate to high metamorphic grade (Wilson & Bell 1979). Muscovite is apparently somewhat stronger than biotite (Wilson & Bell 1979; Bell & Wilson 1981), but little is known about other white mica compositions such as paragonite or phengite. Chlorite behaviour is similar to that of biotite (Bons 1988; Sancheznavas & Galindozaldivar 1993). The rheology of phyllosilicate-rich rocks will probably be largely controlled during deformation by cleavage-parallel slip of favourably oriented grains (Kronenberg *et al.* 1990; Shea *et al.* 1993).

Microstructural features of the Garbenschiefer

The single most striking feature of the Garbenschiefer is the development of large, radiating bundles of hornblende that appear to be postkinematic (Fig. 2). In some samples, hornblende garben occur only on foliation surfaces, but in many cases, hornblende overprints both the foliation and lineation. Despite the postkinematic appearance of hornblende, however, deformation clearly continued after hornblende growth; hornblende is locally fractured and extended parallel to the stretching lineation and, where partially replaced by biotite, the biotite is offset along cleavage planes parallel to the foliation.

In thin section, undeformed hornblende is generally set in a matrix of extremely fine-grained, granoblastic plagioclase, whereas samples with a more prominent foliation generally contain coarser-grained plagioclase with high grain aspect ratios. Regions of fine-grained (typically <30 μm) plagioclase in the Garbenschiefer matrix preserve round to rectangular grain shapes with low grain aspect ratios and only slight lattice-preferred and shape-preferred orientations (Fig. 3a). Plagioclase inclusions in hornblende cores are typically even finer grained than matrix grains, indicating that matrix plagioclase growth occurred during or after hornblende growth. Asymmetric compositional zoning is evident from extinction patterns in fine-grained matrix plagioclase, and is confirmed in back-scattered electron images and quantitative electron microprobe traverses, with compositions ranging from An_{15} to An_{28} in single grains (Fig. 3b and c). Individual grain edges vary from An_{17} to An_{27} in a single sample, with no evidence of equilibration across grain boundaries. Compositional changes as great as An_{15} to An_{25} occur over distances as short as 6 μm in some grains, whereas changes as small as 1–5 mol % occur over entire grains in other cases. These microstructures and varia-

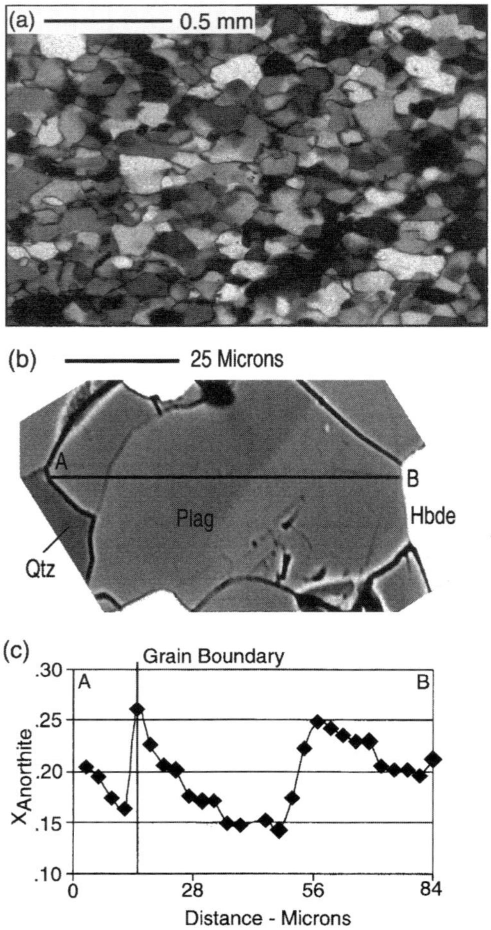

Fig. 3. (**a**) Photomicrograph (crossed polars) of typical grain boundary diffusion creep (GBDC) texture in plagioclase. (**b**) Back-scattered electron image of GBDC plagioclase (note the non-concentric compositional zoning and the compositional breaks across grain boundaries). (**c**) Quantitative electron microprobe traverse along line A–B shown in (**b**).

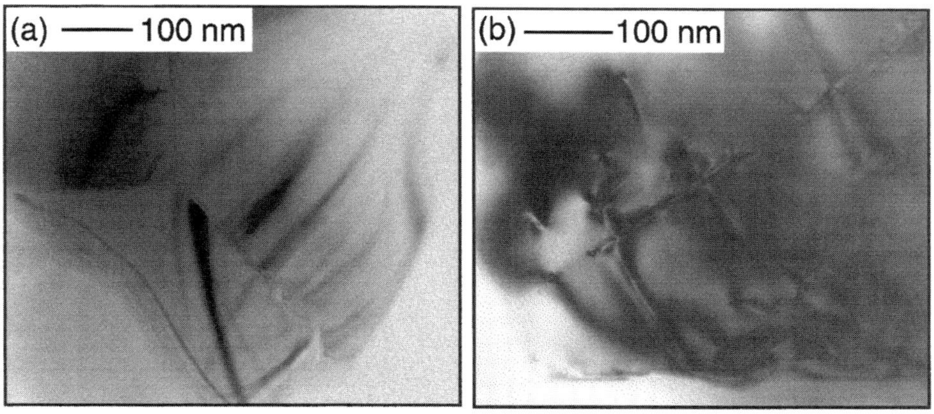

Fig. 4. (**a**) TEM bright field image of low-dislocation-density GBDC plagioclase. (**b**) TEM bright-field image of isolated, moderate-dislocation-density plagioclase grain in GBDC-dominated domain.

tions in composition are consistent with those observed experimentally for GBDC (Tullis & Yund 1991; Yund & Tullis 1991).

Transmission electron microscopy (TEM) bright-field images (Fig. 4a) of fine-grained plagioclase show that >90% of the plagioclase grains examined have extremely low dislocation densities or are dislocation free. The remaining grains have low to moderate dislocation densities (Fig. 4b). The distribution of the strained grains appears to be random: higher dislocation density grains are surrounded by grains with extremely low dislocation densities and are not clustered with other strained grains. When combined with the compositional data, these observations suggest that GBDC was the dominant deformation mechanism in the fine-grained plagioclase aggregates.

Apparent alteration products were observed via TEM along a few grain boundaries, which could indicate that grain boundaries were open during or after deformation; however, no conclusive evidence of open grain boundaries was observed and alteration could have resulted from low-temperature weathering processes. Tullis *et al.* (1996) identified open grain boundaries in experimental samples that preserve evidence of GBDC. However, open grain boundaries were not preserved in experimental samples that were allowed to anneal under hydrostatic conditions for short periods of time, indicating that open boundaries are transient features related to differential stress. Preservation of open grain boundaries is thus unlikely in natural samples, and we can only infer that open grain boundaries (dihedral angle near 0°) existed in the fine-grained plagioclase aggregates during active GBDC deformation.

Not all plagioclase in the Greiner shear zone shows evidence of GBDC. Domains of comparatively coarse grained (>0.5 mm) plagioclase with core-and-mantle structures (Fig. 5) are interspersed on a centimetre to metre scale with domains that show features indicative of GBDC. Compositional zoning in dislocation creep domains is typically smooth and concentric, with edge compositions around An_{28-32} and cores around An_{11-18} (Selverstone *et al.* 1984). These features are consistent with recrystallization- and climb-accommodated dislocation creep (Hirth & Tullis 1992).

Hornblende locally is partially to completely replaced by biotite and other phases in samples that contain white mica (Fig. 6). Selverstone & Munoz (1987) identified three dehydration reactions by which hornblende and white mica can react to form biotite, garnet, plagioclase, quartz, epidote, ± chlorite, ± staurolite in the adjacent graphitic Furtschaglschiefer. At the temperatures of metamorphism of the Furtschaglschiefer (*c.* 550 °C, the same as the Garbenschiefer), graphite locally reduced aH_2O in the metamorphic fluid sufficiently to favour the reaction products over the reactants. Although graphite is rare in the Garbenschiefer, small variations in fluid composition in this unit may none the less have been sufficient to locally destabilize hornblende in the presence of white mica, thereby causing partial to complete pseudomorphing of strong hornblende by weaker phyllosilicate minerals.

Many samples in which hornblende is replaced by biotite ± chlorite are characterized by a prominent foliation. In some cases, biotite is faulted along cleavage planes to such a degree that single grains are extended for long

Fig. 5. Photomicrograph (crossed polars) of dislocation creep features in coarse-grained plagioclase.

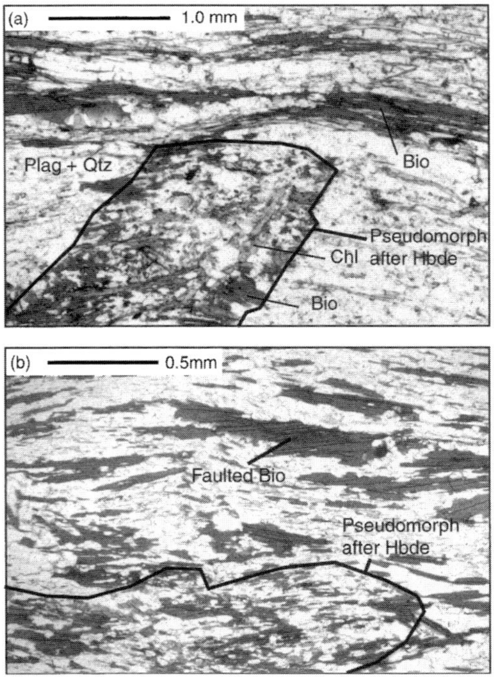

Fig. 6. (**a**) Photomicrograph (plane-polarized light) of biotite and chlorite pseudomorph after hornblende (lower), which grades into a mica-dominated shear foliation (upper). (**b**) Photomicrograph (plane-polarized light) of biotite–plagioclase domain. (Note evidence of slip along biotite cleavage planes and the presence of a pseudomorph after hornblende in the lower left.)

distances (Fig. 6a and b); relict hornblende is typically undeformed in these samples.

Successive shear zone strengthening and weakening

Extreme microstructural variations occur throughout the Garbenschiefer horizon. Matrix minerals, dominated by plagioclase, vary from 10 μm to >1 mm in diameter and show evidence for deformation and recovery via both GBDC and dislocation creep. Hornblende porphyroblasts are undeformed in some regions but in others are fractured, boudinaged, and/or partially replaced by biotite and transposed into the shear foliation. Some horizons preserve an extremely well-developed foliation defined by phyllosilicate minerals, whereas foliation is nearly absent in other horizons. These heterogeneities suggest that considerable strength variations developed in the shear zone during synchronous deformation and metamorphism. On the basis of these heterogeneities, we propose a three-stage history for the spatial and temporal rheological evolution in the Greiner shear zone.

Stage I: softening associated with grain-boundary diffusion creep

Garbenschiefer samples that experienced either cataclasis or production of extremely fine-grained reaction products in the early evolution of the Greiner shear zone underwent subsequent deformation within the fluid-assisted, grain-boundary diffusion creep (GBDC) regime. Within this regime, the rocks experienced grain-size-sensitive behaviour and could have localized large amounts of strain without developing textures, such as subgrains or core-and-mantle structures, that are common in many mylonites. Features indicative of GBDC are preserved only in plagioclase-rich rocks, typically those that now also display well-developed hornblende 'garben'. Horizons that record GBDC range from a few centimetres to several metres in width and extend for metres to hundreds of metres along strike. In many instances these horizons anastomize and connect along strike, whereas in other areas these horizons are separated for great distances by lithologies with different textures. We have no way of determining whether all of these horizons experienced GBDC simultaneously, or whether deformation shifted between them at different times over several million years (*c.* 35 to 30 ± 1 Ma; see above). However, these rocks were almost certainly the weakest rocks within the Greiner shear zone during strain accommodation by GBDC, particularly in comparison with neighbouring horizons dominated by dislocation creep in plagioclase (Tullis *et al.* 1996) or cross-foliation hornblende garben. It is thus likely that GBDC horizons accommodated large amounts of strain within the Greiner shear zone for the duration of GBDC activity.

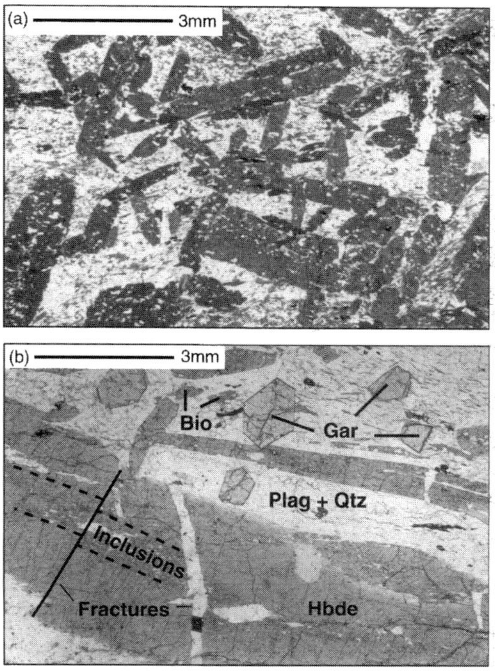

Fig. 7. (**a**) Photomicrograph (plane-polarized light) of undeformed hornblende framework in matrix of fine-grained plagioclase. (**b**) Photomicrograph (plane-polarized light) of fractured hornblende porphyroblasts. Although some hornblende porphyroblasts show evidence of brittle deformation, they do not show evidence for significant translation or rotation.

Stage II: reaction-induced strengthening

In some regions where plagioclase deformed by GBDC, randomly oriented hornblende porphyroblasts up to 20 cm long grew across both foliation and lineation (Garbenschiefer texture, Fig. 7a); much of the hornblende is in the form of delicate radiating bundles. These horizons vary from centimetres to tens of metres in thickness and extend for metres to hundreds of metres along strike. Hornblende nucleation and growth must have been syndeformational, as demonstrated by local faulting and boudinage of some hornblende grains (Fig. 7b), and by slip along cleavage planes of biotite that partially replaces hornblende. However, most hornblende porphyroblasts show no evidence of deformation. In fact, hornblende appears to have established a sufficiently rigid framework that the shear zone became locally locked and deformation shifted to adjacent horizons.

Because large hornblende porphyroblasts developed only in localities where GBDC was the dominant matrix deformation–recovery mechanism before porphyroblast growth (as evidenced by fine-grained, GBDC plagioclase inclusions in hornblende cores), it is likely that there is a causal relationship between GBDC and hornblende nucleation and growth. In these domains, the high surface area to volume ratio of the fine-grained matrix, coupled with rapid grain boundary diffusion rates through an interconnected grain-edge fluid (Tullis *et al.* 1996), would have provided fast bulk diffusion pathways that could have allowed hornblende growth rates to dominate over nucleation rates. Once a critical hornblende nucleus formed, rapid growth could have produced large grains with little to no preferred orientation. Thus, we infer that rapid diffusion rates associated with GBDC were probably responsible for the nucleation and growth of the hornblende porphyroblast framework, although we remain uncertain about the specific reaction(s) that produced hornblende in the Garbenschiefer.

Back-scattered electron images and microprobe traverses taken perpendicular to the *c*-axis

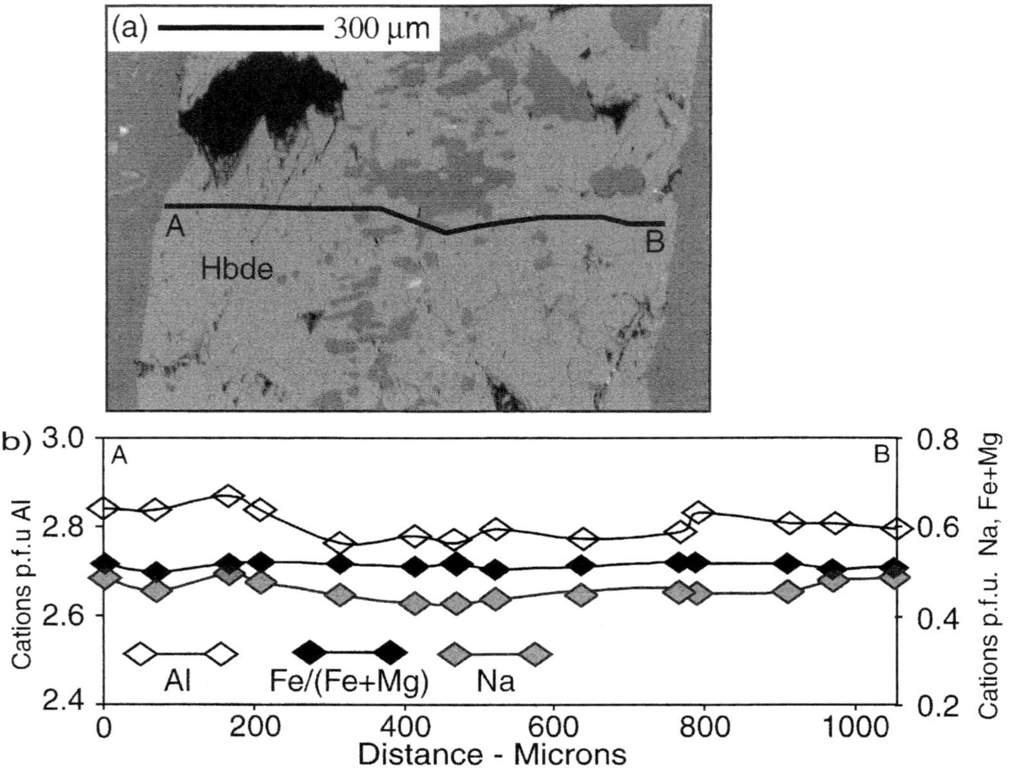

Fig. 8. (**a**) Back-scattered electron image of hornblende porphyroblast showing location of traverse A–B. (**b**) Quantitative electron microprobe traverse along line A–B.

across a hornblende porphyroblast (the orientation that should show maximum chemical zoning) show no appreciable compositional zoning (Fig. 8a and b). Although this is not definitive evidence of rapid growth of hornblende, it is consistent with the growth of hornblende at stable $P-T$ conditions and via a single reaction mechanism throughout the growth cycle.

Partial melting experiments in basalts indicate that sample strength is preserved if at least 35% of interconnected, strong phase grains remain in the sample (Philpotts & Carroll 1996). Deformation in the Garbenschiefer occurred in the solid state, so analogies with partial melting scenarios are imperfect; however, development of a rigid hornblende framework in a fine-grained, superplastic matrix may be somewhat analogous to the melting experiments. Hornblende generally makes up 20–30% of the mode in the Garbenschiefer samples, with some grains in obvious contact with one another in 2D slabs and thin sections. In cases in which the strong phase forms a load-bearing framework, the strength of the bulk aggregate depends primarily on the strength of the strong phase (Handy 1990). Further work is in progress to determine the hornblende abundance and porphyroblast orientations that are required to form a load-bearing framework.

Near the margins of the hornblende-framework zones, the shear foliation is deflected around hornblende porphyroblasts (Fig. 6a), indicating that the porphyroblasts were mechanically fixed and incapable of large amounts of translation or rotation. If the plagioclase matrix had a low strength relative to the hornblende porphyroblasts, there would be nothing to prevent the rotation or translation of hornblende, because the matrix could easily flow into spaces created by hornblende motion. However, if the matrix were more viscous, rotation and translation of elongate porphyroblasts would be difficult. A change from GBDC to any other deformation mechanism (probably Coble creep or regime-2 dislocation creep (Hirth & Tullis 1992) at these $P-T$ conditions), would create a more viscous, less deformable, matrix. The lack of translation of hornblende grains thus suggests that the plagioclase matrix experienced an increase in viscosity coincident with development of the Garbenschiefer texture. This increase was probably related to a change in deformation mechanism in the matrix plagioclase. Observed coarsening of plagioclase from <30 μm inclusions in hornblende cores to >0.5 mm matrix grains in some samples is consistent with an increase in matrix strength during hornblende growth.

It is also possible that the minimal deformation of hornblende reflects a change in stresses applied following development of the hornblende framework. However, it is unlikely that hornblende grew synchronously in every horizon, and that deformation ceased as soon as hornblende grew. Therefore the preservation of the Garbenschiefer domains is probably related to their higher relative strength in comparison with other textural domains observed within the Greiner shear zone.

Stage III: reaction-induced weakening

As described above, partial to complete replacement of hornblende by biotite $\pm$ chlorite $\pm$ plagioclase $\pm$ quartz $\pm$ epidote is evident in many phengite-bearing samples from the Garbenschiefer. In samples in which hornblende is partially altered to biotite, a prominent foliation is typically developed in the biotite-rich portions of the rock but is poorly developed or absent where hornblende is unaltered. Where hornblende was completely replaced by biotite, the earlier hornblende framework is destroyed and large shear strains are evident.

The rheology of the resulting plagioclase and mica assemblage was probably dominated by two mechanisms: (1) sliding along the cleavage planes of mica grains oriented parallel to the foliation; (2) GBDC $\pm$ dislocation creep in the plagioclase matrix. The plagioclase grains in the retrogressed rocks are rectangular, fine grained, and preserve zoning truncations similar to those previously described for GBDC. However, the aspect ratio of these grains is much higher than previously described, often as high as 1:20, and may reflect pinning of plagioclase grain boundaries by micas. An aggregate rheology determined by a combination of sliding along cleavage planes in the micas oriented parallel to foliation and GBDC in the plagioclase matrix will have a composite strength no greater than that of an aggregate determined by one of the mechanisms alone (Tullis & Yund 1991).

Strain localization via slip along cleavage planes and grain boundary sliding in secondary biotite is clearly the youngest of the three strengthening or weakening mechanisms described here, and indicates that shear zone rheology evolved through time in individual horizons of the Garbenschiefer. Adjacent horizons of Garbenschiefer, however, may have experienced weakening via the hornblende-to-biotite reaction at different times, as the reaction appears to have depended on transient fluid infiltration into hornblende + phengite rocks. There are also biotite-rich schists intercalated with the Garbenschiefer that probably never contained hornblende or GBDC plagioclase. These horizons were probably weak zones throughout the entire history of deformation in the Greiner zone.

Discussion: a model for the textural and rheological evolution of the Greiner shear zone

The rheology of the Greiner shear zone varied in both space and time in response to heterogeneities in grain size, bulk-rock composition, fluid composition, and fluid availability. Some layers clearly show a temporal variation from strain softening, through reaction-driven strengthening, to subsequent reaction-controlled weakening. In other cases, variations in rock composition caused simultaneous strain gradients between adjacent horizons. The shear zone as a whole thus had a complex history of shifting strain localization. A model for the textural evolution of the Greiner shear zone (Fig. 9a) and a schematic representation of the changing rheology (Fig. 9b) are discussed in the following section.

Grain size, specifically the amount of grain-size reduction, probably controlled the degree to which GBDC occurred in plagioclase-rich rocks (Fig. 9a). There are three possible mechanisms for grain-size reduction in the Greiner shear zone: cataclasis, dynamic recrystallization in the dislocation creep regime, or the formation of fine-grained products as a result of discontinuous metamorphic reactions. Any of these could result in activation of GBDC and concomitant weakening of the rocks (Fig. 9b). Spatial variations in plagioclase grain size probably represented the dominant control over the activity of GBDC v. dislocation creep in the Greiner shear zone.

Plagioclase samples that appear to accommodate strain by GBDC in laboratory experiments show evidence for deformation-enhanced fluid distribution (Tullis *et al.* 1996), in which metamorphic fluids disperse from isolated droplets located at grain triple junctions to fluids that wet entire grain edges. Deformation-enhanced fluid distribution results in increased surface area for fluid–rock interaction, as well as increasing fluid permeability, diffusion rates, and reaction kinetics. These conditions, if experienced in natural rocks with complex chemical systems (e.g. Fig. 9a), are likely to favour rapid growth of large porphyroblasts, such as the syndeformational hornblende porphyroblasts within the Garbenschiefer (Selverstone 1993). However, because of possible redistribution of fluid in natural samples that remained at hydrostatic conditions after

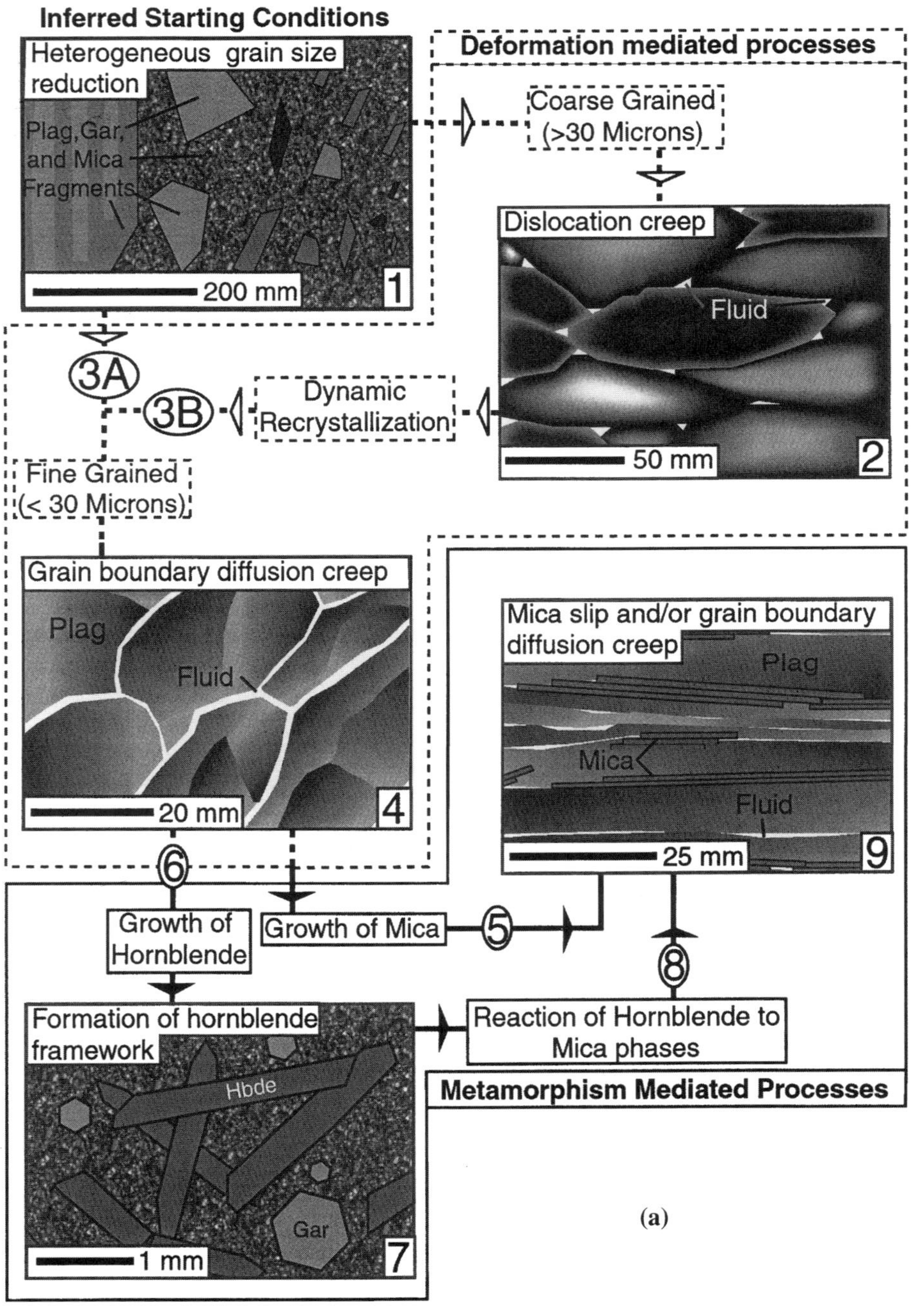

(a)

deformation, it is impossible to confirm if deformation-enhanced fluid distribution was present, other than by analogy with laboratory experiments.

Growth rates of large porphyroblasts could have increased significantly if metastable assemblages existed before deformation. If hornblende was part of the lowest free energy assemblage at the conditions of deformation but had not previously crystallized because of slow reaction kinetics, increased grain boundary and bulk diffusion rates could have released the excess (disequilibrium) energy and led to extremely rapid nucleation and growth of hornblende. Regardless of growth mechanism, the development of the resulting interlocking hornblende porphyroblasts led to localized strengthening of the Greiner shear zone (Fig. 9b). In the most hornblende-rich rocks, strengthening probably occurred via formation of a rigid hornblende framework (Fig. 9a). In other samples, hornblende growth may have shut off GBDC in the plagioclase groundmass via one of three mechanisms. The first of these would have involved decreasing H_2O/CO_2 ratios during hornblende growth, leading to changes in fluid wetting characteristics or diffusion rates. Growth of hornblende would have preferentially consumed H_2O from the fluid present during GBDC, shifting the remaining fluid towards a more CO_2-rich composition (the presence of ankerite in the Garbenschiefer indicates that some CO_2 must have been present in the fluid). CO_2-rich fluids have higher wetting angles than H_2O-rich fluids (Watson & Brenen 1987), and thus are less able to coat grain edges or migrate through the rock (Holness & Graham 1991). A change from lower to higher wetting angles could have diminished diffusion rates sufficiently to shut down GBDC. However, because of the transient nature of fluid distributions, it is difficult to document shifts in wetting angle and concomitant changes in deformation mechanism in naturally deformed samples. By the second of the possible mechanisms, development of strain shadows in matrix plagioclase around hornblende porphyroblasts may have halted deformation-enhanced fluid distribution and also caused an increase in wetting angle (Tullis *et al.* 1996). The third possibility is that grain growth in the fine-grained plagioclase domains may have shut down GBDC by limiting fluid wetting and increasing distances for intracrystalline diffusion.

We believe that grain growth is probably the dominant mechanism that increased matrix strength in hornblende-framework domains. The grain sizes observed in the Garbenschiefer are much larger than the sizes of grains in which the GBDC mechanism was observed experimentally. In addition, the presence of extremely fine-grained (<10 μm) inclusions of plagioclase in the cores of hornblende porphyroblasts, relative to matrix grain sizes of 10–30 μm, indicates that matrix plagioclase grains did indeed coarsen during or subsequent to hornblende growth. This grain growth probably increased the ratio of grain volume to surface area (and hence intracrystalline diffusion distances) to such an extent that strain rates for GBDC became insignificant. Regardless of the mechanism by which GBDC was rendered ineffective,

(b)

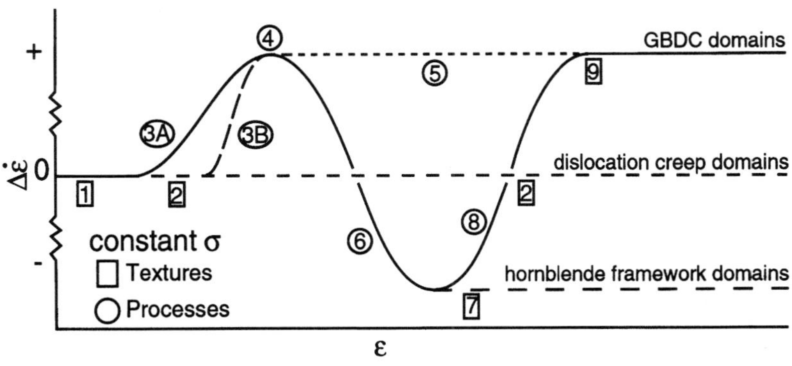

Fig. 9. (**a**) Summary of textural evolution of the Greiner shear zone. The numbers relate textural features and rheological properties of deformational and metamorphic processes observed in the Greiner shear zone. Dashed lines represent deformation-mediated features; continuous lines represent metamorphism-mediated features. (**b**) Schematic portrayal of strain v. change in strain rate during progressive deformation of the Greiner shear zone. Zero point is referenced to strength of a rock dominated by dislocation creep. (Note successive weakening and strengthening in response to changes in deformation mechanism and metamorphic reactions.) Numbers correspond to textures and processes illustrated in (**a**).

its cessation enhanced the strengthening that was afforded by development of the hornblende framework (Fig. 9b). The low dislocation densities preserved in plagioclase indicate that there was little subsequent deformation in the resulting hornblende–plagioclase aggregates, confirming that the strength of these horizons was higher than that of surrounding regions in the shear zone.

In some layers, hornblende porphyroblasts were partially to completely pseudomorphed by biotite-rich assemblages (Fig. 9a) that localized subsequent strain (Fig. 9b). Because areas with stable and pseudomorphed hornblende are finely interspersed, hornblende stability could not have been controlled by variations in $P-T$ conditions. The two possible remaining factors influencing hornblende stability are fluid composition and whole-rock composition. The latter clearly played a role in controlling hornblende stability, as is evident from the fact that hornblende is altered to biotite in rocks containing white mica but is unaffected in rocks that lack white mica. Preliminary whole-rock geochemical data exhibit only slight and largely overlapping compositional variation between different textural domains within the Garbenschiefer (our unpublished data). We are currently unsure if small variations in whole-rock geochemistry provide a major control on the stability of hornblende by controlling the relative stability of white mica and chlorite.

Deformation-enhanced variations in fluid flow may have had an additional effect on hornblende stability by causing localized infiltration of CO_2-rich fluids from the adjacent Furtschaglschiefer, which would in turn may have destabilized hornblende by lowering $a\mathrm{H_2O}$ (Selverstone & Munoz 1987).

Conclusions

The Greiner shear zone shows evidence of several different rheological–textural domains intermingled on a thin-section to outcrop scale, with gradational boundaries between the domains. The inferred stress–strain relationships vary from relatively weak domains deforming dominantly by GBDC and mica slip, to intermediate strength domains where dislocation creep was active, and to strong domains where minimal strain occurred after the development of a hornblende framework (Fig. 9b). The metamorphic and structural evolution of these domains indicates that both spatial and temporal variations in rheology dominated the deformational history of the Greiner shear zone. Modelling the strength of the Greiner shear zone is complicated by small variations in whole-rock chemistry, fluid chemistry, fluid availability, and original grain-size reduction, which all play a role in controlling deformation mechanisms. Furthermore, competition between multiple deformation mechanisms within each polymineralic domain, along with uncertainty regarding the geometrical distribution of the domains at any point in time, make modelling stress–strain relationships for the shear zone as a whole impossible with current techniques.

Garbenschiefer rocks are known from many other areas around the world (Selverstone *et al.* 1984). In most cases, the delicate bundles of hornblende and the common occurrence of other euhedral porphyroblasts (e.g. garnet) suggest that mineral growth was postkinematic and uncomplicated by the effects of shearing. As we point out above, however, Garbenschiefer rocks may in fact *only* develop in high-strain zones because of the ability of fluid-assisted diffusion creep to create domains of high diffusion rates, and the high relative strength of the Garbenschiefer texture may subsequently shift deformation to other horizons.

This study also provides textural evidence for fluid-assisted grain boundary diffusion creep during deep-seated shearing. This is the first study, to our knowledge, that correlates deformation microstructure in natural plagioclase samples with experimental studies in which fluid-assisted GBDC was identified (Tullis & Yund 1991). GBDC may be a rare deformation–recovery mechanism with only limited significance for stress–strain relationships at depth. Alternatively, GBDC may be common during shearing at depth, but the textures that record it may be erased during exhumation or be easily misinterpreted. On the basis of our observations in the Greiner shear zone, we suspect that fluid-assisted GBDC is a more common process than has been recognized previously, implying that mid- to lower-crustal rheologies estimated from power-law creep data will, in many cases, represent vast overestimates of strength during deformation.

We would like to thank J. Wheeler and M. Handy for reviews of this manuscript. Financial support was provided by NSF grant EAR-0000965 (J.S. and A.B.), a Kelley–Silver Graduate Fellowship (K.S.) in the Department of Earth and Planetary Sciences, and by the Regents' Lecturer program (J.S.) at the University of New Mexico. Much of the background work on the Garbenschiefer samples was carried out under the auspices of NSF grants EAR-8658145 and EAR-9219472 to J.S. Transmission electron microscopy was supported by NSF grant EAR-9506481 to A.B. Electron microscopy and electron microprobe analysis was carried out at the Electron Microbeam Analysis Facility at the Department of Earth and Planetary Sciences, University of New Mexico.

References

BEHRMANN, J.H. & FRISCH, W. 1990. Sinistral ductile shearing associated with metamorphic decompression in the Tauern Window, Eastern Alps. *Jahrbuch der Geologischen Bundesanstalt*, **133**, 135–146.

BEHRMANN, J.H. & MAINPRICE, D. 1987. Deformation mechanisms in a high-temperature quartz feldspar mylonite: evidence for superplastic flow in the lower continental-crust. *Tectonophysics*, **140**, 297–305.

BELL, I.A. & WILSON, C.J.L. 1981. Deformation of biotite and muscovite: TEM microstructure and deformation model. *Tectonophysics*, **78**, 201–228.

BIERMANN, C. & VAN ROERMUND, H.L.M. 1983. Defect structures in naturally deformed clinoamphiboles: a TEM study. *Tectonophysics*, **95**, 267–278.

BONS, A.J. 1988. Deformation of chlorite in naturally deformed low-grade rocks. *Tectonophysics*, **154**, 149–165.

BOULLIER, A.M. & GUEGUEN, Y. 1975. SP-mylonites; origin of some mylonites by superplastic flow. *Contributions to Mineralogy and Petrology*, **50**, 93–104.

BRODIE, K.H. & RUTTER, E.H. 1985. On the relationship between deformation and metamorphism, with special reference to the behavior of basic rocks. *In*: THOMPSON, A.B. & RUBIE, D.C. (eds) *Metamorphic Reactions*. Advances in Physical Geochemistry, 138–179.

CARLSON, W.D. 1989. The significance of intergranular diffusion to the mechanism and kinetics of porphyroblast crystallization. *Contributions to Mineralogy and Petrology*, **103**, 1–23.

CHERNOFF, C.B. & CARLSON, W.D. 1997. Disequilibrium for Ca during growth of pelitic garnet. *Journal of Metamorphic Geology*, **15**, 421–438.

CHRISTENSEN, J.N., SELVERSTONE, J., ROSENFELD, J.L. & DEPAOLO, D.J. 1994. Correlation by Rb–Sr geochronology of garnet growth histories from different structural levels within the Tauern window; Eastern Alps. *Contributions to Mineralogy and Petrology*, **118**, 1–12.

COOPER, R.F. & KOHLSTEDT, D.L. 1984. Rheology and structure of olivine-basalt partial melts. *Journal of Geophysical Research*, **91**, 9315–9323.

CUMBEST, R.J., DRURY, M.R., VANROERMUND, H.L.M. & SIMPSON, C. 1989. Dynamic recrystallization and chemical evolution of clinoamphibole from Senja, Norway. *Contributions to Mineralogy and Petrology*, **101**, 339–349.

DE VECCHI, G. & BAGGIO, P. 1982. The Pennine zone of Vizze region of the western Tauern Window (Italian Eastern Alps). *Bollettino della Societa Geologica Italiana*, **101**, 89–116.

DEWEY, J.F., PITMAN, W.C. III, RYAN, W.B.F. & BONNIN, J. 1973. Plate tectonics and the evolution of the Alpine System. *Geological Society of America Bulletin*, **84**, 3137–3180.

DIETRICH, V. 1976. Evolution of the Eastern Alps; a plate tectonics working hypothesis. *Geology*, **4**, 147–152.

DIPPLE, G.M., WINTSCH, R.P. & ANDREWS, M.S. 1990. Identification of the scales of differential element mobility in a ductile fault zone. *Journal of Metamorphic Geology*, **8**, 645–661.

HACKER, B.R. & CHRISTIE, J.M. 1990. Brittle/ductile and plastic/cataclastic transitions in experimentally deformed and metamorphosed amphibolite. *In*: DUBA, A.G., DURHAM, W.B., HANDIN, J.W. & WANG, H.F. (eds) *The Brittle–Ductile Transition in Rocks*. American Geophysical Union, Washington, DC, 127–147.

HANDY, M.R. 1989. Deformation regimes and the rheological evolution of fault zones in the lithosphere: the effects of pressure, temperature, grainsize and time. *Tectonophysics*, **163**, 119–152.

HANDY, M.R. 1990. The solid-state flow of polymineralic rocks. *Journal of Geophysical Research*, **95**, 8647–8661.

HIRTH, G. & TULLIS, J. 1992. Dislocation creep regimes in quartz. *Journal of Structural Geology*, **14**, 145–159.

HOLNESS, M. & GRAHAM, C. 1991. Equilibrium dihedral angles in the system H_2O–CO_2–NaCl–calcite, and the implications for fluid flow during metamorphism. *Contributions to Mineralogy and Petrology*, **108**, 368–383.

KNIPE, R.J. 1989. Deformation mechanisms—recognition from natural tectonites. *Journal of Structural Geology*, **11**, 127–146.

KRONENBERG, A.K., KIRBY, S.T. & PINKSTON, J. 1990. Basal slip and mechanical anisotropy of biotite. *Journal of Geophysical Research*, **95**, 19257–19278.

LAMMERER, B. & WEGER, M. 1998. Footwall uplift in an orogenic wedge: the Tauern Window in the Eastern Alps of Europe. *Tectonophysics*, **285**, 213–230.

LISTER, G.S. & WILLIAMS, P.F. 1983. The partitioning of deformation in flowing rock masses. *Tectonophysics*, **92**, 1–33.

MORTEANI, G. 1974. Petrology of the Tauern Window, Austrian Alps. *Fortschritte der Mineralogie*, **52**, 195–220.

PHILPOTTS, A.R. & CARROLL, M. 1996. Physical properties of partly melted tholeiitic basalt. *Geology*, **24**, 1029–1032.

RATSCHBACHER, L., FRISCH, W., LINZER, H.G. & MERLE, O. 1991. Lateral extrusion in the Eastern Alps; 2. Structural analysis. *Tectonics*, **10**, 257–271.

RUBIE, D.C. 1983. Reaction-enhanced ductility: the role of solid–solid univariant reactions in deformation of the crust and mantle. *Tectonophysics*, **96**, 331–352.

RUTTER, E.H. 1999. On the relationship between the formation of shear zones and the form of the flow law for rocks undergoing dynamic recrystallization. *Tectonophysics*, **303**, 147–158.

RUTTER, E.H. & BRODIE, K.H. 1988. Experimental approaches to the study of deformation metamorphism relationships. *Mineralogical Magazine*, **52**, 35–42.

SANCHEZNAVAS, A. & GALINDOZALDIVAR, J. 1993. Alteration and deformation microstructures of biotite from plagioclase-rich dykes (Ronda Massif, S Spain). *European Journal of Mineralogy*, **5**, 245–256.

SCHMID, S.M. 1982. Microfabric studies as indicators of deformation mechanisms and flow laws operative in mountain building. *In*: HSU, K.J. (ed.) *Mountain Building Processes*. Academic Press, London, 95–110.

SCHMID, S.M. & HANDY, M.R. 1991. Toward a genetic classification of fault rocks: geological usage and tectonophysical implications. *In*: MUELLER, D.W., MCKENZIE, J.A. & WEISSERT, H. (eds) *Controversies in Modern Geology; Evolution of Geological Theories in Sedimentology, Earth History and Tectonics*. Academic Press, London, 339–361.

SELVERSTONE, J. 1988. Evidence for east–west crustal extension in the Eastern Alps: implications for the unroofing history of the Tauern Window. *Tectonics*, **7**, 87–105.

SELVERSTONE, J. 1993. Microscale to macroscale interactions between deformational and metamorphic processes, Tauern Window, Eastern Alps. *Schweizerische Mineralogische und Petrographische Mitteilungen*, **73**, 229–239.

SELVERSTONE, J. & MUNOZ, J.L. 1987. Fluid heterogeneities and hornblende stability in interlayered graphitic and nongraphitic schists (Tauern Window, Eastern Alps). *Contributions to Mineralogy and Petrology*, **96**, 426–440.

SELVERSTONE, J., MORTEANI, G. & STAUDE, J.M. 1991. Fluid channeling during ductile shearing: transformation of granodiorite into aluminous schist in the Tauern Window, Eastern Alps. *Journal of Metamorphic Geology*, **9**, 419–431.

SELVERSTONE, J., SPEAR, F.S., FRANZ, G. & MORTEANI, G. 1984. High-pressure metamorphism in the SW Tauern Window, Austria: *P–T* paths from hornblende–kyanite–staurolite schists. *Journal of Petrology*, **25**, 501–531.

SHEA, W.T., KRUSE, R. & KRONENBERG, A.K. 1993. Strength and anisotropy of foliated rocks with varied mica contents. *Journal of Structural Geology*, **15**, 1097–1121.

SIMPSON, C. 1983. Stain and shape-fabric variations associated with ductile shear zones. *Journal of Structural Geology*, **5**, 61–72.

STÖCKHERT, B., WACHMANN, M., KUSTER, M. & BIMMERMANN, S. 1999. Low effective viscosity during high-pressure metamorphism due to dissolution precipitation creep: the record of HP–LT metamorphic carbonates and siliciclastic rocks from Crete. *Tectonophysics*, **303**, 299–319.

STÜNITZ, H. 1998. Syndeformational recrystallization: dynamic or compositionally induced. *Contributions to Mineralogy and Petrology*, **131**, 219–236.

TULLIS, J. & YUND, R.A. 1987. Transition from cataclastic flow to dislocation creep of feldspar: mechanisms and microstructures. *Geology*, **15**, 606–609.

TULLIS, J. & YUND, R.A. 1991. Diffusion creep in feldspar aggregates: experimental evidence. *Journal of Structural Geology*, **13**, 987–1000.

TULLIS, J., DELLANGELO, L.N. & YUND, R. 1990. Ductile shear zones from brittle precursors of felspathic rocks: the role of dynamic recrystallization. *In*: DUBA, A.G., DURHAM, W.B., HANDIN, J.W. & WANG, H.F. (eds) *The Brittle–Ductile Transition in Rocks*. Geophysical Monograph, American Geophysical Union, **56**, 67–81.

TULLIS, J., YUND, R. & FARVER, J. 1996. Deformation-enhanced fluid distribution in feldspar aggregates and implications for ductile shear zones. *Geology*, **24**, 63–66.

URAI, J.L., MEANS, W.D. & LISTER, G.S. 1986. Dynamic recrystallization of minerals. *In*: HOBBS, B.E. & HEARD, H.C. (eds) *Mineral and Rock Deformation Studies, Paterson Volume*. American Geophysical Union, Washington, DC, 161–179.

WATSON, E.B. & BRENEN, J.M. 1987. Fluids in the lithosphere: 1, Experimentally-determined wetting characteristics of CO_2–H_2O fluids and their implications for fluid transport, host-rock physical properties, and fluid inclusion formation. *Earth and Planetary Science Letters*, **85**, 497–515.

WHITE, S.H. & KNIPE, R.J. 1978. Transformation- and reaction-enhanced ductility in rocks. *Journal of the Geological Society, London*, **135**, 513–516.

WILSON, C.J.L. & BELL, I.A. 1979. Deformation of biotite and muscovite: optical microstructure. *Tectonophysics*, **58**, 179–200.

WINTSCH, R.P. & KNIPE, R.J. 1983. Growth of a zoned plagioclase porphyroblast in a mylonite. *Geology*, **11**, 360–363.

YUND, R.A. & TULLIS, J. 1991. Compositional changes of minerals associated with dynamic recrystallization. *Contributions to Mineralogy and Petrology*, **108**, 346–355.

Geometric controls on the evolution of normal fault systems

J.J. WALSH[1,3], C. CHILDS[1,3], V. MEYER[1,5], T. MANZOCCHI[1,3], J. IMBER[1,3], A. NICOL[1,4], G. TUCKWELL[2], W.R. BAILEY[1,3], C.G. BONSON[1,3], J. WATTERSON[1,6], P.A. NELL[1,7] & J. STRAND[1,3]

[1]*Fault Analysis Group, Department of Earth Sciences, University of Liverpool L69 3GP, UK*

[2]*School of Earth Sciences and Geography, Keele University, Keele ST5 5BG, UK*

[3]*Present address: Fault Analysis Group, Department of Geology, University College Dublin, Belfield, Dublin 4, Ireland (e-mail: fault@fag.ucd.ie)*

[4]*Institute of Geological and Nuclear Sciences, Lower Hutt, New Zealand*

[5]*Institut de Physique du Globe de Paris, Paris, Cedex 05, France*

[6]*Fault Analysis Group, Liverpool University Marine Laboratory, Port Erin, Isle of Man IM9 6JA*

[7]*Badley Earth Sciences Ltd, North Beck Lane, Hundleby, Spilsby PE23 5NB, UK*

Abstract: The growth of normal fault arrays is examined in basins where sedimentation rates were higher than fault displacement rates and where fault growth histories are recorded by thickness and displacement variations within syn-faulting sequences. Progressive strain localization is the principal feature of the growth history of normal faults for study areas from the Inner Moray Firth, a sub-basin of the North Sea, and from the Timor Sea, offshore Australia. The kinematics of faulting are similar in both study areas. Fault displacement rates correlate with fault size, where size is measured in terms of either displacement or length. Small faults have higher mortality rates than larger faults throughout the growth of the fault system. Displacement and strain are progressively localized onto the larger faults at the expense of smaller faults at progressively larger scales. Strain localization and the preferential growth of larger faults are attributed to geometric factors, such as size and location, rather than to the mechanical properties of fault rock in individual faults. This conclusion is supported by numerical models that reproduce the main characteristics of fault system growth established from both study areas.

Details of the growth history of normal fault systems are difficult to establish from outcrop. Onshore neotectonic normal fault systems, for example, are often characterized by the erosion of footwall highs and the deposition of syn-faulting sediments within under-filled hanging-wall basins. In these circumstances, preserved horizons are rarely mappable across individual faults, but occasionally the timing and rates of fault movement on geological time scales can be defined within broad limits (e.g. Petersen & Wesnousky 1994; Morewood & Roberts 1997). Sometimes it may be possible to establish the recent movement history of faults by dating a small number of earthquake slip events from trenches exposed across fault scarps (e.g. Pantosti *et al.* 1996), but these data, although invaluable for earthquake studies, provide limited constraints on the growth of fault systems on longer time scales (greater than *c.* 0.1 Ma). Definition of fault growth on geological time scales is best achieved from datasets providing 3D control on the geometries of both faults and syn-faulting depositional horizons. In the past couple of decades, the acquisition of seismic data in extensional basins has provided such refined datasets.

In this paper we outline the kinematics of two fault systems imaged from 3D offshore seismic datasets in the North Sea and the Timor Sea. The growth of both fault systems is characterized by the progressive localization of displacement and strain onto large faults, a feature we attribute to geometric factors (i.e. fault size, connectivity, position and orientation), rather than to differences between the fault rock mechanical properties of individual faults. A subordinate role for fault rock mechanical properties is

From: HOLDSWORTH, R.E., STRACHAN, R.A., MAGLOUGHLIN, J.F. & KNIPE, R.J. (eds) 2001. *The Nature and Tectonic Significance of Fault Zone Weakening*. Geological Society, London, Special Publications, **186**, 157–170. 0305-8719/01/$15.00

supported by numerical models that reproduce the main characteristics of fault system growth irrespective of the model fault rock rheology. We suggest that fault rock mechanical properties are important for fault growth at certain spatial and temporal scales, but may not exert the prime control over fault system growth on geological time scales.

Kinematic analysis of fault systems

The growth of normal fault systems within extensional basins can be analysed in detail where the faults intersected the Earth's surface and their displacement histories are recorded by thickness and displacement changes within syn-faulting sediments (Childs *et al.* 1993, Childs *et al.* 1995). This is possible only where sedimentation rates exceed fault displacement rates (Nicol *et al.* 1997), a condition that is not satisfied by the many extensional basins characterized by relatively rapid fault displacement rates, with the formation of fault scarps, the erosion of uplifted footwalls or the underfilling of hanging-wall basins (e.g. Contreras *et al.* 2000). Blanketing of fault scarps by sediments and the preservation of fault displacement histories is a feature of our two study areas in the Inner Moray Firth (Underhill 1991*a*, *b*, Nicol *et al.* 1997), a sub-basin of the Late Jurassic North Sea Basin, and in the Cartier Trough, in the Timor Sea, NW Australia (Nicol *et al.* 1995, 1996; Walsh *et al.* 1996). We use two basic analytical approaches; one relies on the comparison of fault maps for different aged syn-faulting depositional horizons to identify the principal changes in fault system structure through time, whereas the other uses a displacement backstripping method (Petersen *et al.* 1992; Childs *et al.* 1993; Clausen *et al.* 1994) to establish the displacement histories of faults.

Inner Moray Firth

The Inner Moray Firth is characterized by a relatively thick sequence (up to *c.* 3 km) of sediments deposited during a *c.* 16 Ma phase of Late Jurassic rifting that affected the entire North Sea. Previous work in this basin suggests that large faults have relatively high displacement rates and that smaller faults have higher mortality rates than larger faults (Nicol *et al.* 1997). Both these features point indirectly to progressive strain localization during fault growth. In this study we highlight some more features of fault system growth from the analysis of a high-quality 3D seismic dataset for a 15 km × 6.5 km area in the Inner Moray Firth. The study area covers a large normal fault (400 m maximum throw) and its associated footwall high, which contains an array of normal faults with maximum throws down to the effective limit of seismic resolution, i.e. 10 m (Fig. 1). The majority of faults show

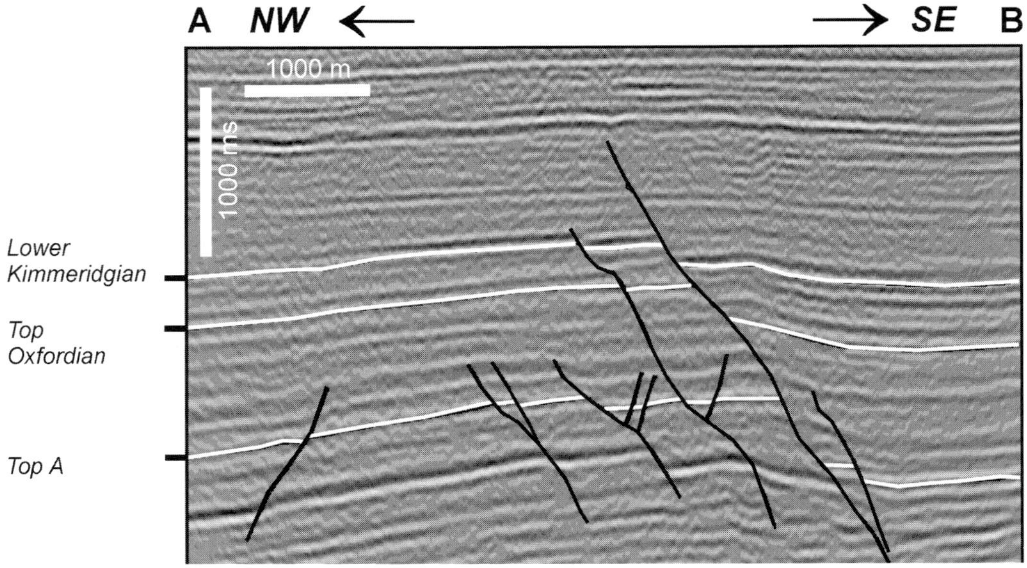

Fig. 1. Seismic section across a fault system in the Inner Moray Firth Basin. The Top A horizon is a Top Callovian reflector, synchronous with the onset of rifting. The uppermost Oxfordian and the Top Lower Kimmeridgian reflectors are syn-rift. (See Fig. 3 for location of seismic section.)

marked upward decreases in displacement, reflecting their synsedimentary nature, with less of the displacement history, and therefore displacements, recorded on younger horizons. Syn-faulting stratigraphic sequences show sympathetic thickening from the upthrown to the downthrown side of faults. The fact that some smaller displacement faults extend through the faulted sequence indicates that the upward loss of displacement on smaller faults is neither a resolution effect nor is it a direct consequence of strains associated with upward propagation through the syn-faulting stratigraphic sequence: some contribution from upward propagation effects cannot, however, be ruled out. Further details of the structural development of this area have been provided elsewhere (Walsh *et al.* in prep.).

Timor Sea

A normal fault system is analysed from a 3D seismic dataset of a *c.* 10 km × 12 km area on the SE edge of the Cartier Trough in the Timor Sea, offshore NW Australia (Fig. 2). Normal faults of the Cartier Trough are assigned to two main phases of extension: Late Jurassic–earliest Cretaceous and Plio-Pleistocene (Pattillo & Nicholls 1990; Woods 1992; Nicol *et al.* 1995, 1996). The most recent phase of extension is represented by a system of NE-striking normal faults, some of which are reactivated Late Jurassic structures. These normal faults dip to the NW and SE in approximately equal numbers within a well-imaged Cenozoic sequence (*c.* 1–1.5 km thick), which is dominated by shelf carbonates (Pattillo & Nicholls 1990). Pliocene and Pleistocene syn-faulting sequence strata thicken by up to 30% across faults, which, in some cases, extend to the sea bed (Fig. 2b). The contemporary horizontal extension is NNW–SSE, i.e. subperpendicular to faulting (Hillis 1991), and is associated with subduction of the Australian continent beneath the Banda Arc (Laws & Kraus 1974). Faulting in the study area is dominated by a central

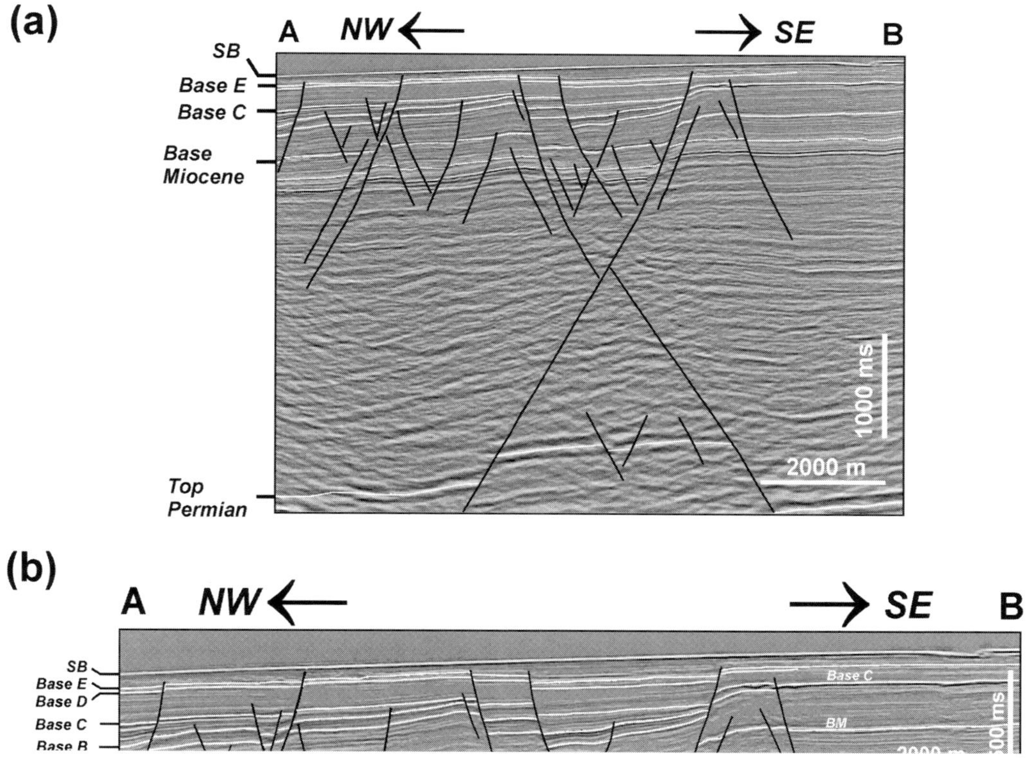

Fig. 2. Seismic section across a fault system in the Timor Sea. (**a**) Entire section; (**b**) uppermost 600 ms. SB, sea bed; Base E is 1.3 Ma, Base D is 1.8 Ma, Base C is 2.7 Ma, Base B is 4.7 Ma, Base A is 6 Ma and Base Miocene horizon (BM in (**b**)) is 23 Ma. The Base A and Base B horizons are restricted to the area in the hanging wall of the largest NW-dipping fault. (See Fig. 4 for location of seismic section.) (Modified from Meyer *et al.* 2001.)

graben bounded by two opposed-dipping, relatively high displacement faults, which are rooted in Late Jurassic–Early Cretaceous normal faults and form a large conjugate structure (Fig. 2a). These large Tertiary faults originated by reactivation and upward propagation of Mesozoic structures from which they inherited their strike. The spatial pattern of pre-existing Mesozoic structures also partly controls the areal distribution of other Tertiary faults, which, on horizons deposited before the onset of Tertiary deformation, are concentrated within, and NW of, the central graben (e.g. Fig. 2a). Further details of the structural development of this area are provided elsewhere (Meyer *et al.* 2001).

Main features of fault growth

Examination of seismic sections across both study areas shows that individual faults generally show upward decreases in displacement accompanied by related across-fault thickness changes of syn-faulting units. These features reflect the synsedimentary nature of the faults, with older horizons recording more of the fault displacement history than younger horizons. There is also a good relationship between the vertical persistence, and therefore longevity, of faults and their displacement (Figs 1 and 2). Smaller displacement faults are often, but not always, restricted to lower parts of the syn-faulting sequence, whereas larger displacement faults most often extend through the entire syn-faulting sequence. As larger displacement faults also generally have larger fault trace lengths, fault maps show that older horizons are characterized by relatively large numbers of smaller faults compared with younger syn-faulting horizons (Figs 3 and 4).

Fault longevity is shown by the colour coding in Figs 3 and 4. The black faults on older horizons are still active at the time of deposition of the next mapped horizon, whereas red faults have become inactive during the period separating the horizons. On each map, red faults are generally the smaller faults, and black faults the larger ones. A large population of smaller, and shorter-lived, faults is a characteristic of older horizons and the earlier stages of faulting. Occasional smaller faults initiated later in the fault system evolution, and most commonly occur in locations adjacent to and between larger faults. These faults are relatively rare and are interpreted to accommodate interactions between larger faults such as relay zones and conjugate fault intersections (e.g. X; Fig. 4b). Larger faults, by contrast, are found on all horizons and their growth has persisted from early in the growth of the system into the later stages of faulting. In both systems an early population of active small to large faults evolves into a system in which only the larger faults are active and the smaller faults have died.

The evolution of the Timor Sea fault system is relatively well represented by the series of maps for different aged horizons (Fig. 4), but in the Inner Moray Firth fault system the degree of localization is very extreme and few faults remain towards the end of rifting. In this case, we highlight the principal changes in fault system evolution by reconstructing the fault system that existed during rifting (Fig. 3c). This reconstruction is performed using displacement back-stripping methods in which displacements on younger horizons are subtracted from those of underlying horizons, to produce the fault map that existed during rifting. The reconstructed fault pattern comprises a broad size range of disconnected faults. The fault system towards the end of rifting is, by contrast, dominated by a few larger faults that extend almost across the entire study area (Fig. 3). A feature best seen in the Inner Moray Firth is the evolution from a symmetrical array with conjugate faults towards a polarized array in which one fault dip direction dominates (Figs. 1–4).

We have described above the spatial evolution of the fault systems by reference to the faults mapped on different horizons. If the ages of the horizons are known, fault growth curves (see Nicol *et al.* 1997) can be constructed to describe the temporal evolution of individual faults. The curves presented in Fig. 5 show the maximum throw of each fault on each mapped horizon, and the curve for any fault that is inactive before the end of rifting (i.e. all the red faults in Figs. 2 and 3) intersects the y-axis at the time when it became inactive. Low gradients on the plot imply high displacement rates, and for the Timor Sea study area, larger displacement faults have higher displacement rates than smaller displacement faults and are more likely to continue to grow throughout the entire rifting phase. For the Inner Moray Firth study area, very few faults extend through the majority of the rifting phase, so we reproduce here the growth curves presented by Nicol *et al.* (1997) for several larger displacement (>100 m) faults from a larger area of the Inner Moray Firth (Fig. 5a). These faults show similar features to those of the Timor Sea, with smaller displacement faults having lower displacement rates and higher mortality rates than larger displacement faults. Even during the early stages of the evolution of the system,

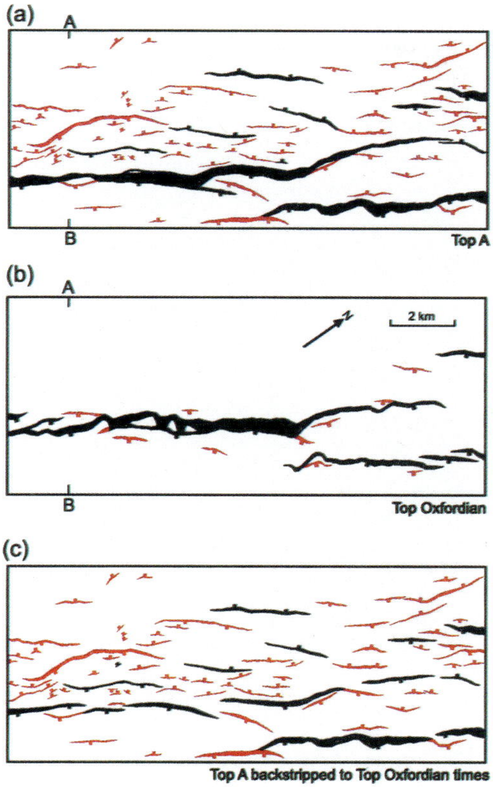

Fig. 3. Fault maps for the Inner Moray Firth fault system. (**a**) Top A. (**b**) Top Oxfordian section. (**c**) Top A restored to Late Oxfordian time. The faults coloured red in (**a**) and (**c**) became inactive by the time of deposition of the Top Oxfordian horizon, and the red faults in (**b**) became inactive by the time of deposition of the Lower Kimmeridgian horizon. AB marks the position of the seismic line shown in Fig. 1.

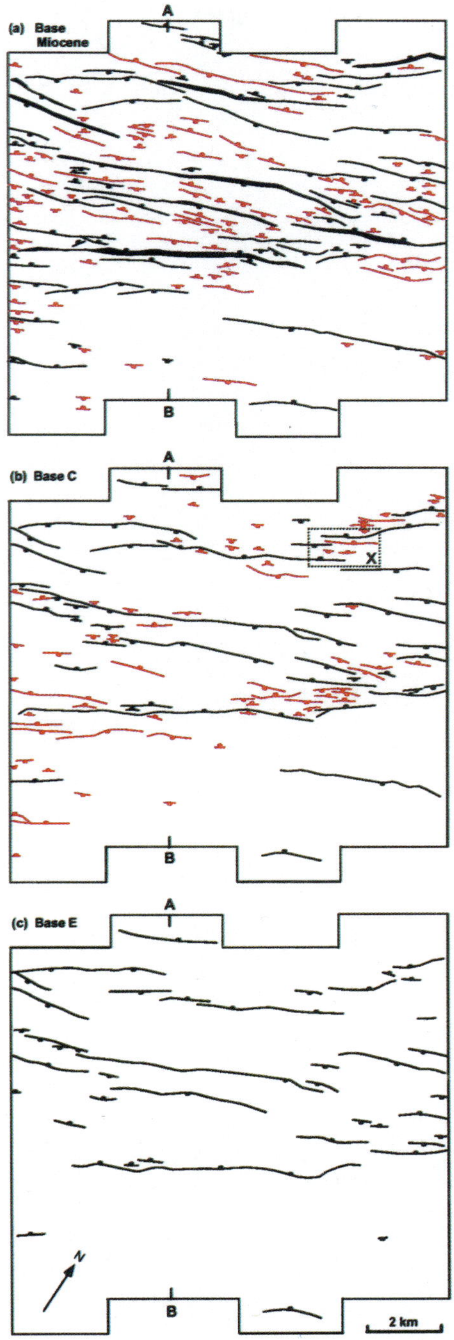

Fig. 4. Fault maps for the Timor Sea fault system. (**a**) Base Miocene section. Red faults are inactive by the time of deposition of lowermost C. (**b**) Base C. Red faults are inactive by the time of deposition of lowermost E. (**c**) Base E. AB marks the position of the seismic line shown in Fig. 2.

when both relatively small and large faults are active, the larger faults grow faster and therefore contribute disproportionately to accommodating extension. Eventually, the large faults accommodate all the displacement and the smaller faults become inactive.

Displacement in the Inner Moray Firth fault system is almost completely localized by the time of the last mapped horizon, on which a hard-linked system in the SW transfers displacement via soft-linkage onto two faults in the NE. The relay zones that accommodate this soft linkage are sites of smaller faults that have remained active. The localization at the scale of observation in the Timor Sea system is less mature than in the Inner Moray Firth fault system, and although it is considerably more localized on later horizons, it remains soft-linked.

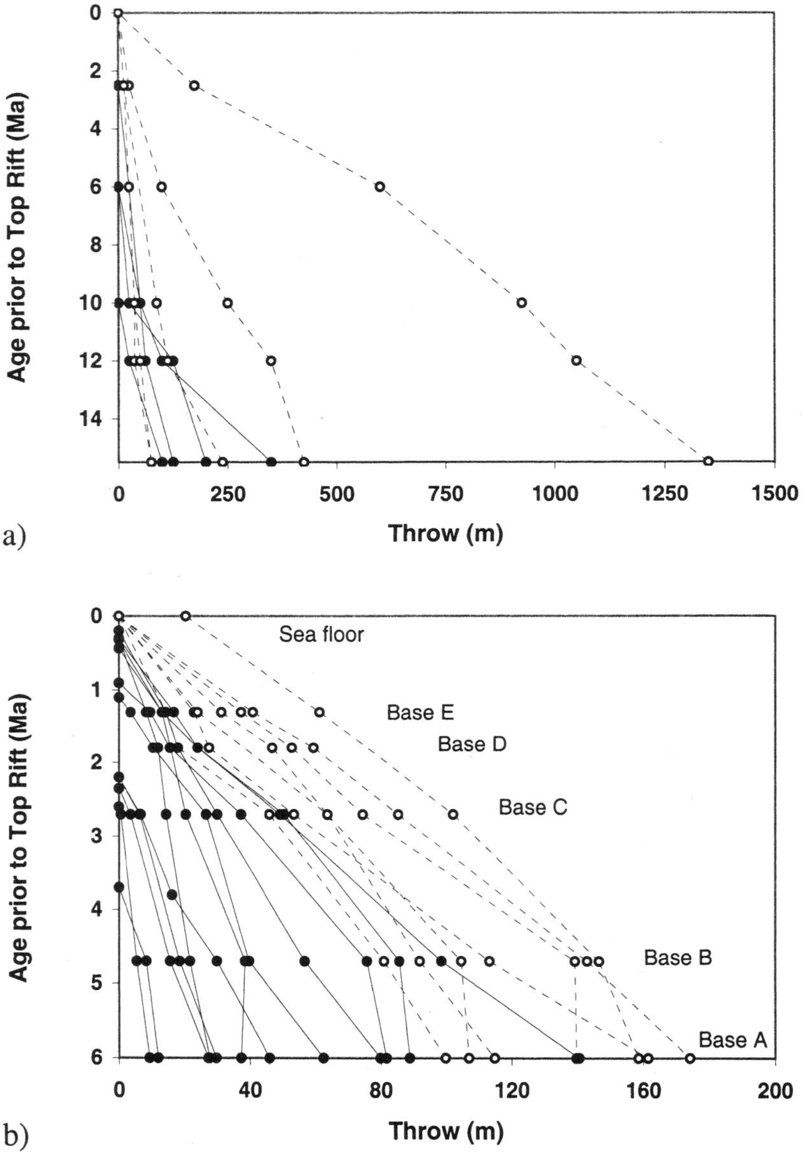

Fig. 5. Fault growth curves for (**a**) faults in the Inner Moray Firth (after Nicol *et al.* 1997), and (**b**) the Timor Sea fault system. ●, faults that became inactive before the end of the rifting episode; ○, faults that were still active.

Summary

The histories of the growth of the two fault systems have several features in common, which either indicate directly or are consistent with fault system geometry providing the main control on displacement localization.

(1) Large faults have higher displacement rates than smaller faults throughout the growth of the fault system, and smaller faults have higher mortality rates. As described by Nicol *et al.* (1997), this observation requires that long-range interactions existed between faults throughout the evolution of the fault system, such that the entire population of active faults operates to accommodate the regional strain. Within such a system the activity of a fault is determined by its position within the population, in terms of relative size and location,

rather than by other intrinsic properties of that fault.

(2) A number of small faults are active at the later stages of evolution of the Timor Sea fault system. These faults are generally confined to the tip regions of large faults or to relay zones (e.g. X in Fig. 4b) between large faults forming part of a soft-linked fault array onto which displacement has become localized. These faults have either initiated or continued to grow by virtue of their locations relative to larger faults.

(3) Within the resolution of the Inner Moray Firth dataset, the death of the large population of faults within the footwall to the large fault is synchronous with the establishment of the through-going fault.

(4) Some larger faults become inactive relatively early in the system evolution, and these faults have opposed dips to the faults upon which displacement becomes localized. These faults were not favourably oriented to efficiently accommodate strain within a progressively more polarized array.

The principal factors that appear to control which faults are important in the localization process are (1) the size of the faults and (2) their locations and orientations. The main characteristics of faulting in both of the study areas suggest that the geometric properties of the fault system are the crucial factors in controlling fault system growth. Fault size correlates with displacement rate and fault systems evolve towards fewer and either better connected larger faults or soft-linked fault arrays. In these natural systems we cannot measure the properties of the fault rocks within the high and low displacement rate faults, but given that each of our observations can be explained by the geometric properties of the fault systems, fault rock material properties may not play a significant role, if any, in determining which faults are active at a particular time. In the next section, we present some simple numerical models that support the evolutionary scheme we propose, and that demonstrate that geometry, rather than fault rock properties, is the dominant control on the displacement rates of individual faults.

Numerical modelling

Numerical models of faulting are presented, which examine the principal relationships established from the natural datasets. These relationships are (1) strain localization onto larger faults and optimally placed smaller faults, and (2) correlations between fault size and displacement rate. Using geometrically identical models we also investigate the relative significance of fault system geometric properties and fault rock mechanical properties in controlling fault system growth.

Discrete element modelling code (Cundall & Strack 1979) is used to simulate the accumulation of displacement on normal faults in a cube of rock represented by a bonded assembly of spherical particles (the code PFC3D is used). The elastic properties of the particles and the strength of the bonds between them are specified to reproduce the bulk properties (Young's modulus, Poisson's ratio, shear modulus and compressive strength) of the material to be modelled: for the models presented in this paper the mechanical properties of a typical fine-grained sandstone are used (Springwell Sandstone; Wu *et al.* 1991; Hazzard 1998). The models are placed under an isotropic confining pressure, and constant strain rate boundary conditions impose a shear strain across the models (Fig. 6). Predefined vertical discontinuities that lie within the plane of shearing and extend between the upper and lower boundaries of the model are defined by particles whose bond strengths and/or elastic stiffness are either higher or lower than those of the matrix particles. In the models presented, 'weak' faults comprise particles with elastic stiffness 50 times lower than those of the host material, and bond strengths of weak fault particles are set to zero. 'Strong' faults have elastic stiffness 10 times higher, and initial bond strengths 100 times greater, than those of the host material. Once a bond is broken, there is no healing of the bonds, and the strength of the faults is controlled by the elastic stiffness of the particles along, and on either side of the discontinuity.

Evolution of model fault systems

Two model fault systems are described. The first model contains eight small faults with lengths up to one-half of the total model width, and the second model contains the same eight faults in addition to a single fault that entirely crosses the model. Figure 7a shows the first fault system modelled with weak faults at a bulk shear strain of 0.03. Contours are of particle depth below a fixed datum (the top of the maps) on the upper surface of the model cube, and are effectively structural contours of an initially planar surface. The structural contour patterns indicate a positive correlation between fault length and displacement: larger faults are responsible for more significant perturbations of the structural contours and for correspondingly greater elevation changes across the faults. Displacements associated with

Fig. 6. Discrete element model cube containing weak faults at a shear strain of 0.03, showing shear direction and particles colour coded according to their displacement. The predefined faults are represented by the white spheres, and extend from top to bottom of the cube. (See text for discussion.)

particular faults arise principally because faults modify the local stress field and interact mechanically as the system evolves (e.g. Pollard & Segall 1987; Ackermann & Schlische 1997; Cowie 1998). A single fault in a fault system may lie in a zone of locally increased or decreased stress. Faults in areas of increased stress will deform to relieve that increased stress, and will thus accumulate greater displacement. Smaller faults located beyond, and adjacent to, the tips of longer faults tend to have higher than average displacements (e.g. faults b and f, Fig. 7a), whereas faults in the stress shadow of larger faults tend to have lower than average displacements (e.g. fault h, Fig. 7a). In this way, the position of a fault relative to the other faults exerts an important control on its development. Fault interaction and enhanced or restricted growth are observed on the same faults in the model containing a single through-going discontinuity (Fig. 7b), but the displacements of all these faults are lower, as the through-going fault accommodates by far the largest portion of localized strain.

The significance of fault rock properties can be examined by comparing the deformation associated with models containing weak faults (Fig. 7a and b) and geometrically identical models containing strong faults (Fig. 7c and d). Qualitatively, the strain fields in the two models appear similar, but quantitative differences between the weak and strong fault models are observed on fault growth curves derived from the models (Fig. 7e and f). These growth curves track the system evolution to a shear strain of 0.1, and show the characteristic features observed in natural fault systems (e.g. Fig. 5; Nicol *et al.* 1997). Large displacement faults have higher displacement rates from the outset of shearing, whereas smaller displacement faults have lower displacement rates and high mortality rates. Fault mortality in the model is indicated by constant throws at successive shear strain intervals (e.g. fault b in all models, Fig. 7e and f). In the early stages of system evolution (shear strain less than *c.* 0.03), the models containing strong faults have lower total fault displacements than the models containing weak faults, hence a larger proportion of the strain is accommodated elastically and by distributed brittle failure. Although localization adjacent to and along the pre-existing discontinuities occurs later if the discontinuities are strong, once it occurs displacement continues to be accommodated along these new planes of failure. At a shear strain of 0.1, in the system containing only small faults (Fig. 7e) about 35% of the total strain in the strong fault model, and 25% in the weak fault model, is accommodated on the soft-linked system comprising faults e, f and g. For the system containing a through-going large fault, about 50% of the strain in the strong fault model and 35% in the weak fault model is accommodated on the through-going fault. Hence not only do strong faults represent geometric heterogeneities that localize deformation, they also appear to be more efficient at doing so than weak faults. Cowie (1998) has argued that the ability of a fault to support stresses is essential for the long-range fault interactions we observe in these numerical models and in the natural fault systems. As strong faults are able to support greater stresses than weak faults, fault interactions are possible over greater distances. Because of their strength, displacements become localized later in the strong fault models, but once localized the systems evolve more rapidly. The models containing strong faults have higher displacement rates on the large faults, fewer faults with intermediate displacements and higher mortality rates of small faults than the models containing weak faults (Fig. 7e and f).

These features suggest that fault size and geometric connectivity have a first-order impact on the growth of fault systems, and that specific fault rock properties have an influence on, but are not a prerequisite to, generating important

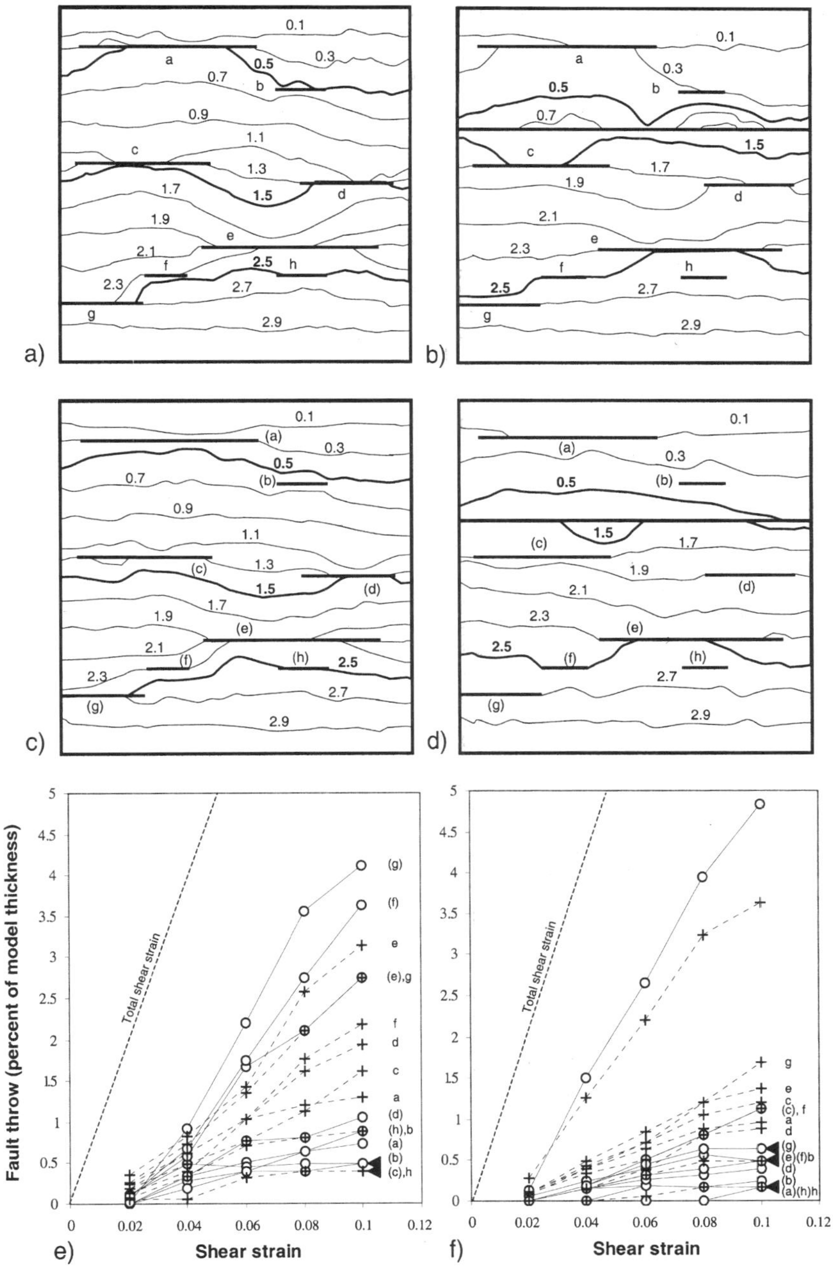

Fig. 7. Contoured maps of particle displacements on the upper surface of the model cubes at a total shear strain of 0.03. Contours are labelled as percentages of the total model thickness. (**a**) Small weak faults. (**b**) Same as (**a**) but including a through-going fault. This map corresponds to the system shown in Fig. 6. (**c**) Small strong faults. (**d**) Same as (**c**) but including a through-going fault. (**e**) Fault growth curves for the two models containing only small faults. Strong faults: ○ and fault identifiers in parentheses. Weak faults: + and fault identifiers without parentheses. (**f**) Fault growth curves for the two models containing small faults and a through-going fault. Symbols as before. For (**e**) and (**f**), shear strain can be taken as a proxy for time. Faults that died during the later stages of deformation are annotated with arrowheads.

characteristics of natural fault systems. Numerical models of faults show similar features to natural fault systems irrespective of the fault rock strength. Larger or favourably positioned faults are mechanically advantaged during deformation and accumulate greater displacements. The models demonstrate that the geometry of faults within a deforming system is the key controlling factor in the accumulation and localization of deformation.

Discussion

Fault system evolution

The geometry and kinematics of natural fault systems suggest that the geometric properties of a fault system, such as fault size, location and potential contribution to connectivity, exert a fundamental control on fault system evolution. Fault systems evolve towards fewer and larger faults, a condition that represents the most efficient means of accommodating tectonic strain. The contribution of a fault to the plan-view connectivity of a fault system scales as a function of the square of its length (e.g. Bour & Davy 1997; Renshaw 1997). Hence a few large faults will contribute significantly more to the connectivity of a fault system than many small faults with the same combined trace length. For purely geometric reasons, therefore, we would expect the larger faults to have a dominant influence on the progression from distributed to localized strain, owing simply to their size. The fundamental role of geometric fault properties in controlling fault system growth are highlighted by the kinematic evidence presented for the Inner Moray Firth and Timor Sea fault systems.

The population systematics of both study areas, presented in detail elsewhere (Meyer *et al.* 2001; Walsh *et al.* in prep.), show progressive changes in scaling properties with fault system evolution. Power-law size populations are characteristic of both areas at the early stages of faulting. With increasing maturity of the fault system, power-law populations are, however, replaced either by power-law populations with lower slopes (and fewer smaller faults), in the case of the Timor Sea, or by non-power-law populations, in the Inner Moray Firth, where strain is mainly localized onto a single fault. These changes in scaling properties support the evidence given in this paper that as a fault system evolves, the strain localizes at progressively larger scales, with the localization at the scale of a given area occurring when through-going faults transect the area (e.g. Main *et al.* 1990). This is particularly clear for the Inner Moray Firth study area, in which the localization of displacements on larger faults is almost complete by the time of the last mapped horizon. A similar evolutionary scheme has been described from laboratory deformation experiments (e.g. Main *et al.* 1990; Lockner *et al.* 1992; Liakopoulou-Morris *et al.* 1994). The evolution of a through-going localized shear fracture in these experiments is charted by monitoring acoustic emissions from the sample. An initially power-law distribution of emission amplitudes evolves into a non-power-law population during the localization phase immediately before sample failure. The natural systems we have studied differ from the laboratory experiments in that boundary conditions do not impose a scale on the system, and these characteristics of fault localization occur at progressively larger scales and presumably over progressively longer intervals. For example, although the initiation and earliest stages (i.e. the first *c.* 1–2 Ma) of the fault systems cannot be established from existing data, the power-law seismically imaged populations suggest they were each characterized by a large population of small, subseismic, faults that localized to form the mappable fault systems. The scale invariance of strain localization in fault systems is consistent with many previous qualitative descriptions of fault systems (e.g. Riedel 1929; Tchalenko 1970).

A model for the evolution of fault systems proposed recently by Cowie (1998), based on numerical modelling studies (Cowie *et al.* 1993), resembles closely the evolution of fault systems described in this paper. In this model, fault interactions enhance displacement rates on optimally positioned faults. An essential feature of this model (Cowie 1998) is that the faults recover some or all of their strength between displacement increments. If the faults remain weak, the long-range organization and feedback controlling displacement accruement on longer and optimally positioned smaller faults are dampened, and the characteristic localized system is not formed.

'Weak' faults

The concept of fault strength, i.e. whether a fault is weak or strong, is currently used in a variety of contexts. Fault strength can be an attribute of either fault rocks or individual faults, although direct equivalence is often assumed. Weak faults are generally considered to have weak fault rocks, and weak fault rocks are taken to be characteristic of weak faults. Whereas the strength of fault rocks can either be measured experimentally or be estimated qualitatively from microstructural considerations, definitions of the strength of individual faults vary with context. In a geological context, 'weak' faults can be those that are more likely to reactivate or those that are characterized by high displacement rates. From an earthquake seismology perspective, 'weak' faults are those that are characterized by low stress drops and are more earthquake prone. In all cases, faults that are weaker are often considered to have weaker fault rocks. Below, we suggest that acceptance of this equivalence is, at best, oversimplified, and, at worst, wrong.

Our analysis of natural fault systems combined with numerical modelling studies (see also Cowie *et al.* 1993; Cowie 1998), suggests that weaker fault rocks are not a prerequisite for either high displacement rates or longevity. Geometric properties of the fault system control fault system growth, with displacement progressively localized onto larger and better connected faults. For the same reasons, weaker fault rocks do not have to be a feature of those faults that happen to be reactivated in a given area. Again, geometric properties, such as size, position, orientation and connectivity, could be the principal control on the reactivation histories of individual faults within a fault system and could provide a basis for the common observation that large faults are preferentially reactivated and generally longer lived than smaller faults (e.g. Kelly *et al.* 1999). This view is supported by numerical models showing that large faults even with strong fault rock can localize displacements. Later fault localization may be at the boundary to the earlier strong fault zone, but for most geological purposes it is one and the same fault that is reactivated. Zones of deformation bands are classic examples of progressive localization of displacement within faults zones that are characterized by fault rocks stronger than the surrounding country rock (Aydin & Johnson 1978; Underhill & Woodcock 1987; Antonellini *et al.* 1994). In natural fault systems the propensity of large faults or connected networks of faults to localize later displacements is further increased by the geometric effects of earlier faulting. Faults are responsible for geometric offsets of layers or bodies within the host rock volume. In the same way that strong faults can localize displacements, the geometric effects of faulting can alter the stress, and related strain, distributions within a deforming volume of rock and be responsible for localizing displacements on earlier, particularly larger, faults.

Although the distribution of fault rocks within fault zones is highly variable and complex, particularly along normal faults and strike-slip faults that usually cut the dominant mechanical layers at high angles, the importance of weak fault rocks in fault reactivation has been suggested from metre- to decametre-scale fault zone outcrop studies (White *et al.* 1986; Holdsworth *et al.* 1997; Imber *et al.* 1997). These studies show that later displacement reactivation of faults with large map traces (i.e. several kilometres to hundreds of kilometres) is often preferentially localized within weaker fault rocks of a fault zone. Weaker fault rocks were, on that basis, taken to be a prerequisite for the preferential reactivation of some faults compared with others. Our contention is that this is a small-scale, within fault zone, localization phenomenon that cannot be extrapolated to the larger-scale problem of whether a fault is more or less likely to be reactivated. From the perspective of fault reactivation, little significance can be attached to metre- to decametre-scale outcrop observations of fault zones with lateral and vertical extents of several kilometres to hundreds of kilometres. We believe that the geometric properties of faults, and their geometric consequences in terms of offsetting the rock volume, could dominate the reactivation process. Geometric properties and effects will generate associated stress heterogeneities that can, on their own, be responsible for reactivating larger, and better connected, faults in preference to smaller faults.

All faults are zones of concentrated strain and, by definition, at the time of faulting are planes of weakness. The key question is whether faults are weak by virtue of the mechanical properties of the fault rocks or their capacity to efficiently relieve elastic strains. Both the geometric features of the natural systems examined here and numerical models presented here and elsewhere indicate that fault system geometry exerts the primary control on fault activity during the development of a fault system. Where a fault system is reactivated by extension oblique to the initial extension direction, it is well established that only those faults or parts of faults that are optimally oriented with respect to the new stress regime will reactivate (e.g. Sibson 1985). Where the new extension direction is subparallel to the initial extension direction, we suggest that the factors that determine which fault moves will be the same as those that determined which faults localized displacement during the original phase of extension.

Our study of natural fault systems has little direct bearing on earthquake seismological arguments about weak and strong faults (e.g. Scholz 2000; Townend & Zoback 2000; Zoback 2000). However, our examination of fault zones (Childs *et al.* 1995, 1996, 1997; Foxford *et al.* 1998), suggests a level of complexity about fault zone structure and content, and fault surface geometry (e.g. fault surface rugosity and segmentation), that argues against the direct use of laboratory derived frictional properties in earthquake modelling. The scaling properties of geometric characteristics of the fault system could again be pre-eminent (e.g. Stirling *et al.* 1996).

Conclusions

(1) Large faults generally have higher displacement rates than smaller faults, throughout the growth of a fault system.

(2) Smaller faults have higher mortality rates and large faults grow at their expense.

(3) Progressive localization of displacement and strain onto large faults is accompanied by an increase in the connectivity of large faults at the scale of observation.

(4) Geometric factors of a fault system (i.e. fault size, connectivity, position, orientation) may exert the dominant control on fault system evolution.

(5) Progressive strain localization is a fundamental characteristic of fault system evolution, with the preferential growth of larger faults and the formation of better connected fault systems at progressively larger scales.

(6) Fault rock mechanical properties may play a subordinate role in the evolution of the fault systems studied.

We thank other members of the Fault Analysis Group, past and present, and BP Amoco and BHP Petroleum for access to data. This work was partly funded by DTI/OGPSO Programme (Project 7234), EU Non-Nuclear Energy Programme (contract no. JOF3-CT97-0036), EU Hydrocarbon Reservoirs Programme Project 'Secondary Migration of Petroleum through Caprock and Carrier Sequences' (contract no. JOF3-CT95-0014) and by an EU Marie-Curie Postdoctoral Fellowship (contract no. ENV4-CT97-5087).

Thanks to Gerald Roberts, David Ferrill and Bob Holdsworth for helpful reviews and discussion.

References

ACKERMANN, R.V. & SCHLISCHE, R.W. 1997. Anticlustering of small normal faults around larger faults. *Geology*, **25**, 1127–1130.

ANTONELLINI, M.A., AYDIN, A. & POLLARD, D.D. 1994. Microstructure of deformation bands in porous sandstones at Arches National Park, Utah. *Journal of Structural Geology*, **16**, 941–959.

AYDIN, A. & JOHNSON, A.M. 1978. Development of faults as zones of deformation bands and as slip surfaces in sandstone. *Pure and Applied Geophysics*, **116**, 931–942.

BOUR, O. & DAVY, P. 1997. Connectivity of random fault networks following a power-law length distribution. *Water Resources Research*, **33**, 1567–1583.

CHILDS, C., EASTON, S.J., VENDEVILLE, B.C., JACKSON, M.P.A., LIN, S.T., WALSH, J.J. & WATTERSON, J. 1993. Kinematic analysis of faults in a physical model of growth faulting above a viscous salt analogue. *Tectonophysics*, **228**, 313–329.

CHILDS, C., WATTERSON, J. & WALSH, J.J. 1995. Fault overlap zones within developing normal fault systems. *Journal of the Geological Society, London*, **152**, 535–549.

CHILDS, C., WATTERSON, J. & WALSH, J.J. 1996. A model for the structure and development of fault zones. *Journal of the Geological Society, London*, **153**, 337–340.

CHILDS, C., WATTERSON, J. & WALSH, J.J. 1997. Discussion on a model for the structure and development of fault zones. *Journal of the Geological Society, London*, **154**, 366–368.

CLAUSEN, O. R., KORSTGåRD, J. A., PETERSEN, K., & 7 OTHERS 1994. Systematics of faults and fault arrays. *In*: HELBIG, K. (ed.) *Modelling the Earth for Oil Exploration. Final report of the CEC's Geoscience Program 1990–1993.* Pergamon, Oxford, 205–316.

CONTRERAS, J., ANDERS, M.H. & SCHOLZ, C.H. 2000. Growth of a normal fault system: observations from the Lake Malawi basin of the east African rift. *Journal of Structural Geology*, **22**, 159–168.

COWIE, P.A. 1998. A healing–reloading feedback control on the growth rate of seismogenic faults. *Journal of Structural Geology*, **8**, 1075–1087.

COWIE, P.A., VANNESTE, C. & SORNETTE, D. 1993. Statistical physics model for the spatiotemporal evolution of faults. *Journal of Geophysical Research*, **98**, 21809–21821.

CUNDALL, P.A. & STRACK, O.D.L. 1979. A discrete numerical model for granular assemblies. *Géotechnique*, **29**, 47–65.

FOXFORD, K.A., WALSH, J.J., WATTERSON, J., GARDEN, I.R., GUSCOTT, S.C. & BURLEY, S.D. 1998. Structure and content of the Moab Fault Zone, Utah, U.S.A., and its implications for fault seal prediction. *In*: JONES, G., FISHER, Q.J. & KNIPE, R.J. (eds) *Faulting, Fault Sealing and Fluid Flow in Hydrocarbon Reservoirs.* Geological Society, London, Special Publications **147**, 87–103.

HAZZARD, J. F. 1998. Numerical modelling of acoustic emissions and dynamic rock behaviour. Ph.D. thesis, University of Keele.

HILLIS, R.R. 1991. Australia–Banda Arc collision and *in situ* stress in the Vulcan Sub-basin (Timor Sea) as revealed by borehole breakout data. *Exploration Geophysics*, **22**, 189–194.

HOLDSWORTH, R.E., BUTLER, C.A. & ROBERTS, A.M. 1997. The recognition of reactivation during continental deformation. *Journal of the Geological Society, London*, **154**, 73–78.

IMBER, J., HOLDSWORTH, R.E., BUTLER, C.A. & LLOYD, G.E. 1997. Fault-zone weakening processes along the reactivated Outer Hebrides Fault Zone, Scotland. *Journal of the Geological Society, London*, **154**, 105–109.

KELLY, P.G., PEACOCK, D.C.P., SANDERSON, D.J. & MCGURK, A.C. 1999. Selective reverse-reactivation of normal faults, and deformation around reverse-reactivated faults in the Mesozoic of the Somerset coast. *Journal of Structural Geology*, **21**, 493–509.

LAWS, R.A. & KRAUS, G.P. 1974. The regional geology of the Bonaparte Gulf Timor Sea area. *Australian Petroleum Exploration Association Journal*, **14**, 77–84.

LIAKOPOULOU-MORRIS, F., MAIN, I.G., CRAWFORD, B.R. & SMART, B.G.D. 1994. Microseismic properties of a homogeneous sandstone during fault nucleation and frictional sliding. *Geophysical Journal International*, **119**, 219–230.

LOCKNER, D.A., BYERLEE, J.D., KUKSENKO, V., PONOMAREV, A. & SIDORIN, A. 1992. Observations of quasistatic fault growth from acoustic emissions. *In*: EVANS, B. & WONG, T.-F. (eds) *Fault Mechanics and Transport Properties of Rocks*. Academic Press, London, 3–31.

MAIN, I.G., MEREDITH, P.G., SAMMONDS, P.R. & JONES, C. 1990. Influence of fractal flaw distributions on rock deformation in the brittle field. *In*: KNIPE, R.J. & RUTTER, E.H. (eds) *Deformation Mechanisms, Rheology and Tectonics*. Geological Society, London, Special Publications **54**, 71–79.

MEYER, V., NICOL, A., CHILDS, C., WALSH, J.J. & WATTERSON, J. 2001. Progressive localisation of strain during the evolution of a normal fault system in the Timor Sea. *Journal of Structural Geology*, in press.

MOREWOOD, N.C. & ROBERTS, G.P. 1997. Geometry, kinematics and rates of deformation in a normal fault segment boundary, central Greece. *Geophysical Research Letters*, **24**, 3081–3084.

NICOL, A., WALSH, J.J., WATTERSON, J. & BRETAN, P.G. 1995. Three dimensional geometry and growth of conjugate normal faults. *Journal of Structural Geology*, **17**, 847–862.

NICOL, A., WATTERSON, J., WALSH, J.J. & CHILDS, C. 1996. The shapes, major axis orientations and displacement patterns of fault surfaces. *Journal of Structural Geology*, **18**, 235–248.

NICOL, A., WALSH, J.J., WATTERSON, J. & UNDERHILL, J.R. 1997. Displacement rates of normal faults. *Nature*, **390**, 157–159.

PANTOSTI, D., D'ADDEZIO, G. & CINTI, F.R. 1996. Paleoseismicity of the Ovindoli–Pezza fault, central Apennines, Italy: a history including a large, previously unrecorded earthquake in the Middle Ages (860–1300 AD). *Journal of Geophysical Research*, **101**, 5937–5959.

PATTILLO, J. & NICHOLLS, P.J. 1990. A tectonostratigraphic framework for the Vulcan Graben, Timor Sea region. *Australian Petroleum Exploration Association Journal*, **30**, 27–51.

PETERSEN, M.D. & WESNOUSKY, S.G. 1994. Fault slip rates and earthquake histories for active faults in Southern California. *Bulletin of the Seismological Society of America*, **84**, 1608–1649.

PETERSEN, K., CLAUSEN, O.R. & KORSTGÅRD, J.A. 1992. Evolution of a salt-related listric growth fault near the D-1 well, block 5605, Danish North Sea: displacement history and salt kinematics. *Journal of Structural Geology*, **14**, 565–577.

POLLARD, D.D. & SEGALL, P. 1987. Theoretical displacement and stresses near fractures in rock: with application to faults, joints, veins, dikes, and solution surfaces. *In*: ATKINSON, B.K. (ed.) *Fracture Mechanics of Rock*. Academic Press, London, 277–347.

RENSHAW, C.E. 1997. Mechanical controls on the spatial density of opening-mode fracture networks. *Geology*, **25**, 923–926.

RIEDEL, W. 1929. Zur Mechanik geologischer Brucherscheingen. *Zentralblatt für Mineralogie, Geologie und Paläontologie, Abhandlung B*, **1929**, 354–368.

SCHOLZ, C.H. 2000. Evidence for a strong San Andreas fault. *Geology*, **28**, 163–166.

SIBSON, R.H. 1985. A note on fault reactivation. *Journal of Structural Geology*, **7**, 751–754.

STIRLING, M.W., WESNOUSKY, S.G. & SHIMAZAKI, K. 1996. Fault trace complexity, cumulative slip, and the shape of the magnitude–frequency distribution for strike-slip faults: a global survey. *Geophysical Journal International*, **124**, 833–868.

TCHALENKO, J.S. 1970. Similarities between shear zones of different magnitudes. *Geological Society of America Bulletin*, **81**, 1625–1640.

TOWNEND, J. & ZOBACK, M.D. 2000. How faulting keeps the crust strong. *Geology*, **28**, 399–402.

UNDERHILL, J.R. 1991*a*. Controls on Late Jurassic seismic sequences, Inner Moray Firth, UK North Sea: a critical test of a key segment of Exxon's original global cycle chart. *Basin Research*, **3**, 79–98.

UNDERHILL, J.R. 1991*b*. Implications of Mesozoic–Recent basin development in the western Inner Moray Firth, UK. *Marine and Petroleum Geology*, **8**, 359–369.

UNDERHILL, J.R. & WOODCOCK, N.H. 1987. Faulting mechanisms in high-porosity sandstones; New Red Sandstone, Arran, Scotland. *In*: JONES, M.E. & PRESTON, R.M.F. (eds) *Deformation of Sediments and Sedimentary Rocks*. Geological Society, London, Special Publications **29**, 91–105.

WALSH, J.J., WATTERSON, J., CHILDS, C. & NICOL, A. 1996. Ductile strain effects in the analysis of seismic interpretations of normal fault systems. *In*: BUCHANAN, P.G. & NIEUWLAND, D.A. (eds) *Modern Developments in Structural Interpretation, Validation and Modelling*. Geological Society, London, Special Publications **99**, 27–40.

WHITE, S.H., BRETAN, P.G. & RUTTER, E.H. 1986. Fault-zone reactivation: kinematics and mechanics. *Philosophical Transactions of the Royal Society, London, Part A*, **317**, 81–97.

WOODS, E.P. 1992. Vulcan sub-basin fault styles—implications for hydrocarbon migration and entrapment. *Australian Petroleum Exploration Association Journal*, **32**, 138–158.

WU, B., KING, M.S. & HUDSON, J.A. 1991. Stress-induced ultrasonic wave velocity anisotropy in a sandstone. *International Journal of Rock Mechanics, Mineral Science and Geomechanical Abstracts*, **28**, 101–107.

ZOBACK, M.D. 2000. Strength of the San Andreas. *Nature*, **405**, 31–32.

The nature and origin of asymmetric arrays of shear surfaces in fault zones

STEVEN F. WOJTAL

Department of Geology, Oberlin College, Oberlin, OH 44074-1044, USA
(e-mail: steven.wojtal@oberlin.edu)

Abstract: Mid- to upper-crustal fault zones often possess arrays of shear surfaces whose traces on sections perpendicular to the fault surface and parallel to the *ac*-plane conform with one or more of the 'Riedel shear' orientations. These shear surfaces often are oblique to the transport plane, however, so arrays exhibit monoclinic rather than orthorhombic symmetry. In a mudstone-dominated mélange in Humber Arm Supergroup strata in the Bay of Islands, Newfoundland, and in serpentinites from the base of the Bay of Islands complex, shear surfaces have orientations inclined to major fault-zone boundaries and an inferred *ac*-plane for macroscopic fault-related deformation. Deformation in these zones exhibits an overall monoclinic symmetry. The 3D, asymmetric character of shear surface fabrics suggests that a factor other than stress or the symmetric strain rate tensor controlled their formation. The velocity gradient tensor in a steady, non-coaxial shearing flow possesses a symmetry consistent with monoclinic fabrics. Shear surfaces in asymmetric arrays may initiate with predictable orientations relative to the velocity gradient tensor and then rotate toward flow apophyses, which identify stable positions in steady, 3D flows.

A general aim of studies of ancient, inactive fault zones and their fault rocks is the determination of the strain rate and stress distributions that prevailed in the zones during fault slip. This determination is a prerequisite in assessing whether fault zones softened or hardened. The question of the fault-zone strength is critical to understanding the emplacement of thrust sheets and ophiolites and the development of orogenic belts. Fault slip may constitute a significant component of the resistance to emplacement for individual thrust or ophiolite sheets if fault zones are strong (Hubbard & Rubey 1959; Moores 1982). If fault zones are weak, fault slip may require an order of magnitude less energy than work against gravity or the work of deforming the sheet (Elliott 1976*a*, *b*; Mitra & Boyer 1986). The taper values typical of natural orogenic wedges suggest that basal detachments are relatively weak (Chapple 1978; Davis *et al.* 1983), but it is not always clear why fault zones are weak. Further, temporal changes in fault-zone strength can affect the emplacement of individual sheets (Coward 1982; Wojtal & Mitra 1986), and may alter the architecture of a deforming orogenic wedge (DeCelles & Mitra 1995; Mitra 1997). Therefore, it is appropriate to study fault-zone behaviour and evolution.

Often, structural geologists studying fault-zone behaviour follow an approach that mimics laboratory studies of deformation, where one envisions that the applied stresses are independent parameters and material velocities, incremental displacements, or total displacements are a dependent material response. This approach can be used to examine the origin of any number of fabric elements, but it is especially common in studying the arrays of fractures or minor faults typically found in large fault zones. For example, by viewing fault arrays as a dependent material response, Erickson & Wiltschko (1991) inferred, using the Mohr–Coulomb criterion, constraints on the stress distribution that prevailed in the Lewis thrust fault zone during portions of its slip history. The work by Erickson & Wiltschko is a particularly elegant and successful example of such an analysis, but many others have taken essentially the same approach (Harris & Milici 1977; Platt & Leggett 1986; Platt *et al.* 1987; Roberts 1994; Teixell *et al.* 2000). One especially common argument made using this general approach is that minor faults are Riedel shears whose geometries reflect the orientations of stresses in a fault zone (e.g. Logan *et al.* 1979; Stearns *et al.* 1981; Rutter *et al.* 1986; Price & Cosgrove 1990, p. 145; Yeats *et al.* 1997, p. 189).

From: HOLDSWORTH, R.E., STRACHAN, R.A., MAGLOUGHLIN, J.F. & KNIPE, R.J. (eds) 2001. *The Nature and Tectonic Significance of Fault Zone Weakening*. Geological Society, London, Special Publications, **186**, 171–193. 0305-8719/01/$15.00

'Kinematic approaches', in which parameters such as incremental displacements, displacement gradients, or velocity gradients are inferred from rock fabrics, are preferred to such 'stress approach' treatments of fault-slip data (Twiss *et al.* 1991, 1993; Twiss & Unruh 1998). This preference arises primarily because fault-slip data, i.e. slickenlines on individual shear surfaces, directly record local incremental displacements, not shear stresses on planes (Twiss *et al.* 1991; Wojtal & Pershing 1991). Using one of several inversion techniques (e.g. Angelier 1994), fault-slip data from collections of shear surfaces may fix the principal values of an incremental displacement gradient tensor (Twiss & Unruh 1998). By defining a parameter that measures the progress of deformation, a measure of 'material time' (Hsu 1967), one can interpret incremental displacement gradient tensor components as 'material' deformation rates that are scalar multiples of 'temporal' or 'true' deformation rates. The principal values of such a deformation rate tensor need not, and in general will not, correlate directly with (i.e. be scalar multiples of) principal values of an inferred stress tensor, especially in non-coaxial flows (Twiss & Unruh 1998). Preference for kinematic approaches arises also because material anisotropies often control the orientation of individual faults in fault zones. Finally, this preference arises because stresses may, depending upon the boundary conditions for fault-zone deformation, be a dependent material response (Twiss *et al.* 1991; Tikoff & Wojtal 1999). Although one may analyse minor faults in a fault zone by attempting to define stresses in the fault zone, one must recognize that the stresses need not be truly independent. In this paper, I describe and analyse arrays of shear surfaces (minor faults and bounding surfaces for small shear zones) from two large fault zones using 'kinematic approaches' as defined by Twiss & Unruh (1998).

Velocity gradients across boundary layers provide insight into the bulk behaviour of a fluid undergoing shearing flow (Batchelor 1967). Velocity gradients in glaciers or ice sheets constrain inferences on the bulk flow properties of glaciers and ice sheets (e.g. Paterson 1994, pp. 238–288 (but see also pp. 173–203, 301–311 and 317–354); Harper *et al.* 1998). Kehle (1970); Elliott (1976*a*); Ramberg (1977); Sanderson (1982); Merle (1984, 1989), Wojtal (1992*a*) and many others have examined the deformation at the base of thrust sheets as a kind of boundary layer flow, and sought to determine velocity gradients to constrain the bulk material behaviour of the moving sheets. It is relatively straightforward in theory, albeit difficult in practice, to determine profiles of strain or profiles of rock fabric elements that reflect strains in the vicinity of a major thrust fault (see Hossack 1968, 1978; Ramberg 1974, 1975, 1977; Elliott 1976*b*; Mitra & Elliott 1980; Coward & Kim 1981; Sanderson 1982; Gilotti & Kumpulainen 1986; Geiser 1988; Handschy 1998; and many others). Deriving a velocity gradient from a strain or fabric profile is more difficult (Wojtal 1992*a*, *b*). Further, it is rarely clear to what extent changes in rock fabric, and therefore inferred changes in material velocity, reflect changes in flow parameters in a material that follows a single constitutive relationship, or changes in constitutive relationships caused by changes in rock structure or microstructure. In this context, the shear surface arrays from one location described here are interesting. The orientations of shear surfaces, both minor faults and the bounding surfaces of small shear zones, change with distance from the major fault, whereas the general deformation mechanism, i.e. shearing on discrete surfaces, remains the same. Thus, the changes in the shear surface arrays may reflect changes in the velocity field in a material exhibiting uniform behaviour. Deformation in the two fault zones described here exhibits monoclinic kinematic character, like the fault-zone deformation described by Twiss *et al.* (1991); Twiss & Unruh (1998). Moreover, shear surfaces within these arrays nucleated in orientations that yield asymmetrical geometric patterns. Taken together, the shear surface arrays apparently accommodated shearing flow with concomitant thickening or flattening, corresponding to transtensional or transpressive shearing. That such distinctive arrays exhibiting distinctive deformation kinematics occur in highly localized zones may indicate significant fault-zone weakening.

For the purposes of visualizing and analysing strain and displacement gradient data or strain rate and velocity gradient data adjacent to major thrust faults, it is customary to focus on a plane that is perpendicular to the intermediate principal axis for emplacement-related strain and/or perpendicular to the axis of internal vorticity associated with fault-parallel shearing (Christie 1963; McLeish 1971; Chapman *et al.* 1979; Coward & Kim 1981; Milton & Williams 1981; Platt & Leggett 1986; Wojtal 1986; Merle 1989; and many others). Many analyses of wall-rock deformation during faulting presume that deformation is plane strain. From the arcuate nature of thrust belts (e.g. Marshak 1988) and fanning of strain principal directions, this restriction cannot, strictly speaking, hold. In fact, the heterogeneous deformation associated with slip on thrust faults is often demonstrably non-plane (Wojtal 1986, 1992*a*; Unruh & Twiss 1998; Strine & Wojtal 1999). This paper also describes fabrics that developed during slip on major thrusts where wall-rock deformation cannot readily be reconciled with plane-strain flow: the deformation associated with faulting was general 3D shearing.

This paper will then (1) describe arrays of shear surfaces found in fault rocks from two major thrust zones, (2) relate these arrays of shear surfaces to 3D velocity gradient fields, and (3) use the inferred overall velocity field to draw inferences on the relative strength of the major thrust zones.

Geological setting

Both fault zone examples described here are from the Bay of Islands area in western Newfoundland (Fig. 1a). This area is well known for the Bay of Islands complex, which consists of

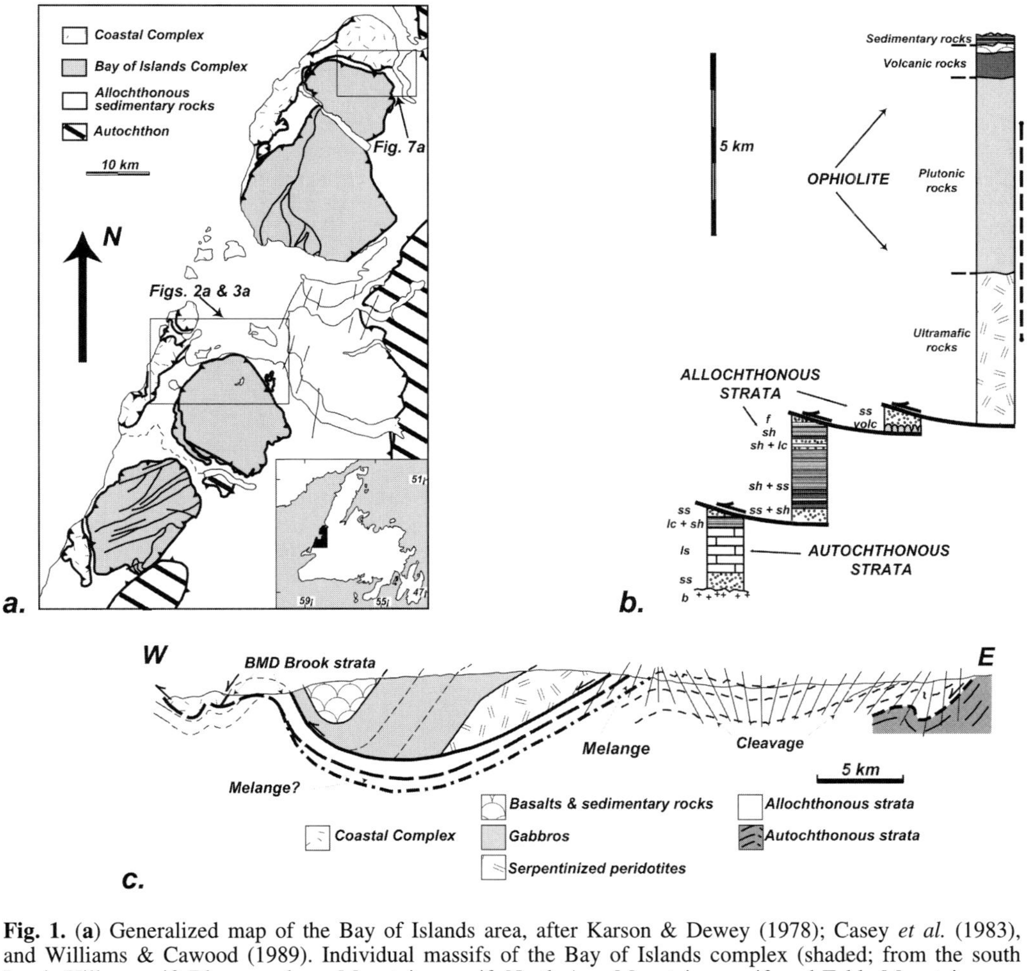

Fig. 1. (**a**) Generalized map of the Bay of Islands area, after Karson & Dewey (1978); Casey *et al.* (1983), and Williams & Cawood (1989). Individual massifs of the Bay of Islands complex (shaded; from the south Lewis Hills massif, Blow-me-down Mountain massif, North Arm Mountain massif, and Table Mountain massif) and the Coastal complex (with random dashes) occupy structural depressions in complexly folded allochthonous sedimentary rocks (in white) and authochthonous sedimentary strata and basement (with diagonal ruling). (**b**) Schematic 'stratigraphic' columns, after Casey & Karson (1981); Waldron (1985), indicating the character of the structural units. b, basement; ss, sandstone; sh, shale; ls, limestone; lc, limestone conglomerate; f, flysch; volc, basaltic volcanic rocks. The dashed line at right of the ophiolite column gives the approximate 'stratigraphic' position of the inferred transform fault rocks in the Lewis Hills massif (Karson 1984). (**c**) Generalized structural section across the Blow-me-down Mountain massif, based on local down-structure projections (Wojtal & Mitra 1988); east end of the section after Waldron (1985).

serpentinized peridotites, variably altered gabbroic rocks, and associated basaltic volcanic rocks and marine sedimentary rocks, and the related Coastal complex, which consists of altered, mainly mafic intrusive and extrusive rocks and their metamorphosed equivalents. The Bay of Islands complex is inferred to be obducted sea floor and upper mantle (Casey *et al.* 1983; Casey & Dewey 1984; Cawood & Suhr 1992; Suhr & Cawood 1993). The Coastal complex contains the altered remnants of an oceanic fracture zone (Karson & Dewey 1978; Karson 1984; compare Suhr & Cawood 1993; McCaig & Rex 1995). Both Bay of Islands complex and Coastal complex rocks sit, at present, on top of lower Palaeozoic sedimentary rocks of the Humber Arm Supergroup (Williams 1975; Waldron 1985; Williams & Cawood 1989; Cawood & Suhr 1992). Humber Arm Supergroup strata are themselves allochthonous, and are underlain by autochthonous carbonates and shales that accumulated along the edge of a lower Palaeozoic continental shelf (Williams 1975; Waldron 1985; Williams & Cawood 1989) (Fig. 1b and c). Humber Arm Supergroup units probably represent the slope and/or rise of the palaeo-North American continental margin (James & Stevens 1982; Lindholm & Casey 1989; Waldron *et al.* 1998). The Bay of Islands complex and Coastal complex together constitute the highest of several structural slices (Williams 1975); they were emplaced as a single, laterally extensive sheet that must be highly allochthonous (Casey *et al.* 1983; Casey & Dewey 1984; Wojtal & Mitra 1988; Waldron *et al.* 1998).

After emplacement, the region was subjected to folding about NNE-trending axes followed by folding about ESE-trending axes; the Bay of Islands–Coastal complex sheet now consists of four separate massifs (the Lewis Hills, Blow-me-down Mountain, North Arm Mountain, and Table Mountain) found in structural depressions (Waldron 1985) (Fig. 1a). The folding provides access to major faults at several locations across and along strike (Wojtal & Mitra 1988). This study examines minor structures in the Companion Mélange exposed at Frenchman's Cove and in serpentinites along the basal detachment beneath the Table Mountain massif.

Companion Mélange at Frenchman's Cove

Immediately below much of the Bay of Islands–Coastal complex sheet is a sequence of grey to buff sandstones and variegated argillites assigned to the Blow-me-down Brook Formation (Williams 1972, 1975; Williams & Cawood 1989) (Fig. 2). These sandstones and argillites are Late Proterozoic to Cambrian in age, derived from the west, and probably are rift-related sedimentary rocks that accumulated at the base of the North American shelf-slope–rise prism (Lindholm & Casey 1989; Williams & Cawood 1989). Near the Blow-me-down Mountain massif, Blow-me-down Brook Formation strata generally dip toward and under the massif (Figs 2 and 3). Numerous minor faults cut these strata. In investigating deformation within the Blow-me-down Brook Formation, I endeavoured to measure the orientations of all faults visible on vertical and horizontal surfaces in exposures along the rugged coastline and sinuous streams. The overall trends of coasts and streams dictate the typical attitudes of exposure faces. Further, weathering may preferentially remove evidence for faulting on surfaces parallel to the exposure face. As a result, there may be biases in measured fault-slip data. In this portion of the Bay of Islands, however, the local attitudes of sedimentary strata reflect the two folding episodes and, in turn, control the orientation of the coast and streams. Inasmuch as variations in the orientations in minor faults correlate with changes in bedding attitude, I infer that observed variations reflect real changes in minor fault populations.

Most minor faults cut bedding at high angles. Bed thicknesses change abruptly across some high-angle faults, suggesting that they are growth faults originally related to rifting. Most high-angle faults lack growth geometries, have down-dip or steeply raking slip mineral-fibre lineations, and normal offsets of bedding (Fig. 3b). These minor faults extend layering along WNW-trending to –NNW-trending axes. A small number of minor faults, most of which occur in strata higher in the section (closer to the Bay of Islands–Coastal complex sheet), cut bedding at low angles, have generally NW-trending lineations, and show reverse offsets. These minor faults shorten bedding along a NW-trending axis. Mineral-coated faults that cut bedding at high angles also cut and offset faults at low angles to bedding. Low-angle faults may have formed during the emplacement of the Bay of Islands–Coastal complex sheet; most high-angle faults probably formed during folding about NNE-trending axes.

Near Frenchman's Cove, NE of the Blow-me-down Mountain massif (Fig. 3a), strata have NNW strikes and dip steeply to the west (Fig. 3b). Units are folded about a kink fold with a gently SSE-plunging axis (Fig. 3b), so that the base of the Blow-me-down Brook Formation, which here consists of massive sand-

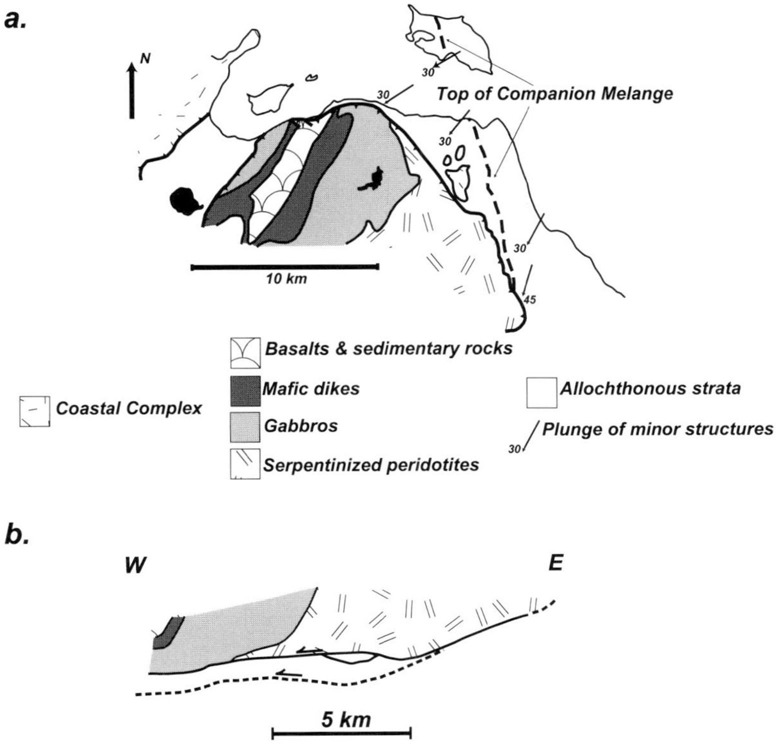

Fig. 2. (**a**) Generalized geological map of the NE portion of the Blow-me-down massif, modified from Williams & Cawood (1989). (**b**) Generalized down-plunge projection of the Blow-me-down massif showing the approximate geometric character of the top of the Companion Mélange.

stones with minor shales or argillites, locally dips gently to the south. As strata return to NNW strikes and steep dips, Blow-me-down Brook Formation sandstones give way to a zone of 100–200 m thickness of discontinuous, lens-shaped blocks of sandstone in chaotically distorted argillite, siltstone, mudstone, and shale. The transition from well-layered sandstone to chaotically deformed sandstone and argillite is abrupt, but there is no single discontinuity marking the base of the sandstone. Numerous faults cut interlayered sandstone and shale or argillite in this transition zone. Faults in the transition zone cut layering at low angles, and define irregular contractional duplexes. The stratigraphically lower, chaotically deformed rocks are part of the Companion Mélange.

The Companion Mélange separates the Blow-me-down Brook Formation from structurally lower and stratigraphically younger limestone conglomerates, sandstones and shale of the Curling Group (Williams 1976; Waldron 1985; Williams & Cawood 1989). On Woods Island, the mélange zone includes chaotically deformed mafic volcanic rocks from the base of the Blow-me-down Brook Formation. The stratigraphic and structural position of the mélange indicates that it is the locus of regional thrusting, carrying older strata over younger strata (Figs. 1c and 2). In deformed strata from the transition zone, minor structures are clearly consistent with regional shortening.

Characteristics of mélange zone deformation

Deformation within the mélange proper is not so straightforward. To assess deformation patterns, I collected data from a shoreline exposure on Woods Island, east–west-trending and north–south-trending shoreline exposures on the mainland, and from a series of road cuts a short distance south of those shoreline exposures. Figures 3b and 4a present stereographic projections of those measured deformation elements. At this position in a cross-strike transect, most Humber Arm Supergroup strata lack a mappable cleavage. Spaced solution cleavage occurs occasionally in impure sandstones, and, in thin section, quartz exhibits undulose extinction and some subgrain development. The primary planar

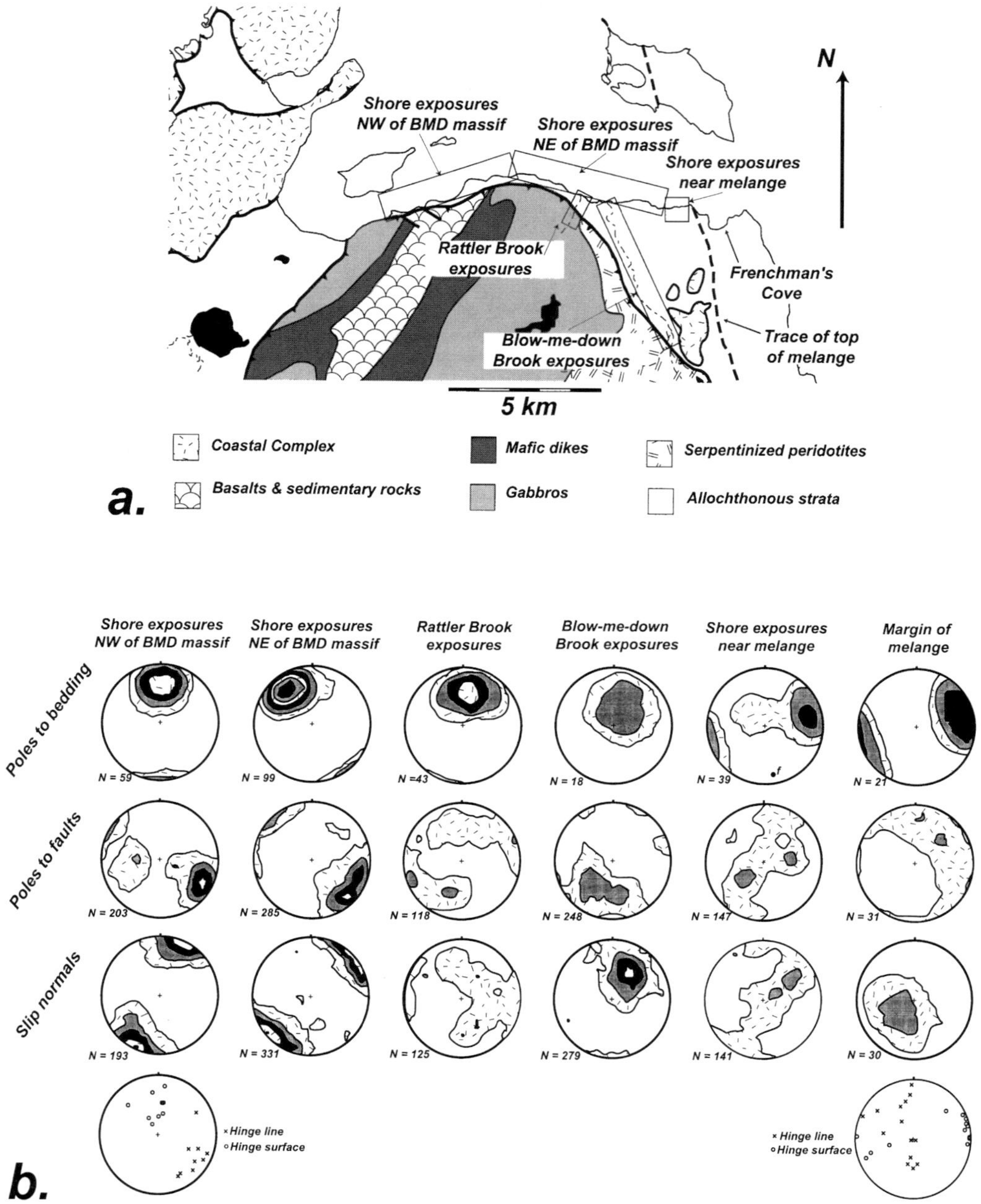

Fig. 3. (**a**) Generalized geological map of NE portion of the Blow-me-down massif showing the location of exposures of Blow-me-down Brook Formation. (*b*) Contoured, equal-area stereographic projections of structural elements in the Blow-me-down Brook exposures at locations shown in (**a**) and from the top of the Companion Mélange. Contours drawn by Allmendinger's STEREONET program using the Kamb method, with contours at 3σ intervals.

fabric in the mélange zone is compositional layering. Layering is often clearly bedding. Individual beds are often thinned or truncated by faults, but sometimes they neck or pinch out with no discernible faulting. Fine-grained layers often possess a fissility parallel or nearly parallel to the compositional layering; this fissility is, at locations, a secondary fabric composed of clo-

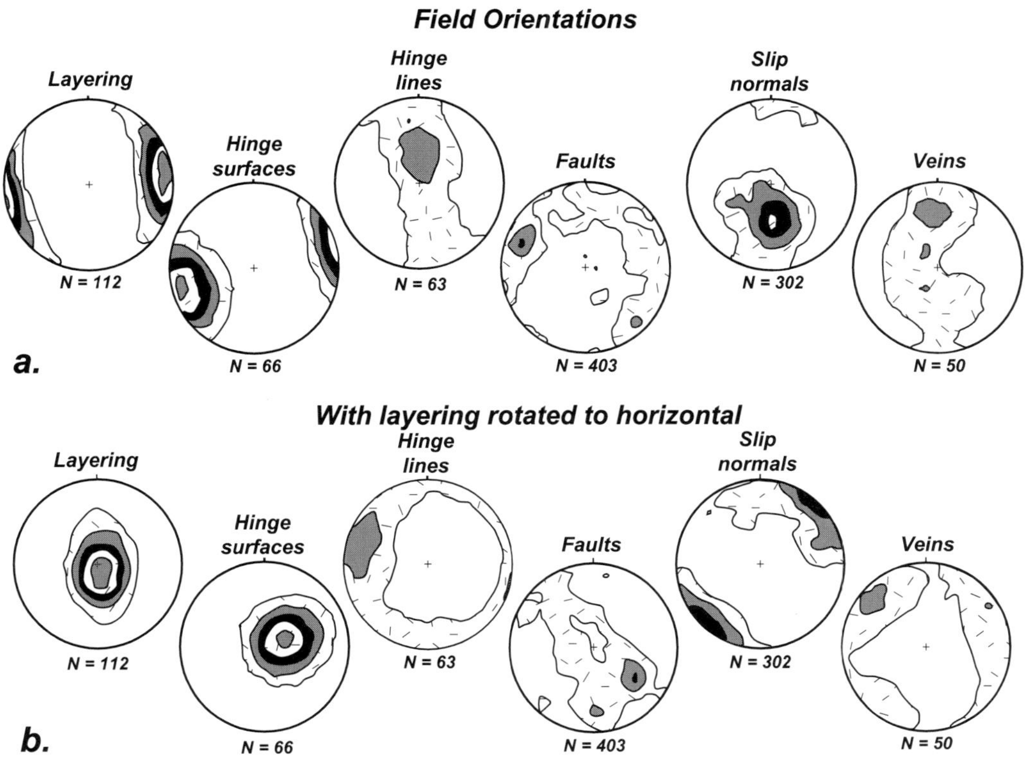

Fig. 4. (**a**) Contoured, equal-area stereographic projections of structural elements in the Companion Mélange in the vicinity of Frenchman's Cove. Contours drawn by Allmendinger's STEREONET program using the Kamb method, with contours at 3σ intervals. Plots labelled 'layering', 'hinge surfaces', 'faults' and 'veins' give poles to those planar features. (**b**) Contoured, equal-area stereographic projections of rotated structural elements in the Companion Mélange in the vicinity of Frenchman's Cove. All structural elements were rotated through the same angle required to bring the mélange zone layering to horizontal. Contours drawn by Allmendinger's STEREONET program using the Kamb method, with contours at 3σ intervals. Plots labelled 'layering', 'hinge surfaces', 'faults' and 'veins' again give poles to those planar features.

sely spaced shear surfaces or minor faults. Layering in the Companion Mélange at Frenchman's Cove, whether it is bedding or foliation, uniformly strikes NNW and dips steeply to the west (Figs. 3b and 4a).

Layering is often folded, but isolated fold hinges are more common than antiform–synform pairs. Fold hinge surfaces strike more westerly than the prominent layering, and fold hinge lines generally plunge steeply (Fig. 4a). Faults and veins are very common within the mélange zone. Poles to mineral-filled veins define a great circle girdle, with a diffuse point maximum at one point on the girdle (Fig. 4a). Faults have a range of orientations, although poles to faults cluster about two point maxima amongst significant numbers of faults with a variety of orientations (Fig. 4a). Faults often truncate sandstones, creating discontinuous sandstone stringers in argillite. In other cases, faults juxtapose isolated antiforms against other antiforms with no intervening synforms, or juxtapose isolated synforms against other synforms with no intervening antiforms. Mutually cross-cutting relationships suggest that faults with different orientations were active concurrently. It is not possible to determine the magnitudes of offsets on most faults, and many faults lack clear indications of even the sense of relative offset. It is, then, not possible to determine finite or incremental bulk strains in the mélange zone. Likewise, the paucity of unfaulted antiform–synform pairs yields few data (n = 23) giving the sense of overturning of folds. The geometry and kinematics of the faults hold the few keys available to understand the deformation in these rocks. Slickenlines exist on many of the fault surfaces, and slip normals, determined by taking the pole to the great circle that contains the fault pole and the slickenline (Price 1967;

Wojtal 1986), cluster at a prominent point maximum (Fig. 4a).

The SSE-plunging kink fold is related to the late, regional folding. It formed after the formation of the mélange, and reoriented deformation elements in it. Some slickenlines on faults are most easily related to interlayer slip associated with this late folding. I endeavoured to remove the effects of late folding by rotating all stereographic data through the angle required to bring the point maximum for poles to layering to the centre of the stereonet (Fig. 4b). Such a rigid rotation does not undo all effects of folding, but it does provide some insight into the pre-folding structure of the mélange zone. After rotation, poles to faults define a diffuse girdle corresponding to a NNW-striking vertical plane. Rotation through the same angle brings the mean orientation of slip normals to horizontal, normal to this vertical plane. A slip normal is equivalent to the 'local' tectonic *b*-axis for a fault, or is parallel to the axis of internal vorticity associated with fault-parallel shearing. The point maximum of slip normals perpendicular to the vertical plane defined by poles to faults suggests that the NW-striking vertical plane is the *ac*-plane of the deformation. With the mélange zone boundaries returned to horizontal, a NE-trending tectonic *b*-axis is consistent with analyses of the regional shortening in the Bay of Islands (e.g. Waldron *et al.* 1998).

Inferred kinematics of deformation in mélange zone

In detail, it is not clear how to interpret deformation in the mélange zone. The diffuse girdle defined by poles to faults consists of two girdles, each inclined to the NW-striking vertical great circle (Fig. 5a). Tangent lineations (Twiss *et al.* 1991) on most minor faults possess mirror symmetry about a comparably oriented, NW-striking plane (Fig. 5b); the remaining tangent lineations are consistent with layer-parallel shearing associated with the late folding. Together, the fault data suggest that this NW-striking plane is the *ac*-plane for the deformation. With the layering, and thus the inferred boundaries of the mélange zone, rotated to horizontal, minor faults dip NW and are inclined to the *ac*-plane, or dip SE and are inclined to the *ac*-plane (Fig. 5c). Faults rarely have traces on the mélange zone boundary parallel to the inferred tectonic *b*-axis, and are rarely parallel to the mélange zone boundaries. The sense of shearing in this zone is not clear. Striated faults in critical orientations have ambiguous indicators of the sense of offset. In the absence of clear data on the sense of offset on these critical faults, the tangent lineation pattern could fit either shear sense (Figs. 5b and 6). Rotated fold hinge lines are roughly parallel to the inferred *a*-axis, but rotated fold hinge surfaces do not correlate readily with other deformation elements (Fig. 4b). The sense of overturning of folds is also ambiguous. Nine out of 23 asymmetric folds are consistent with top-to-the-SE shearing after being rotated through the angle needed to bring layering to horizontal. Six asymmetric folds are, after rotation, consistent with top-to-the-NW shearing. The remaining eight asymmetric folds have, after rotation, NW–SE-trending hinge lines, and so are not diagnostic of shear sense. The rotated poles to mineral-filled veins define a point maximum generally parallel to the inferred tectonic *a*-axis (Fig. 4b), consistent with an inference of elongation parallel to the *a*-axis. The point maximum of poles to veins suggests a weak preferred orientation of veins inclined steeply to the NW (with layering rotated to horizontal), but significant numbers of poles to veins also define a diffuse horizontal great circle parallel to the primitive. The point maximum may indicate top-to-the-SE shearing, or, if veins rotated during deformation, top-to-the-NW shearing. In any case, the orientations of veins suggest elongation in a range of directions normal to the inferred tectonic *c*-axis and parallel to the mélange zone boundaries.

Deformation in this mélange zone resembles that in other major fault zones (e.g. Wojtal 1986) in some important ways: poles to layering, fault poles, and poles to veins define an *ac*-plane, and slip normals define a tectonic *b*-axis for the bulk deformation. The general trends of tangent lineations for mélange zone faults exhibit monoclinic symmetry about a NNW-striking plane. By inspection, the tangent lineations resemble the pattern expected for SE-directed shearing combined with fault-zone flattening or NW-directed shearing combined with fault-zone constriction (Twiss *et al.* 1991) (Fig. 6). The orientations of minor faults in the Companion Mélange at Frenchman's Cove exhibit monoclinic symmetry compatible with the monoclinic symmetry of the tangent lineation pattern. Faults are arrayed at a variety of angles relative to the mélange zone boundaries. Many faults are inclined to the SE relative to horizontal mélange zone boundaries, and a moderate number of faults are inclined 50–60° toward the NW relative to horizontal mélange zone boundaries. In foreland fold-and-thrust-belt thrust zones (e.g. Wojtal 1986), a pattern of

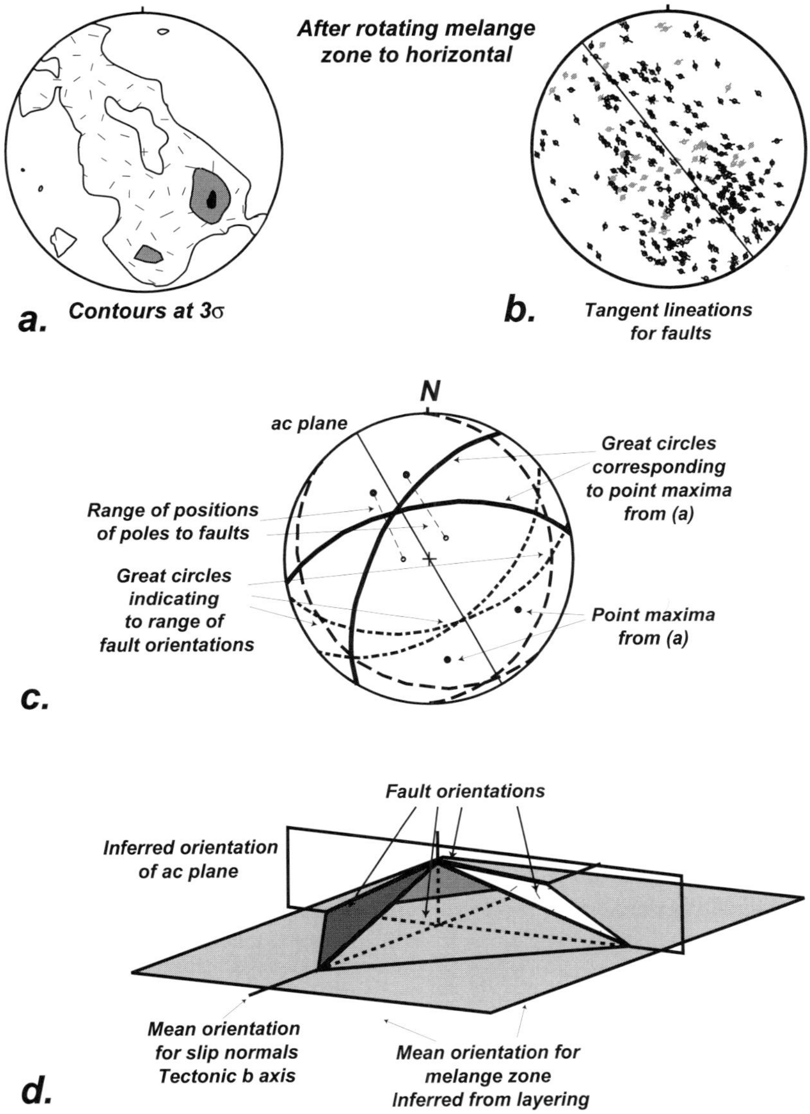

Fig. 5. (**a**) Contoured, equal-area stereographic projections of poles to faults in the Companion Mélange in the vicinity of Frenchman's Cove, after rotation through angle required to bring mélange zone layering to horizontal. Contours drawn by Allmendinger's STEREONET program using the Kamb method, with contours at 3σ intervals. (**b**) Equal-area stereographic projection of tangent lineations for faults in the Companion Mélange in the vicinity of Frenchman's Cove. Tangent lineations plotted in black are inferred to relate to thrust-related shearing; tangent lineations plotted in grey are inferred to relate to later folding. (**c**) Stereographic projection showing the relationships between deformation elements. (**d**) Schematic diagram showing the orientations of faults, the inferred *ac*-plane, and tectonic *b*-axis relative to the mélange zone layering.

shallowly SE-dipping faults and steeply NW-dipping faults would suggest top-to-the-NW shearing. The evidence for such an inference here is more ambiguous, but I prefer this interpretation for deformation in the mélange zone. Faults in the mélange zone rarely have strikes normal to the inferred transport direction, as is often the case for thrust-related arrays of minor faults (e.g. Wojtal 1986). Given the lack of a consistent cross-cutting pattern among the faults, I infer that faults of all orientations within this monoclinic array were simul-

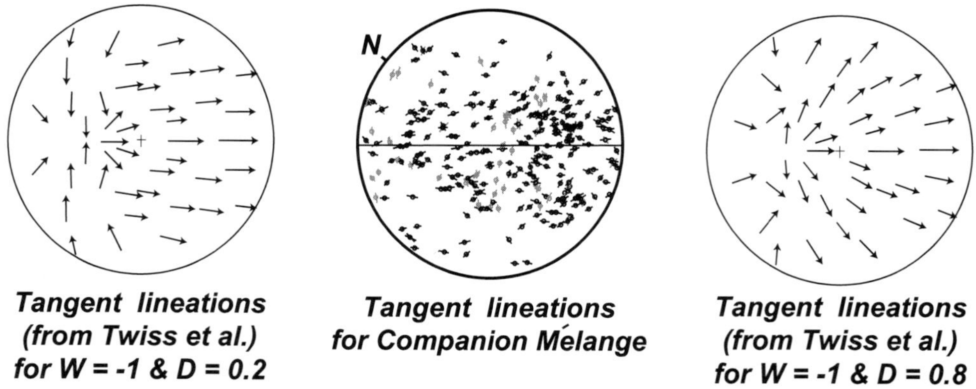

Fig. 6. Comparison of tangent lineation plots for shearing plus flattening and shearing plus constriction non-coaxial flows, taken from Twiss *et al.* (1991), and tangent lineations for faults in the Companion Mélange in the vicinity of Frenchman's Cove. D denotes the differences between ratios of the principal values of the total or macroscopic deformation rate tensor. $D = 0.2$ corresponds to shearing plus constriction and $D = 0.8$ corresponds to shearing plus flattening. W denotes the ratio of the net vorticity to the maximum difference between the principal values of the deformation rate matrix. In the Twiss *et al.* (1991) formulation, W depends upon the relative magnitudes of the vorticity of the macroscopic deformation and the relative vorticity or 'microspin' of individual fault-bounded blocks. $W = -1$ may correspond to pure shearing with relative microspin of fault-bounded blocks or simple shearing with no relative microspin of fault-bounded blocks.

taneously active. The tangent lineation pattern suggests shearing parallel to the mélange-zone boundaries and shortening or elongation normal to the shearing direction; this minor fault array formed during a 3D general shearing flow.

Winterhouse Brook exposure of the Bay of Islands complex sheet detachment

At Winterhouse Brook, on the NE side of the Table Mountain massif (Figs. 1 and 7), serpentinized peridotites at the base of the Bay of Islands complex sheet lie on variably metamorphosed shales and mudstones. At the contact, a rodingite mylonite of 20 cm thickness, with discrete, grooved upper and lower boundaries, separates weakly metamorphosed mudstones and shales from a flaggy serpentinite. The contact strikes WNW at present, and dips steeply to the south (Fig. 7b). Grooves on the contact separating weakly metamorphosed mudstones and shales from rodingites, grooves on the contact separating rodingites and a flaggy serpentinite, and slickenlines and slickenfibres in serpentinites all plunge gently to the ESE. This suggests a WNW movement of the Bay of Islands complex sheet relative to the underlying metasedimentary rocks, and yields a slip normal for the major fault that plunges steeply to the SSW. Rotating the fault surface to horizontal yields a NNE-trending tectonic b-axis and a WNW-trending tectonic a-axis (Fig. 7b).

Characteristics of serpentinite deformation

More than 10 m above the contact with metasedimentary rocks, serpentinized peridotites are generally massive. In thin section, the massive serpentinites exhibit an undeformed mesh texture (Maltman 1978; fig. 5a of Wojtal & Mitra 1988). Numerous planar to smoothly curved fractures cut the massive serpentinites. Fractures are usually filled by cross-fibre serpentine. In many cases, fractures also have serpentine slickenfibre coatings. The poles to faults in massive serpentinites >35 m from the contact with metasedimentary rocks cluster along a great circle girdle (Fig. 8a). Rotating fault poles through the angle needed to bring the major fault and its lineation to horizontal yields a diffuse great circle roughly perpendicular to the major fault (Fig. 8b). Fault poles in the great circle girdle cluster at two point maxima. One point maximum corresponds to faults inclined slightly to the SE relative to the major fault; the other corresponds to faults inclined at moderate angles to the NW relative to the major fault. Between 10 and 35 m from the contact, poles to faults define a diffuse great circle oblique to the major fault (Fig. 8b). Faults here generally are inclined to the SE relative to a horizontal major fault, but some are perpendicular to the major fault (Fig. 8b).

About 10 m from the contact with metasedimentary rocks, massive serpentinite gives way to a mesoscopic 'block in fine-grained matrix'

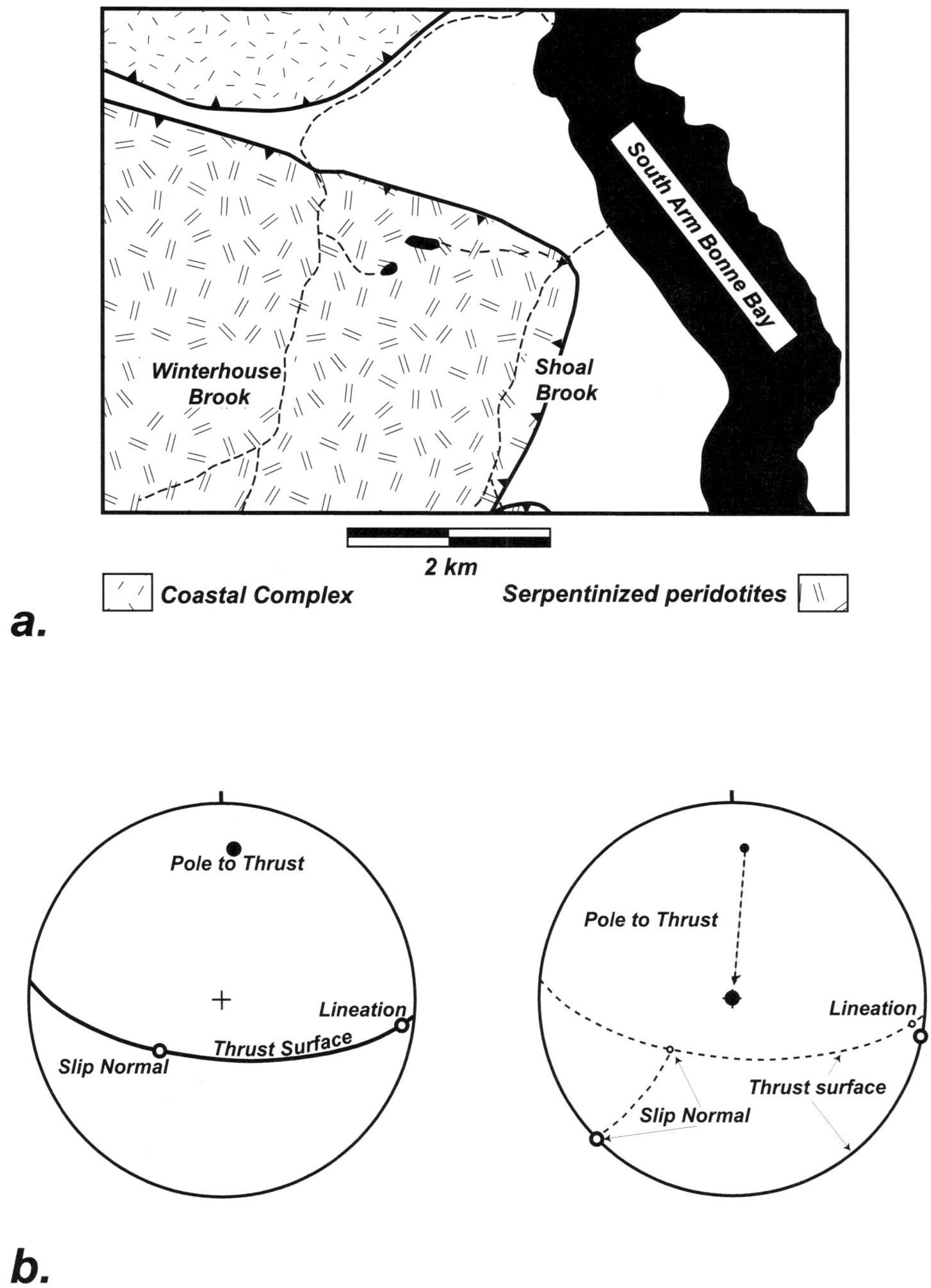

Fig. 7. (**a**) Generalized geological map of the NE portion of the Table Mountain massif showing the location of the Winterhouse Brook exposure of the basal detachment of the Bay of Islands complex sheet. (**b**) Stereographic projection showing the present geometry of the contact between serpentinized peridotite and metasedimentary rocks, the lineations on that surface, and the slip normal for the fault, which is the best viewing axis for fault-zone structures, and showing orientation of structural elements after rotating thrust to horizontal.

serpentinite fabric. O'Hanley (1996) called this mesoscopic fabric 'foliate serpentine breccia'. The rounded to lenticular 'blocks' are undeformed to weakly deformed, fault-bounded bodies of massive serpentinite. In thin section, rounded blocks exhibit undeformed mesh-

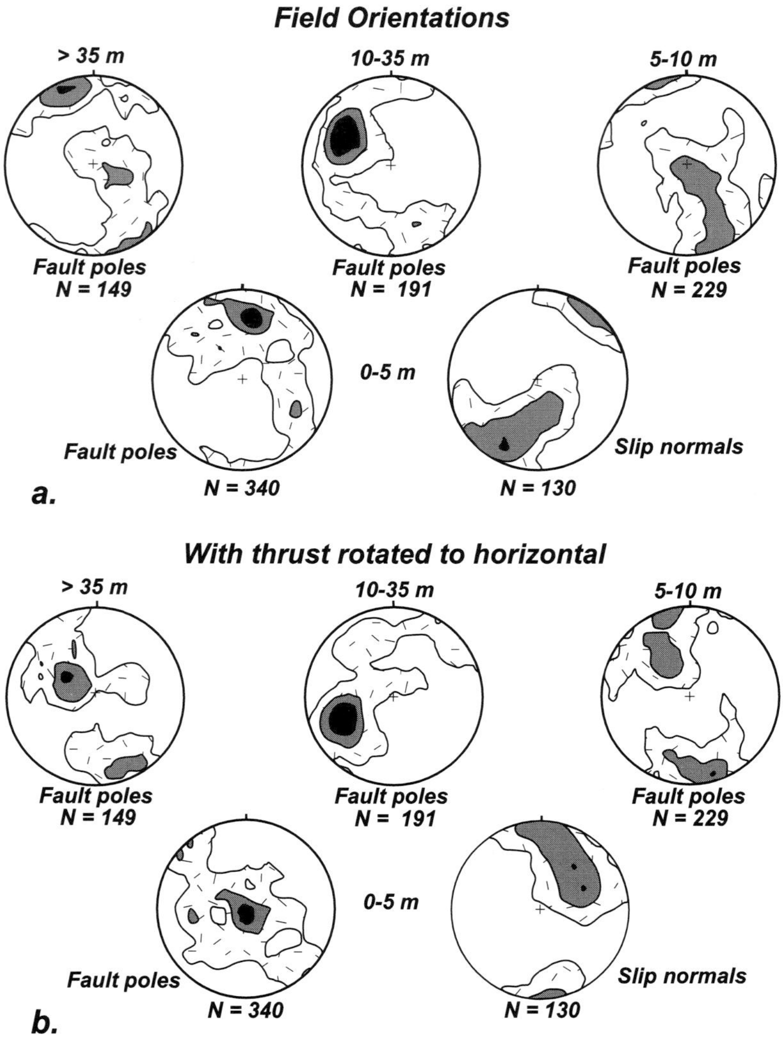

Fig. 8. (**a**) Contoured, equal-area stereographic projections of structural elements in the Winterhouse Brook exposure of the basal detachment of the Bay of Islands complex sheet. Contours drawn by Allmendinger's STEREONET program using the Kamb method, with contours at 3σ intervals. Plots give poles to shear surfaces in rock farther than 35 m from the contact with metasedimentary rocks, at 10–35 m from the contact, 5–10 m from the contact, and 0–5 m from the contact, and slip normals on shear surfaces within 5 m of the contact. (**b**) Contoured, equal-area stereographic projections of structural elements in the Winterhouse Brook exposure of the basal detachment of the Bay of Islands complex sheet. All structural elements were rotated through the same angle required to bring the fault-zone layering to horizontal. Contours drawn by Allmendinger's STEREONET program using the Kamb method, with contours at 3σ intervals. Plots give poles to shear surfaces in rock farther than 35 m from the contact with metasedimentary rocks, at 10–35 m from the contact, 5–10 m from the contact, and 0–5 m from the contact, and slip normals on shear surfaces within 5 m of the contact.

texture serpentine, like that seen in the overlying rocks. In lenticular 'blocks' the mesh texture is often significantly distorted. Bounding surfaces for the blocks are striated surfaces with orientations that, in part, are parallel to the orientations of faults in the overlying massive serpentinite (Fig. 8b). These surfaces often meet at low angles to form anastomosing systems. Usually, individual segments in the anastomosing systems are no longer discrete fault surfaces. Rather, wall-rock adjacent to segments is highly distorted, and in some cases exhibits the fine-grained bladed texture of Maltman (1978). Thus, many of the striated bounding surfaces of blocks are actually bounding surfaces of shear zones. Where anastomosing faults or shear surfaces intersect, distorted mesh texture and/or fine-grained bladed texture serpentine occurs even farther from the cores of small faults or shear zones (fig. 5c of Wojtal & Mitra 1988). This creates, in any outcrop-scale profile, braided bands of foliated 'fine-grained' serpentine 'matrix' that separate mesoscopic 'blocks' of undeformed to weakly deformed serpentinized peridotite. The thickness of small faults or shear zones with different orientations suggests that many faults accrued significant offset. Apparently, fault- or shear-zone orientations were such that intersecting faults or shear zones could transfer slip to one another effectively, thereby leading to slip being partitioned on many fault or shear zones simultaneously (see Jones *et al.* 1997). Slip on these anastomosing systems probably accommodated larger offsets than slip on discrete faults in the overlying rocks.

There are more small fault or shear zones per unit volume 5 m from the contact than there are 10 m from the contact. Fault- or shear-zone intersections are also more common. As a result, there is more foliated 'fine-grained matrix' in rock 5 m from the contact than there is in rock 10 m from the contact. Correspondingly, the median size of mesoscopic 'blocks' is smaller nearer the contact.

Within 5 m of the contact with the metasedimentary rocks, the mesoscopic character of the rock, and the geometry of the fault array, changes subtly. The number of faults or shear zones increases dramatically, so that the mean size of fault-bounded 'blocks' decreases by an order of magnitude from what it was *c.* 10 m from the contact. Many more of the shear surface orientations are at low angles to or are nearly parallel to the major fault. As a result of the orientations of shear surfaces in the anastomosing array, individual fault-bounded blocks are highly flattened parallel to the fault zone and, if the major fault were rotated back to horizontal, elongate in a NW-trending line. The 'fine-grained' serpentine 'matrix' itself exhibits a strong foliation and usually exhibits a strong lineation parallel to the long axes of the fault-bounded blocks. Thus, the rock is a serpentinite mylonite, with an incipient to well-defined S–C–C' fabric. At thin-section scale, serpentine is nearly entirely recrystallized, exhibiting fine-grained fibrous to fine-grained bladed textures (Maltman 1978; fig. 5d of Wojtal & Mitra 1988). Only remnant 'porphyroclasts' retain their mesh texture. In these foliated serpentinites and serpentinite S–C–C' mylonites, mutually cross-cutting relationships suggest that shear surfaces with different orientations were active at the same time; Jones *et al.* (1997) drew similar inferences from mesoscopic structures in sheared serpentinites associated with the Troodos complex.

Structures similar to those seen at this exposure occur at other locations along the detachment at the base of the Bay of Islands complex sheet. For example, along a tributary of Blow-me-down Brook on the NE side of the Blow-me-down Mountain massif are strongly sheared serpentinites like those found at the base of the Winterhouse Brook section. Similarly, on the SW side of the Blow-me-down massif is a succession from massive serpentinized peridotite to strongly sheared, mylonitic serpentinite entirely comparable with that observed in the Winterhouse Brook exposures. Wojtal & Mitra (1988) argued that this sequence of mesoscopic structures is typical of the base of the Bay of Islands complex sheet at comparable cross-strike locations.

Inferred kinematics of deformation in serpentinites

After rotating all structural data through the angle required to bring the major fault to horizontal, tangent lineations (Twiss *et al.* 1991) on many shear surfaces, especially those oriented at low angles to the thrust-zone boundaries, are symmetrically oriented relative to a WNW-trending plane (Fig. 9b). Tangent lineations on the remaining shear surfaces, including many oriented at higher angles to the thrust-zone boundaries, are consistent with the late folding of the fault zone. This suggests that a WNW-trending plane is the *ac*-plane for the thrust-related bulk deformation. With the fault-zone boundaries horizontal, a WNW-trending tectonic *a*-axis is consistent with analyses of the regional deformation patterns (e.g. Waldron *et al.* 1998). After a comparable rotation, slip normals cluster along a great circle that is perpendicular to this plane and inclined toward the east relative to the major fault. This pattern resembles that observed in foreland fold-and-thrust belt thrust zones (e.g. Wojtal 1986).

Rotating the poles to shear surfaces within 5 m of the contact with the underlying metasedimentary rocks through the same angle yields a diffuse, NW-striking great circle girdle. Shear surfaces inclined steeply relative to a horizontal major fault have orientations similar to those of shear surfaces in the overlying rocks. Unique to this portion of the fault zone are shear surfaces

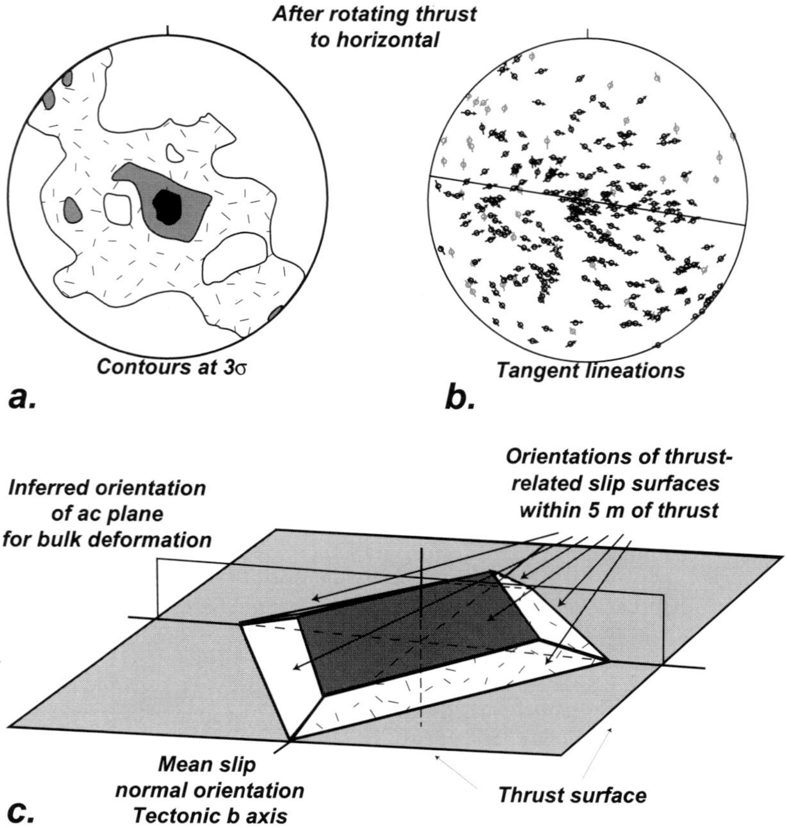

Fig. 9. (**a**) Contoured, equal-area stereographic projections of poles to shear surfaces in serpentinites within 5 m of the contact with metasedimentary rocks after rotation through the angle required to bring the fault zone to horizontal. Contours drawn by Allmendinger's STEREONET program using the Kamb method, with contours at 3σ intervals. (**b**) Equal-area stereographic projection of tangent lineations for shear surfaces within 5 m of the contact with metasedimentary rocks. Tangent lineations plotted in black are inferred to relate to thrust-related shearing; tangent lineations plotted in grey are inferred to relate to later folding. (**c**) Schematic diagram showing idealized orientations of thrust-related shear surfaces, the inferred *ac*-plane, and inferred tectonic *b*-axis relative to the mélange zone layering.

at low angles to the major fault. With the thrust horizontal, the shear surfaces at low angles to the thrust may be oriented parallel to the thrust, inclined to the west and oblique to the *ac*-plane, or inclined to the east and oblique to the *ac*-plane (Fig. 9a). Figure 9c is a schematic representation of the orientations of shear surface nucleated within 5 m of the underlying metasedimentary rocks. Shear surfaces in serpentinites within 5 m of the contact with meta-sedimentary rocks at Winterhouse Brook exhibit a distinctive monoclinic symmetry, consistent with the monoclinic symmetry of the tangent lineations for these faults.

Discussion

The Companion Mélange at Frenchman's Cove and the Winterhouse Brook serpentinites share significant characteristics. In both cases, deformation is concentrated in a layer tens to hundreds of metres thick that is cut by a 3D array of shear surfaces. Tangent lineations for most shear surfaces conform to monoclinic patterns, suggesting deformation kinematics consisting of shearing with concurrent fault-zone flattening or shearing with concurrent fault-zone thickening; tangent lineations on the remaining shear surfaces are appropriately oriented to record interlayer slip during later folding. Mutually cross-cutting relationships suggest that shear surfaces with different orientations were active at the same time, so shearing across either fault zone would have required deformation of the blocks between shear surfaces. In the most highly deformed regions of these two fault zones, shear surfaces are oriented at low angles to the inferred shear plane, are inclined toward and away from the inferred shearing direction, and are oblique to the inferred *ac*-plane for the deformation. In the Companion Mélange at Frenchman's Cove, the array itself exhibits monoclinic symmetry. In the Winterhouse Brook serpentinites, the shear surfaces most readily correlated with thrust-related deformation also exhibit monoclinic symmetry. The kinematic similarities of these two fault zones suggest that similar processes operated in both. Two aspects of the fault array at Frenchman's Cove and the shear surface array at Winterhouse Brook require explanation. The first concerns the 3D, asymmetric character of deformation at both of these locations. The second concerns the similarity of orientations for faults in the Companion Mélange on one hand and shear zones in the Winterhouse Brook serpentinites on the other hand.

Origin of faults and/or shear zones

The anastomosing faults in the Companion Mélge at Frenchman's Cove resemble the 'mudstone-dominated brittle fault zones' or 'Type IV mélanges' of Cowan (1985). Cowan & Brandon (1994) argued that many faults in anastomosing fault systems in mudstone-dominated brittle fault zones are Riedel composite structures. In profiles parallel to the inferred *ac*-plane, anastomosing faults at Frenchman's Cove certainly resemble Riedel shears (see Rutter *et al.* 1986). In a strict sense, however, Riedel shears form in plane flows, and should therefore be cylindrical features that intersect along lines normal to the *ac*-plane. Riedel shears should, to use a term from crystallography, have a common zone axis parallel to the tectonic *b*-axis. Faults in the Companion Mélange at Frenchman's Cove, and in the author's experience faults in many other mélange zones, lack such a zone axis. This mélange zone presents comparable anastomosing patterns in sections parallel, normal, or oblique to the *ac*-plane. This pattern suggests 3D, not plane, strain. Tangent lineations on faults (Fig. 5c) support such an inference, as does the range of orientations of mineral-filled veins (Fig. 3a and b). For this reason, it is not appropriate to interpret the anastomosing fault pattern using 2D Mohr–Coulomb theory, as is typical of analyses of Riedel shears (e.g. Tchalenko 1970; Price & Cosgrove 1990, p. 144). Neither do the observed faults have orientations consistent with 3D Mohr–Coulomb theory (see Reches 1983; Reches & Dieterich 1983). For any reasonable orientation for principal stresses, shear surfaces are asymmetrically disposed relative to these principal directions. De Paor (1994) argued that asymmetric arrays of shear surfaces, some of which would have atypical orientations relative to the most compressive and most tensile stresses, may form in response to transient, asymmetric stress fields. De Paor formulated this criterion in two dimensions, but his approach could conceivably be extended to three dimensions. Considering the two examples described in this paper, a problem arises in addressing the nucleation and growth of the shear surfaces in serpentinites. The formation of small shear zones in serpentinites would seem to require asymmetric stresses to persist for some time. A separate issue arises in comparing the geometries of faults in the Companion Mélange with the geometries of shear zones in the serpentinites. Faults in unmetamorphosed sedimentary rock are usually inferred to indicate pressure-sensitive failure criteria such as the Mohr–Coulomb criterion, whereas the shear zones in the metamorphosed serpentinites are inferred to indicate pressure-insensitive plastic flow (see fig. 9.9 of Twiss & Moores 1992). Differences in the angles of internal friction inferred for the two failure criteria predict different orientations for faults and shear zones, generally yielding steeper inclinations (relative to the zone boundaries) for significant numbers of shear zones. Comparing the arrays of faults in the mélange with the orientations of shear surfaces in the serpentinite suggests that the latter have shallower inclinations relative to the zone boundaries (compare Figs. 5c and 9d). Finally, as noted above, there are significant impediments to defining independently stresses in fault zones (see Fig. 10; Twiss *et al.* 1991; Twiss & Unruh 1998), and these lead to significant problems in drawing conclusions based on the stresses in fault zones.

Wojtal & Mitra (1988); Shimamato (1989) argued that minor faults, shear zones, or other slip surfaces result from the localization of deformation in fine-grained rocks during semi-ductile flow in fault zones. This proposal cannot apply in the fault zones described here. Most minor faults in the Companion Mélange have slickenfibre coatings and lack fine-grained fault gouge or cataclasite. Further, not all fault-

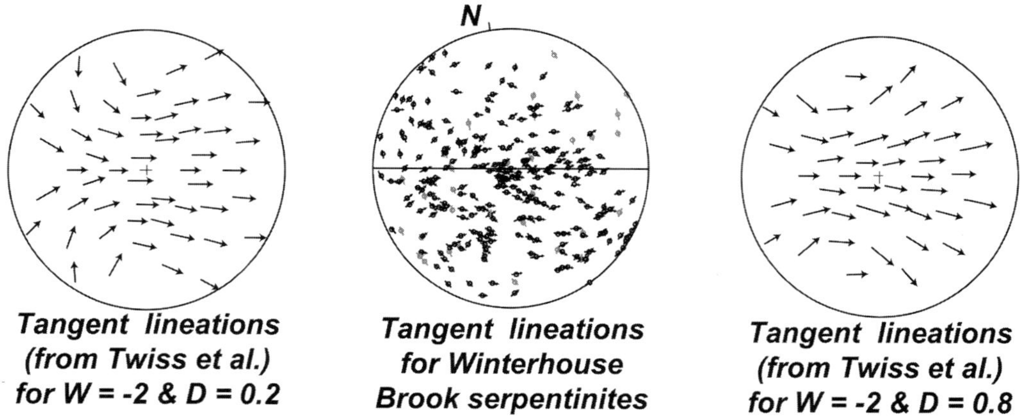

Fig. 10. Comparison of tangent lineation plots for shearing plus flattening and shearing plus constriction non-coaxial flows, taken from Twiss *et al.* (1991), and tangent lineations for shear surfaces in serpentinites within 5 m of the contact with underlying metasedimentary rocks at Winterhouse Brook. *D* denotes the differences between ratios of the principal values of the total or macroscopic deformation rate tensor. $D = 0.2$ corresponds to shearing plus constriction, and $D = 0.8$ corresponds to shearing plus flattening. *W*denotes the ratio of the net vorticity to the maximum difference between the principal values of the deformation rate matrix. In the Twiss *et al.* (1991) formulation, *W* depends upon the relative magnitudes of the vorticity of the macroscopic deformation and the relative vorticity or 'microspin' of individual fault-bounded blocks. $W = -2$ corresponds to deformations in which the microspin is equal in magnitude but opposite in direction to the relative macro-spin.

bounded blocks in the Companion Mélange are intensely deformed. The localization of semi-ductile deformation seems unlikely to have occurred here. In the case of the deformed serpentinites at Winterhouse Brook, the generation of fine-grained rock, the 'matrix' in these foliate breccias, results from the formation of shear surfaces, not the other way around.

Watterson (1999) argued that shear surfaces form along surfaces that experience no finite elongation in coaxial deformation. Watterson's criterion would seem to apply to kinematically equivalent deformation elements such as small shear zones. However, there is a difficulty in defining and using finite strain parameters, which refer to the symmetric portion of the displacement gradient field, in the strongly non-coaxial flow observed in fault zones. Furthermore, finite strain parameters are not sufficiently time independent to use as reliable measures of general shearing flows.

Twiss & Unruh (1998) have argued that fault-zone structures are most readily and effectively interpreted as indicators of velocity gradient fields. Velocity gradient fields in fault zones, unlike incremental strain or strain rate tensors, are asymmetrical. Asymmetrical shear surface arrays like those described here are additional evidence of the difficulty of interpreting fault-zone character using incremental strains or the symmetric portion of the velocity gradient field. It is especially clear in the Winterhouse Brook serpentinites that new shear surfaces developed during fault formation. These *neoformed* faults did not develop with the orthorhombic geometry expected for faults (see Angelier 1994). The question is how do shear surfaces initiate in asymmetrical arrays.

Let us consider a non-coaxial flow corresponding to a non-spinning fault zone, in which the velocity gradient matrix has an upper-triangular form (Fossen & Tikoff 1993; Tikoff & Fossen 1993) corresponding to a combination of 3D coaxial deformation (i.e. 'pure shear') and simple shear (see Figs 11 and 12). It is not clear whether or how a material surface would distinguish the component of the transverse velocity gradient across it ($\partial v_i/\partial x_j$ in Fig. 11) ascribed to the symmetric strain rate tensor from that component ascribed to the deformation-induced vorticity. Thus, it is sensible to suggest that both components of the velocity gradient field will affect the initiation of shear surfaces in a non-coaxial deformation. Shear surfaces nucleated in response to gradients of the total velocity field in general shearing could yield shear surface orientations asymmetrical to fault-zone boundaries (Simpson & De Paor 1993). Figure 10 suggests such a failure criterion, and illustrates how it differs from criteria tied to the symmetric portion of the velocity gradient field. Shear surfaces formed in

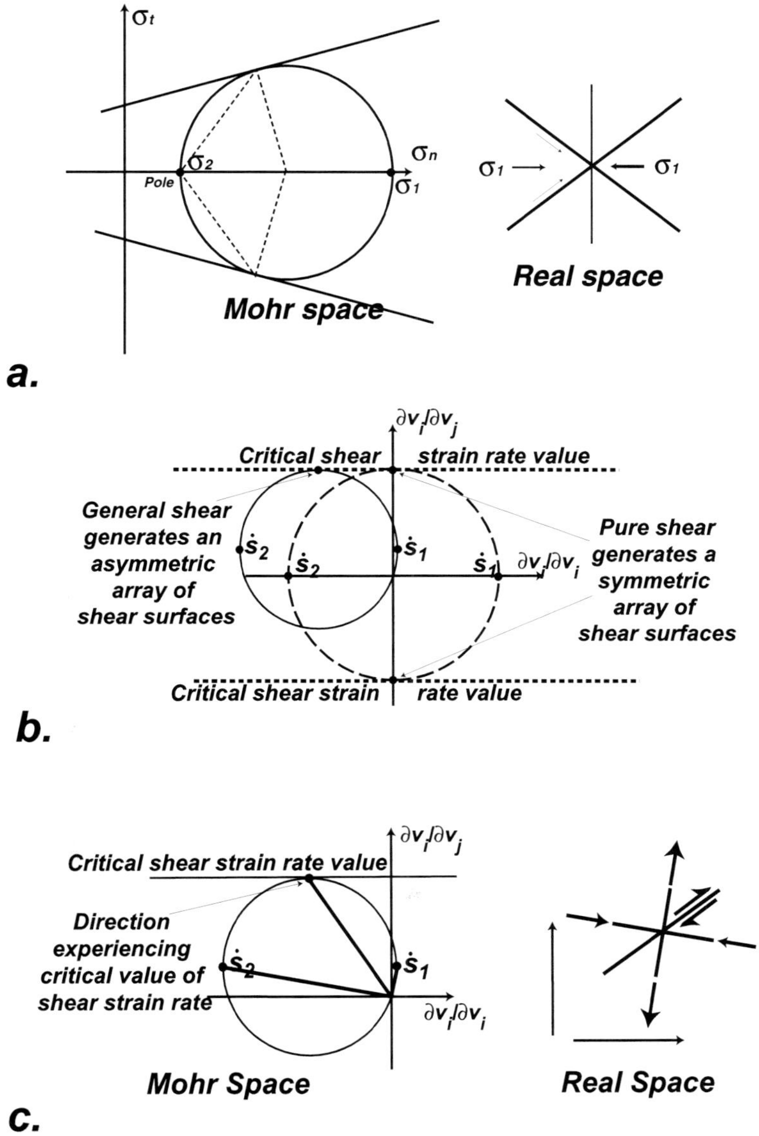

Fig. 11. (**a**) Mohr–Couloumb criterion for faulting, with 2D stress Mohr circle tangent to failure envelope, at left, and diagrammatic cross-section showing the geometry of faults relative to most compressive principal stresses at right. (**b**) Diagrammatic representation of the hypothesis that shear surfaces form when shearing velocity gradient magnitudes reach a critical value. In $\partial v_i/\partial x_i - \partial v_i/\partial x_j$ space, a Mohr circle centred at the origin (dashed) depicts a pure shear flow. In this flow, formation of shear surface at a critical value of transverse velocity gradient leads to two, orthogonal conjugate shear surfaces. The off-axis, off-centre Mohr circle depicts a ductile deformation zone undergoing simultaneous shearing and shortening. (**c**) At left is an off-axis, off-centre Mohr circle depicting a ductile deformation zone undergoing simultaneous shearing and shortening. At right is a schematic representation showing that the faults or shear zones with critical value of shearing velocity gradient are inclined to the directions of maximum instantaneous shortening and elongation.

response to such a criterion would take orientations that are asymmetric to the large-scale deformation-zone boundaries in shearing flows, and could yield shear surface arrays that possess overall symmetries that conform to the monoclinic symmetry of the velocity field. It is not clear, however, that such a criterion would lead to shear surfaces that lack a zone axis. An alternative hypothesis is that shear surfaces originate along numerous surfaces whose trans-

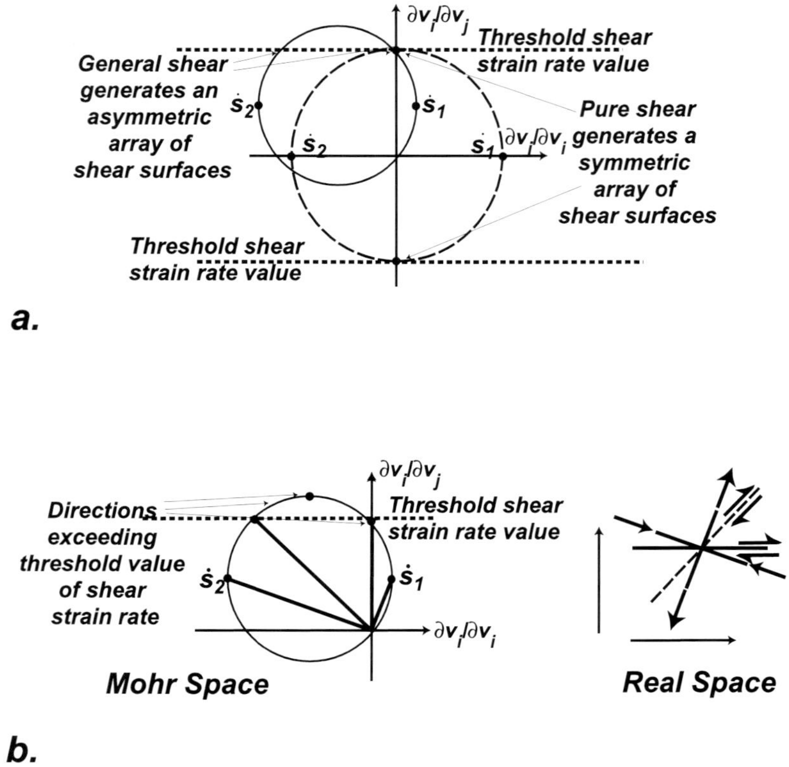

Fig. 12. (**a**) Diagrammatic representation of the hypothesis that shear surfaces form when transverse velocity gradient magnitudes are above a threshold value, and that well-defined surfaces form parallel to planes that experience no instantaneous shortening or elongation. In $\partial v_i/\partial x_i - \partial v_i/\partial x_j$ space, a Mohr circle centred at the origin (dashed) depicts a pure shear flow. In this flow, formation of shear surfaces at a critical value of shearing velocity gradient leads to two, orthogonal conjugate faults or shear surfaces. The off-axis, off-centre Mohr circle depicts a ductile deformation zone undergoing simultaneous shearing and shortening. (**b**) At left is an off-axis, off-centre Mohr circle depicting a ductile deformation zone undergoing simultaneous shearing and shortening. At right is a schematic representation showing that the faults or shear zones formed parallel to planes that experience no instantaneous shortening or elongation are inclined to the directions of maximum instantaneous shortening and elongation.

verse velocity gradients exceed some threshold value. Of the shear surfaces that nucleate under these conditions, those material surfaces that are neither shortened (buckled) nor elongated (boudinaged) might persist long enough to accrue finite shear displacement without disruption, thereby leading to the development of a shear surface array in which shear surfaces exhibited a limited range of orientations. Clearly, there is no proof that either of these criteria pertained in the deformation of Companion Mélange at Frenchman's Cove or in the serpentinites at the Winterhouse Brook exposure of the Bay of Islands complex basal detachment. Still, it is clear that existing failure criteria are inadequate to explain the general character of deformation in these fault zones or to rationalize the geometric patterns of shear surfaces that nucleated in these and other natural fault zones.

Evolution of fault- and/or shear-zone arrays in non-coaxial flows

The array of shear surfaces in the serpentinized peridotites at Winterhouse Brook provides two insights into the history and evolution of fault zones. First, two observations indicate temporal changes in the orientations of shear surfaces relative to the incremental shortening and elongation directions. In massive serpentinites >10 m from the contact with metasedimentary rocks, individual surfaces possess both cross-fibre serpentine vein fillings and serpentine slickenfibres. These material surfaces changed their orientation relative to the incremental shortening and elongation directions. This change in relative orientation could occur by solid-body rotation of massive serpentinite with respect to relatively fixed shortening and elongation directions, or by rotation of the principal directions with respect to relatively fixed massive serpentine. Such reorientations should be expected in non-coaxial flows. In serpentinites closer to the base of the sheet, shear surfaces are cut and offset by other shear surfaces as shear surface densities increased. This is most apparent in rocks 5–10 m from the contact with the underlying metasedimentary rocks. Shear surfaces there often bend in the vicinity of intersections with other shear surfaces. In some cases, bending is due to deformation of wall-rock adjacent to the later shear surface. In other cases, bending results from flattening of mesoscopic blocks. In any general shearing flow, shear surfaces, like any material surfaces, will tend to reorient as a result of the flow. The suggestion that shear surfaces reorient in response to bulk deformation is not new. Mitra (1979) used the reorientation of shear zones in bulk pure shear to estimate bulk strains in deformed basement in the Blue Ridge of Virginia. In non-coaxial flows, the flow apophyses (or *stable eigenvectors*) are *fabric attractors* (Passchier & Trouw 1998). They are the orientations of stable lines in the flow, and all other material lines will rotate toward one of the flow apophyses during deformation. In general shears, the flow apophyses are not three mutually perpendicular lines but directions oblique to the shearing plane (Fossen *et al.* 1994), and in flows undergoing shearing concurrent with fault-zone flattening all three may be at low angles to the fault-zone boundaries. The exact significance of this effect is not clear, but it might be locally significant. Second, for fault numbers to increase, new faults must nucleate. Whatever the mechanism, shear surfaces within 10 m of the contact with underlying metasedimentary rocks nucleated with orientations at lower angles to fault-zone boundaries. Both factors are likely to lead to slip being preferentially partitioned into these rocks. The distinctive geometry of shear surfaces shallowly inclined relative to fault-zone boundaries readily accommodates significant bulk shearing. It appears, then, that, a kind of macroscopic 'geometric softening' occurs.

It is harder to assess the temporal evolution of the fault array in the Companion Mélange. Faults in the mélange probably reoriented like those in the serpentinites. Some faults are distorted or deflected where they intersect other faults. Moreover, the hinges of many folds in fault-bounded blocks are now aligned parallel to the inferred transport direction, suggesting significant reorientation of bounding faults and deformation of fault-bounded blocks. However, there is little direct evidence for continuous strains within individual fault-bounded blocks. Fault densities are high in some segments of the mélange, and in these regions faults have consistent orientations. This too suggests that these regions experienced strain localization reflecting a macroscopic geometric softening. In this fault zone, there is no single, discrete fault surface on which much or most of the slip was partitioned.

Conclusions

In two major fault zones from the Bay of Islands area in western Newfoundland, arrays of shear surfaces occur in distinctive arrays. In these arrays, (1) shear surfaces are at low angles to the fault-zone boundaries, (2) shear surfaces are oblique to the *ac*-plane for the bulk deformation inferred from tangent lineation plots, and (3) the intersections of shear surfaces and the fault surface are not parallel to the tectonic *b*-axis inferred for the bulk deformation. This distinctive geometry probably reflects minor fault- or shear-zone formation in a general, 3D, shearing flow.

In sections parallel to the *ac*-plane, arrays of shear surfaces may resemble Riedel shears. One model commonly used to understand the formation of shear surfaces within fault zones is the formation of Riedel shears by Mohr–Coulomb failure. The Riedel shear model predicts the formation of conjugate sets of shear surfaces in fault zones that conform with plane strain deformation of the fault zone. Shear surface arrays in natural fault zones rarely fit fully the patterns predicted for Riedel shears, however. Shear surfaces usually are oblique to the inferred *ac*-plane and often exhibit symmetries other than the predicted orthorhombic symmetry.

The character of deformation in these fault zones suggests that some factor other than stresses controls the formation of shear surfaces. Gradients of the total velocity fields (excluding external vorticity) in steady shearing flows could lead to deformation characteristics more in keeping with those observed in natural fault zones. This study suggests two alternative ways to envision the formation of shear surfaces in non-coaxial flows. Both predict some characteristics for shear surface arrays that match natural fault zones, but neither explains or rationalizes all aspects of fault-zone deformation. These alternatives cannot, then, fully explain the formation of shear surface arrays. Still, the line of reasoning that shear surfaces form with a predictable relationship to the velocity gradient field holds, in the view of the author, promise. Another factor rarely considered in stress-based models for the formation of shear surface arrays is the effect of subsequent deformation. Once formed, shear surfaces will rotate toward flow apophyses, which identify stable positions in steady 3D flows. This factor may help to explain the distinctive geometric character of shear surfaces seen in many high-strain zones within major fault zones.

I wish to thank H. Fossen, R. Holdsworth, and especially R. Twiss for thoughtful reviews of this manuscript. B. Tikoff has, through innumerable conversations, helped me to clarify my thinking on deformation in non-coaxial flows. Acknowledgment is made to the donors of the Petroleum Research Fund, administered by the American Chemical Society, for partial support of this research. I also wish to acknowledge the Oberlin College Research and Development Committee for partial support of this research.

References

ANGELIER, J. 1994. Fault slip analysis and paleostress reconstruction. *In*: HANCOCK, P.L. (ed.) *Continental Deformation*. Pergamon Press, Oxford, 53–100.

BATCHELOR, G.K. 1967. *An Introduction to Fluid Dynamics*. Cambridge University Press, Cambridge.

CASEY, J.F. & DEWEY, J.F. 1984. Initiation of subduction zones along transform and accreting plate boundaries, triple-junction evolution and forearc spreading centers—implications for ophiolitic geology and obduction. *In*: GASS, I.P., LIPPARD, S.J. & SHELTON, A.W. (eds) *Ophiolites and Oceanic Lithosphere*. Geological Society, London, Special Publications, **13**, 267–290.

CASEY, J.F. & KARSON, J.A. 1981. Magma chamber profiles from the Bay of Islands ophiolite complex. *Nature*, **292**, 295–301.

CASEY, J.F., KARSON, J.A., ELTHON, D.L., ROSENCRANTZ, E. & TITUS, M. 1983. Reconstruction of the geometry of accretion during formation of the Bay of Islands ophiolite complex. *Tectonics*, **2**, 509–528.

CAWOOD, P.A. & SUHR, G. 1992. Generation and obduction of ophiolites: constraints from the Bay of Islands complex, western Newfoundland. *Tectonics*, **11**, 884–897.

CHAPMAN, T.J., MILTON, N.J. & WILLIAMS, G.D. 1979. Shape fabric variation in deformed conglomerates at the base of the Laksefjord Nappe, North Norway. *Journal of the Geological Society, London*, **136**, 683–691.

CHAPPLE, W.M. 1978. Mechanics of thin-skinned fold-and-thrust belts. *Geological Society of America Bulletin*, **81**, 1189–1198.

CHRISTIE, J.M. 1963. The Moine thrust zone in the Assynt region, northwest Scotland. *University of California Publication of the Geological Sciences*, **40**, 345–440.

COWAN, D.S. 1985. Structural styles in Mesozoic and Cenozoic melanges of North America. *Geological Society of America Bulletin*, **96**, 451–462.

COWAN, D.S. & BRANDON, M.T. 1994. A symmetry-based method for kinematic analysis of large-slip brittle fault zones. *American Journal of Science*, **294**, 257–306.

COWARD, M.P. 1982. Surge zones in the Moine thrust zone of NW Scotland. *Journal of Structural Geology*, **4**, 247–256.

COWARD, M.P. & KIM, J.H. 1981. Strain within thrust sheets. *In*: MCCLAY, K.R. & PRICE, N.J. (eds) *Thrust and Nappe Tectonics*. Geological Society, London, Special Publications, **9**, 275–292.

DAVIS, D., SUPPE, J. & DAHLEN, F.A. 1983. Mechanics of fold-and-thrust belts and accretionary wedges. *Journal of Geophysical Research*, **88**, 1153–1172.

DECELLES, P.G. & MITRA, G. 1995. History of the Sevier orogenic wedge in terms of critical taper models, Northeast Utah and Southwest Wyoming. *Geological Society of America Bulletin*, **107**, 454–462.

DE PAOR, D.G. 1994. The role of asymmetry in the formation of structures. *Journal of Structural Geology*, **16**, 467–477.

ELLIOTT, D.W. 1976*a*. The motion of thrust sheets. *Journal of Geophysical Research*, **81**, 949–963.

ELLIOTT, D.W. 1976*b*. The energy balance and deformation mechanisms of thrust sheets. *Philosophical Transactions of the Royal Society of London, Part A*, **283**, 289–312.

ERICKSON, S.G. & WILTSCHKO, D.V. 1991. Spatially heterogeneous strength in thrust fault zones. *Journal of Geophysical Research*, **96**, 8427–8439.

FOSSEN, H. & TIKOFF, B. 1993. The deformation matrix for simultaneous simple shearing, pure shearing and volume change, and its application to transpression–transtension tectonics. *Journal of Structural Geology*, **15**, 413–422.

FOSSEN, H., TIKOFF, B. & TEYSSIER, C. 1994. Strain modeling of transpressional and transtensional deformation. *Norsk Geologisk Tidsskrift*, **74**, 134–145.

GEISER, P.A. 1988. The role of kinematics in construction and analysis of geological cross sections in deformed terranes. *In*: MITRA, G. & WOJTAL, S. (eds) *Geometries and Mechanisms of Thrusting, with Special Reference to the Appalachians*. Geological Society of America, Special Papers, **222**, 47–76.

GILOTTI, J. & KUMPULAINEN, R. 1986. Strain-softening induced ductile flow in the Sarv thrust sheet, Scandinavian Caledonides. *Journal of Structural Geology*, **8**, 44–56.

HANDSCHY, J.W. 1998. Spatial variation in structural style, Endicott Mountains allochthon, central Brooks Range, Alaska. *In*: OLDOW, J.S. & AVÉ LALLEMANT, H.G. (eds) *Architecture of the Central Brooks Range Fold and Thrust Belt, Arctic Alaska*. Geological Society of America, Special Papers, **324**, 33–50.

HARPER, J.T., HUMPHREY, N.F. & PFEFFER, W.T. 1998. Three-dimensional deformation measured in an Alaskan glacier. *Science*, **281**, 1340–1342.

HARRIS, L.D. & MILICI, R.C. 1972. *Characteristics of thin-skinned style of deformation in the southern Appalachians, and potential hydrocarbon traps*. US Geological Survey Professional Papers **1018**.

HOSSACK, J.R. 1968. Pebble deformation and thrusting in the Bygdin area (South Norway). *Tectonophysics*, **5**, 315–339.

HOSSACK, J.R. 1978. The correction of stratigraphic sections for tectonic strain in the Bygdin area, Norway. *Journal of the Geological Society, London*, **135**, 229–241.

HSU, T.C. 1967. Velocity field and strain rates in plastic deformation. *Journal of Strain Analysis*, **2**, 196–206.

HUBBARD, M.K. & RUBEY, W.W. 1959. Role of fluid pressure in mechanics of overthrust faulting. *Geological Society of America Bulletin*, **70**, 115–166.

JAMES, N. P. & STEVENS, R. K. 1982. *Excursion 2B: Anatomy and evolution of a lower Paleozoic continental margin, western Newfoundland*. International Association of Sedimentologists Field Excursion Guidebook. 11th International Congress on Sedimentology, McMaster University, Hamilton, Ont., 22–27 August 1982.

JONES, R.R., HOLDSWORTH, R.E. & BAILEY, W. 1997. Lateral extrusion in transpression zones: the importance of boundary conditions. *Journal of Structural Geology*, **19**, 1201–1218.

KARSON, J.A. 1984. Variations in structure and petrology in the Coastal complex, Newfoundland: anatomy of an oceanic fracture zone. *In*: GASS, I.P., LIPPARD, S.J. & SHELTON, A.W. (eds) *Ophiolites and Oceanic Lithosphere*. Geological Society, London, Special Publications, **13**, 131–144.

KARSON, J.A. & DEWEY, J.F. 1978. Coastal complex western Newfoundland: an early Ordovician fracture zone. *Geological Society of America Bulletin*, **89**, 1037–1049.

KEHLE, R.O. 1970. Analysis of gravity sliding and orogenic translation. *Geological Society of America Bulletin*, **81**, 1641–1664.

LINDHOLM, R.M. & CASEY, J.F. 1989. Regional significance of the Blow Me Down Formation, western Newfoundland: new fossil evidence for an Early Cambrian age. *Geological Society of America Bulletin*, **101**, 1–13.

LOGAN, J.M., FRIEDMAN, J., HIGGS, N., DENGO, C. & SHIMAMOTO, T. 1979. Experimental studies of simulated gouge and their application to studies of natural fault zones. *US Geological Survey Open-File Report*, **79-1239**, 305–343.

MALTMAN, A.J. 1978. Serpentinite textures in Anglesey, North Wales, United Kingdom. *Geological Society of America Bulletin*, **89**, 972–980.

MARSHAK, S. 1988. Kinematics of orocline and arc formation in thin-skinned orogens. *Tectonics*, **7**, 73–86.

MCCAIG, A. & REX, D. 1995. Argon-40/argon-39 ages from the Lewis Hills massif, Bay of Islands Complex, west Newfoundland. *In*: HIBBARD, J.P., VAN STAAL, C.R. & CAWOOD, P.A. (eds) *Current Perspectives in the Appalachian–Caledonian Orogen*. Geological Association of Canada, Special Papers **41**, 137–146.

MCLEISH, A.J. 1971. Strain analysis of deformed pipe rock in the Moine thrust zone, Northwest Scotland. *Tectonophysics*, **12**, 469–503.

MERLE, O. 1984. Déplacement et déformation des nappes superficielles. *Revue de Géologie Dynamique et de Géographie Physique*, **25**, 3–17.

MERLE, O. 1989. Strain models within spreading nappes. *Tectonophysics*, **165**, 57–71.

MILTON, N.J. & WILLIAMS, G.D. 1981. The strain profile above a major thrust fault, Finnmark, Norway. *In*: MCCLAY, K.R. & PRICE, N.J. (eds) *Thrust and Nappe Tectonics*. Geological Society, London, Special Publications **9**, 235–240.

MITRA, G. 1979. Ductile deformation zones in Blue Ridge basement rocks and estimations of finite strains. *Geological Society of America Bulletin*, **90**, 935–951.

MITRA, G. 1997. Evolution of salients in a fold-and-thrust belt: the effects of sedimentary basin geometry, strain distribution and critical taper. *In*: SENGUPTA, S. (ed.) *Evolution of Geological Structures from Macro- to Micro- Scales*. Chapman and Hall, London, 59–90.

MITRA, G. & BOYER, S.E. 1986. Energy balance and deformation mechanisms of duplexes. *Journal of Structural Geology*, **8**, 291–304.

MITRA, G. & ELLIOTT, D.W. 1980. Deformation of basement in the Blue Ridge and the development of the South Mountain cleavage. *In*: WONES, D.R. (ed.) *The Caledonides in the USA, IGCP Project 27*. Department of Geological Sciences, Virginia Polytechnic Institute and State University Memoir, **2**, 307–311.

MOORES, E.M. 1982. Origin and emplacement of ophiolites. *Reviews of Geophysics and Space Physics*, **20**, 735–760.

O'HANLEY, D.S. 1996. *Serpentinites: Records of Tectonic and Petrological History*. Oxford Monographs on Geology and Geophysics, **34**. Oxford University Press, Oxford.

PASSCHIER, C.W. & TROUW, R.A.J. 1998. *Microtectonics*. Springer, Berlin.

PATERSON, W.S.B. 1994. *The Physics of Glaciers* (3rd edition). Pergamon Press, Oxford.

PLATT, J.P. & LEGGETT, J.K. 1986. Stratal extension in thrust footwalls, Makran accretionary prism: implications for thrust tectonics. *AAPG Bulletin*, **70**, 191–203.

PLATT, J.P., LEGGETT, J.K. & ALAM, S. 1987. Slip vectors and fault mechanics in the Makran accretionary wedge, southern Pakistan. *Journal of Geophysical Research*, **93**, 7955–7973.

PRICE, R.A. 1967. The tectonic significance of mesoscopic subfabrics in the southern Rocky Mountains of Alberta and British Columbia. *Canadian Journal of Earth Sciences*, **4**, 39–70.

PRICE, N.J. & COSGROVE, J.W. 1990. *Analysis of Geological Structures*. Cambridge University Press, Cambridge.

RAMBERG, H. 1974. Superposition of homogeneous strain and progressive deformation in rocks. *Bulletin of the Geological Institutions of the University of Uppsala, N.S.*, **6**, 35–67.

RAMBERG, H. 1975. Particle paths, displacement and progressive strain applicable to rocks. *Tectonophysics*, **28**, 1–37.

RAMBERG, H. 1977. Some remarks on the mechanism of nappe movement. *Geologiska Föreningen i Stockholm Förhandlingar*, **99**, 110–117.

RECHES, Z. 1983. Faulting of rocks in three-dimensional strain fields. II. Theoretical analysis. *Tectonophysics*, **95**, 133–156.

RECHES, Z. & DIETERICH, J. 1983. Faulting of rocks in three-dimensional strain fields. I. Failure of rocks in polyaxial, servo-controlled experiments. *Tectonophysics*, **95**, 111–132.

ROBERTS, G.P. 1994. Displacement localization and palaeo-seismicity of the Rencurel Thrust Zone, French Sub-Alpine Chains. *Journal of Structural Geology*, **16**, 633–646.

RUTTER, E.H., MADDOCK, R.H., HALL, S.H. & WHITE, S.H. 1986. Comparative microstructures of natural and experimentally produced clay-bearing fault gouges. *Pure and Applied Geophysics*, **124**, 3–30.

SANDERSON, D.J. 1982. Models of strain variation in nappes and thrust sheets; a review. *Tectonophysics*, **88**, 201–233.

SHIMAMATO, T. 1989. The origin of S–C mylonites and a new fault zone model. *Journal of Structural Geology*, **11**, 51–64.

SIMPSON, C. & DE PAOR, D.G. 1993. Strain and kinematic analysis in general shear zones. *Journal of Structural Geology*, **15**, 1–20.

STEARNS, D., COUPLES, G.D., JAMISON, W.R. & MORSE, J.D. 1981. Understanding faulting in the shallow crust: contributions from selected experimental and theoretical studies. *In*: CARTER, N.L., FRIEDMAN, M., LOGAN, J.M. & STEARNS, D. (eds) *Mechanical Behavior of Crustal Rocks; the Handin Volume*. Geophysical Monograph, American Geophysical Union, **24**, 215–229.

STRINE, M. & WOJTAL, S. 1999. Implications of quartz *c*-axis patterns and grain shapes in the Moine thrust zone, Loch Strath Nan Aisinnin, Northwest Scotland. *Geological Society of America, Abstracts with Programs*, **31/7**, A109.

SUHR, G. & CAWOOD, P.A. 1993. Structural history of ophiolite obduction, Bay of Islands, Newfoundland. *Geological Society of America Bulletin*, **105**, 399–410.

TCHALENKO, J.S. 1970. Similarities between shear zones of different magnitudes. *Geological Society of America Bulletin*, **81**, 1625–1640.

TEIXELL, A., DURNEY, D.W. & ARBOLEYA, M.-L. 2000. Stress and fluid control on décollement within competent limestone. *Journal of Structural Geology*, **22**, 349–371.

TIKOFF, B. & FOSSEN, H. 1993. Simultaneous pure and simple shear: the unifying deformation matrix. *Tectonophysics*, **217**, 267–283.

TIKOFF, B. & WOJTAL, S.F. 1999. Displacement control of geologic structures. *Journal of Structural Geology*, **20**, 959–968.

TWISS, R.J. & MOORES, E.M. 1992. *Structural Geology*. W.H. Freeman, New York.

TWISS, R.J. & UNRUH, J.R. 1998. Analysis of fault-slip inversions: do they constrain stress or strain rate. *Journal of Geophysical Research*, **103**, 12205–12222.

TWISS, R.J., PROTZMAN, G.M. & HURST, S.D. 1991. Theory of slickenline patterns based on velocity gradient tensor and microrotation. *Tectonophysics*, **186**, 215–239.

TWISS, R.J., SOUTER, B.J. & UNRUH, J.R. 1993. The effect of block rotations on global seismic moment tensor and the patterns of *P* and *T* axes. *Journal of Geophysical Research*, **98**, 645–674.

UNRUH, J.R. & TWISS, R.J. 1998. Coseismic growth of basement-involved anticlines: the Northridge–Laramide connection. *Geology*, **26**, 335–338.

WALDRON, J.W.F. 1985. Structural history of continental margin sediments beneath the Bay of Islands ophiolite. *Canadian Journal of Earth Sciences*, **22**, 1618–1632.

WALDRON, J.W.F., ANDERSON, S.D., CAWOOD, P.A. & 6 OTHERS 1998. Evolution of the Appalachian Laurentian margin: Lithoprobe results in western Newfoundland. *Canadian Journal of Earth Sciences*, **35**, 1271–1287.

WATTERSON, J. 1999. The future of failure: stress or strain? *Journal of Structural Geology*, **20**, 939–948.

WILLIAMS, H. 1972. *Bay of Islands Map-Area, Newfoundland (12G) 1:125 000*. Geological Survey of Canada Paper **72-34**.

WILLIAMS, H. 1975. Structural succession, nomenclature, and interpretation of transported rocks in western Newfoundland. *Canadian Journal of Earth Sciences*, **12**, 1874–1894.

WILLIAMS, H. & CAWOOD, P.A. 1989. *Geology, Humber Arm Allochthon, Newfoundland, scale 1:250 000*. Geological Survey of Canada Map **1678A**.

WOJTAL, S. 1986. Deformation within foreland thrust sheets by populations of minor faults. *Journal of Structural Geology*, **8**, 341–360.

WOJTAL, S. 1992. One-dimensional models for plane and non-plane power-law flow in shortening and elongating thrust zones. *In*: MCCLAY, K.R. (ed.) *Thrust Tectonics*. Chapman & Hall, London, 41–52.

WOJTAL, S. 1992. Shortening and elongation of thrust zones within the Appalachian fold–thrust belt. *In*: MITRA, S. & FISHER, G.W. (eds) *Structural Geology of Fold and Thrust Belts*. Johns Hopkins University Press, Baltimore, MD, 93–103.

WOJTAL, S. & MITRA, G. 1986. Strain hardening and strain softening in fault zones from foreland thrusts. *Geological Society of America Bulletin*, **97**, 674–687.

WOJTAL, S. & MITRA, G. 1988. Nature of deformation in fault rocks from Appalachian thrusts. *In*: MITRA, G. & WOJTAL, S. (eds) *Geometries and Mechanisms of Thrusting, with Special Reference to the Appalachians*. Geological Society of America, Special Papers, **222**, 17–33.

WOJTAL, S. & PERSHING, J. 1991. Paleostresses associated with faults of large offset. *Journal of Structural Geology*, **13**, 49–62.

YEATS, R.S., SIEH, K. & ALLEN, C.R. 1997. *The Geology of Earthquakes*. Oxford University Press, Oxford.

A quantitative study of the influence of pre-existing compositional and fabric heterogeneities upon fracture-zone development during basement reactivation

L.E. BEACOM[1,2,3], R.E. HOLDSWORTH[2], K.J.W. McCAFFREY[2] & T.B. ANDERSON[1]

[1]*Department of Geology, Queen's University of Belfast, Belfast BT7 1NN, UK*

[2]*Reactivation Research Group, Department of Geological Sciences, University of Durham, Durham DH1 3LE, UK (e-mail: R.E.Holdsworth@durham.ac.uk)*

[3]*BP Norge, Godesetdalen 8, 4065 Stavanger, Norway*

Abstract: In common with many other regions of exposed continental basement, the Late Archaean to Palaeoproterozoic Lewisian Complex, NW Scotland, preserves numerous examples of faults that appear to reactivate pre-existing compositional and structural heterogeneities in the host gneisses. A regionally recognized set of late Laxfordian sinistral strike-slip faults and fractures are spatially associated with pre-existing NW–SE-trending ductile shear zones of Inverian and Laxfordian age. Field observations suggest that most of the sinistral displacements have been accommodated along laterally persistent faults (here termed principal displacement zones (PDZ)) that lie sub-parallel to the pre-existing foliation in the shear zones. Geometric and orientation data collected during structural logging of the PDZ faults have been used to quantitatively test the influence of lithology and pre-existing structural geometry on the spatial patterns of fault development. Stereographic analysis shows a strong geometrical correspondence between the intensity and form of the pre-existing anisotropy and the alignment of the PDZ brittle faults. Spatial clustering of PDZ faults varies depending on lithology (amphibolite v. acid gneiss v. quartz–mica schist). A close correlation exists between the geometry and intensity of the pre-existing foliation and fault spatial clustering. The results demonstrate that reactivation of pre-existing anisotropies in typical continental basement gneisses exert a significant control on brittle fault development and growth in the upper crust.

It is very widely believed that pre-existing mechanical discontinuities such as faults, shear zones, compositional layering and grain boundaries undergo reactivation, therefore influencing fault-zone localization and development (e.g. Holdsworth *et al.* 1997). The influence of pre-existing structures has been suggested at plate scales (e.g. Sykes 1978; Daly *et al.* 1989; Snyder *et al.* 1997; Tikoff *et al.* 2001), regional scales (e.g. Doré *et al.* 1997; Needham & Morgan 1997; Pinheiro & Holdsworth 1997), outcrop scales (e.g. Laubach & Marshak 1987; Shail & Alexander 1997), grain scales and microscopic scales (e.g. Hippler & Knipe 1990; Lloyd & Knipe 1992; Knipe & Lloyd 1994). Much of the evidence presented to support these hypotheses is circumstantial and there have to date been few attempts to quantitatively demonstrate that pre-existing heterogeneities do affect fracture-zone development. This has become possible with the emergence of quantitative techniques to describe the spatial attributes and evolution of fracture zones in rocks (Barton 1983; Gillespie *et al.* 1993, Gillespie *et al.* 1999).

The influence of pre-existing structures is particularly important in continental regions as the relatively buoyant quartzo-feldspathic crust is rarely subducted, thereby imparting a long-lived architecture of inheritance that is not preserved in (mainly younger) oceanic lithosphere (e.g. Sutton & Watson 1986). Exposed regions of high-grade metamorphic basement rocks are generally believed to be representative of the continental crust that underlies most deformation zones and rifted basins at depth. Most basement complexes exhibit widespread circumstantial evidence for reactivation of pre-existing structures and, if they are well exposed, are therefore good locations to quantify the influence of pre-existing structures upon fracture-zone development. In this paper the control of

From: HOLDSWORTH, R.E., STRACHAN, R.A., MAGLOUGHLIN, J.F. & KNIPE, R.J. (eds) 2001. *The Nature and Tectonic Significance of Fault Zone Weakening*. Geological Society, London, Special Publications, **186**, 195–211. 0305-8719/01/$15.00

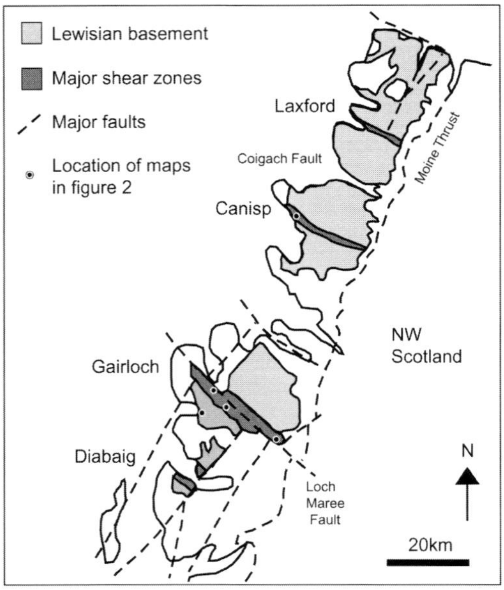

Fig. 1. Location of major shear zones in the Lewisian Basement Complex, NW Scotland.

protolith heterogeneities upon fault-zone attributes is examined in typical continental basement gneisses from the Lewisian Complex of NW Scotland (Fig. 1).

Geometric properties of fault zones

Faults and fractures occupy 3D volumes of rock, but it is difficult to characterize the spatial characteristics within that volume without taking serial sections as happens during a 3D seismic survey. Commonly, analyses of fracture parameters are limited to 1D and 2D samples that may be obtained easily from microstructural, outcrop, air-photo and satellite data.

There are numerous studies that attempt to quantify the spatial and scaling relationships of fault and fracture attributes such as length, spacing, displacement, linkage and orientation (e.g. Thorp *et al.* 1983; DeMarsily 1985; Barton & Heish 1989; Korvin 1989; Scholz & Cowie 1990; Sornette *et al.* 1990; Gillespie *et al.* 1993; Walsh & Watterson 1993; Hatton *et al.* 1994; Cowie *et al.* 1996; Johnston & McCaffrey 1996; Needham *et al.* 1996; Nicol *et al.* 1996b; Peacock & Sanderson 1996; Yielding *et al.* 1996; Walsh *et al.* 2001). These studies demonstrate that there is a wide range of variability in fault population attributes that may depend on local fault systematics or variation of fault mechanisms with scaling. They also suggest that overall fault populations may not exhibit simple or uniform scaling relations over the entire range of observable scales. A limited number of these workers have directly attributed the scaling behaviours to the physical processes operating during faulting or to the influence of pre-existing mechanical and lithological anisotropies (e.g. Walsh & Watterson 1991; Hatton *et al.* 1994; Nicol *et al.* 1996a; Wojtal 1996). Most studies have only examined fracture systems in sedimentary or igneous rocks containing relatively simple fracture sets.

Consistent patterns between the scaling relations of different datasets depend on the tectonic setting and the kinematics of the deformation (e.g. Wojtal 1994). These factors interact to determine the strain distribution in a region and the contribution of fractures of different sizes to the nucleation and coalescence of faults (e.g. Little 1995). Clustering of smaller faults around the major structures evolves as the strain accumulates in the deforming rock (e.g. Knott *et al.* 1996).

Quantifying the clustering behaviour of fault and fracture populations has typically been achieved by measuring the spacing between fractures along 1D transects. Several studies examine fracture spacing along line samples, addressing subsurface borehole data (e.g. Kulander *et al.* 1979; Velde *et al.* 1990; Barton & Zoback 1992) and surface studies of fracture spacing in various geological environments (e.g. Barton 1983; Gillespie *et al.* 1993, Gillespie *et al.* 1999; Manning 1994). Using 1D analysis, faults have been found to show power-law spacing distributions and unrefined joint datasets commonly give log-normal or exponential with negative slope frequency distributions (e.g. Gillespie *et al.* 1993). The results may in part reflect the sparseness of some datasets, but also the effect of preferred orientations in fracture patterns measured along differently trending line samples across 2D fracture surfaces.

In this study, the spatial attributes of fracture populations are investigated in late-Laxfordian-age, sinistral, strike-slip fault zones from metamorphic basement rocks of the Lewisian Complex, NW Scotland (Fig. 1). The aim of this study is to quantitatively assess the influence of pre-existing compositional and structural heterogeneities in the basement rocks upon two main fracture attributes: orientation and spacing. To our knowledge, no such analysis has been attempted previously in metamorphic basement rocks of this age and complexity.

Geological setting, NW Scottish mainland

The Lewisian Complex of mainland NW Scotland (Fig. 1) has experienced a long and complex history of deformation and reactivation, from Archaean time to the present day (Table 1). The oldest Badcallian structures (*c.* 2.5 Ga) preserved are NE–SW-trending fabrics that developed under granulite-facies metamorphic conditions in the lower crust. With exhumation to amphibolite-facies conditions, continued mid-crustal deformation (*c.* 2.5–2.4 Ga) produced NW–SE-trending Inverian shear zones such as the Canisp and (possibly) Gairloch shear zones (Fig. 1; Park *et al.* 1994). Further exhumation occurred during emplacement of the regional Scourie dyke swarm and deposition of the Loch Maree Group (*c.* 2.4–1.9 Ga), the latter forming a localized assemblage of metasedimentary and metavolcanic rocks thought to have been deposited unconformably upon older Lewisian gneisses in the Loch Maree–Gairloch region (Park *et al.* 1994, and references therein). The Scourie dykes and Loch Maree Group are postdated by the main Laxfordian events (*c.* 1.9–1.6 Ga) that heterogeneously rework a large part of the Lewisian Complex. Deformation occurred in mid- and upper-crustal environments forming new NW–SE shear zones and reactivating pre-existing Inverian shear zones. Successive amphibolite- to greenschist-facies transpressional and transtensional deformation events produced mylonitic and blastomylonitic fault rocks with dextral strike-slip displacement components being dominant (Table 1; Park *et al.* 1994). These fault rocks form a heterogeneously distributed, but interconnected network of NW–SE-trending high-

Table 1. *Summary of geological, structural and metamorphic chronology of the Lewisian Complex of mainland NW Scotland (from Beacom 1999)*

Ma (*c.*)	Event	Kinematics
2900–2490	Badcallian granulite-facies metamorphism and deformation NE–SW fabrics and shear zones	Folding and subhorizontal thrusting
2490–2400	Inverian amphibolite-facies metamorphism and deformation Steep NW–SE shear zones	Dextral transpression (north-up overthrusting with small dextral component)
2400–1900	Emplacement of Scourie dyke swarm NW–SE- to east–west-trending structures Deposition of Loch Maree Group	Dextral transtension
1900–1800	Early Laxfordian amphibolite-facies metamorphism and deformation NW–SE-trending structures	Dextral transtension (on oblique shears and asymmetric shear folds)
1600	Mid-Laxfordian upper greenschist-facies metamorphism and deformation NW–SE-trending structures	Dextral transpression (north-up overthrusting and upright folds)
1400–1200	Late Laxfordian lower greenschist-facies metamorphism and deformation NW–SE-trending structures	Sinistral strike-slip (steeply plunging asymmetric folds and crush belts)
1200	Pre- to early Torridonian brittle deformation North–south-, NE–SW- and ENE–WSW-trending structures Deposition of Stoer Group	Dextral transtension (WNW–ESE extension)
1100	Deposition of the Torridon Group	–
1100–present	Vendian rifting, Caledonian thrusting, Late Palaeozoic, Mesozoic and Tertiary rifting related to formation of nearby offshore basins North–south-, NW–SE-, NE–SW- and ENE–WSW-trending structures	Mainly extensional (mainly dip-slip extension)

strain fabrics that anastomose around lens-shaped lower-strain domains (e.g. Attfield 1987; Park *et al.* 1987).

A later suite of interconnected, low greenschist-facies mylonitic shear zones, brittle faults and localized folds formed sub-parallel to the pre-existing high-strain fabrics in the Laxfordian and Inverian shear zones. NW–SE-trending subhorizontal mineral and slickenline lineations are widely associated with sinistral shear sense indicators. The brittle fracture systems formed during this strike-slip event are recognized throughout the region (e.g. Figs 1–4) and are the subject of the current paper. They are widely associated with the development of linked fracture arrays (e.g. Fig. 3), often with infills of cataclasite and pseudotachylyte forming fault-bounded domains referred to as 'crush belts' by some workers (e.g. Fig. 4; Shihe & Park 1993; Park *et al.* 1994). The precise age of the sinistral faulting is poorly constrained, but these structures are unconformably overlain by the Torridonian Stoer Group in the central region of the Lewisian Complex (Beacom 1999). These sedimentary rocks are thought to have been deposited during Riphean (*c.* 1.2 Ga) intracontinental rifting of Laurentia (Soper & England 1995). Thus the sinistral faulting in the basement is generally ascribed to a late Laxfordian event (*c.* 1.4–1.2 Ga; see also Park *et al.* 1994).

The deposition of the Stoer Group is directly associated with a brittle deformation involving faulting along north–south-, NE–SW- and ENE–WSW-trending structures that are recognized widely throughout the Lewisian outcrop on the NW Scottish mainland (Beacom 1999). Palaeostress inversions suggest an overall WNW–ESE direction of extension, possibly during regional dextral transtension (Beacom *et al.* 1999). The faults cross-cut and are generally discordant to the earlier Laxfordian structures in the basement gneisses, although oblique extensional reactivation of pre-existing NW–SE faults is recognized locally (Beacom *et al.* 1999). All of these structures are unconformably overlain by the Torridon Group of the Torridonian succession, a unit thought to have been deposited no earlier than 1.1 Ga (Park *et al.* 1994).

A number of post-Torridonian brittle deformation episodes are known to affect the Lewisian Complex, although the locations, number and timing of individual faulting events are often poorly constrained. They include a Vendian phase of rifting thought to be associated with the opening of Iapetus (e.g. Soper & England 1995), faulting during the Caledonian orogeny (*c.* 430 Ma; e.g. Butler 1997) and Late Palaeozoic, Mesozoic and Tertiary rifting events related to the formation and development of nearby offshore basins such as the West Orkney, Minch and Sea of the Hebrides basins (e.g. Laubach & Marshak 1987; Morton 1992; Fyfe *et al.* 1993; Stoker *et al.* 1993; Roberts & Holdsworth 1999). Locally, extensional or strike-slip reactivation of pre-existing structures occurs during these later events, but it is often very difficult to assign precise ages to individual movements. Regional-scale reactivation of the Coigach and Loch Maree faults seems likely, based on indirect geological evidence, but the actual fault zones are largely unexposed (Beacom 1999).

Late Laxfordian fault zones

Field observations

The late Laxfordian, brittle sinistral fault zones were studied in two of the regional Inverian–Laxfordian high-strain zones: the Gairloch and Canisp shear zones. Four sub-areas exhibiting different dominant host-rock lithologies were studied in the Gairloch shear zone at Tollie (Fig. 2a, acid gneisses), Kerry Beach (Fig. 2b; mafic gneisses), Meall Aundrary (Fig. 2c; amphibolites) and Slioch, Loch Maree (Fig. 2d; quartz–mica schists). The more lithologically uniform acid gneisses of the Canisp shear zone were also studied along the shores of Achmelvich Bay (Fig. 2e). Lithological heterogeneities such as thin mafic bands within the Tollie and Canisp acid gneisses are up to a few tens of centimetres in size. Mica-rich and quartz-rich bands within the quartz–mica schist at Slioch are interbanded on scales of a millimetre to a few tens of centimetres. Quartz–feldspar bands and ultramafic units in the mafic gneisses at Kerry Beach occur on scales of tens of centimetres up to a few metres. At Meall Aundrary lithological banding is defined by interlayered feldspar- and amphibole-rich layers on scales of a millimetre to tens of centimetres.

Systematic structural measurements and observations made during logging of the late Laxfordian fault arrays across and along the Gairloch and Canisp fault zones demonstrate that the fault kinematics are similar in all the localities (e.g. Figs 3 and 5). Brittle fractures associated with the fault zones occur in a wide range of orientations, but the longest faults lie mainly sub-parallel to the earlier, NW–SE-trending, steeply dipping foliation in the host gneisses and mylonites (e.g. Figs 3 and 4; compare Fig. 5a and b). The NW–SE-trending fault

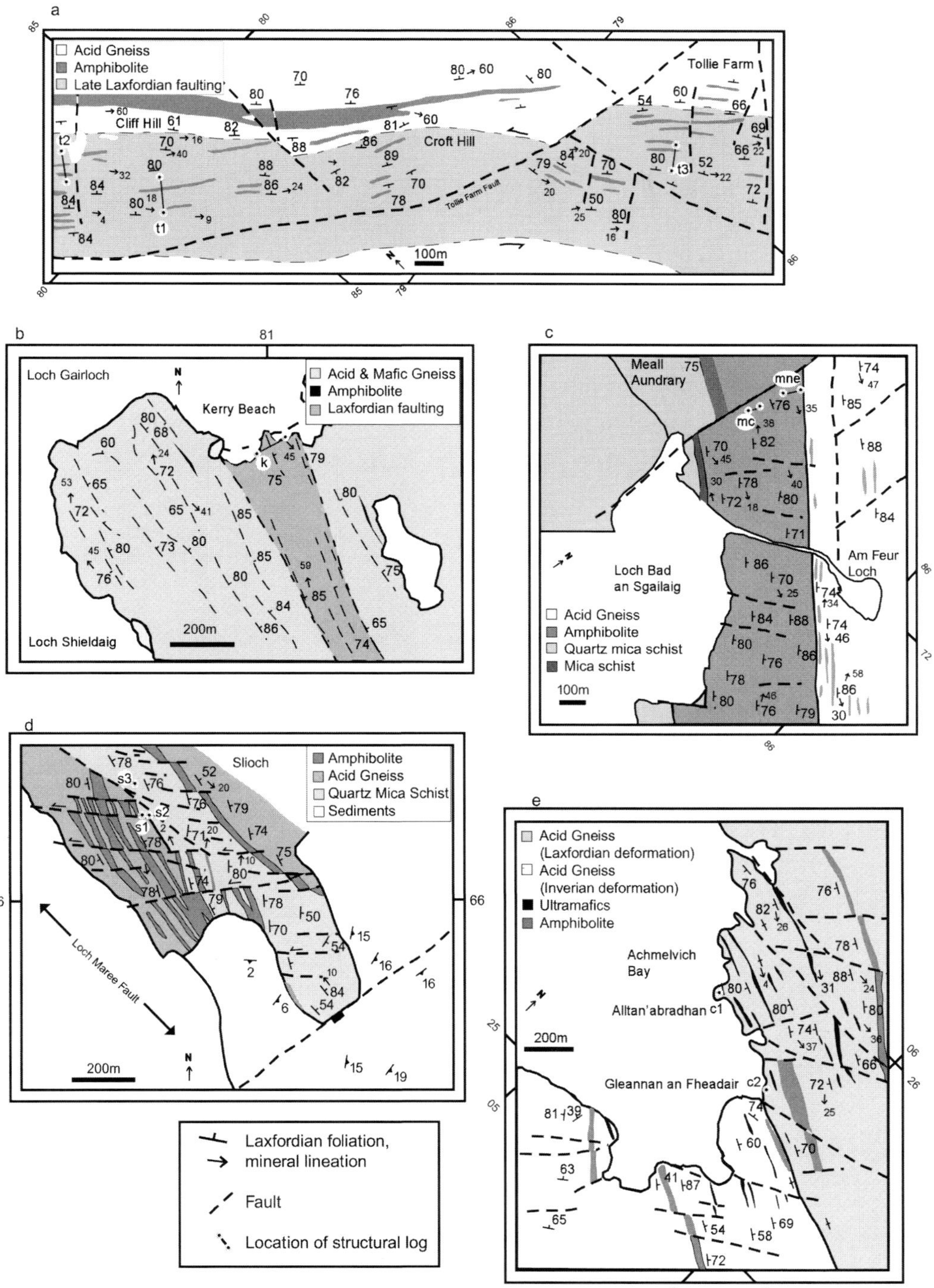

Fig. 2. Geological maps of the Gairloch shear zone at (**a**) Tollie, (**b**) Kerry Beach, (**c**) Meall Aundrary and (**d**) Slioch, and the Canisp shear zone at (**e**) Achmelvich Bay. The transect locations are indicated where detailed fault and fracture characterisitics were logged. Regional locations are shown in Fig. 1.

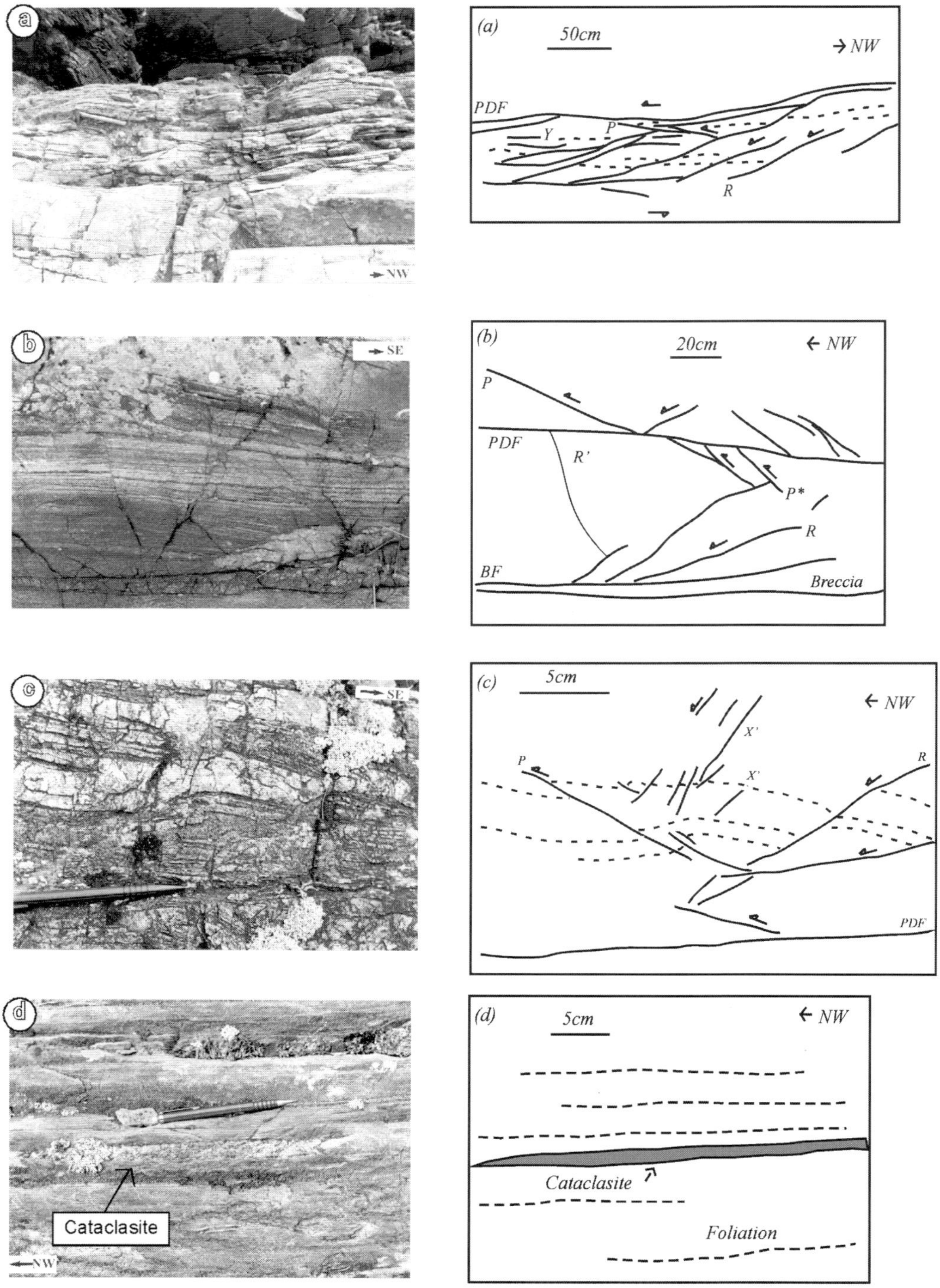

Fig. 3. Foliation and fracture characteristics within the sinistral strike-slip fault zones. (**a**) Canisp (NC05722558): acid gneiss with simple shear fracture configurations (*R*, *P* and *Y* shears). PDZ, principal displacement zone faults. (**b**) Kerry Beach mafic gneiss (NG81037370): simple shear fractures (*R*, *R'* and *P*) and foliation-parallel compressional, pure shear (*P**) fractures. (**c**) Slioch quartz–mica schist (NG00736633): simple shear (*R* and *P*) fractures with foliation-parallel extensional shears (X') fractures. (**d**) Meall Aundrary amphibolite (NG85107231): foliation-parallel cataclasite zones and fractures.

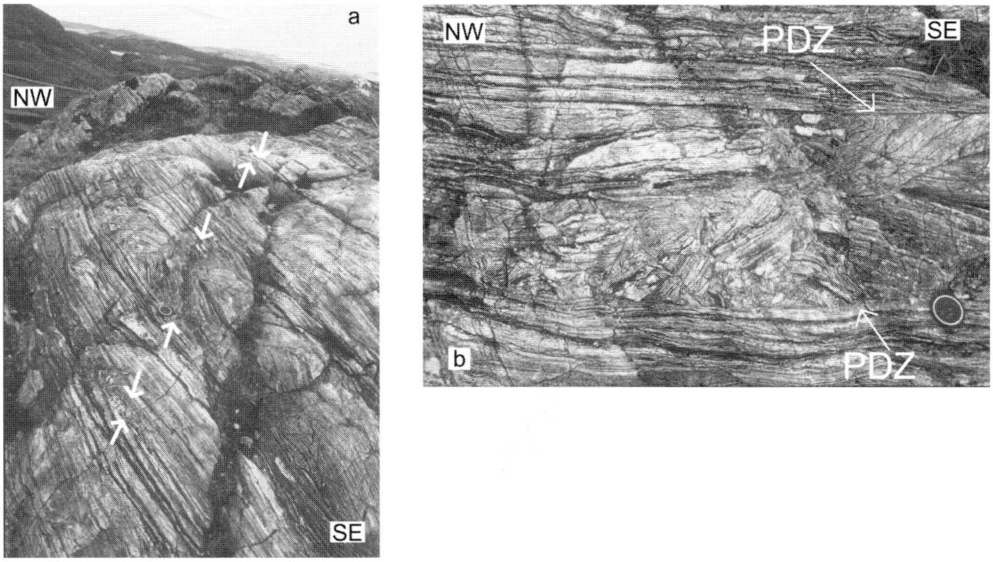

Fig. 4. Styles of late Laxfordian faulting in acid gneiss at Tollie (NG84708122). (**a**) Oblique view of foliation-parallel 'damage zones' bounding brecciated gneiss with cemented cataclasite matrix; bounding faults (PDZ, see text) indicated by arrows. (**b**) Plan-view close-up of central part of the thickest damage zone. The lens cap (50 mm wide) is in same position in both photographs.

zones carry predominantly subhorizontal or shallowly plunging slickenline lineations (Fig. 5b) and display offsets and associated asymmetric secondary fracture arrays consistent with sinistral displacements (Fig. 3; see below). Where offsets of pre-existing marker horizons and granitic veins can be measured, the largest displacements occur along individual faults at low angles or sub-parallel to the foliation in the gneisses. Thus, qualitative field observations suggest that the NW–SE-trending, foliation-parallel faults represent the principal displacement zones (PDZs) during brittle sinistral strike-slip deformation. Variously oriented sets of shorter, predominantly shear fractures and subordinate tensile fractures are spatially associated with, and often linked to the foliation-parallel faults. The asymmetry of these arrays is generally consistent with sinistral senses of shear and can be assigned to standard secondary shear and tensile fracture groups (e.g. R, R′, P, X, X′, etc.) with a reasonable degree of confidence (Fig. 3). Secondary fault arrays and associated brittle fault rocks (mainly cataclasite and pseudotachylyte) are often enclosed by foliation-parallel faults to form planar or lenticular domains here informally termed 'damage zones' (e.g. Figs. 3 and 4). Enclosed fault-bounded blocks may exhibit considerable angular rotations about subvertical axes, predominantly in a clockwise direction (e.g. the large fault-bounded blocks in Fig. 2d, see also Fig. 4a and b).

Field mapping demonstrates that the foliation-parallel PDZs are often preferentially developed along major pre-existing shear-zone boundaries (e.g. between the Inverian and Laxfordian shear domains in the Canisp shear zone, Fig. 2e), along compositional boundaries or where a pervasive, sub-planar, mylonitic foliation is particularly intense. In the Gairloch shear zone, the majority of PDZs lie within domains of brittle faulting 100–500 m thick (Fig. 2a–d), whereas at Canisp, the PDZ domain does not exceed 60 m and lies on the southern side of the pre-existing Laxfordian shear zone (Fig. 2e).

Geometric characteristics of foliation and PDZ faults

A visual inspection of the stereoplot compilations (Fig. 5a and b) suggests (1) that there is a close correspondence between the orientations of the poles to the pre-existing foliation and the PDZ faults, and (2) that the degree and form of preferred orientation, or geometric clustering, of the foliation and fault planes varies between each of the localities studied. Such qualitative assessments can be quantitatively tested using the eigenvector ratios *C* (strength of preferred orientation) and *K* ('shape' of preferred

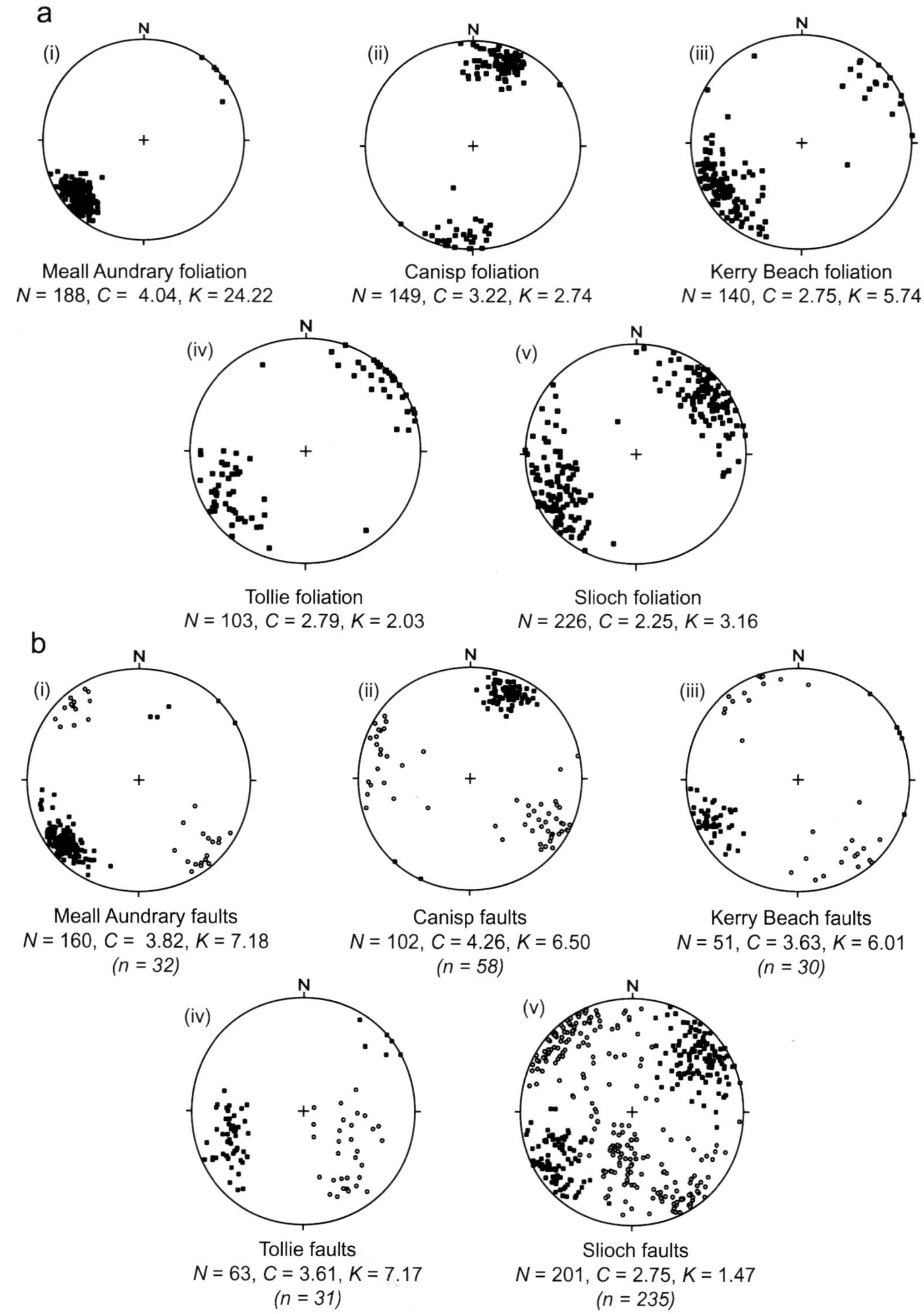

Fig. 5. Stereoplots of (**a**) poles to foliation and (**b**) poles to PDZ fault planes and slickenline lineations for each of the named localities studied. Number of readings, together with *C* and *K* values for the foliation or fault plane data are also shown. In (**b**), the number of lineation measurements is indicated in parentheses.

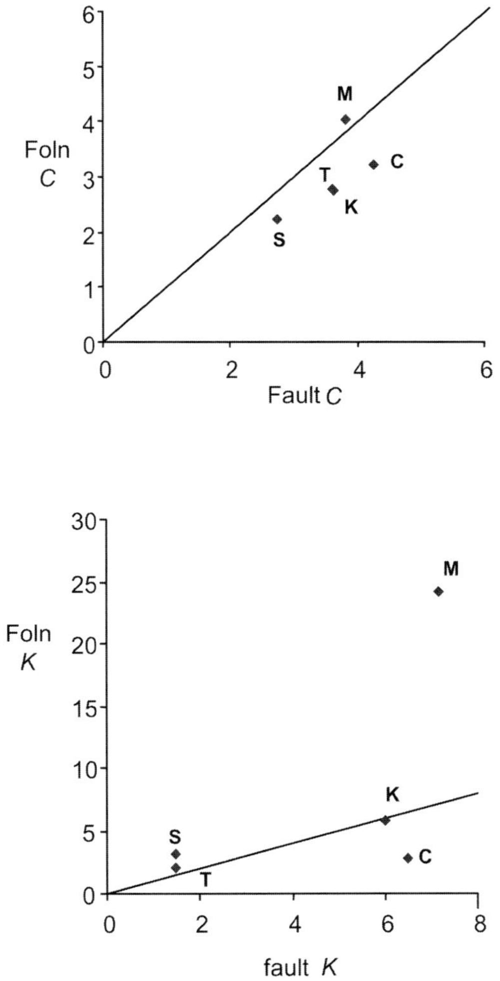

Fig. 6. Plots of (**a**) foliation C v. PDZ fault C, and (**b**) foliation K v. PDZ fault K. M, Meall Aundrary; C, Canisp; T, Tollie; K, Kerry Beach; S, Slioch. The inclined line on each graph is line of 100% geometric correspondence (i.e. foliation parameter = PDZ fault parameter).

orientation). For any given dataset the calculated eigenvectors λ_1, λ_2 and λ_3 may be normalized to give S_1, S_2 and S_3 and $C = \ln(S_1/S_3)$ and $K = \ln(S_1/S_2)/\ln(S_2/S_3)$ (Woodcock & Naylor 1983).

The C and K values for foliation can be plotted against those for PDZ faults (Fig. 6a and b) and show a strong positive correlation confirming that the foliation and later brittle fault zones are geometrically coincident. The C values at Meall Aundrary and Canisp indicate a narrow range of preferred orientation for both poles to foliation and foliation-parallel faults, whereas those at Kerry Beach, Tollie and Slioch are less strongly aligned (Fig. 5a). The K value for foliation at Meall Aundrary is markedly higher than that for the PDZ faults, whereas the other localities show equivalent and lower values (Fig. 5b). High K values indicate a strong degree of planar preferred orientation, whereas lower values suggest a greater tendency towards a girdle distribution (Woodcock & Naylor 1983). In general, the lowest values of K (and C) seem to come from those localities where the foliation and faults are locally reoriented by rotation of fault-bounded blocks about steeply plunging axes and/or by metre-scale brittle–ductile folds, e.g. Slioch and Tollie.

Spatial characteristics of PDZ faults

Method. The spatial clustering and linkage of the foliation-parallel PDZ fault set were investigated at each of the Gairloch and Canisp shear-zone localities. The spatial geometry was quantified by measuring the spacing between adjacent PDZ faults along several traverses oriented perpendicular to the major faults (Table 2; locations of traverses shown in Fig. 2a–e).

Gillespie *et al.* (1993, 1999) demonstrated that the fracture spacing population technique of Harris *et al.* (1991) is the only reliable method for characterizing the spatial attributes of fracture populations measured in 1D samples. In this technique, the spacing between immediately adjacent fractures is represented on plots of spacing (s) v. cumulative number of fracture spacing values greater than or equal to s (Harris *et al.* 1991). Power-law distributions plot as straight lines on log–log axes and indicate fractal-type clustering with the value of the slope D_s the fractal dimension of the 1D spacing distribution (Gillespie *et al.* 1999). Lower values of D_s indicate higher clustering, i.e. the distribution is dominated by large spaces, and are thought to be typical of tectonic fault systems (Gillespie *et al.* 1993). Exponential distributions with negative slope plot as straight lines on log cumulative number v. linear spacing axes and are thought to be typical of joint data that incorporate more than one joint set. Log-normal distributions form straight lines on linear cumulative number v. log spacing and a normal distribution plots as a straight line on linear–linear axes; both of these may be produced by a single fracture set in which anti-clustering or regular spacing predominates (Gillespie *et al.* 1993). The coefficient of variation (Cv) may be used as a numerical measure of clustering in 1D datasets and is valid for any

Table 2. *Summary of locations, lithologies and fracture data collected during structural logging along traverses shown in Fig. 2*

Name	Locality	Lithology	Traverse length (m)	Number of fractures	Average damage zone thickness (cm)	Fracture density (m^{-1})	Coefficient of variation
t1	Tollie NG84797986	acid gneiss	134.29	158	37.61	1.17	1.47
t2	Tollie NG84708024	acid gneiss	94.31	132	31.81	1.4	1.68
t3	Tollie 859785	acid gneiss	92.02	187	28.42	2.03	1.22
mne	Meall Aundrary NG852724	amphibolite	54.65	179	2.32	3.27	1.63
mc	Meall Aundrary NG85107231	amphibolite	21.82	105	2.53	4.8	1.83
s1	Slioch NH00736625	quartz–mica schist	18.87	53	34.16	2.81	0.64
s2	Slioch NH00746625	quartz–mica schist	23.44	73	28.11	3.11	0.46
s3	Slioch NH00736633	quartz–mica schist	30.83	81	36.06	2.62	0.44
k	Kerry Beach NG81037370	mafic gneiss	99.32	53	147.63	0.53	0.75
c1	Canisp NC053257	acid gneiss	27.51	48	56.24	1.74	0.95
c2	Canisp NC05722558	acid gneiss	31.25	76	29.95	2.43	1.24

type of distribution (McCaffrey & Johnston 1996). *Cv* is defined as the ratio of the standard deviation (*SD*(*s*)) to the mean (*s*) (Cox & Lewis 1986). If fractures are clustered then $Cv > 1$, if anticlustered then $Cv < 1$. $Cv = 0$ means regular spacing and $Cv = 1$ reflects a random distribution (exponential with negative slope) (Huang & Angelier 1989; McCaffrey & Johnston 1996).

Individual traverse lengths in the present study are generally at least two orders of magnitude larger than the average space between adjacent PDZ faults, indicating that the datasets are a reasonable sample of the spacing within the decimetre- to centimetre-scale range. Sample lines were limited by exposure and ranged from *c.* 18 m to 134 m and thus did not span the entire fault zones. As a result, clustering at larger scales cannot be addressed.

Results. The spatial distributions of the foliation-parallel PDZ fault datasets are shown in Fig. 7 and display power-law, negative exponential or log-normal regression lines fitted as appropriate. The spatial distributions are similar between samples from each locality, but differ between localities (Fig. 7a–e). The fault spacing determined along traverses t1, t2 and t3 in the acid gneiss at Tollie (Fig. 7a), the acid gneisses at Canisp (c1 and c2) and the mafic gneiss at Kerry Beach (k) are best described by a power-law distribution above 10–15 cm spacing resolution. The log-normal fault spacing distributions along traverses in the amphibolite at Meall Aundrary (Fig. 7e) are very different from those within any other lithology. The spacing data for faults in the quartz–mica schist from Slioch (s1, s2 and s3) are best described by a negative exponential distribution (Fig. 7b). The coefficient of variation (*Cv*) calculated for each dataset (Table 2) is also consistent for datasets from the same locality and is dissimilar between different localities. The *Cv* varies between spatially clustered datasets (1.83 for Meall Aundrary) to relatively anticlustered (0.44 for Slioch).

Analysis. As all the localities have the same fault kinematics, the variability in the magnitude of spacing of NW–SE-trending PDZ faults and the shape of the cumulative number–fracture spacing curves in the different lithologies imply that other factors influence the location of the faults and their spatial distributions. Fault spacing and spatial clustering parameters can be used to quantitatively investigate whether differences in basement lithology and/or pre-existing foliation geometry significantly influence the spacing of the late Laxfordian PDZ faults.

A degree of power-law fault clustering is evident in the acid gneisses (Tollie, Canisp; Fig. 7a and c), confirmed by *Cv* values >1 (Figs. 8 and 9). The quartz–mica schist (Slioch) is most closely approximated by a negative exponential curve and has *Cv* <1 or is anticlustered. The mafic gneiss (Kerry Beach) is poorly described by a power law over a short range of spacing values and the $Cv = 0.75$ shows that the fault spacings are anticlustered. The amphibolite (Meall Aundrary) yielded the highest *Cv* values, implying that it has the most clustered distribution, but the spacing distribution is log-normal, implying a characteristic spacing or anticlustering. Overall, the *Cv* values appear to provide a quantitative measure of fault clustering and do appear to be significantly dissimilar for different lithologies (Fig. 8). In the case of the Meall Aundrary dataset, vein systems with similar highly clustered log-normal spacing distributions have been explained by either a poor resolution of the smallest faults (Gillespie *et al.* 1999), or the possibility of a change in power-law exponents with scale (Johnston & McCaffrey 1996). Without displacement data for the faults, it is not possible to resolve either of these possibilities.

Further characterization of lithological controls is possible by plotting *Cv* against other fault system attributes. For example, a plot of *Cv* v. fault density (Table 2) illustrates that the different basement lithologies plot in different fields (Fig. 8). The acid gneiss shows low-density clustering, whereas the amphibolite shows high-density clustering, the mafic gneiss displays low-density anticlustering whereas the quartz–mica schist has a medium-density anticlustered distribution. The fact that the acid gneiss data from Tollie and Canisp plot in the same field further supports the hypothesis that fault clustering is influenced by lithology.

A preliminary investigation of the influence of pre-existing foliation is assessed by examining the relationship between the spatial clustering of the faults and the intensity and 'shape' of preferred orientations of the pre-existing foliation in the host rocks (Fig. 9a and b). The strength of clustering of poles to foliation (given by *C*) and the shape factor of the distribution (given by *K*) are plotted versus the average *Cv* values for each locality. The positive relationship between *C* and *Cv* (Fig. 9a) indicates that localities with stronger preferred orientation or more planar pre-existing foliation display more strongly clustered faults. The relationship between *K* and *Cv* (Fig. 9b) is less clear, but

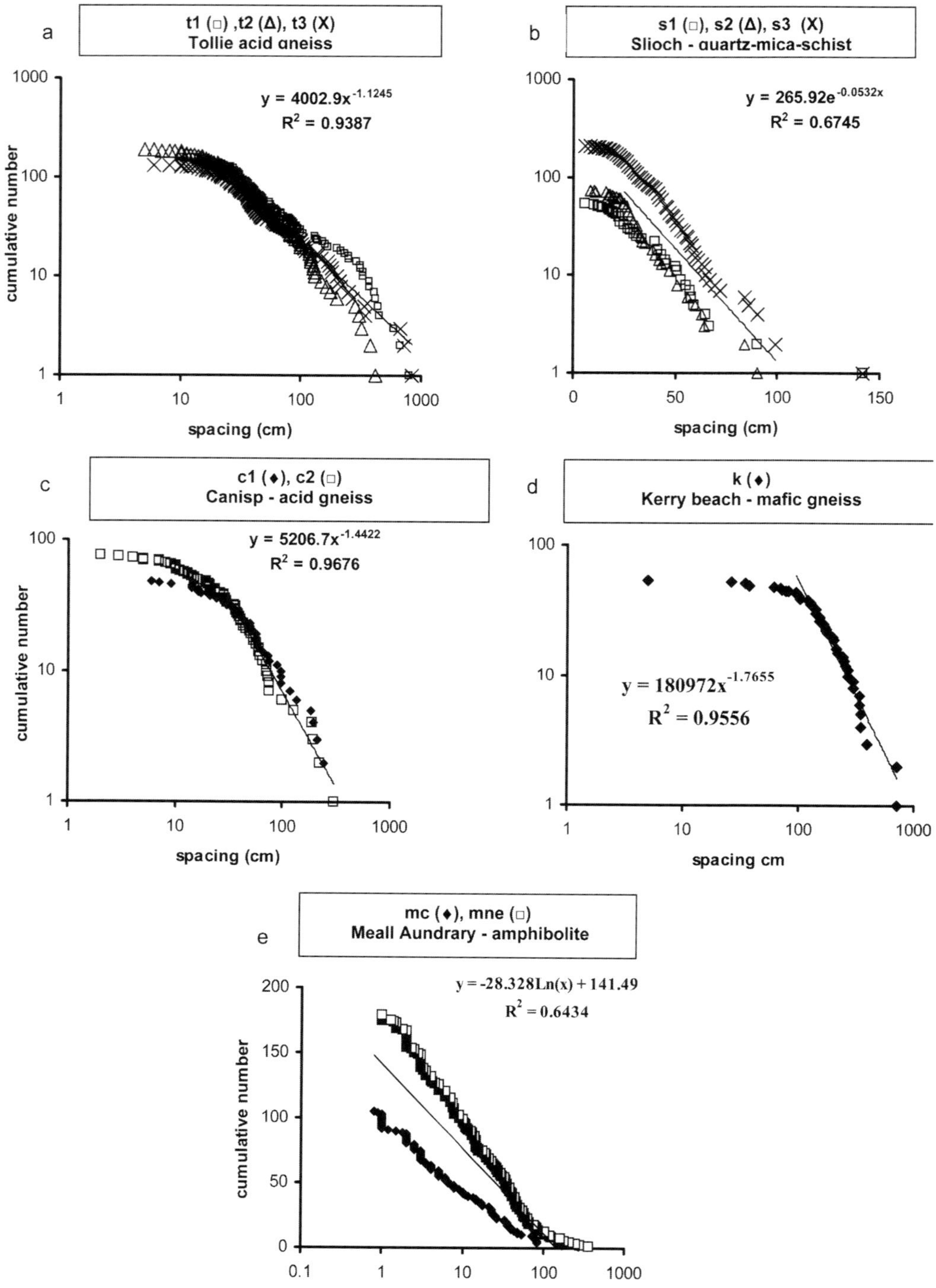

Fig. 7. Cumulative number v. spacing plots for PDZ faults at the named localities. The fitted line in (**b**) and (**e**) is an average fit, hence the low R^2 value. (See text and Table 2 for further details and key to traverse datasets.)

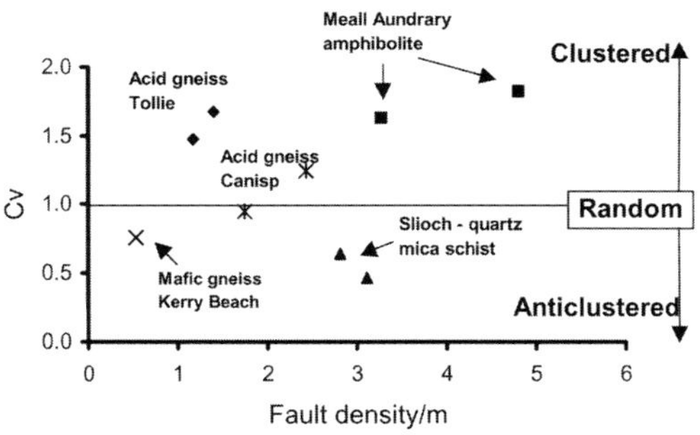

Fig. 8. Plot of *Cv* v. PDZ fault density for the named localities and basement gneiss lithologies.

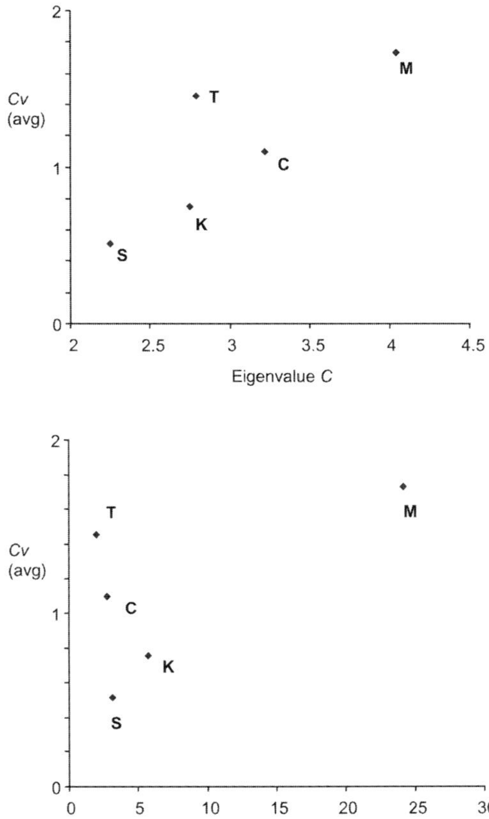

Fig. 9. Plots of (a) *Cv* v. foliation *C* value and (**b**) *Cv* v. foliation *K* value. Locality letters as in Fig. 6.

may be related to differences in the degree of linkage of the brittle faults (see below).

Discussion

In our view, the analysis presented above supports the qualitative assessment made in the field that both lithology and the geometry of the foliation in the pre-existing NW–SE Inverian–Laxfordian shear zones significantly control the attributes of the late Laxfordian sinistral strike-slip PDZ faults. All localities show a high degree of geometric correspondence between the faults and the pre-existing foliation, with significant differences depending on the lithology and the geometric properties of the pre-existing foliation (Figs 5 and 6). Acid and mafic gneisses, the dominant lithologies in the Lewisian Complex, display power-law fault spacing distributions (Fig. 7a, c and d). The non-power-law distributions occur in regionally subordinate lithologies of the Loch Maree Group that display the most and least spatially clustered PDZ faults, respectively in the amphibolites (Meall Aundrary; Figs 7e and 8) and the quartz–mica schists (Slioch; Figs 7b and 8). In addition, it is these same lithologies that display respectively the highest and lowest values of *C* and *K*, i.e. they have the strongest–most planar and the weakest–least planar geometric clustering of poles to foliation and PDZ faults (Fig. 9a and b). Qualitative field and thin-section observations suggest that the amphibolites and quartz–mica schists are, respectively, the least and most compositionally (and by implication mechanically) heterogeneous units, and this may have exerted a significant control on fracture development.

Fracture linkage may form an additional factor influencing the spatial clustering data. In the field, linkage of fractures related to the PDZ sinistral faulting appears to be most limited in the amphibolites of Meall Aundrary and greatest in the quartz–mica schists at Slioch. If we assume that the brittle fault displacements at each locality were the same, the inhibition of linkage would require more closely clustered PDZ faults to form in the amphibolites; this closer clustering is observed (e.g. Fig. 8). Linkage may also be important as it can influence fault-hosted processes that lead to changes in the long-term mechanical properties of the fault zones (Holdsworth *et al.* 2001). Previous research demonstrates that reactivated fault zones typically consist of a highly interconnected network of fractures in which the fault rocks display textural and mineralogical changes related to the syntectonic influx of fluids (e.g. Wintsch *et al.* 1995; Imber *et al.* 1997; Stewart *et al.* 2000). In the late Laxfordian fault zones, subsequent reactivation is recognized based on the presence of later, mainly dip-slip or oblique-slip slickenlines on individual fault planes and/or cross-cutting relationships (Beacom 1999). Irrespective of the age of these later events, field and thin-section observations show that the effects of fluid alteration are mostly restricted to PDZs where fault linkage is abundant, e.g. Slioch. Faults at this locality exhibit numerous later dip-slip and oblique-slip slickenlines (see Fig. 5b, stereoplot v) and lie adjacent to the Loch Maree Fault for which there is considerable indirect regional evidence of reactivation during both Palaeozoic and Mesozoic time. Conversely, all PDZ fault zones that demonstrate limited fault linkage show little evidence of fluid influx and commonly have not experienced any significant reactivation. An intermediate group of PDZ faults display highly interconnected fracture networks, but show little evidence of reactivation (e.g. those shown in Fig. 4a and b). Typically, faults in this group are filled with pseudotachylytes and/or highly cemented cataclasites that may have prevented fluid access and therefore effectively inhibited any weakening and later reactivation. In the absence of quantitative data concerning linkage attributes these suggestions remain speculative, but they do suggest that once fracture sets and infills have formed, they influence the siting and mechanisms of other fault-zone-related processes such as fluid influx and alteration. This in turn can lead to changes in the physical properties of the deformation zone that influence the subsequent distribution of later reactivation events.

Further work is required in a variety of tectonic settings to fully characterize the relationships between the spatial distributions, geometry and scaling of fault populations attributes in continental basement gneiss complexes. This will allow a more rigorous assessment of the relative importance of pre-existing heterogeneities in influencing the location and architecture of structures formed during regional brittle deformation events in the continental crust. Having assessed the distribution of later reactivation events, it should then be possible to incorporate microstructural information concerning the nature and textural evolution of associated fault rocks. This will lead to a clearer understanding of the possible feedback mechanisms that exist between internal fault-zone processes and the growth of fault networks at different scales throughout the upper crust.

Conclusions

(1) The long-lived structural inheritance exhibited by the Late Archaean to Palaeoproterozoic Lewisian Complex of NW Scotland is representative of the deformation histories preserved in many regions of continental crust. Before any quantitative analysis is attempted in such settings, the complexity and longevity of the brittle deformation history usually requires that the relative and, if possible, absolute ages of faulting episodes are known, together with the geometric and kinematic characteristics of individual fault arrays. The late Laxfordian faults studied here have been selected as the relative age and nature of their movements are particularly well constrained.

(2) Field observations suggest that the late Laxfordian sinistral strike-slip PDZ faults reactivate pre-existing ductile fabrics in the Inverian and earlier Laxfordian Canisp and Gairloch ductile shear zones. Geometric and orientation data collected during structural logging of the fault zones have been used to quantitatively test the influence of lithology and pre-existing structural geometry on the spatial patterns of fault development.

(3) Stereographic analysis shows a strong geometrical correspondence between the intensity and form of the pre-existing anisotropy and the alignment of the PDZ brittle faults.

(4) There is a close correspondence between the geometric and spatial clustering attributes of the PDZ faults, which also show characteristically different spatial clustering patterns in certain lithologies (amphibolite v. acid gneiss v. quartz–mica schist).

(5) The present study provides the first quantitative analysis of the spatial patterns of faults that reactivate pre-existing anisotropies in typical continental basement gneisses. The results demonstrate a clear relationship between fault development and pre-existing lithology and structure.

The initial fieldwork was carried out whilst L.E.B. was in receipt of a DENI Distinction Award at the Queen's University of Belfast and the University of Durham. Many thanks go to T. Needham, G. Yielding and J. Walsh for their constructive and forthright reviews, and to J. Sleight for comments on the revised version. R.E.H. would like to acknowledge the continued financial assistance of Statoil (UK) Ltd.

References

ATTFIELD, P. 1987. The structural history of the Canisp Shear Zone. *In*: PARK, R.G. & TARNEY, J. (eds) *Evolution of the Lewisian and Comparable Precambrian High Grade Terrains*. Geological Society, London, Special Publications **27**, 165–173.

BARTON, C.C. 1983. Systematic jointing in the Cardium Sandstone along the Bow River, Alberta, Canada. Ph.D. thesis, Yale University, New Haven, CT.

BARTON, C.C. & HEISH, P.A. 1989. *Physical and Hydrologic Flow Properties of Fractures. Field-trip Guidebook T385, 28th International Geological Congress*. American Geophysical Union, Washington, DC.

BARTON, C.C. & ZOBACK, M.D. 1992. Self-similar distribution and properties of a macroscopic fractures at depth in crystalline rocks in the Cajon Pass scientific drill hole. *Journal of Geophysical Research*, B4, **97**, 5181–5200.

BEACOM, L.E. 1999. The kinematic evolution of reactivated and non-reactivated faults in basement rocks, NW Scotland. Ph.D. thesis, Queen's University of Belfast.

BEACOM, L.E., ANDERSON, T.B. & HOLDSWORTH, R.E. 1999. Using basement hosted clastic dykes as syn-rifting palaeostress indicators: an example from the basal Stoer Group, Scotland. *Geological Magazine*, **136**, 301–310.

BUTLER, R.W.H. 1997. Late Proterozoic rift faults and basement–cover relationships within the Ben More thrust sheet, NW Scotland. *Journal of the Geological Society, London*, **154**, 761–764.

COWIE, P.A., KNIPE, R.J. & MAIN, I.G. 1996. Introduction to the Special Issue. *Journal of Structural Geology*, **18**, v–xi.

COX, D.R. & LEWIS, P.A.W. 1986. *The Statistical Analysis of a Series of Events*. Methuen, London.

DALY, M.C., CHOROWICZ, J. & FAIRHEAD, J.D. 1989. Rift basin evolution in Africa: the influence of reactivated steep basement shear zones. *In*: COOPER, M.A. & WILLIAMS, G.D. (eds) *Inversion Tectonics*. Geological Society, London, Special Publications **44**, 309–334.

DEMARSILY, G. *Flow and Transport in Fractured Rocks: Connectivity and Scale Effect*. International Association of Hydrologists, Memoir **17**, 267–277.

DORÉ, A.G., LUNDIN, E.R., FICHLER, C. & OLESEN, O. 1997. Patterns of basement structure and reactivation along the NE Atlantic margin. *Journal of the Geological Society, London*, **154**, 85–92.

FYFE, J.A., LONG, D. & EVANS, D. 1993. *United Kingdom Offshore Regional Report: the Geology of the Malin–Hebrides Area*. British Geological Survey, Keyworth.

GILLESPIE, P.A., HOWARD, C.B., WALSH, J.J. & WATTERSON, J. 1993. Measurement and characterisation of spatial distributions of fractures. *Tectonophysics*, **226**, 113–141.

GILLESPIE, P.A., JOHNSTON, J.D., LORIGA, M.A., MCCAFFREY, K.J.W., WALSH, J.J. & WATTERSON, J. 1999. Influence of layering on vein systematics in line samples. *In*: MCCAFFREY, K.J.W., LONERGAN, L. & WILKINSON, J.J. (eds) *Fractures, Fluid Flow and Mineralization*. Geological Society, London, Special Publications **155**, 35–56.

HARRIS, C., FRANSSEN, R. & LOOSVELD, R. 1991. Fractal analysis of fractures in rocks: the Cantors Dust method—comment. *Tectonophysics*, **198**, 107–115.

HATTON, C.G., MAIN, I.G. & MEREDITH, P.G. 1994. Non-universal scaling of fracture length and opening displacement. *Nature*, **367**, 160–162.

HIPPLER, S.J. & KNIPE, R.J. 1990. The evolution of cataclastic fault rocks from a pre-existing mylonite. *In*: KNIPE, R.J. & RUTTER, E.H. (eds) *Deformation Mechanisms, Rheology and Tectonics*. Geological Society, London, Special Publications **54**, 71–79.

HOLDSWORTH, R.E., BUTLER, C.A. & ROBERTS, A.M. 1997. The recognition of reactivation during continental deformation. *Journal of the Geological Society, London*, **154**, 73–78.

HOLDSWORTH, R.E., STEWART, M., IMBER, J. & STRACHAN, R.A. 2001. The structure and rheological evolution of reactivated continental fault zones: a review and case study. *In*: MILLER, J.A., HOLDSWORTH, R.E., BUICK, I.S. & HAND, M. (eds) *Continental Reactivation and Reworking*. Geological Society, London, Special Publications, **184**, 115–137.

HUANG, Q. & ANGELIER, J. 1989. Fracture spacing and its relation to bed thickness. *Geological Magazine*, **126**, 355–362.

IMBER, J., HOLDSWORTH, R.E., BUTLER, C.A. & LLOYD, G.E. 1997. Fault-zone weakening processes along the reactivated Outer Hebrides Fault Zone, Scotland. *Journal of the Geological Society, London*, **153**, 105–109.

JOHNSTON, J.D. & MCCAFFREY, K.J.W. 1996. Fractal geometries of vein systems and the variation of scaling relationships with mechanisms. *Journal of Structural Geology*, **18**, 349–358.

KNIPE, R.J. & LLOYD, G.E. 1994. Deformation history during faulting: the Skiag Bridge backthrust, Assynt, NW Scotland. *Pure and Applied Geophysics*, **142**, 229–254.

KNOTT, S.D., BEACH, A., BROCKBANK, P.J., BROWN, J.L., MCCALLUM, J.E. & WELBON, A.I. 1996. Spatial and mechanical controls on normal fault populations. *Journal of Structural Geology*, **18**, 359–372.

KORVIN, G. 1989. Fractured but not fractal: fragmentation of the Gulf Suez basement. *Pure and Applied Geophysics*, **131**, 290.

KULANDER, B. K., BARTON, C. C. & DEAN, S. L. 1979. *The Application of Fractography to Core and Outcrop Fracture Investigations*. Morgantown Energy Technology Center. US Department of Energy, METC/SP-79/3..

LAUBACH, S.E. & MARSHAK, S. 1987. Fault patterns generated during extensional deformation of crystalline basement, N.W. Scotland. *In*: COWARD, M.P., DEWEY, J.F. & HANCOCK, P.L. (eds) *Continental Extensional Tectonics*. Geological Society, London, Special Publications **28**, 495–499.

LITTLE, T.A. 1995. Brittle deformation adjacent to the Awatere strike-slip fault in New Zealand: faulting patterns, scaling relationships, and displacement partitioning. *Geological Society of America Bulletin*, **11**, 1255–1271.

LLOYD, G.E. & KNIPE, R.J. 1992. Deformation mechanisms accommodating faulting of quartzite under upper crustal conditions. *Journal of Structural Geology*, **14**, 127–143.

MCCAFFREY, K.J.W. & JOHNSTON, J.D. 1996. Fractal analysis of a mineralized vein deposit; Curraghinalt gold deposit, County Tyrone. *Mineralium Deposita*, **31**, 52–58.

MANNING, C.E. 1994. Fractal clustering of metamorphic veins. *Geology*, **22**, 335.

MORTON, N. 1992. Dynamic stratigraphy of the Triassic and Jurassic of the Hebrides basin, NW Scotland. *In*: PARNELL, J. (ed.) *Basins on the Atlantic Seaboard: Petroleum Geology, Sedimentology and Basin Evolution*. Geological Society, London, Special Publications **62**, 97–110.

NEEDHAM, T. & MORGAN, R. 1997. The East Irish Sea and adjacent basins: new faults or old? *Journal of the Geological Society, London*, **154**, 145–150.

NEEDHAM, T., YIELDING, G. & FOX, R. 1996. Fault population description and prediction using examples from the offshore U.K. *Journal of Structural Geology*, **18**, 155–167.

NICOL, A., WALSH, J.J., WATTERSON, J. & CHILDS, C. 1996*a*. The shapes, major axis orientations and displacement patterns of fault surfaces. *Journal of Structural Geology*, **18**, 235–248.

NICOL, A., WALSH, J.J., WATTERSON, J. & GILLESPIE, P.A. 1996*b*. Fault-size distributions—are they really power law? *Journal of Structural Geology*, **18**, 191–197.

PARK, R.G., CLIFF, R.A., FETTES, D.J. & STEWART, A.D. 1994. Precambrian rocks in northwest Scotland west of the Moine Thrust: the Lewisian Complex and the Torridonian. *In*: GIBBONS, W. & HARRIS, A.L. (eds) *A Revised Correlation of Precambrian Rocks in the British Isles*. Special Report of the Geological Society, London **22**, 6–22.

PARK, R.G., CRANE, A. & NIAMATULLAH, M. 1987. Early Proterozoic structure and kinematic evolution of the southern mainland Lewisian. *In*: PARK, R.G. & TARNEY, J. (eds) *Evolution of the Lewisian Complex and Comparable High Grade Terrains*. Geological Society, London, Special Publications **27**, 139–151.

PEACOCK, D.C.P. & SANDERSON, D.J. 1996. Effects of propagation rate on displacement variations along faults. *Journal of Structural Geology*, **18**, 311–320.

PINHEIRO, R.V.L. & HOLDSWORTH, R.E. 1997. Reactivation of Archaean strike-slip fault systems, Amazon region, Brazil. *Journal of the Geological Society, London*, **154**, 99–104.

ROBERTS, A.M. & HOLDSWORTH, R.E. 1999. Linking onshore and offshore structures: Mesozoic extension in the Scottish Highlands. *Journal of the Geological Society, London*, **156**, 1061–1064.

SCHOLZ, C.H. & COWIE, P.A. 1990. Determination of total strain from faulting using measurements. *Nature*, **346**, 837–839.

SHAIL, R.K. & ALEXANDER, A.C. 1997. Late Carboniferous to Triassic reactivation of Variscan basement in the western English Channel: evidence from onshore exposures in south Cornwall? *Journal of the Geological Society, London*, **154**, 163–168.

SHIHE, L. & PARK, R.G. 1993. Reversals of movement sense in Lewisian brittle–ductile shear zones at Gairloch, NW Scotland, in the context of Laxfordian kinematic history. *Scottish Journal of Geology*, **29**, 9–19.

SNYDER, D.B., ENGLAND, R.W. & MCBRIDE, J.H. 1997. Linkage between mantle and crustal structures and its bearing on inherited structures in northwestern Scotland. *Journal of the Geological Society, London*, **154**, 79–84.

SOPER, N.J. & ENGLAND, R.W. 1995. Vendian and Riphean rifting in NW Scotland. *Journal of the Geological Society, London*, **152**, 11–14.

SORNETTE, A., DAVY, Ph. & SORNETTE, D. 1990. Growth of fractal fault patterns. *Physics Reviews Letters*, **65**, 2266–2269.

STEWART, M.A., HOLDSWORTH, R.E. & STRACHAN, R.A. 2000. Deformation processes and weakening mechanisms within the frictional–viscous transition zone of major crustal faults: insights from the Great Glen Fault Zone Scotland. *Journal of Structural Geology*, **22**, 543–560.

STOKER, M.S., HITCHEN, K. & GRAHAM, C.C. 1993. *United Kingdom Offshore Regional Report: the Geology of the Hebrides and West Shetland Shelves, and Adjacent Deep-water Areas*. British Geological Survey, Keyworth.

SUTTON, J. & WATSON, J.V. 1986. Architecture of the continental lithosphere. *Philosophical Transactions of the Royal Society, London, Series A*, **317**, 5–12.

SYKES, L.R. 1978. Intraplate seismicity, reactivation of pre-existing zones of weakness, alkaline magmatism and other tectonism post-dating continental fragmentation. *Reviews of Geophysics and Space Physics*, **16**, 621–688.

THORP, R.K., WALTERS, B.J. & RALPH, W.E. 1983. Strength and permeability of an ultra-large specimen of granitic rock. *In*: MATHEWSON, C.C. (ed.) *Rock Mechanics Theory—Experiment—Practice. Proceedings of the 24th US Rock Mechanics Symposium*. 511–518. Texas A&M University Press, College Station, TX.

TIKOFF, B., KELSO, P., MANDUCA, C., MARKELY, M.J. & GILLASPY, J. 2001. Lithospheric and crustal reactivation of an ancient plate boundary: the assembly and disassembly of the Salmon River suture zone, Idaho, USA. *In*: HOLDSWORTH, R.E., STRACHAN, R.A., MAGLOUGHLIN, J.F. & KNIPE, R.J. (eds) *The Nature and Tectonic Significance of Fault Zone Weakening*. Geological Society, London, Special Publications, **186**, 213–231.

VELDE, B., DUBOIS, J., TOUCHARD, G. & BADRI, A. 1990. Fractal analysis of fractures in rocks: the Cantors Dust method. *Tectonophysics*, **179**, 345–352.

WALSH, J.J. & WATTERSON, J. 1991. Geometric and kinematic coherence and scale effects in normal fault systems. *In*: ROBERTS, A.M., YIELDING, G. & FREEMAN, B. (eds) *The Geometry of Normal Faults*. Geological Society, London, Special Publications **56**, 193–203.

WALSH, J.J. & WATTERSON, J. 1993. Fractal analysis of fracture patterns using the standard box counting technique: valid and invalid methodologies. *Journal of Structural Geology*, **15**, 1509–1512.

WALSH, J.J., CHILDS, C., MEYER, V. & 9 OTHERS. 2001. Geometrical controls on the evolution of normal fault systems. *In*: HOLDSWORTH, R.E., STRACHAN, R.A., MAGLOUGHLIN, J.F. & KNIPE, R.J. (eds) *The Nature and Tectonic Significance of Fault Zone Weakening*. Geological Society, London, Special Publications, **186**, 157–170.

WINTSCH, R.P., CHRISTOFFERSEN, R. & KRONENBERG, A.K. 1995. Fluid–rock reaction weakening of fault zones. *Journal of Geophysical Research*, **100**, 13021–13032.

WOJTAL, S. 1994. Fault scaling laws and temporal evolution of fault systems. *Journal of Structural Geology*, **16**, 603–612.

WOJTAL, S.F. 1996. Changes in fault displacement populations correlated to linkage between faults. *Journal of Structural Geology*, **18**, 265–279.

WOODCOCK, N.H. & NAYLOR, M.A. 1983. Randomness testing in three-dimensional orientation data. *Journal of Structural Geology*, **5**, 539–548.

YIELDING, G., NEEDHAM, T. & JONES, H. 1996. Sampling of fault populations using sub-surface data: a review. *Journal of Structural Geology*, **18**, 135–146.

Lithospheric and crustal reactivation of an ancient plate boundary: the assembly and disassembly of the Salmon River suture zone, Idaho, USA

B. TIKOFF[1], P. KELSO[2], C. MANDUCA[3], M.J. MARKLEY[4] & J. GILLASPY[1]

[1]*Department of Geology and Geophysics, University of Wisconsin–Madison, 1215 W Dayton Street, Madison, WI 53706, USA (e-mail: basil@geology.wisc.edu)*

[2]*Department of Geology and Physics, Lake Superior State University, Sault St Marie, MI 49783, USA*

[3]*DLESE Outreach Coordinator, Carleton University, Boulder, CO 80307, USA*

[4]*Department of Geology, Mt Holyoke College, North Hampton, MA 01075, USA*

Abstract: The Salmon River suture zone, western Idaho, is a fundamental lithospheric boundary between the North American craton and the accreted terranes of the Cordilleran margin. The initial juxtaposition along this north–south-oriented structure occurred during Early Cretaceous time. This zone was potentially reactivated twice by subsequent tectonism, once during Cretaceous time and once during Miocene time. The Late Cretaceous western Idaho shear zone formed along the Salmon River suture zone, as denoted by a sharp gradient in the isotopic signature of the granitoids that intruded the lithospheric boundary zone. The reconstructed Late Cretaceous orientation of the western Idaho shear zone contains subvertical fabrics (lineation, foliation). The same boundary also acted as a locus for subsequent Miocene Basin and Range extensional deformation. Domino-style normal faulting and deep (2100 m) basin formation accommodated the motion between the extending accreted terranes to the west and the unextended Idaho batholith to the east. Whereas either the mantle boundary or a crustal-scale structuring controls the regional extent of the extensionally reactivated zone, locally crustal basement faults and lithological contacts control the orientation and precise location of faults that accommodate reactivation. The multiple reactivation of the Salmon River suture zone is critical for several reasons. The Early Cretaceous suture zone apparently created a fundamental lithospheric flaw, which was reactivated after terrane accretion. Whether this zone was a fracture or a shear zone, the fabric in the mantle lithosphere was apparently not 'healed' during orogenesis. Thus, juxtaposition of mantle lithosphere, which is inferred to occur by faulting in the uppermost mantle, acts as a weakness during later tectonism. Second, the paucity of strike-slip plate boundaries in the geological record makes sense in the context of reactivation. The vertical, lithospheric-scale nature of these structures makes them particularly susceptible to lithospheric-scale reactivation during both transcurrent and/or extensional deformation. These reactivations both overprint the earlier deformation and modify the original geometry. Steeply dipping fabrics, rather than vertical fabrics, may be the general signature of major, ancient strike-slip faults.

Most field geologists understand the importance of reactivation of earlier structures. Reactivation controls much of the heterogeneity of deformation within orogenic belts, and results in very complex displacements, rotations, and internal strains arising from the applied boundary conditions (e.g. Dewey *et al.* 1986). Although the crustal reactivation is often difficult to recognize, because of the necessary overprinting of some of the earlier fabrics during later tectonism, it has received recent attention in the literature (e.g. Holdsworth *et al.* 1997, and references therein). The presence of reactivated crustal structures is commonly invoked to explain the patterns of deformation, such as reactivation of late Palaeozoic Ancestral Rockies trends during the Late Cretaceous–Paleogene Laramide orogeny (e.g. Oldow *et al.* 1989).

A more complex reactivation issue involves reactivation applied to the whole lithosphere. Strike-slip fault systems, in particular, are

From: HOLDSWORTH, R.E., STRACHAN, R.A., MAGLOUGHLIN, J.F. & KNIPE, R.J. (eds) 2001. *The Nature and Tectonic Significance of Fault Zone Weakening*. Geological Society, London, Special Publications, **186**, 213–231. 0305-8719/01/$15.00

thought to extend through the entire lithosphere (e.g.Teyssier & Tikoff 1998). Recent seismic work along the San Andreas fault in northern California supports this contention, as discrete offsets in the Moho are observed under the San Andreas and two parallel faults (Henstock *et al.* 1997). As the mantle is considered to be the strongest of the lithospheric layers (e.g. Molnar 1992), the presence of previously active faults and/or shear zones in the uppermost mantle may control the lithospheric reactivation. The notion of the lithospheric mantle deforming in a brittle manner at these pressure and temperature conditions is consistent with experimental deformation of olivine (Brace & Kohlstedt 1980). Consequently, if fractures in the mantle do not anneal during the intervening tectonic quiescence, it is possible that they are also the locus of reactivation during later tectonism (e.g. Vauchez *et al.* 1997). Suggestions of multiple reactivation of lithospheric structures, including the existence of a proto-San Andreas fault (Suppe 1970; Nilsen 1978) and reactivation of the Taura fault, New Zealand (Sutherland & Melhuish 2000), suggests that mantle faults can experience reactivation.

Although the lithospheric mantle can be imaged geophysically, this provides only an instantaneous picture of the current geometry of the lithospheric layers. Geologically, reactivation of the lithospheric mantle is a difficult process to document, requiring: (1) a field location in which the crust holds a record, through igneous intrusions, of the underlying mantle; (2) juxtaposed lithospheric mantles that both contain distinctive, but differing, geochemical signatures; (3) reactivation of the earlier structure by recognizable later tectonic events. The Salmon River suture zone, Idaho, meets these criteria and is an ideal place to study lithospheric reactivation.

The Salmon River suture zone juxtaposes accreted material of oceanic origin against the North American margin during an Early Cretaceous suturing event. These two tectonic blocks contain distinctive Sr and O isotopic signatures, which are recorded in the voluminous plutons that subsequently rose through this region as part of a magmatic arc. Moreover, the Salmon River suture zone was potentially reactivated twice. The first reactivation occurred in Late Cretaceous time, as an intra-arc shear zone (McClelland *et al.* 2000), and a second possible reactivation is associated with Basin and Range extension. In both cases, the lithospheric flaw resulting from the initial suturing played a critical role in localizing deformation. We first summarize the geological history of the Salmon River suture zone and then discuss the crustal and lithospheric reactivation that occurred in the Salmon River suture zone.

Geological evolution

Overview

The Salmon River suture zone of west–central Idaho (Figs. 1 and 2) separates accreted oceanic-arc (Seven Devils–Wallowa) terranes and the North America craton, as denoted by a wide variety of geological, geochemical, and geophysical techniques (e.g. Fleck & Criss 1985; Lund & Snee 1988; Manduca *et al.* 1993). Wall rocks and pendants in plutons of the western part of the Idaho batholith change within 5 km from amphibolites, quartzofeldspathic gneisses and lesser calc-silicate material representing metamorphosed accreted terranes in the west, to quartzites, schists, and gneisses formed from metamorphosed continentally derived sedimentary rocks in the east (Myers 1982; Lund 1984; Onasch 1987; Manduca *et al.* 1993). These eastern metaclastic rocks belong to the Belt Supergroup, part of the North American craton, and presumably overlie basement of Archaean or Proterozoic age (Hyndman *et al.* 1988). The

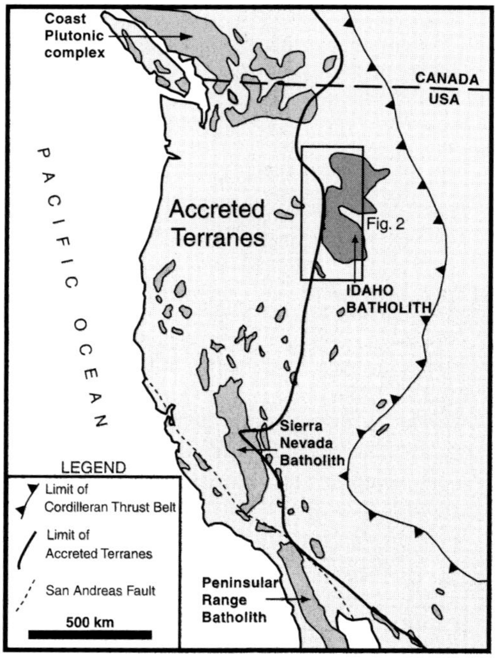

Fig. 1. Mesozoic batholiths of the central North American Cordillera.

depositional age of these rocks is poorly constrained but considered to be Proterozoic (1.6–1.3 Ga; e.g. Burchfiel *et al.* 1992). A distinct chain of ultramafic pods is found along the suture zone, but always on the western, accreted side (Godchaux & Bonnichsen 1993). There is also a distinct contrast in the petrology and geochemistry of igneous rocks on each side of the suture zone, representing differences in lithospheric mantle and lower crust through which the magmas rose (Fleck & Criss 1985; Manduca *et al.* 1993). The plutons on the eastern side contain magmatic epidote, indicating crystallization at >7 kbar (Zen 1988; Manduca *et al.* 1993).

The most obvious regional demarcation of the Salmon River suture zone is a prominent change in the $^{87}Sr/^{86}Sr$ ratio (Fig. 2; Armstrong *et al.* 1977; Fleck & Criss 1985; Criss & Fleck 1987). The rocks of the Seven Devils–Wallowa terranes typically have a $^{87}Sr/^{86}Sr$ ratio <0.7045 and the intrusive rocks within the North American craton have a $^{87}Sr/^{86}Sr$ ratio >0.707. The narrowness of the zone over which this change occurs is unique in the North American Cordillera. A similarly dramatic variation is found in the $\delta^{18}O$ values across this boundary, grading from *c.* +7 on the western, accreted side to >+10 on the continental side (Fleck & Criss 1985; Criss & Fleck 1987; Manduca *et al.* 1992). A steep dip for this boundary is supported by the isotopic systematics, which indicates a change in source rocks across a zone of 10–15 km width juxtaposing fundamentally different lithospheres (Manduca *et al.* 1992).

This distinct break in lithological and geochemical data can be explained by the truncation of the continent before collision of the Seven Devils–Wallowa terranes, either by strike-slip faulting or by rifting (e.g. Davis *et al.* 1978; Hamilton 1978; Lund 1984; Wernicke & Klepacki 1988; Selverstone *et al.* 1992). This idea is most clearly supported by the lack of any Palaeozoic shelf or slope deposits (Hamilton 1969), or an accretionary complex (Myers 1982; Lund 1984).

Early Cretaceous time: suturing of accreted terranes

Although the accretion of the Seven Devils–Wallowa terrane to the continental margin was a major tectonic event, the details of the collision are both poorly preserved and poorly understood (Fig. 3a). Very little evidence remains from the Seven Devils–Wallowa terrane accretion (e.g. Avé Lallemant 1995; Vallier 1995), except for minor calc-alkaline magmatism at 145 Ma (Vallier 1995). No evidence of the accretion is preserved on the North American side of the suture zone, as a result of subsequent emplacement of the Idaho batholith. Two major thrust faults (the Pollock Mountain and Rapid River faults) occur within the accreted terrane (Fig. 4). Both faults dip to the SE at 45–60°, juxtapose rock packages of differing metamorphic grade, and are interpreted as syn- to

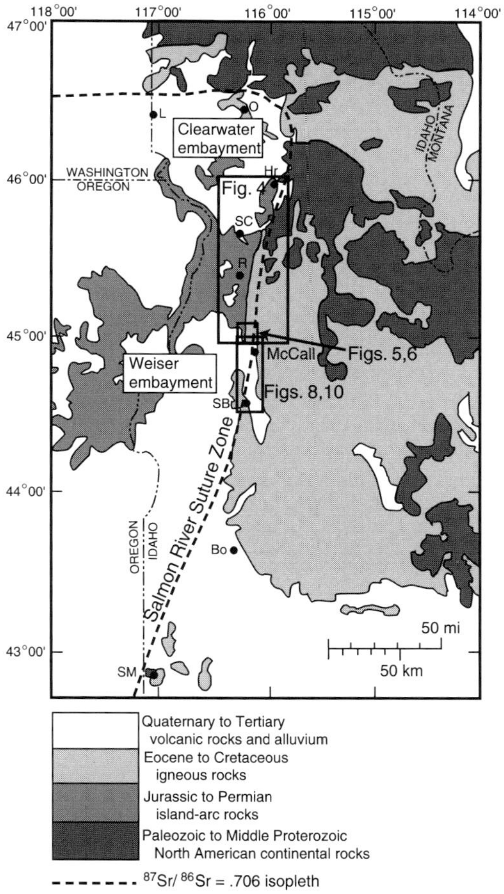

Fig. 2. The Salmon River suture zone in western Idaho. The dashed line indicates the abrupt break in the $^{87}Sr/^{86}Sr$ isotopic line, from <0.7045 in the accreted terranes (western side) to >0.707 in the Precambrian basement of North American (eastern side). The 'gneissic border' extends from Orofino south to Snowbank Mountain (>250 km) and is found on the south side of the Snake River plain in the South Mountain (>400 km). The Weiser and Clearwater embayments are eastward extensions of Columbia River basalts. Relevant landmarks: McCall; L, Lewiston; O, Orofino; Hr, Harpster; SC, Slate Creek; R, Riggins; SB, Snowbank Mountain; Bo, Boise; SM, South Mountain. Modified from Lund (1995).

late metamorphic (Selverstone *et al.* 1992). The Rapid River thrust fault, located west of the Pollock Mountain thrust fault, places rocks of the Rapid River plate (metamorphosed phyllites, volcanic metasediments, and amphibolites of the Riggins group) over the Wallowa terrane. The Pollock Mountain thrust places the Pollock Mountain amphibolite and intrusive granodiorites over the Rapid River plate. Potentially equivalent faults (Downey Creek, Castle Creek faults) occur along strike to the north, in the Harpster area (Myers 1982).

In general, there is an increase in metamorphic grade from the western accreted terranes into the western Idaho shear zone, located east of the Pollock Mountain thrust (Fig. 4; Hamilton 1963). The surface expression of the Pollock Mountain thrust occurred at 9–11 kbar at 600–650 °C, whereas the Rapid River thrust occurred at 6–8 kbar at 475–500 °C (Selverstone *et al.* 1992). Selverstone *et al.* (1992) dated the deformation using Sm–Nd dating, arriving at a mineral isochron age of 128 Ma and a garnet–whole-rock age of 136–144 Ma for material of the Rapid River plate. Thus, the best constraint on the initiation of the collision of the accreted terranes with North America is *c.* 144 Ma. This date is consistent with $^{40}Ar/^{39}Ar$ dating of hornblende and biotite within the shear zone, which suggests cooling ages as early as 130 Ma in the western part of the Salmon River suture zone (Lund & Snee 1988; Snee *et al.* 1995).

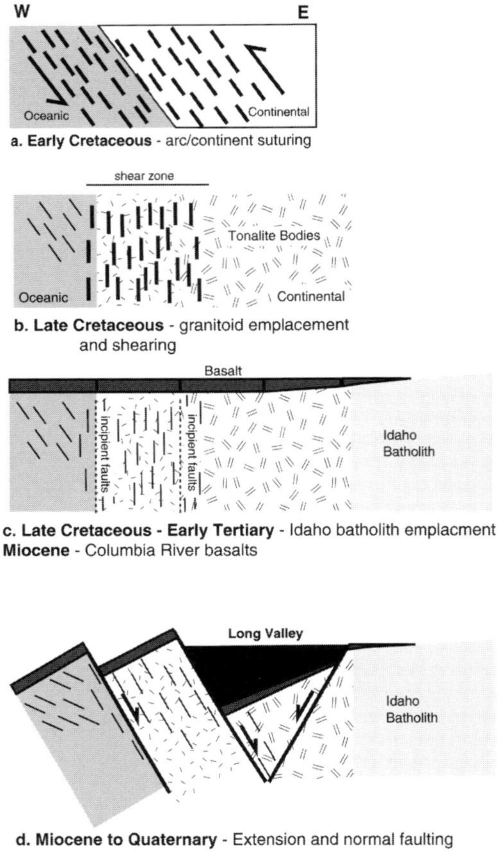

Fig. 3. Inferred 2D geological history of the Salmon River suture zone shown in cross-section. An Early Cretaceous suture is reactivated by both Late Cretaceous shearing and Miocene–Quaternary extension. Bold lines indicate deformation that occurred during the given time interval.

Late Cretaceous time: pluton emplacement and shearing

After suturing, a series of tabular intrusive bodies were emplaced within a zone of 10–15 km width east of the Rapid River and Pollock Mountain thrusts (Fig. 3b). The bodies are elongated north–south, and are concordant to the surrounding foliation (i.e. sills). These intrusions are well foliated and lineated, and were referred to in the early literature as the western 'gneissic' zone of the Idaho batholith (Hamilton 1963; Taubeneck 1971). This gneissic fabric results from a late-stage, Late Cretaceous deformation in the western Idaho shear zone, discussed below. As opposed to the common ilmenite-bearing, two-mica, S-type granites of the main Idaho batholith, the western units are magnetite bearing, granodioritic to tonalitic in composition, and are I-type granitoids (e.g. Piccoli & Hyndman 1985; Manduca *et al.* 1993).

The intrusions of the western gneissic zone of the Idaho batholith near McCall, ID, were split into three subdivisions (Hazard Creek complex, Little Goose Creek complex, and Payette River tonalite), based on distinct petrological, structural, geochronological, and geochemical characteristics (Fig. 5; Manduca *et al.* 1992). The westernmost Hazard Creek complex is tonalite to trondhjemite in composition, with uncommon alkali-feldspar-bearing units, and contains magmatic epidote. Only the easternmost part of the Hazard Creek complex is involved in the western Idaho shear zone. A concordant U–Pb age on zircon dates the unit to 118 $\pm$ 5 Ma (Manduca *et al.* 1992). The Hazard Creek complex displays low $\delta^{18}O$ (<+9) and low $^{87}Sr/^{86}Sr$ values (<0.704) (Fig. 6; Manduca *et al.* 1992). The central Little Goose Creek complex is dominantly dioritic to granodioritic in composition, and there is no magmatic epidote. The

Fig. 4. Regional geological map of the Salmon River suture zone, between the Weiser and Clearwater embayments. The Rapid River and Pollock Mountain thrusts are Early Cretaceous contractional structures related to initial accretion of the Blue Mountain arc terranes to cratonic North America. To the east, the Late Cretaceous western Idaho shear zone is an amphibolite-facies shear zone with steeply east-dipping fabrics. R_i indicates the values of the $^{87}Sr/^{86}Sr$ ratio across the zone. The main part of the Idaho batholith was intruded during Late Cretaceous–early Tertiary time.

entire Little Goose Creek complex is affected by solid-state deformation associated with the western Idaho shear zone. A slightly discordant U–Pb age on zircon of 110 $\pm$ 5 Ma is reported from this unit (Manduca *et al.* 1993). The Little Goose Creek complex displays, systemically increasing from west to east, $\delta^{18}O$ values from +8 to +10 and $^{87}Sr/^{86}Sr$ values from 0.704 to 0.708. The youngest unit occurs in the east, and is the Payette River tonalite. This varies from tonalitic to granitic in composition, but is generally tonalitic along its western border, and locally contains magmatic epidote. Only the western margin of the tonalite is affected by the solid-state deformation of the western Idaho shear zone. The Payette River tonalite in the McCall area is considered broadly syntectonic, as its western edge exhibits solid-state foliation, which grades eastward into subparallel magmatic foliation. The tonalite body appears regionally continuous and is always associated with the eastern edge of the western Idaho shear zone. Two U/Pb ages on zircon are reported from the tonalite: a discordant 90 $\pm$ 5 Ma age in the McCall area (Manduca *et al.* 1993) and 90 $\pm$ 3 Ma in the Slate Creek area (80 km north of McCall; Snee *et al.* 1995). $\delta^{18}O$ values in the Payette River tonalite are generally >+10, although there is a large range and no spatial organization to the pattern, and $^{87}Sr/^{86}Sr$ values are >0.708 in the McCall area (Manduca *et al.* 1992).

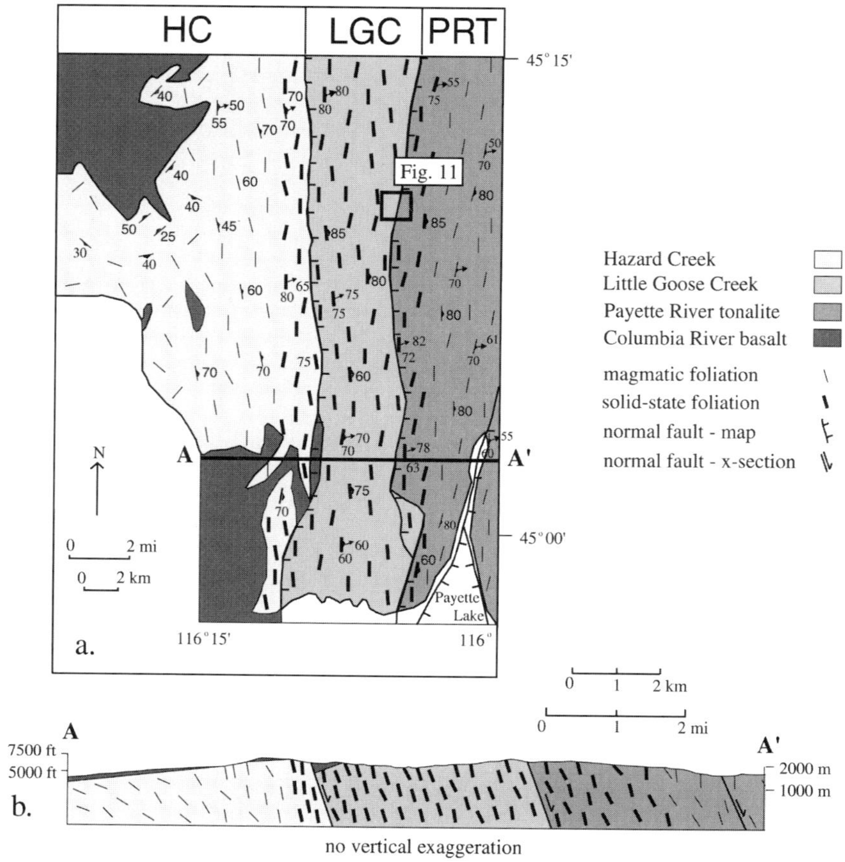

Fig. 5. (**a**) Geological map of the area immediately north of McCall. The Hazard Creek complex, Little Goose Creek complex, and Payette River tonalite are distinguishable in the field and discussed in the text. Partially based on Manduca *et al.* (1992). (**b**) Cross-section of the same region, indicating the steep (60–70°) dip of the Cretaceous fabric. Normal faulting occurs parallel to the Cretaceous foliation and generally along lithological contacts.

Sr and O isotopic data are thought to reflect interaction between the magmas and the mantle and crustal rocks through which they ascended. As a result, isotopic compositions of plutonic rocks are often used to demarcate the western limit of cratonic North America. $^{87}Sr/^{86}Sr$ initial ratios <0.706 typically are taken to indicate the absence of North American lithosphere and to reflect material added to the craton since Palaeozoic time. The sharp gradient of the isotopic systematics requires a steep dip for the structure that juxtaposes the accreted Wallow terrane and North American lithospheres (Manduca *et al.* 1992).

The western Idaho shear zone (Fig. 3b), which affects the igneous units, is an amphibolite-facies shear zone characterized by steeply east-dipping foliation and a down-dip lineation (Figs. 5 and 7). The kinematics of the shear zone is interpreted as either contractional (Blake 1991; Manduca *et al.* 1993) or transpressional (Tikoff & Manduca 1997; McClelland *et al.* 2001). This issue is not discussed here, as it is not relevant to the question of reactivation.

The western Idaho shear zone is centred within the Little Goose Creek unit, although it also affects the adjacent Hazard Creek complex and Payette River tonalite. As the Little Goose Creek complex displays the largest gradient in Sr and O isotopic values (e.g. records the largest variation in the shortest east–west distance), it is the unit located structurally above the magmatic source region along the juxtaposition of the lithospheric mantle between the accreted terranes and North America. This spatial relationship suggests that the inherited Early Cretaceous accretionary boundary was preferentially reactivated during Late Cretaceous time.

The uplift and cooling of the Late Cretaceous shear zone is documented using $^{40}Ar/^{39}Ar$

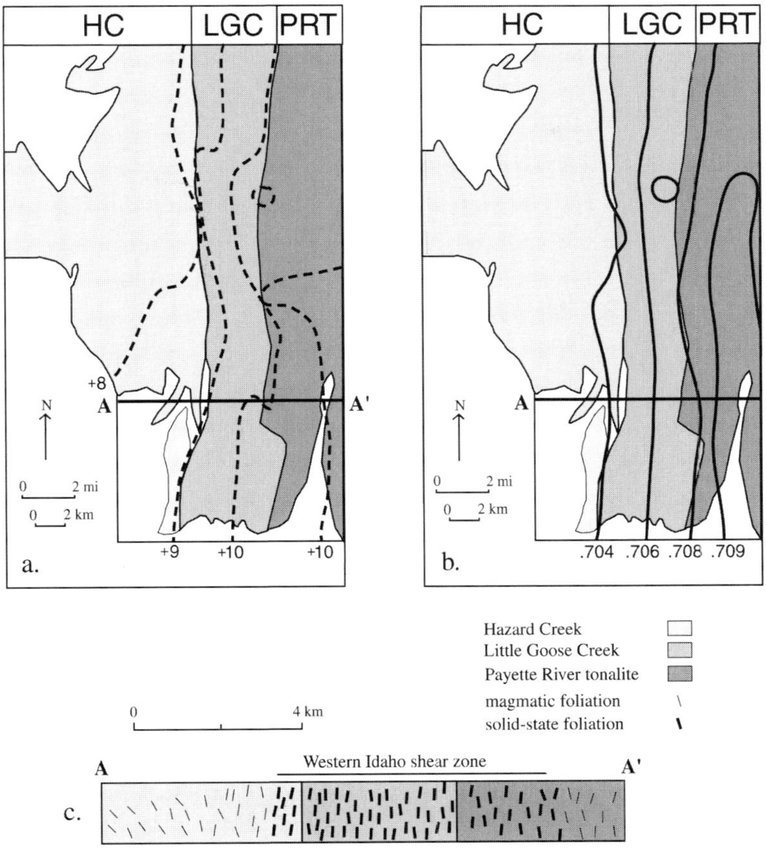

Fig. 6. Pre-Miocene geological maps for same area as in Fig. 5, constructed by restoring extensional deformation. (**a**) Distribution of $\delta^{18}O$, which shows a sharp rise across the Little Goose Creek complex. (**b**) Distribution of $^{87}Sr/^{86}Sr$ across the same area, with a similar steep gradient across the Late Cretaceous western Idaho suture zone. (**c**) Cross-section of the same region, illustrating the original vertical orientation of the Cretaceous fabric. (**a**) and (**b**) are based on Manduca *et al.* (1992).

results. These plutons and wall rocks yield $^{40}Ar/^{39}Ar$ dates from hornblende and biotite of 140 Ma along the western part of the Salmon River suture zone and 75 Ma within the Late Cretaceous shear zone (slightly east of the $^{87}Sr/^{86}Sr = 0.706$ line; Snee *et al.* 1995). Thus, the fabrics are time transgressive, with generally older ages in the west, and deformation ceased by *c.* 75 Ma in this area.

Latest Cretaceous to mid-Tertiary time: Idaho batholith emplacement

The area east of the western Idaho shear zone is largely underlain by the Idaho batholith, a composite Cretaceous–Tertiary diorite to granodiorite complex that intrudes Triassic rocks west of the western Idaho shear zone and rocks ranging from Pennsylvanian–Permian to Proterozoic age to the east (Fig. 3c). Characteristic two-mica granites of the Idaho batholith intrude the first phase of tonalite bodies east of McCall. Conventional U–Pb data on zircon indicate that plutonism initiated at *c.* 90 Ma in the western batholith and continued to *c.* 50–55 Ma in the east (Bickford *et al.* 1981; Shuster & Bickford 1985; Toth & Stacey 1992; Manduca *et al.* 1993). $^{40}Ar/^{39}Ar$ studies on biotite and hornblende (Lund & Snee 1988; Snee *et al.* 1995) and SHRIMP U–Pb analyses on zircon (Foster & Fanning 1997) indicate a two-phase emplacement of the batholith. Quartz diorite and tonalite emplacement occurred to the west at *c.* 70–95 Ma followed by granite emplacement to the east and within the Bitteroot core complex at 53–70 Ma. Moreover, the younger plutons are intruded at higher structural levels, suggesting that uplift occurred during batholith emplace-

Fig. 7. Prominent steeply east-dipping foliation and small-scale lithological contacts within the Little Goose Creek complex.

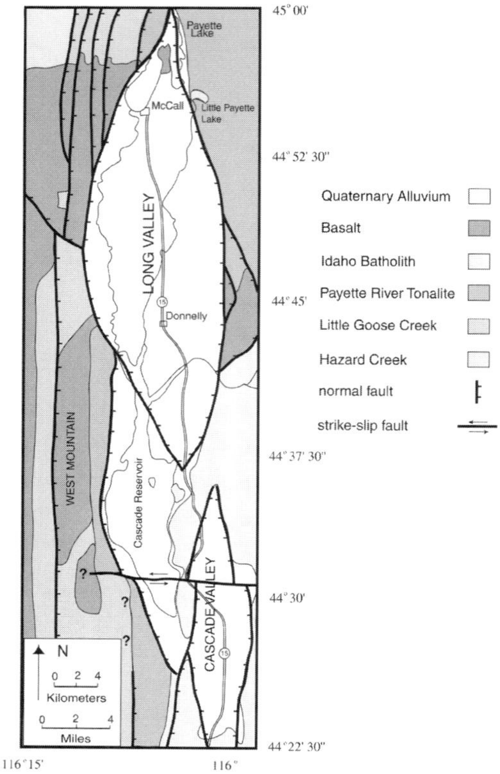

Fig. 8. Generalized geological map of Long Valley. Long Valley forms on the eastern side of rocks affected by the western Idaho shear zone. The half-grabens with east-dipping normal faults NW of Long Valley follow the fabric of western Idaho shear zone.

ment. After batholith emplacement, magmatism continued to migrate eastward as represented by the Eocene (43–51 Ma) Challis arc. These rocks are commonly extrusive and compositionally variable but predominantly of intermediate composition (e.g. McIntyre *et al.* 1982; Norman & Leeman 1989).

Miocene time: Columbia River basalts and extensional deformation

During Miocene time, the Columbia River basalts covered much of the area (Fig. 3c). These basalts are regionally extensive, occurring throughout eastern Oregon, eastern Washington, and western Idaho. The Columbia River basalts are dated at 6–17 Ma, although they were primarily erupted from 17 to 14 Ma (e.g. Watkins & Baksi 1974; Christiansen & Yeats 1992). The Columbia plateau has major lobes or embayments that extend eastward into Idaho, including the Weiser embayment to the south and the Clearwater embayment to the north (Fig. 2; Fitzgerald 1982). The flows in these embayments are significantly thinner than the massive basalt flows of eastern Oregon. The McCall area is at the northern end of the Weiser embayment and, within the embayment, basalt flows extend eastward to the western edge of the Idaho batholith. Northwest of McCall is the best exposure of the Seven Devils–Wallowa terranes, including the Rapid River and Pollock Mountain thrusts, between the Weiser and Clearwater embayments (Fig. 2). In the McCall area and southward (e.g. Long Valley), basalt flows locally overlie the western Idaho shear zone and the westernmost part of the Idaho batholith.

Extensional deformation appears to have commenced after extrusion of the basalt flows. This deformation is attributed to Basin and Range style extension and block faulting that seems to have initiated during Miocene time in the Great Basin and Oregon–Idaho region (Oldow *et al.* 1989). In general, faulting occurs up to the edge of the Idaho batholith, but does not extend pervasively into the batholith (Hamilton 1963). The western 'gneissic' edge a.

b.

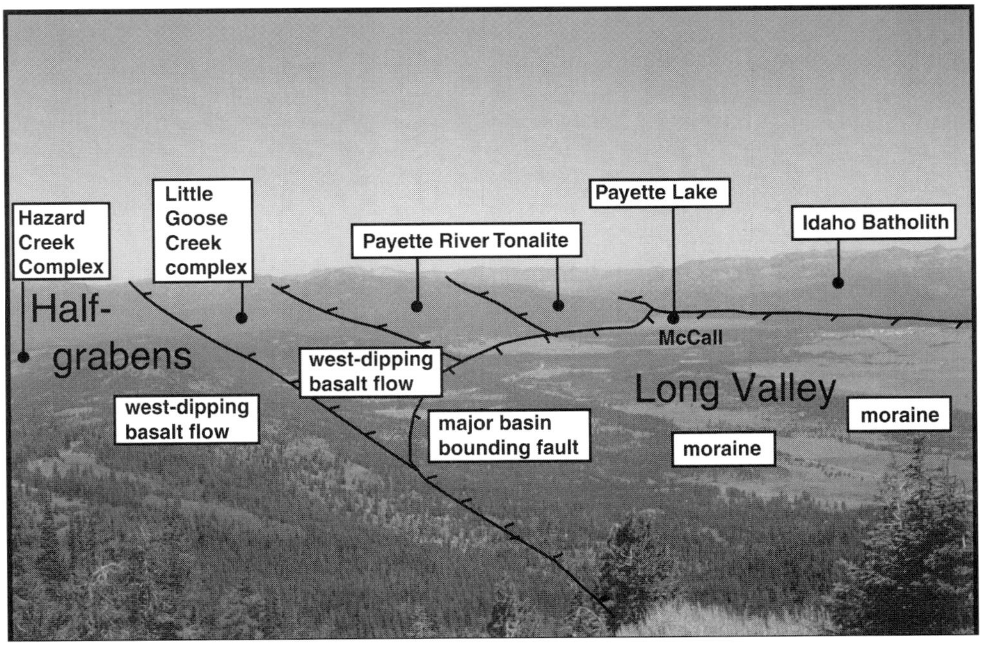

Fig. 9. (**a**) Photograph to the NNE of Long Valley from No Business Mountain. (**b**) Modification of the photograph, showing geological features. Payette Lake, in the northernmost part of Long Valley, is located within a graben. Half-grabens NW of Long Valley separate the lithological units of the Salmon River suture zone. These half-grabens are truncated by the major bounding fault on the west side of Long Valley.

of the batholith has a distinct style of normal faulting. Within the McCall area, the Columbia River basalts are typically tilted westward 20–30° on steep, east-dipping normal faults (Fig. 3d; Hamilton 1963; Bonnichsen 1987; Manduca 1988). In these cases, the 20–30° west dip of the basalt flows is subperpendicular to the 60–70° east dip of the western Idaho shear-zone fabric in the underlying rocks. The throws on some of the faults are obtainable using the offset basalt flows, assuming that the basalts flowed over a subhorizontal surface. In Fig. 5, the throw on the fault that separates the Hazard Creek complex from the Little Goose Creek complex is calculated at *c.* 100 m. Faults are more easily recognized south of McCall within the Weiser embayment, because of the presence of the Columbia River basalt units (Figs. 8 and 9).

Normal faults contained within the rocks of the western Idaho shear zone have a consistent east-dipping orientation, with the east side moving down relative to the west side (Fig. 3d). Farther west in the Weiser embayment, there are domains of both NW- and NE-trending normal faults, but neither area shows a consistent sense of rotation (Fitzgerald 1982).

An exception to the consistent domino-style rotation of the normal faulting along the Salmon River suture zone is Long Valley, which forms a graben. Long Valley is the largest (11 km by 40 km) in a series of north–south-trending valleys that formed within or immediately east of the western Idaho shear zone (Fig. 10). Regional gravity surveys suggest that up to 1524 m of sediment exists in Long Valley, implying 2134 m of throw (e.g. Kinoshita 1962). The gravity data suggest an asymmetric graben geometry to the valley, in which the west side of the valley has significantly more sediment than the east side (Fig. 10). This interpretation is consistent with the presence of deep (>186 m) water wells in the valley that do not penetrate basement. Thus, although normal faulting occurs regionally, Miocene and younger extensional deformation is concentrated in the Long Valley region. Relatively minor normal faulting is observed east of Long Valley, in the main part of the Idaho batholith.

Lithospheric reactivation

The spatial coincidence of the Early Cretaceous suturing, Late Cretaceous shearing, and Miocene extension suggests that the earlier events affected the younger events. We interpret the Late Cretaceous reactivation, and possibly the Miocene reactivation, to occur as a result of the Early Cretaceous juxtaposition of differing lithospheres, which created a fundamental lithospheric flaw. The record of this lithospheric juxtaposition is the isotopic signatures of igneous rocks that rose through lithosphere and crust after the initial suturing (e.g. Fleck & Criss 1985; Manduca 1988; Leeman *et al.* 1992). Differentiating a crustal v. mantle source of the isotopic variations in the granitoid plutons is difficult, as significant crustal contamination occurs as the viscous magmas move upward. Thus, the possibility exists that the plutons, because of their long residence time in the crust, record a crustal juxtaposition rather than a lithospheric one. This difficulty is allayed by the presence of voluminous basalt magmatism

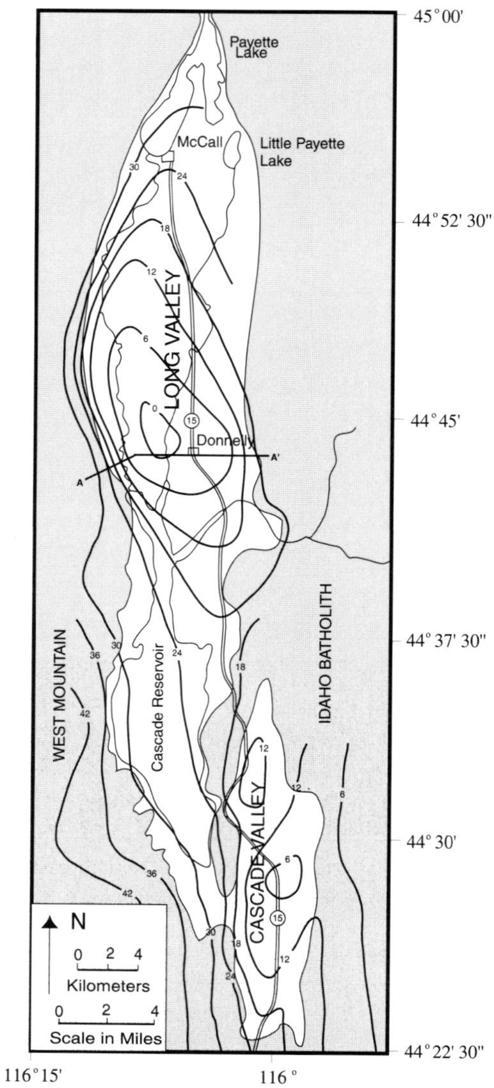

in the Snake River plain, which erupts on both sides of the southward extension of the Salmon River suture zone in southwestern Idaho and southeastern Oregon (surrounding South Mountain; Fig. 2). In this location, Miocene–Quaternary tholeiitic basalts demonstrate the same distinctive isotopic break as observed further north along the suture zone (Hart 1985; Leeman *et al.* 1992). Tholeiitic basalt magmas have low viscosity and move quickly through the crust with a minimum amount of contamination, and are derived from the mantle. Consequently, they are good isotopic indicators of the upper mantle, and indicate that indeed a major lithospheric juxtaposition of differing mantles occurred underneath the Salmon River suture zone. Consequently, there is clear indication that the currently exposed structures occur in relation to a break in the mantle lithosphere. However, whereas the upper-mantle 'flaw' controls the location of the reactivation on a regional scale, crustal basement faults and lithological contacts control the orientation and precise location of faults that accommodate reactivation.

Miocene reactivation

Crustal scale. Hamilton (1963, p. 785) wrote: 'The orientation and competence of the pre-Tertiary rock complexes have to a large extent controlled the character of late Cenozoic structures. The massive interior of the Idaho batholith is not broken by large normal faults; fault blocks trending toward the batholith from the north and from the southeast dwindle and disappear as they reach the batholith. The gneissic western border zone of the batholith is broken by many large normal faults that are parallel in strike and probably also in dip to the foliation of the zone.'

Although the presence of Tertiary normal faults was noticed in the 1960s (e.g. Hamilton 1963), they were not studied in detail. We undertook a study to map these faults, document their movement, and investigate their implications for the geological history of the region. Figure 5 shows the north–south strike of the normal faults parallel to the strike of the foliation in the Late Cretaceous plutons (Manduca *et al.* 1993; our mapping). The dip of the

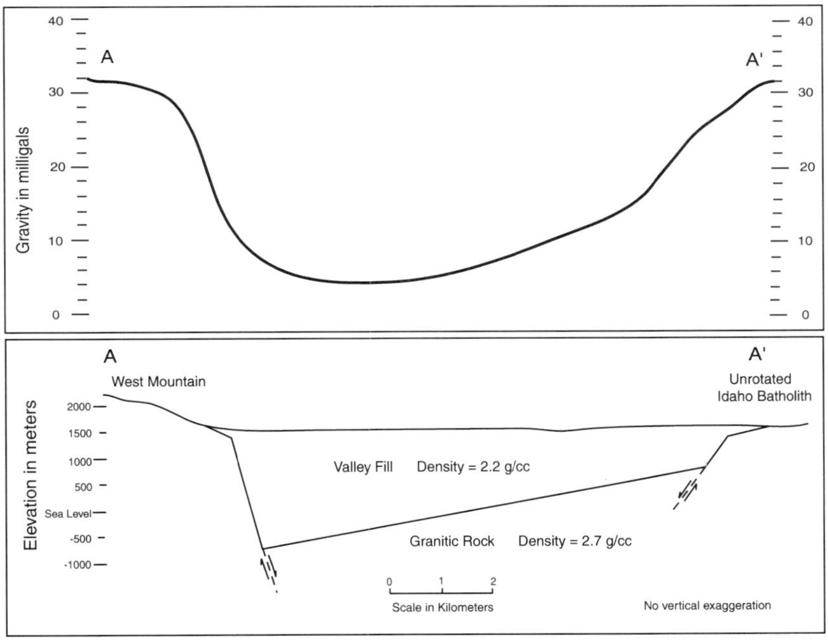

Fig. 10. Gravity data from Long Valley from Kinoshita (1962). (**a**) Bouguer anomaly map of Long Valley, with 6 mgal contours. This survey was not tied to a known reference point, thus there are no absolute measurements on the contours. A closed gravity low of 27 mgal exists in the northern end of Long Valley and the southern end does not exhibit a closed gravity low. (**b**) Gravity profile and cross-sectional interpretation of the gravity data for line A–A′. Gravity profile results from subtraction of a regional gradient. Interpretation assumes a density difference of 0.5 g cm^{-3}. The maximum basin depth in Long Valley is 1500 m. The estimated fault throw is 2100 m on the west side of the valley and 460 m on the east side.

normal faults, however, is difficult to determine. North of McCall, north–south-oriented river drainages (e.g. Goose Creek, Fisher Creek, Payette River, Little French Creek) are subparallel to the strike of both the normal faults and the basement foliation. Exposures of these normal faults tend to be poor. In Fisher Creek saddle, a glaciated cirque at the southern headwaters of Little French Creek exposes an anastomosing gouge zone that separates the Cretaceous Little Goose Creek complex from the Cretaceous Payette River tonalite (Fig. 11). Individual gouge layers are decimetres in thickness and defined by reduced grain size and dark layering. These gouge zones are deeply weathered and anastamose on a metre scale. A global positioning system (GPS) survey of five points along the gouge defined the fault as striking 348° and dipping 66° to the east. The strike and dip of the fault approximately parallel the local foliation (Fig. 11). These data suggest that normal faulting initiated by reactivation of the structural grain of the western Idaho shear zone (Hamilton 1963).

The orientation of the Columbia River basalts also provides additional information about the normal faulting. Assuming that the basalts originally covered a relatively flat erosional surface, reconstruction of the original orientation of the basalts, which at present dip 20–30° westward, restores the pre-Miocene orientation of foliation and lineation in the Cretaceous basement plutons with the western Idaho shear-zone fabric to a subvertical position (Fig. 12). It should be noted that we are assuming horizontal axis rotations only. Different domains of the western Idaho shear zone all reconstruct to this same geometry, although they have different present orientations. For example, lineation in the tectonites north of McCall generally has an ENE bearing, whereas the lineation in the Snowbank Mountain area (Fig. 2) has an east bearing. This difference is due to a difference in normal fault geometry. The basalt flows covering the McCall rocks have a slight southward component of tilt in addition to the dominant western tilt. This southward component of tilt is observed on the regional scale by the increasing amount of basalt covering the granitoids southward toward McCall, despite the decreasing elevation. In contrast, the basalts on Snowbank Mountain have only a westward tilt.

The pre-Miocene reconstruction of the subvertical Cretaceous fabric also requires that the Miocene normal faults initiated in a vertical orientation. This observation is not consistent with Andersonian theory (Anderson 1951), which, based on the assumption of isotropic

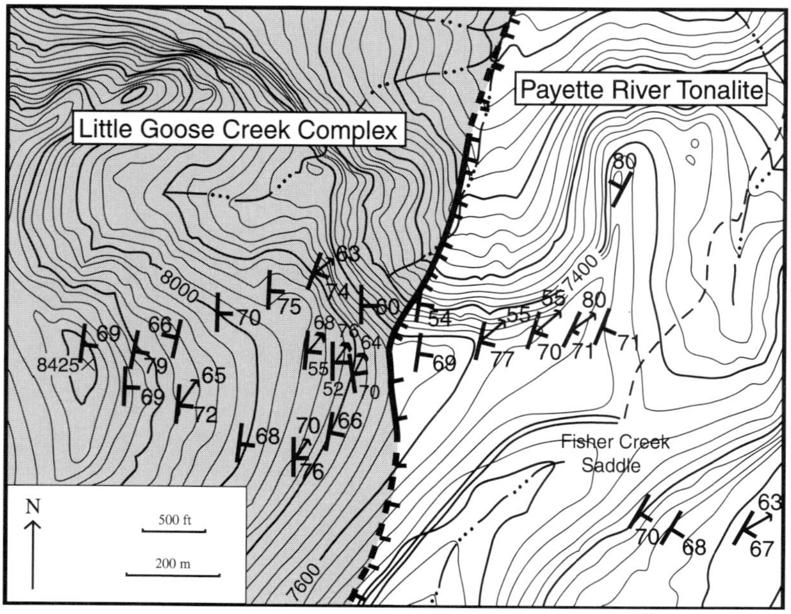

Fig. 11. Geological map of cirque locality on Fisher Creek saddle, separating south-flowing Fisher Creek from north-flowing Little French Creek. (See Fig. 5 for location.) The strike and dip of the fault calculated using GPS measurements are 348° and 66°E, subparallel to the Late Cretaceous foliation. This parallelism suggests that the normal faulting reactivates the foliation.

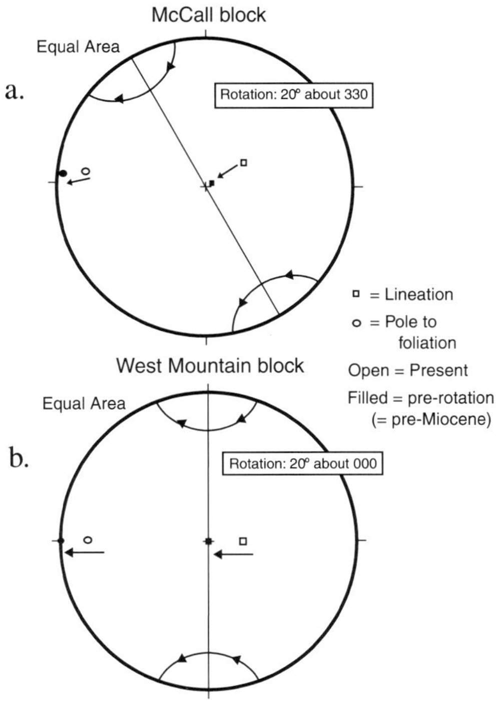

Fig. 12. Equal-area, lower hemispheric projections of fabric in the western Idaho shear zone. □, current orientations; ■, rotation of fabric to a first-order, pre-Miocene orientation by rotating the overlying basalt flows to horizontal. (a) Measurements from north of McCall. (b) Measurements from West Mountain area. Both measurements were taken from the Little Goose Creek complex and equivalent units and the measured basalt flows overlie those units.

material, predicts an initial 60° dip of normal faults. The strong gneissic layering apparently acts as a material anisotropy, which guides the formation of the normal faults (e.g. Donath 1961).

Normal faulting also reactivates lithological contacts. The Cretaceous intrusions in the Salmon River suture zone have a vertical tabular shape, elongated in a north–south direction. Normal faults occur between each of the major intrusive types (Hazard Creek, Little Goose Creek, Payette River). Normal faulting does locally occur within each of the complexes; for example, it occurs within the Payette River tonalite (Fig. 5). In the absence of overlying basalt flows, it is difficult to determine the offset on the normal faults, or the presence of minor faults.

In summary, Miocene normal faulting is parallel to and reactivated the Late Cretaceous foliation. Both the normal faults and the western Idaho shear-zone fabric (foliation, lineation) reconstruct to a vertical orientation at the time of fault initiation. In addition, the domino-style fault blocks in the McCall area contain a distinct set of intrusive rocks (Hazard Creek, Little Goose Creek, Payette River). This coincidence suggests that Miocene normal faults also reactivated north–south-oriented lithological contacts.

Lithospheric scale. Along the western margin of the Idaho batholith, coincident with the trace of the $^{87}Sr/^{86}Sr$ 0.706 isopleth, are a series of Late Cenozoic basins. These include, from south to north, Squaw Valley, Cascade Valley, Long Valley, Upper Payette Lake area, and Big Hazard Lake area. The largest of these basins is Long Valley, which contains Payette Lake at its north end and Cascade Reservoir at its south end. The basin is steeply bounded on the west by West Mountain, with up to 900 m relief. In the east, the valley is gently bounded by the granites of the Payette River tonalite in its northern end, the Idaho batholith in its southern end, and the overlying Columbia River basalts.

A regional gravity study (Kinoshita 1962) reported a closed gravity low of 27 mgal located in the northern end of Long Valley (Fig. 10). The southern half of the valley, centred on Cascade Reservoir, does not have a closed gravity low. Instead, gravity contours run south into the Cascade Valley. A subsurface fault, probably striking NNW and dipping steeply, must connect these basins. Assuming the density difference between the basin sediment and the granitoids is 0.5 g cm^{-3}, the maximum basin depth in Long Valley is 1500 m at the western edge of the valley (Fig. 10). The range-bounding normal fault on the west side of the valley has an estimated throw of 2100 m, and the fault on the east side of the valley has a throw of 460 m (Fig. 10). The east-dipping normal fault on the west side of the valley parallels, to a first approximation, the steep Late Cretaceous fabrics. The largely subsurface fault on the east side of the valley has a shallow (*c.* 30°), westward dip, which does not follow the pre-existing foliation or a lithological contact. These fault offsets are significantly greater than offsets on faults associated with half-grabens further north in the McCall block.

Long Valley occupies a structural position between the extended Basin and Range province and the Idaho batholith (Fig. 13). Although detailed mapping of faults within the Idaho batholith is not complete, geomorphological and structural evidence suggests that if large normal faults exist within the batholith, the fault spacing is much wider and less consistent in direc-

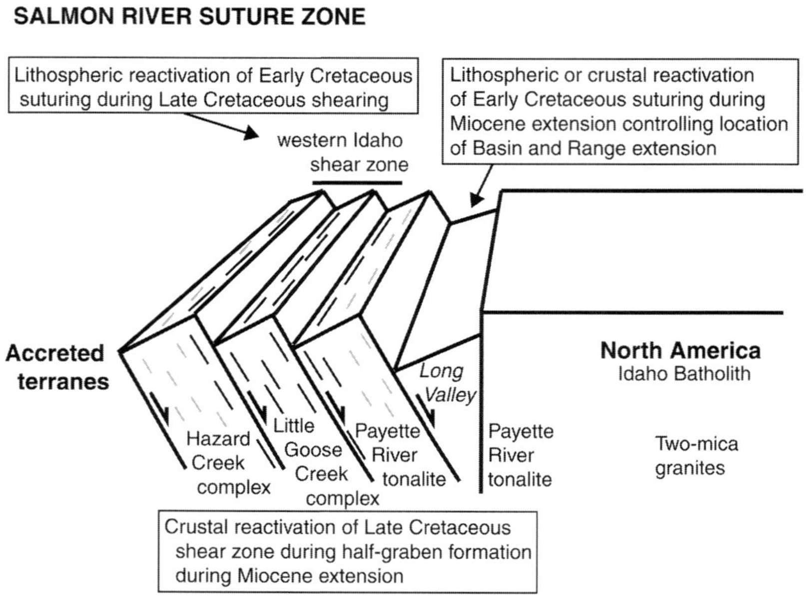

Fig. 13. Schematic illustration of the current geometry of the Salmon River suture zone as the result of reactivation during Late Cretaceous and Miocene time. Bold dashes indicate solid-state foliation; light dashes indicate magmatic foliation.

tion of offset (Hamilton 1963; Lund *et al.* 1997). Consequently, the Salmon River suture zone also corresponds to the transition zone from Basin and Range extension to the undeformed batholith. Long Valley may be a Miocene structural 'hole' that opened as an accommodation space between these two different structural domains (Fig. 13).

The spatial coincidence of Long Valley and the $^{87}Sr/^{86}Sr$ 0.706 isopleth suggests that the difference in structural style is associated with heterogeneities in the mantle lithosphere. The lithospheric mantle of the oceanic terrane was reactivated by the Basin and Range extension, whereas the North American lithosphere was not. Unfortunately, the Idaho batholith has received very little geophysical or geochemical attention. Studies of the Sierra Nevada batholith of California (Ducca & Saleeby 1998) suggest the presence of a lithospheric keel, consisting of eclogite and granulite, which grew during formation of the batholith. We speculate that the presence of a lithospheric keel under the Idaho batholith may explain this difference in response to extensional deformation.

The idea of Miocene lithospheric reactivation is inconsistent with the model proposed by Leeman *et al.* (1992). They also considered the Salmon River suture zone as a vertical structure that cut through the entire lithosphere during Early Cretaceous time. However, during the Tertiary thrusting, the vertical structure in the crust moved eastward *c.* 100 km on a subhorizontal detachment relative to the vertical structure in the mantle (orogenic float model; Oldow *et al.* 1989). The model proposed by Leeman *et al.* (1992) does not alter the Late Cretaceous reactivation (described below), but does suggest that the Miocene reactivation is a result of only crustal heterogeneities, rather than mantle heterogeneities. At present, there are too few data to distinguish between the two models, although the existence of the deep valleys (e.g. Long Valley) along the trend of the $^{87}Sr/^{86}Sr$ 0.706 isopleth favours a mantle explanation.

Late Cretaceous reactivation

Reconstruction of the normal faulting, using the Columbia River basalts, returns the fabrics of the western Idaho shear zone to subvertical orientation before Miocene time. We assume that there is no prior rotation of the fabrics, and that the pre-Miocene orientation is also the Late Cretaceous geometry of the shear zone. Consequently, the present horizontal width (*c.* 7 km) of the western Idaho shear zone is greater than its original width of *c.* 6 km (Figs. 5 and 6). This reconstruction also suggests that the isotopic (Sr and O) anomalies had steeper gradients during Late Cretaceous time. Consequently, the dip of the mantle structure below the western

Idaho shear zone is even steeper (e.g. subvertical) than reported by Manduca *et al.* (1992) using an unreconstructed geometry.

The key to the Late Cretaceous reactivation is the position of the western Idaho shear zone with respect to the isotopic values. Solid-state deformation is concentrated in the Little Goose Creek complex, and affects only the western edge of the Hazard Creek complex and the eastern edge of the Payette River tonalite. The Little Goose Creek complex lies along the steepest gradient in the isotopic values, indicating that these magmas were intruded along the boundary between differing lithospheric mantles at depth. Alternatively stated, the Little Goose Creek complex (and thus the western Idaho shear zone) lies directly above the lithospheric suture zone. The concentration of deformation in this zone indicates that the lithospheric fault or shear zone below the western Idaho shear zone moved during Late Cretaceous time.

Two major models for Late Cretaceous tectonism have been proposed. Contractional models emphasize the eastward dip of the foliation and the down-dip lineations (e.g. Manduca *et al.* 1993). Strike-slip models emphasize the locally steep orientation of the boundary, the opposite vergence of structures away from the shear zone, and the clear break of lithologies across the zone (e.g. Lund & Snee 1988). McClelland *et al.* (2001) provided evidence for dextral structures locally preserved in the zone and argued for dextral transpression. Although the reconstruction using the Columbia River basalt flows does not eliminate either model, the vertical nature of the boundary does favour a strike-slip or transpressional model.

Discussion

Recognition of old suture zones or strike-slip zones

Suture zones that separate distinct lithospheric types are good candidates for reactivation (Holdsworth 1994; Vauchez *et al.* 1997). In particular, vertical structures (e.g. strike-slip and transpression features) that penetrate the mantle lithosphere (e.g. Henstock *et al.* 1997) provide a lithospheric-scale weakness. Such zones are readily reactivated in a wide range of tectonic environments, including contraction, extension, or strike-slip.

This reactivation complicates the interpretation of the geological history of a zone. In western Idaho, there is a debate concerning the location and presence of a suture zone or shear zone (Lund & Snee 1988; Strayer *et al.* 1989). In its present orientation, the Late Cretaceous shearing is coincident with the steep isotopic gradient that is the boundary between the accreted terrane and North America. However, as the age of the rocks within the shear zone postdates the accretion, the western Idaho shear zone cannot record the accretion event. No discrete suture zone, representing the initial Early Cretaceous accretion, exists in western Idaho. Regardless of whether the exact relationship between the accretion and the Late Cretaceous reactivation is gradual (e.g. Blake 1991), punctuated (e.g. Snee *et al.* 1995), or discrete (McClelland *et al.* 2001), the accretion event is not preserved because of later magmatism and tectonism. Thus, major structures may currently occupy the position of a lithospheric boundary, although the initial juxtaposition occurred much earlier.

With respect to crustal reactivation, it is solely the presence of the Columbia River basalts and the geomorphology of the actively deforming region (e.g. prominent north–south-oriented river valleys) that allow us to make inferences about the initial orientation of the fault geometries and highlight the location of the faults. High-grade, highly anisotropic igneous terranes generally lack stratigraphic offset markers, particularly when the faults parallel the dominant fabric. The faults within the Salmon River suture zone are best recognized in the areas of some basalt coverage (e.g. the NW corner of Long Valley; Fig. 8).

The combination of crustal and lithospheric weakness explains the paucity of major strike-slip zones recognized in ancient orogens (Sylvester 1988). The consequence of the crustal reactivation of these well-foliated zones above lithospheric boundaries is a reorientation of fabric. In west–central Idaho, this demonstrably reoriented fabric is extremely consistent on a regional scale. Consequently, if these zones are also typified by lithospheric weakness, it is extremely unlikely that unmodified crustal exposures of major strike-slip and transpressional faults or shear zones will retain their original geometry. The presence of steeply dipping fabric (rather than vertical fabric) may be the key to recognizing ancient strike-slip or transpressional zones (e.g. McClelland *et al.* 2001).

Anisotropy

Quantitative models of crustal deformation generally assume that rocks are isotropic. Workers use this assumption, although it is often unrealistic, because it renders simple the geometric relation between the principal stresses applied to rocks and resulting principal strain rates. However, the effect of anisotropy on the distribution and style of crustal-scale structures such as faults is difficult to quantify because the geometric relation between stress and strain is complex. For example, in experiments on natural rocks with strong planar fabrics, faults develop in orientations not accounted for by Mohr–Coulomb or von Mises failure criteria (Donath 1961). Moreover, deformation may intensify anisotropy through the formation of new fabrics and structures (Elliott 1972), or it may reactivate older structures (Holdsworth *et al.* 1997).

The strong lithological and structural fabric in the Late Cretaceous Western Idaho shear zone clearly controls the Tertiary normal faulting, as the faults are parallel to the older foliation. This geometry also requires that the present-day normal fault initiated in a vertical orientation. Using the analysis of Sibson (1985) and the experiments of Donath (1961), we can evaluate these relations. First, the initiation of the normal faults is *c.* 30° from that predicted by Mohr–Coulomb behaviour. The presence of the overlying Columbia River basalts implies that the rocks deformed by the western Idaho shear zone were at the surface during Miocene time. Sibson (1985) contended that faults or pre-existing weaknesses that are up to 30° out of optimal orientation for faulting may reactivate in extensional settings. The physical experiments of Donath (1961) demonstrated that schistosity is reactivated, rather than the formation of new faults, up to 30° off orientation of faults predicted by Mohr–Coulomb theory. Thus, the inferred pattern of faulting by reactivating foliation is generally consistent with theoretical and experimental studies.

The only exception is the fault on the east side of Long Valley, which is inferred from the gravity data. Although this fault lies entirely within the subsurface, it is not parallel to the inferred magmatic fabric. In some settings, lithospheric control, which requires structural separation between the distinct tectonic provinces of the extending Basin and Range and non-extending Idaho batholith, apparently takes precedence over local crustal anisotropy.

Lithospheric implications

The main result of our study is that the process of mantle reactivation occurs and can be studied using crustal rocks. In western Idaho, the distinctive geochemical signature between the accreted Blue Mountain terranes and North America provides the necessary condition to document this occurrence (e.g. Armstrong *et al.* 1977; Fleck & Criss 1985). The crustal record of the underlying mantle is carried by the Cretaceous granitoid intrusions in the field area (e.g. Manduca *et al.* 1992), and corroborated by Miocene and younger basaltic volcanism farther south along the same lithospheric boundary (Hart 1985). The reactivation of the lithospheric mantle occurred despite voluminous magmatism within the Salmon River suture zone, which occupied an intra-arc setting during Late Cretaceous time. Consequently, it appears that mantle discontinuities are long-lived geological features and are not healed at elevated temperatures found in arc settings.

A mechanism for weakening of the lithospheric mantle was explored by Vauchez *et al.* (1997). They speculated that the mechanical anisotropy of the mantle defined by the crystallographic alignment of olivine is preferentially reactivated, in analogy with the reactivation of the foliation in the western Idaho shear zone by the Tertiary normal faulting. Thus, the lithospheric-scale weakness of the Salmon River suture zone is also the effect of anisotropy. Unfortunately, the geological, geophysical, and geochemical studies of Idaho are not as comprehensive as those in other regions of the US Cordillera. Consequently, more work is required before we fully appreciate the mechanisms and ramifications of lithospheric reactivation in western Idaho.

Conclusions

The Salmon River suture zone is an ideal place to study the role of mantle reactivation, because (1) the crust contains igneous rocks that contain a record of the underlying mantle, and (2) the juxtaposition of the Precambrian North American craton and the Mesozoic accreted terranes provides a distinct break in the isotopic signatures. The steepest isotopic (Sr, O, Nd) gradients preserved in igneous rocks define a zone that was reactivated at least twice. These reactivations include amphibolite-facies shearing during Late Cretaceous time (western Idaho shear zone) and brittle Basin and Range extension during Miocene time.

Reactivation of faults is commonly attributed to an inherent weakness of the fault zones themselves. However, if the mantle lithosphere is strong relative to the crust, reactivation of the mantle is potentially a more significant control on a plate scale. In particular, reactivation of the entire lithosphere, crust and upper mantle is probably common in palaeo-suture zones.

Steeply eastward-dipping fabrics characterize the western edge of the Idaho batholith, for

>200 km along strike. Despite the regional consistency of this fabric, the presence of the overlying and tilted Columbia River basalt flows clearly demonstrates that these fabrics were rotated from an originally vertical orientation. Thus, the regionally consistent nature of fabrics does not imply that they are unmodified. The faults that rotated the fabric follow the pervasive fabric of the Late Cretaceous western Idaho shear zone, and fault spacing is locally controlled by the thickness of the Late Cretaceous igneous units, indicating an additional crustal control on reactivation. Experimental and theoretical studies suggest that pre-existing weaknesses up to 30° out of optimal orientations for faulting may be reactivated. Consequently, crustal reactivation of vertical fabrics by later extensional deformation is likely. The combination of mantle and crustal reactivation, such as seen in the Salmon River suture zone, suggests that major strike-slip zones are unlikely to be preserved in their original vertical geometry. Rather, ancient strike-slip and transpressional zones may predominantly contain steeply dipping fabrics as a result of later reactivation and reorientation.

We wish to thank W. McClelland and D. Blake for numerous helpful conversations about Idaho geology. C. Johnson is thanked for relevant discussions about geochemistry. B. Gustavson, E. Horsman, J. Minott, and A. Kolker assisted with portions of the fieldwork. J. Magloughlin, J. B. Murphy, and E. Tavarnelli are acknowledged for extremely helpful reviews of the manuscript. This work was partially funded by NSF EAR 0001092.

References

ARMSTRONG, R.L., TAUBENECK, W.H. & HALES, P.O. 1977. Rb–Sr and K–Ar geochronometry of Mesozoic granitic rocks and their Sr isotopic composition: Oregon, Washington, and Idaho. *Geological Society of America Bulletin*, **88**, 397–411.

ANDERSON, E.M. 1951. *The Dynamics of Faulting*. Oliver and Boyd, Edinburgh.

AVÉ LALLEMANT, H.G. 1995. Pre-Cretaceous tectonic evolution of the Blue Mountains province, northeastern Oregon. *In*: VALLIER, T.L. & BROOKS, H.C. (eds) *Geology of the Blue Mountains Region of Oregon, Idaho, and Washington: Petrology and Tectonic Evolution of Pre-Tertiary Rocks of the Blue Mountains Region*. US Geological Survey, Professional Papers **1438**, 271–304.

BICKFORD, M.E., CHASE, R.B., NELSON, B.K., SHUSTER, R.D. & ARRUDA, E.C. 1981. U–Pb studies of zircon cores and overgrowths, and monazite: implications for age and petrogenesis of the northeastern Idaho batholith. *Journal of Geology*, **89**, 433–457.

Blake, D. E. 1991. Geology of the western Idaho suture zone in the Salmon River gorge, west–central Idaho. Ph.D. thesis, Washington State University, Pullman..

BONNICHSEN, B. 1987. Pre-Cenozoic geology of the West Mountain–Council Mountain–New Meadows area, west–central Idaho. *In*: VALLIER, T.L. & BROOKS, H.C. (eds) *Geology of the Blue Mountains Region of Oregon, Idaho, and Washington: Petrology and Tectonic Evolution of Pre-Tertiary Rocks of the Blue Mountains Region*. US Geological Survey, Professional Papers **1436**, 151–170.

BRACE, W.F. & KOHLSTEDT, D.L. 1980. Limits on lithospheric stress imposed by laboratory experiments. *Journal of Geophysical Research*, **85**, 6248–6252.

BURCHFIEL, B.C., COWAN, D.S. & DAVIS, G.A. 1992. Tectonic overview of the Cordilleran orogen in the western United States. *In*: BURCHFIEL, B.C., LIPMAN, P.W. & ZOBACK, M.L. (eds) *The Cordilleran Orogen: Conterminous US*. Geological Society of America, Boulder, CO, 407–479.

CHRISTIANSEN, R.L. & YEATS, R.S. 1992. Post-Laramide geology of the U.S. Cordillera region. *In*: BURCHFIEL, B.C., LIPMAN, P.W. & ZOBACK, M.L. (eds) *The Cordilleran Orogen: Conterminous US*. Geological Society of America, Boulder, CO, 261–406.

CRISS, R.E. & FLECK, R.J. 1987. Petrogenesis, geochronology, and hydrothermal systems of the northern Idaho batholith and adjacent areas based on $^{18}O/^{16}O$, D/H, $^{87}Sr/^{86}Sr$, K–Ar and $^{40}Ar/^{39}Ar$ studies. *In*: VALLIER, T.L. & BROOKS, H.C. (eds) *Geology of the Blue Mountains Region of Oregon, Idaho, and Washington: Petrology and Tectonic Evolution of Pre-Tertiary Rocks of the Blue Mountains Region*. US Geological Survey, Professional Papers **1436**, 95–138.

DAVIS, G.A., MONGER, J.W.H. & BURCHFIEL, B.C. 1978. Mesozoic construction of the Cordilleran 'collage', central British Columbia to central California. *In*: HOWELL, D.G. & MCDOUGALL, K.A. (eds) *Mesozoic Paleogeography of the Western United States*. Society of Economic Paleontology and Mineralogy, Pacific Section, Pacific Coast Paleogeography Symposium **2**, 1–32.

DEWEY, J.F., HEMPTON, M.R., KIDD, W.S.F., SAROGLU, F. & SENGÖR, A.M.C. 1986. Shortening of continental lithosphere: the neotectonics of Eastern Anatolia—a young collision zone. *In*: COWARD, M.P. & RIES, A.C. (eds) *Collision Tectonics*. Geological Society, London, Special Publications **19**, 3–36.

DONATH, F.A. 1961. Experimental study of shear failure in anisotropic rocks. *Geological Society of America Bulletin*, **72**, 985–989.

DUCCA, M.N. & SALEEBY, J.B. 1998. The age and origin of a thick ultramafic keel from beneath the Sierra Nevada Batholith. *Contributions to Mineralogy and Petrology*, **133**, 169–185.

ELLIOTT, D. 1972. Deformation paths in structural geology. *Geological Society of America Bulletin*, **83**, 2621–2638.

FITZGERALD, J.F. 1982. Geology and basalt stratigraphy of the Weiser Embayment, west–central Idaho. *In*: BONNICHSEN, B. & BRECKENRIDGE, R.M. (eds) *Cenozoic Geology of Idaho*. Idaho Bureau of Mines and Geology Bulletin **26**, 103–128.

FLECK, R.J. & CRISS, R.E. 1985. Strontium and oxygen isotopic variations in Mesozoic and Tertiary plutons of central Idaho. *Contributions to Mineralogy and Petrology*, **90**, 291–308.

FOSTER, D.A. & FANNING, C.M. 1997. Geochronology of the northern Idaho batholith and the Bitterroot metamorphic core complex: magmatism preceding and contemporaneous with extension. *Geological Society of America Bulletin*, **109**, 379–394.

GODCHAUX, M.M. & BONNICHSEN, B. 1993. Ultramafic rocks of the western Idaho suture zone; Asbestos Peak and Misery Ridge. *Geological Society of America, Abstracts with Programs*, **25**, 5, 43.

HAMILTON, W.R. 1963. Metamorphism in the Riggins Region, Western Idaho. *US Geological Survey, Professional Papers*, **436**, 1–94.

HAMILTON, W.R. 1969. *Reconnaissance Geologic Map of the Riggins Quadrangle, West–Central Idaho, 1:125 000*. US Geological Survey Miscellaneous Geological Investigation Map **I-579**.

HAMILTON, W.R. 1978. Mesozoic tectonics of the western United States. *In*: HOWELL, D.G. & MCDOUGALL, K.A. (eds) *Mesozoic Paleogeography of the Western United States*. Society of Economic Paleontology and Mineralogy, Pacific Section, Pacific Coast Paleogeography Symposium **2**, 33–70.

HART, W.K. 1985. Chemical and isotopic evidence for mixing between depleted and enriched mantle, northwestern U.S.A. *Geochimica et Cosmochimica Acta*, **49**, 131–144.

HENSTOCK, T.J., LEVANDER, A. & HOLE, J.A. 1997. Deformation in the lower crust of the San Andreas fault system in Northern California. *Science*, **278**, 855–858.

HOLDSWORTH, R.E. 1994. Structural evolution of the Gander–Avalon terrane boundary: a reactivated transpression zone in the NE Newfoundland Appalachians. *Journal of the Geological Society, London*, **151**, 629–646.

HOLDSWORTH, R.E., BUTLER, C.A. & ROBERTS, A.M. 1997. The recognition of reactivation during continental deformation. *Journal of the Geological Society, London*, **154**, 73–78.

HYNDMAN, D.W., ALT, D. & SEARS, J.W. 1988. Post-Archean metamorphic and tectonic evolution of western Montana and northern Idaho. *In*: ERNST, W.G. (ed.) *Metamorphism and Crustal Evolution of the Western United States, Rubey Volume VII*. Prentice–Hall, Englewood Cliffs, NJ, 332–361.

KINOSHITA, W.T. 1962. *A gravity study of part of the Long Valley district, Idaho*. US Geological Survey Survey Open-File Report, **62-73**.

LEEMAN, W.P., OLDOW, J.S. & HART, W.K. 1992. Lithosphere-scale thrusting in the western US Cordillera as constrained by Sr and Nd isotropic transitions in Neogene volcanic rocks. *Geology*, **20**, 63–66.

LUND, K. 1984. Tectonic history of a continent–island arc boundary: west–central Idaho. Ph.D. thesis, Pennsylvania State University, State College.

LUND, K. 1995. Metamorphic and structural development of island-arc rocks in the Slate Creek–John Day Creek area, west–central Idaho. *In*: VALLIER, T.L. & BROOKS, H.C. (eds) *Geology of the Blue Mountains Region of Oregon, Idaho, and Washington: Petrology and Tectonic Evolution of Pre-Tertiary Rocks of the Blue Mountains Region*. US Geological Survey, Professional Papers **1438**, 517–540.

LUND, K. & SNEE, L.W. 1988. Metamorphism, structural development, and age of the continent–island arc juncture in west–central Idaho. *In*: ERNST, W.G. (ed.) *Metamorphism and Crustal Evolution of the Western United States, Rubey Volume VII*. Prentice–Hall, Englewood Cliffs, NJ, 296–331.

LUND, K., KUNTZ, M.A., MANDUCA, C.A., GAMMONS, C.H., EVANS, K.V., TYSDAL, R.G., WINKLER, G.T. & CONNOR, J.J. 1997. *Geologic Map of the Western Salmon River Mountains, Valley and Idaho Counties, West–Central Idaho*. US Geological Survey Map, **I-2599**.

MANDUCA, C. A. 1988. Geology and geochemistry of the oceanic arc–continent boundary in the western Idaho batholith near McCall. Ph.D. thesis, California Institute of Technology, Pasadena.

MANDUCA, C.A., KUNTZ, M.A. & SILVER, L.T. 1993. Emplacement and deformation history of the western margin of the Idaho batholith near McCall, Idaho: Influence of a major terrane boundary. *Geological Society of America Bulletin*, **105**, 749–765.

MANDUCA, C.A., SILVER, L.T. & TAYLOR, H.P. 1992. $^{87}Sr/^{86}Sr$ and $^{18}O/^{16}O$ isotopic systematics and geochemistry of granitoid plutons across a steeply-dipping boundary between contrasting lithospheric blocks in western Idaho. *Contributions to Mineralogy and Petrology*, **109**, 355–372.

MCCLELLAND, W.C., MANDUCA, C.A. & TIKOFF, B. 2000. Two-phase evolution of accretionary margins: examples from the North American Cordillera. *Tectonophysics*, **326**, 37–55.

MCINTYRE, D.H., EKREN, E.B. & HARDYMAN, R.F. 1982. Stratigraphic and structural framework of the Challis Volcanics in the eastern half of the Challis 1° × 2° Quadrangle. *In*: BONNICHSEN, B. & BRECKENRIDGE, R.M. (eds) *Cenozoic Geology of Idaho*. Idaho Bureau of Mines and Geology Bulletin **26**, 3–22.

MOLNAR, P. 1992. Brace–Goetze strength-profiles, the partitioning of strike-slip and thrust faulting at zones of oblique convergence, and the stress–heat flow paradox of the San Andreas fault. *In*: EVANS, B. & WONG, T.-F. (eds) *Fault Mechanics and Transport Properties of Rocks*. Academic Press, London, 435–459.

MYERS, P.E. 1982. Geology of the Harpster area. *Idaho County, Idaho. Idaho Bureau of Mines and Geology Bulletin*, 25.

NILSEN, T.H. 1978. Late Cretaceous geology of California and the problem of the proto-San Andreas Fault. *In*: HOWELL, D.G. & MCDOUGALL, K. (eds) *Mesozoic Paleogeography of the Western United States*. Pacific Coast Paleogeography Symposium **2**, 559–573.

NORMAN, M.D. & LEEMAN, W.P. 1989. Geochemical evolution of Cenozoic–Cretaceous magmatism and its relation to tectonic setting, southwestern Idaho, USA. *Earth and Planetary Science Letters*, **94**, 78–96.

OLDOW, J.S., BALLY, A.W., AVÉ LALLEMANT, H.G. & LEEMAN, W.P. 1989. Phanerozoic evolution of the North American Cordillera: United States and Canada. *In*: BALLY, A.W. & PALME, A. (eds) *The Geology of North America—an Overview*. Geological Society of America, Boulder, CO, 139–232.

ONASCH, C.M. 1987. Temporal and spatial relations between folding, intrusion, metamorphism and thrust faulting in the Riggins area, west–central Idaho. *In*: VALLIER, T.L. & BROOKS, H.C. (eds) *Geology of the Blue Mountains Region of Oregon, Idaho, and Washington: Petrology and Tectonic Evolution of Pre-Tertiary Rocks of the Blue Mountains Region*. US Geological Survey, Professional Papers **1436**, 139–150.

PICCOLI, P.M. & HYNDMAN, D.W. 1985. Magnetite/ilmenite boundary in the western Atlanta lobe of the Idaho batholith. *Northwest Geology*, **14**, 1–5.

SELVERSTONE, J., WERNICKE, B. & ALIBERTI, E. 1992. Intracontinental subduction and hinged uplift along the Salmon River suture zone in west central Idaho. *Tectonics*, **11**, 124–144.

SHUSTER, R.D. & BICKFORD, M.E. 1985. Chemical and isotopic evidence for the petrogenesis of the northeastern Idaho Batholith. *Journal of Geology*, **93**, 727–742.

SIBSON, R.H. 1985. A note on fault reactivation. *Journal of Structural Geology*, **7**, 751–754.

SNEE, L.W., LUND, K., SUTTER, J.F., BALCER, D.E. & EVANS, K.V. 1995. An $^{40}Ar/^{39}Ar$ chronicle of the tectonic development of the Salmon River suture zone, western Idaho. *In*: VALLIER, T.L. & BROOKS, H.C. (eds) *Geology of the Blue Mountains Region of Oregon, Idaho, and Washington: Petrology and Tectonic Evolution of Pre-Tertiary Rocks of the Blue Mountains Region*. US Geological Survey, Professional Papers **1438**, 359–414.

STRAYER, L.M., HYNDMAN, D.W., SEARS, J.W. & MYERS, P.E. 1989. Direction and shear sense during suturing of the Seven Devils–Wallowa terrane against North America in western Idaho. *Geology*, **17**, 1025–1029.

SUPPE, J. 1970. Offset of late Mesozioc terranes by the San Andreas fault system. *Geological Society of America Bulletin*, **81**, 165–188.

SUTHERLAND, R. & MELHUISH, A. 2000. Formation and evolution of the Solander Basin, southwestern South Island, New Zealand, controlled by a major fault in the continental crust and upper mantle. *Tectonics*, **19**, 44–61.

SYLVESTER, A.G. 1988. Strike-slip faults. *Geological Society of America Bulletin*, **100**, 1666–1703.

TAUBENECK, W.H. 1971. Idaho batholith and its southern extension. *Geological Society of America Bulletin*, **82**, 1899–1928.

TIKOFF, B. & MANDUCA, C. 1997. Late Cretaceous dextral transpressional shearing in the Idaho suture zone. *Geological Society of America, Abstracts with Programs, Cordilleran Section*, **29**, 5, 70.

TEYSSIER, C. & TIKOFF, B. 1998. Strike-slip partitioned transpression of the San Andreas fault system: a lithospheric scale approach. *In*: HOLDSWORTH, R.E., STRACHAN, R.A. & DEWEY, J.F. (eds) *Continental Transpression and Transtension Tectonics*. Geological Society, London, Special Publications **135**, 143–158.

TOTH, M.I. & STACEY, J.S. 1992. Constraints on the formation of the Bitterroot lobe of the Idaho batholith, Idaho and Montana, from U–Pb zircon geochronology and feldspar Pb isotopic data. *US Geological Survey Bulletin*, 2008.

VALLIER, T.L. 1995. Petrology of pre-Tertiary igneous rocks in the Blue Mountains region of Oregon, Idaho, and Washington: implications for the geologic evolution of a complex island arc. *In*: VALLIER, T.L. & BROOKS, H.C. (eds) *Geology of the Blue Mountains Region of Oregon, Idaho, and Washington: Petrology and Tectonic Evolution of Pre-Tertiary Rocks of the Blue Mountains Region*. US Geological Survey, Professional Papers **1438**, 125–209.

VAUCHEZ, A., BARRUOL, G. & TOMMASI, A. 1997. Why do continents break-up parallel to ancient orogenic belts? *Terra Nova*, **9**, 62–66.

WATKINS, N.D. & BAKSI, A.K. 1974. Magnetostratigraphy and oroclinal folding of the Columbia River, Steens, and Owyhee basalts in Oregon, Washington, and Idaho. *American Journal of Science*, **274**, 148–189.

WERNICKE, B. & KLEPACKI, D.W. 1988. Escape hypothesis for the Stikine block. *Geology*, **16**, 461–464.

ZEN, E. 1988. Tectonic significance of high-pressure plutonic rocks in the Western Cordillera of North America. *In*: ERNST, W.G. (ed.) *Metamorphism and Crustal Evolution of the Western United States, Rubey Volume VII*. Prentice–Hall, Englewood Cliffs, NJ, 41–67.

Sequential ductile to brittle reactivation of major fault zones along the accretionary margin of Gondwana in Central Argentina

CAROL SIMPSON[1], STEVEN J. WHITMEYER[1], DECLAN G. DE PAOR[1], L. PETER GROMET[2], ROBERTO MIRO[3], MICHAEL A. KROL[1] & HEATHER SHORT[1]

[1]*Department of Earth Sciences, Boston University, Boston, MA 02215, USA (e-mail: csimpson@bu.edu)*

[2]*Department of Geological Sciences, Brown University, Providence, RI 02912, USA*

[3]*Geological Survey of Argentina, Córdoba Branch, Córdoba, Argentina*

Abstract: Metamorphic and plutonic basement rocks and cover sequences of the Eastern Sierras Pampeanas, Argentina, have undergone multiple episodes of fault reactivation. Faults take advantage of mid- to late Cambrian, NW–SE-striking, steeply east-dipping foliations in Vendian-aged accretionary prism metasedimentary rocks. Foliations in peraluminous schists, paragneisses and migmatites are deflected into late Cambrian amphibolite-grade high-strain zones. Greenschist-grade mylonite zones and thick retrogressed ultramylonite zones with mainly NNW strikes, easterly dips, and east-over-west movement, affect the metasedimentary rocks and Ordovician-aged intrusive rocks and are presumably related to early Devonian accretion of terranes to the west of Gondwana. pseudotachylyte veins occur in nearly all mylonite zones. Brittle deformation during Carboniferous to Triassic time produced major pull-apart basins located above terrane boundaries. Outcrop patterns of Triassic to Cretaceous sedimentary rocks are consistent with transtensional pull-apart basins followed by Andean transpressional deformation. The theoretical basis for fault reactivation and production of 'short cuts' is examined in the context of Tertiary to Recent basin inversion faults. The inversion faults follow the Palaeozoic trends and produce the present-day NNW-oriented, deep sedimentary basins and intervening ranges of basement rocks.

Multiple reactivation of zones of crustal weakness is a common feature of continental regions that have been influenced by several tectonic events. Plate margin orogens contain many occurrences of reactivated faults. For example, the Norumbega fault zone in the northern Appalachians is an Acadian dextral transcurrent fault zone that was reactivated during the Alleghanian orogeny and again during Mesozoic extension (Ludman *et al.* 1999; West 1999). Seismic profiles across the zone reveal deep subvertical reflectors that offset the Moho, suggesting that the fault is linked to an original subsurface suture between Avalon and ancestral North America (Doll *et al.* 1996). The Brevard fault zone in the southern Appalachians also has a history of several episodes of deformation, from mid–late Palaeozoic thrust and dextral strike-slip movement to late Alleghanian brittle thrusting (Bobyarchick 1998). Although the style of deformation often changes from one faulting event to the next, the location of each new faulting event appears to be constrained by the location and orientation of the original deep-rooted zones of crustal weakness.

Fault reactivation zones also occur well inboard of plate margins, especially in regions of active shallow subduction such as the Rocky Mountain region of Colorado (Thomas & Sims 1996) and northern New Mexico (Naeser *et al.* 1989; Karlstrom & Daniel 1993), and the Sierras Pampeanas region of central Argentina (this paper). The fault zones in the Sierras Pampeanas (Fig. 1) occur in basement blocks well to the east of the Andes. NW–SE-oriented ductile and brittle fault zones are prevalent throughout this region and demonstrate evidence of Cambrian to Devonian reverse faulting, Carboniferous to Cretaceous normal and (?)strike-slip faulting, and Tertiary high-angle reverse fault movement. It has been suggested that the present-day NW strike of these reactivated fault

From: HOLDSWORTH, R.E., STRACHAN, R.A., MAGLOUGHLIN, J.F. & KNIPE, R.J. (eds) 2001. *The Nature and Tectonic Significance of Fault Zone Weakening*. Geological Society, London, Special Publications, **186**, 233–255. 0305-8719/01/$15.00

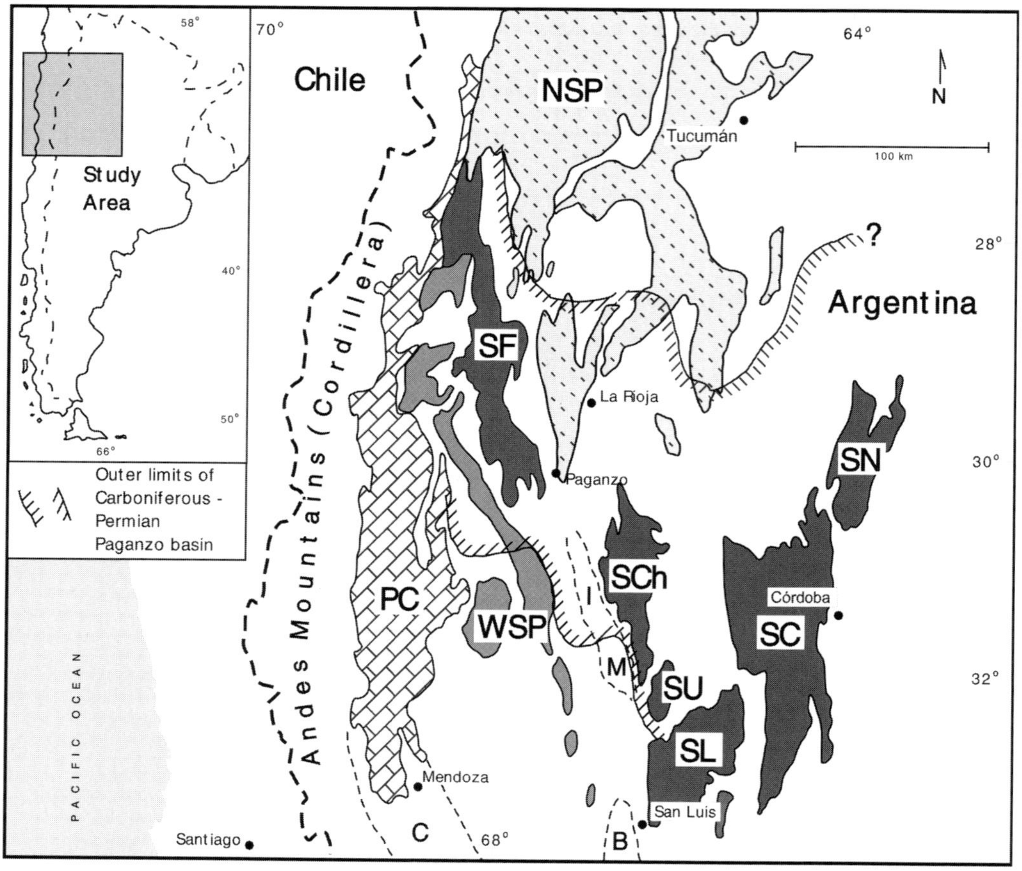

Fig. 1. Location map of the Sierras Pampeanas and Precordillera tectonic provinces of Argentina. Modified from Achilli *et al.* (1997). Boundaries of Paganzo Basin modified from Fernandez-Seveso & Tankard (1995). Location of Beazley (B), Cuyo (C), Ischigualasto (I) and Marayes (M) basins from Ramos & Kay (1991). PC, Precordillera; SC, Sierras de Córdoba; SCh, Sierras de Chepes; SF, Sierra de Famatina; SL, Sierra de San Luis; SN, Sierra del Norte; SU, Sierras de Ulapes and Las Minas; NSP, Northern Sierras Pampeanas; WSP, Western Sierras Pampeanas.

zones is related to Palaeozoic suturing of one or more terranes to this region of the western margin of Gondwana (Ramos & Kay 1991), a suggestion strengthened by our field studies in the Eastern Sierras Pampeanas. We observe that the location and orientation of many of the later Palaeozoic ductile and semi-brittle faults, and subsequent Mesozoic and Tertiary brittle fault zones, are controlled by fabrics that first formed in early Palaeozoic ductile fault zones. We concentrate here on the several reactivation events that are found within the basement fault blocks and that occurred during the Palaeozoic evolution of this segment of the Gondwana margin. We also discuss the Tertiary fault reactivation and inversion of Carboniferous to Cretaceous rift basins that are located on or adjacent to the Palaeozoic terrane boundaries in the region.

Tectonic setting

The Sierras Pampeanas (Pampean Ranges) in central Argentina have been uplifted on east-dipping, high-angle reverse faults since Tertiary time as a consequence of stress transfer by the relatively flat portion of the Nazca Plate as it subducts under this part of South America (Jordan & Allmendinger 1986). The Eastern Sierras Pampeanas (Fig. 1), comprising the Sierras de Córdoba in the east, the Sierra de San Luis to the SW, and the Sierras de Chepes and Las Minas in the NW (La Rioja province), are a system of north–south basement uplifts, up to

3000 m in elevation, separated by extremely deep and asymmetrical, sediment-filled valleys.

In the Eastern Sierras Pampeanas, multiple magmatic, deformation, and metamorphism events are preserved in low- to high-grade metasedimentary and plutonic rocks. The eastern and northern parts of the Sierras de Córdoba contain well-exposed Cambrian I-type calc-alkaline plutons that are thought to represent part of a Pampean continental arc (Lira *et al.* 1996; Pérez *et al.* 1996; Rapela *et al.* 1998*a*, *b*, *c*) that developed along the western edge of the Rio de La Plata craton on the margin of Gondwana by early to mid-Cambrian time.

The plutonic belt of the Sierras de Córdoba is flanked to the SW by complexly deformed metapelitic and psammitic schists, para- and ortho-gneisses, and migmatites, with discontinuous lenses of closely associated calc-silicate rocks and amphibolites (Caminos 1988; Lucero Michaut *et al.* 1995). To the north of the Sierras de Córdoba there is a section of 2000 m thickness of turbiditic, volcanic, and calcareous rocks of the Puncoviscana Formation, which contains ichnofauna of Vendian to Lower Cambrian (Tommotian) age (Durand 1996). These rocks are the presumed time and depositional equivalents to the higher-grade schists and gneisses of the Sierras de Córdoba (Willner & Miller 1986; Caminos 1988) and are thought to represent part of an early Cambrian accretionary prism that formed along this segment of the Gondwana margin (Lyons *et al.* 1997; Northrup *et al.* 1998*a*, *b*). Locally significant volumes of peraluminous S-type granites, derived by partial melting of the metasedimentary rocks at *c.* 520 Ma (Rapela *et al.* 1998c), are present in the west and south of the Sierras de Córdoba. Foliations and ductile deformation zones of the Sierras Pampeanas are thought to have formed during the suturing of a Pampean arc system (Northrup *et al.* 1998*a*) or a Pampean microcontinent (e.g. Kraemer *et al.* 1995; Durand 1996; Rapela *et al.* 1998*a*, *b*, *c*) to the Rio de La Plata craton during the Cambrian Pampean orogeny. The Pampean plutonic and metasedimentary belts are strikingly similar in composition, structural style, and age, to the Ross and Delamerian orogenic belts in Antarctica and Australia, which together may have formed an extensive palaeo-Pacific margin to Gondwana (Northrup *et al.* 1998a).

Evidence for juxtaposition of Laurentian crust against western South America during or after Early Ordovician time comes from the presence of Laurentian basement and cover rocks in the Precordillera and Pie de Palo ranges of the Western Sierras Pampeanas (Fig. 1). The Precordillera terrane contains Grenville-aged basement (Abbruzzi *et al.* 1993; McDonough *et al.* 1993; Ramos *et al.* 1998; Vujovich & Kay 1998) and lower Ordovician fossils of Laurentian affinity (Astini *et al.* 1995; Astini 1998), and is considered by many workers to be the 'missing' part of the southern Appalachians (e.g. Dalla Salda *et al.* 1992a; 1992b; Dalziel *et al.* 1994; Thomas & Astini 1996; Dickerson & Keller 1998).

The presence of an Early Ordovician arc to the west of the Sierras de Córdoba, between Gondwana and the accreted Precordillera (Fig. 1), was suggested by Coira *et al.* (1982), Astini *et al.* (1995), Huff *et al.* (1998) on the basis of Celtic faunal affinity in shallow marine sedimentary rocks, and tuffs and volcanic rocks of Arenig age that now crop out east of the Precordillera. Astini *et al.* (1995) also noted a regionally diachronous unconformity that has a maximum gap in rocks of the central Precordillera in the Arenig–Ludlow interval, which coincides well with the peak of magmatic activity in the Sierra de Chepes (Fig. 1) and surrounding regions of the Eastern Sierras Pampeanas (Pankhurst *et al.* 1996). The development of this arc may represent the early stages of the Ordovician-aged Famatinian orogen of Dalla Salda *et al.* (1992*a*); and Pankhurst *et al.* (1996), during which the Precordillera terrane was accreted.

Ramos (1984), Ramos *et al.* (1986) and Ramos *et al.* (1998) suggested that the Chilenia Terrane docked to the NW of the Precordillera during a late Devonian to Early Carboniferous orogenic event, named the 'Achalian' by Sims *et al.* (1997, 1998). There is little direct evidence for a major late Devonian orogenic or suturing event in the region of the Sierras Pampeanas; however, rocks of the Eastern Sierras Pampeanas were intruded by a suite of mainly post-tectonic, late Devonian granites that Sims *et al.* (1997, 1998) considered related to this event.

From late Devonian to early Carboniferous time, thousands of metres of turbidites were deposited in the region of what is now the Cordillera Principal, or main Andean range (Caminos 1979), and the locus of the volcanic arc shifted to the west (Ramos & Kay 1991). Subduction-related magmatic activity ceased in early Permian time. Brittle deformation during a regional Carboniferous to Triassic transtensional event initiated major pull-apart basins throughout central and western Argentina. In the region of the Sierras Pampeanas, the Carboniferous–Permian Paganzo basin developed in conjunction with a major peneplain. The basin initiated as a series of isolated pull-apart basins along a dextral strike-slip system that utilized the earlier, NNW-trending basement structures; these pull-apart basins eventually amalgamated to form the main Paganzo basin (Fernandez-Seveso & Tankard 1995).

Transition to the modern Andean arc-related volcanism and deformation began to the west of the accreted terranes in mid-Jurassic time (e.g. Coira *et al.* 1982; Mpodozis & Ramos 1990), and continues to the present (Ramos *et al.* 1986; Ramos & Kay 1991). In several areas, latest Jurassic to Early Cretaceous alkaline volcanism and rift basins were superimposed on the earlier Permo-Triassic rift structures, and their bounding faults followed the earlier fabrics in the basement rocks (Schmidt *et al.* 1995). Tertiary to Recent high-angle reverse faulting resulted in the present-day north- to NNW-oriented deep sedimentary basins and intervening ranges that provide access to the basement rocks in the Eastern Sierras Pampeanas.

High-strain events

Early Palaeozoic fabric formation

The earliest observable deformation in the Eastern Sierras Pampeanas rocks is a pressure solution cleavage preserved in psammite lenses within paragneisses in the northern Sierras de Córdoba (Simpson & Northrup 1998; Simpson *et al.* 1998*b*) and in phyllites of the westernmost Sierras de Córdoba and central Sierra de San Luis (Figs 2 and 3a). The pressure solution cleavage is axial planar to tight, upright chevron folds and strikes NW to NNW with a moderate to steep easterly dip (Fig. 4a). Local crenulation and minor folding of the pressure solution cleavage and schistosity occur throughout the area, but the predominant fabric dips to the east and exhibits little or no mineral elongation lineation. Buckle-folding of bedding-parallel quartz veins indicates a minimum of 48% shortening perpendicular to cleavage with only minor extension in the cleavage plane (Simpson & Northrup 1998). Large-scale, upright chevron folds with north to NE axial surfaces are also seen in the unmetamorphosed sedimentary rocks of the Puncoviscana Formation to the west of Salta (Omarini 1991), well north of the Sierras de Córdoba, and indicate that this style of deformation and shortening of the accretionary prism sequence may have been common all along the Gondwana margin.

The pressure solution cleavage formed a pervasive structural fabric that was subsequently metamorphosed, at high temperatures and relatively low pressures, to form biotite–sillimanite ± cordierite schists and gneisses (Fig. 3b). The ubiquitous NNW-striking, mostly steeply east-dipping foliation in the schists and gneisses (Fig. 4a) appears to be inherited from the earlier diffusional mass transfer process. Peak metamorphic mineral assemblages in the highest-grade metapelitic rocks contain garnet + biotite ± cordierite ± sillimanite ± K-feldspar (±hypersthene locally) (Gordillo 1984; our own observations), indicating upper-amphibolite to lower-granulite facies conditions. Quantitative thermobarometry gives estimates of peak metamorphic temperatures of 650–850 °C (e.g. Cerredo 1996; Otamendi & Rabbia 1996; Rapela *et al.* 1998*b*), consistent with the observed mineral assemblages. Paragneisses, with the exception of the psammites, were variably migmatized during this event and the entire metasedimentary section was intruded by pulses of peraluminous, cordierite-bearing S-type granite magmas, some of which reach batholithic proportions (Fig. 2; Caminos 1988; see also Lyons *et al.* 1997).

In contrast to the strong penetrative foliation developed before migmatization, there was little new development of penetrative structure associated with the high-temperature metamorphic event. The S-type granites are largely undeformed, although they locally contain weak to moderately developed foliations that are subparallel to foliations in the surrounding gneisses. Early-formed chevron fold hinges are preserved in amphibolite-grade gneisses, in garnet–biotite–sillimanite schists, and in metasedimentary inclusions within the undeformed S-type granite massifs throughout the Sierras de Córdoba. Chevron fold hinges are mimetically overgrown by prograde metamorphic mineral assemblages, indicating that the intense shortening of the metasedimentary section ceased before peak metamorphic conditions were reached.

Rapela *et al.* (1998c) obtained a 520–525 Ma age for the migmatization event in the Sierras de Córdoba, consistent with data of Lyons *et al.* (1997); Rapela *et al.* (1998*b*); Sims *et al.* (1998) for peraluminous bodies in other parts of the Sierras de Córdoba. U–Pb ages of 515–520 Ma for metamorphic monazite from paragneisses (e.g. Gromet & Simpson 1999) indicate the synchronous development of high-grade metamorphism and peraluminous magmatism. A U–Pb zircon age of 474 ± 5 Ma (Gromet & Simpson 1999) from an undeformed granitic pluton that cross cuts the *c.* 515 Ma paragneisses provides a firm younger age limit for the development of the regionally pene-

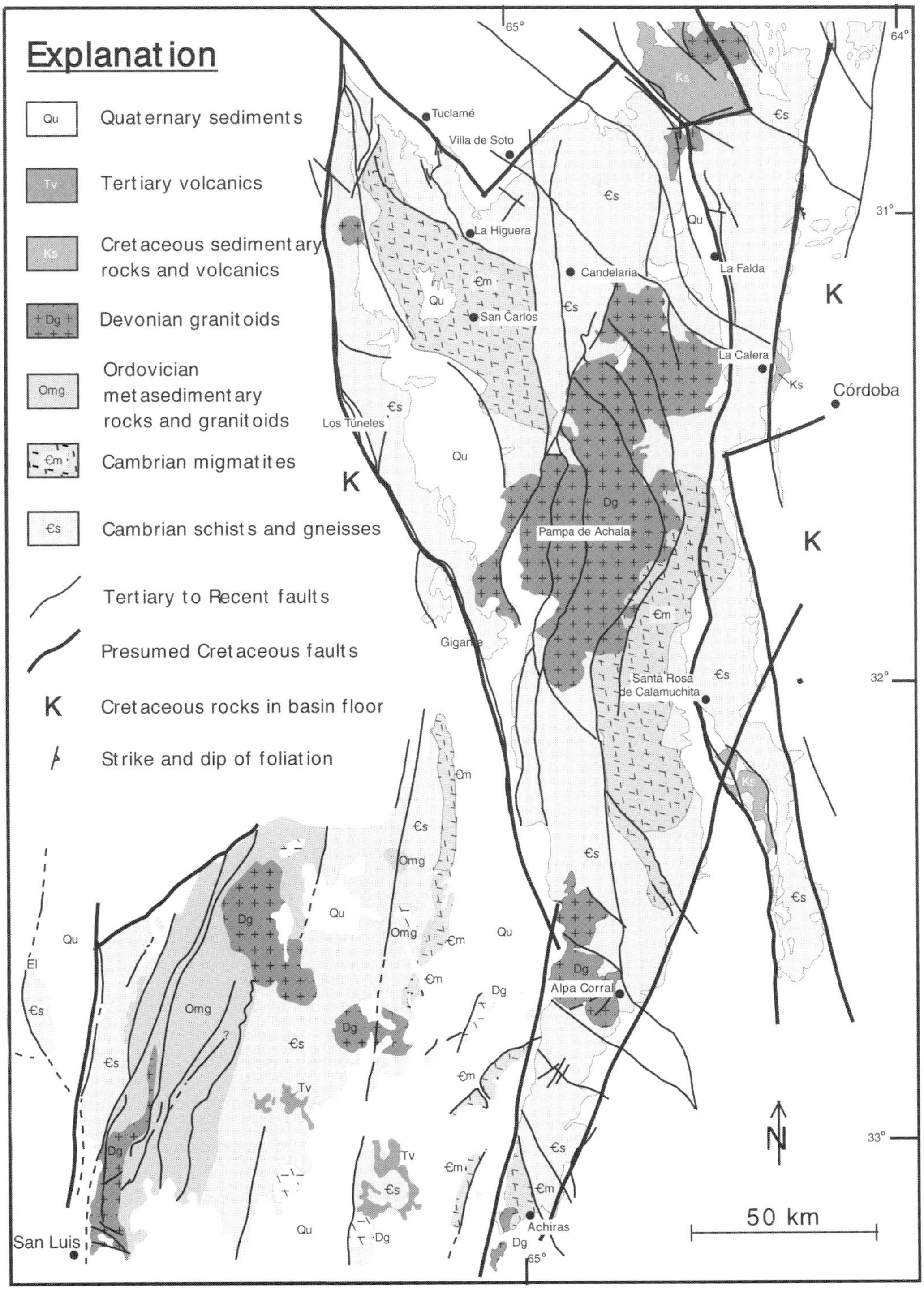

Fig. 2. Simplified geological map of the Sierras de Córdoba (eastern block) and Sierra de San Luis (western block), modified from Lucero Michaut *et al.* (1995).

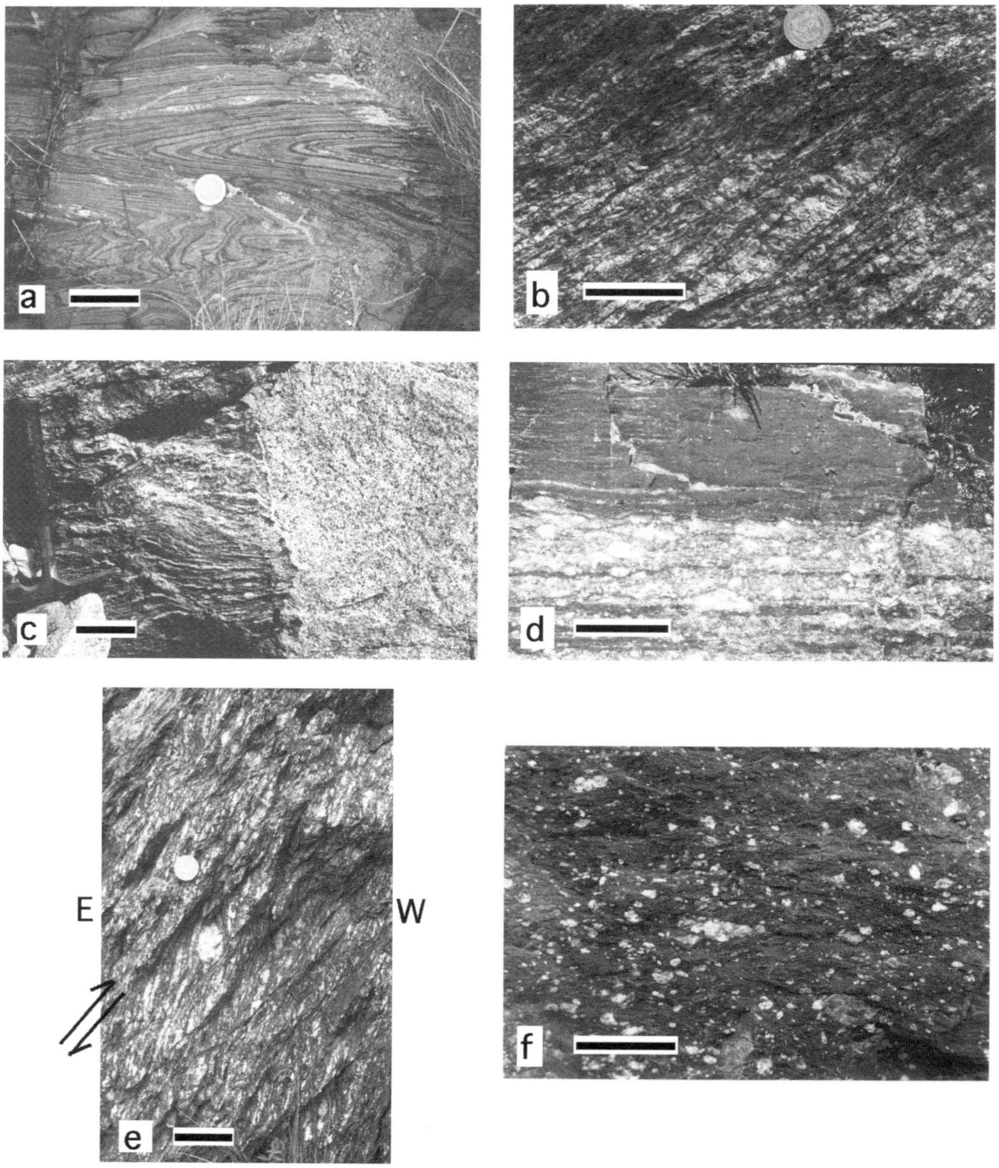

Fig. 3. (**a**) Folded pressure solution cleavage seams in psammites of the northwestern Sierras de Córdoba, near Tuclame. (**b**) Garnet–biotite–sillimanite–cordierite gneisses of central Sierras de Córdoba showing well-developed and regular compositional banding. (**c**) Undeformed 474 $\pm$ 5 Ma Passo del Carmen granite cross-cuts cordierite-bearing paragneisses. (**d**) View perpendicular to lineation and foliation in mylonites of Sierras de Córdoba. Pre-existing pegmatite vein and garnetiferous gneiss are sheared into parallelism. (**e**) East-over-west chlorite-grade shear bands in the Los Túneles section, westernmost Sierras de Córdoba. (**f**) Ultramylonite from the south central Sierras de Córdoba. (Note lack of a strong foliation and preservation of rounded clasts of feldspar and quartz in a predominantly chloritic matrix.) Scale bars for (**a**), (**b**), (**d**) and (**f**) represent 5 cm; in (**c**) and (**e**), 10 cm.

trative Pampean deformation. In addition, $^{40}Ar/^{39}Ar$ cooling ages for muscovite from the San Carlos migmatite massif in the Sierras de Córdoba give an integrated apparent age of 502 $\pm$ 5 Ma (Krol & Simpson 1999), indicating that cooling of the S-type granites and their host gneisses was relatively rapid after emplacement.

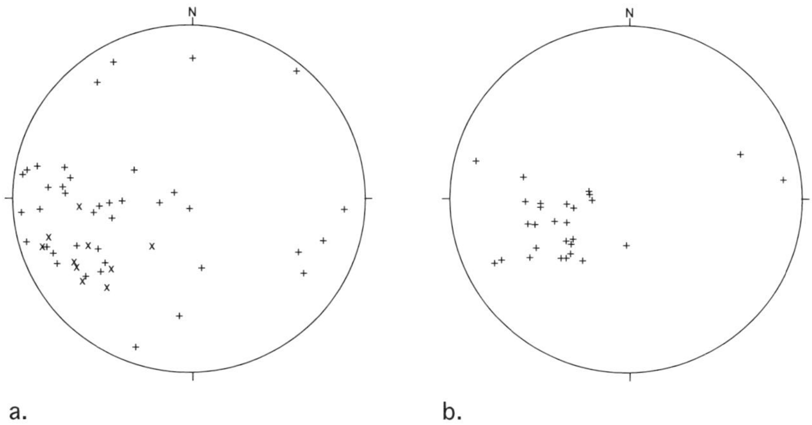

Fig. 4. Lower hemisphere stereographic projections of (**a**) poles to foliation in schists and gneisses (+) and poles to pressure solution cleavage (×), central Sierras de Córdoba, and (**b**) poles to foliation in chlorite-grade shear bands from Los Túneles and in ultramylonite, Ambul region, Sierras de Córdoba.

Pampean high-strain zones

The early schistose and gneissic fabrics are deflected into several north–south-trending, upper-amphibolite to upper-greenschist facies mylonite zones of hundred-metre to kilometre scale (Fig. 5a; Martino *et al.* 1993*a*, *b*). Outside these zones, foliations in the gneisses and schists are generally steep, mineral lineations are rare, and early formed chevron folds in the paragneisses and schists are common. Inside the shear zones, foliations are planar, steeply east-dipping, and contain strong down-dip lineations. Almost all of these shear zones contain shear sense indicators such as sigma grains (e.g. Passchier & Simpson 1986) or rotated boudins (e.g. Hanmer & Passchier 1991) that indicate west-directed reverse movement (Martino *et al.* 1993*a*, *b*; Simpson *et al.* 1998*a*). However, in most areas, structural blocks that are juxtaposed across the shear zones contain similar metamorphic mineral assemblages, indicating little post-metamorphic differential vertical movement. One of these high-strain zones north of Candelaria (Figs 2 and 5a) yielded 509 ± 2 Ma U–Pb metamorphic titanite ages for neocrystallized grains and a 490 ± 2 Ma U–Pb apatite cooling age (Fantini *et al.* 1998). These results indicate that the deformation in the eastern margin of the Sierras Pampeanas occurred at a late stage of the Cambrian-aged Pampean orogenic event.

Greenschist-grade shear zones

Many of the high-grade deformation zones in the Sierras de Córdoba are locally overprinted by younger, lower- to mid-greenschist facies mylonite zones (Figs 3d and 5a) with similar NNW strike to the earlier zones (Fig. 4b). The younger zones contain closely spaced foliations, dynamically recrystallized grains typical of greenschist-facies mylonites, and a strong down-dip mineral lineation defined by quartz ribbons, new muscovite, and recrystallized feldspar trails. Although foliations in these mylonites dip steeply to the east or, less commonly, steeply to the west, preserved shear bands, sigma and delta grains, and grain shape preferred orientations indicate that movement sense in almost all cases was again east over west (Simpson *et al.* 1998*a*). The greenschist mylonite zones commonly have nucleated on the margins of aplite or pegmatite dykes. They also occur within 468–486 Ma (Pankhurst *et al.* 1996; Pieters & Skirrow 1997) calc-alkaline plutonic rocks of the Sierras de Chepes and San Luis that are thought to be part of the Ordovician (Famatinian) arc (Toselli *et al.* 1993; Sims *et al.* 1997, 1998).

On the eastern flank of the Sierra de Chepes (Fig. 1), the calc-alkaline granitoids are deformed into metre-scale, east-side-up, lower-greenschist grade mylonites with S–C structure (Simpson *et al.* 1998a). Pankhurst *et al.* (1998); and Sims *et al.* (1998) suggested that this phase of deformation in the Sierra de Chepes was contemporaneous with Ordovican pluton emplacement during a Famatinian orogenic event. However, the low metamorphic grade and presence of many thin ultracataclasite and related pseudotachylyte veins in these rocks suggest that a significant cooling period must have occurred before the deformation event. A biotite cooling age of 428 ± 12 Ma from pseudotachylyte in the La Calera region suggests that the high-grade basement rocks of the eastern Sierras de Córdoba had reached upper-crustal levels by this time, consistent with a regionally rapid post-Cambrian exhumation and cooling. In the San Luis ranges, the El Volcán muscovite granite (Fig. 2), thought to be Ordovician in age by correlation with similar lithologies elsewhere in the region (Sims *et al.* 1997), is both crenulated and sheared at lower- to middle-greenschist grade, with an east-over-west shear sense (Simpson *et al.* 1998*a*). Lyons *et al.* (1997) obtained 354 Ma Ar cooling ages on muscovites from this granite. However, all of the micas in this rock are cataclastically deformed or kinked, therefore the actual age of shear-zone formation, although probably post-470 Ma, remains poorly constrained.

The clear east-over-west, dip-slip movement in almost all areas of the Eastern Sierras Pampeanas (Martino *et al.* 1993*b*; Simpson *et al.* 1998*a*) is difficult to reconcile with any significant strike-slip component of terrane emplacement at this time. Strong down-dip stretching lineations can form within predominantly transpressional regimes (e.g. Simpson & De Paor 1993; Tikoff & Teyssier 1994), but lack of any evidence for partitioning of the deformation into accompanying strike-slip zones argues against this as the predominant deformation mechanism for the entire Eastern Sierras Pampeanas.

Semi-brittle fault reactivation

A major low-angle thrust zone with significant westward displacement crops out south of the Achala batholith (Fig. 5b) and may be related to a second west-directed thrust zone along the western border of the Sierras de Córdoba, well exposed in the Los Túneles road section (Fig. 2 and 5b). Steeply east-dipping foliations in sillimanite-grade gneiss (Gordillo 1984) are retrogressed and deformed into a zone of several hundred metres thickness of ultramylonite, mylonite, and chloritic shear bands. In the Los Túneles section, a zone of 300 m thickness contains 5–10 cm spaced chlorite-grade shear bands (Fig. 3e) that dip moderately to the east. The shear banded and retrogressed gneiss is here thrust westward on a series of discrete, moderately east-dipping brittle fault planes over almost unmetamorphosed, kinked phyllites in which bedding and pressure solution cleavage are still preserved; these phyllites were considered by Rapela *et al.* (1998*b*) to be part of the Cambrian accretionary prism section. Metamorphic grade in the Los Túneles thrust zone decreases downward, toward the contact with the phyllites.

At a distance of about 200 m structurally above the Los Túneles thrust plane, relict lenses of unretrogressed amphibolite preserve tight chevron folds in their cores and steeply dipping amphibolite-grade shear-zone foliations on their margins. However, for a vertical distance of *c.* 100 m above the thrust plane, all rocks are completely retrogressed to chlorite grade. Rocks immediately above the brittle thrust contact with the phyllites, and for a vertical distance of about 50 m, contain numerous centimetre-thick pseudotachylyte veins. The veins become more abundant as the thrust plane is approached, but are absent in the phyllites below the thrust.

pseudotachylyte veins are commonly associated with the semi-brittle faulting event. Northwest-striking ultracataclasite bands (Fig. 6a) and ductile to semi-brittle zones (Fig. 6b) occur within 1 km of a NW-striking Neogene fault in otherwise massive S-type granite migmatite in the San Carlos massif, near La Higuera (Fig. 2). There are thin (<1 cm) pseudotachylyte veins within many of these ultracataclasite zones. pseudotachylyte veins up to 10 cm thick occur parallel to foliation in paragneisses from the La Puerta area (Fig. 6c), and occasional veins of 1–5 cm thickness occur in almost all of the mylonite and chlorite-grade fault zones (Fig. 5b). In the Los Túneles and La Calera regions, centimetre-thick pseudotachylyte veins are numerous (Simpson *et al.* 1998a). The orientations of semi-brittle, brittle, and pseudotachylyte zones are generally subparallel to earlier greenschist- or higher-grade shear zones. $^{40}Ar/^{39}Ar$ spot fusion analyses of a pseudotachylyte sample from La Calera (Fig. 6d) gave a formation age of 428 ± 12 Ma (mean square weighted deviation (MSWD) 0.71; Northrup *et al.* 1998*b*); however, a pseudotachylyte from a 20 cm vein in the La Puerta region in the central Sierras de Córdoba (Fig. 2) yielded a 365 ± 10 Ma (MSWD 2.5) formation age (Northrup *et al.* 1998*b*), which is consistent with its formation during the Achalian event.

Other examples of chlorite- to greenschist-grade fault reactivation within the Sierras de Córdoba include at least three major zones of severely retrogressed mylonite and ultramylonite (Figs. 3f and 5a and b) that are many tens of metres thick, have a gentle to moderate dip to the ENE, and contain only a weak foliation with almost no discernible lineation; sense of shear indicators are rare but where present are consistent with east-over-west movement. These retro-

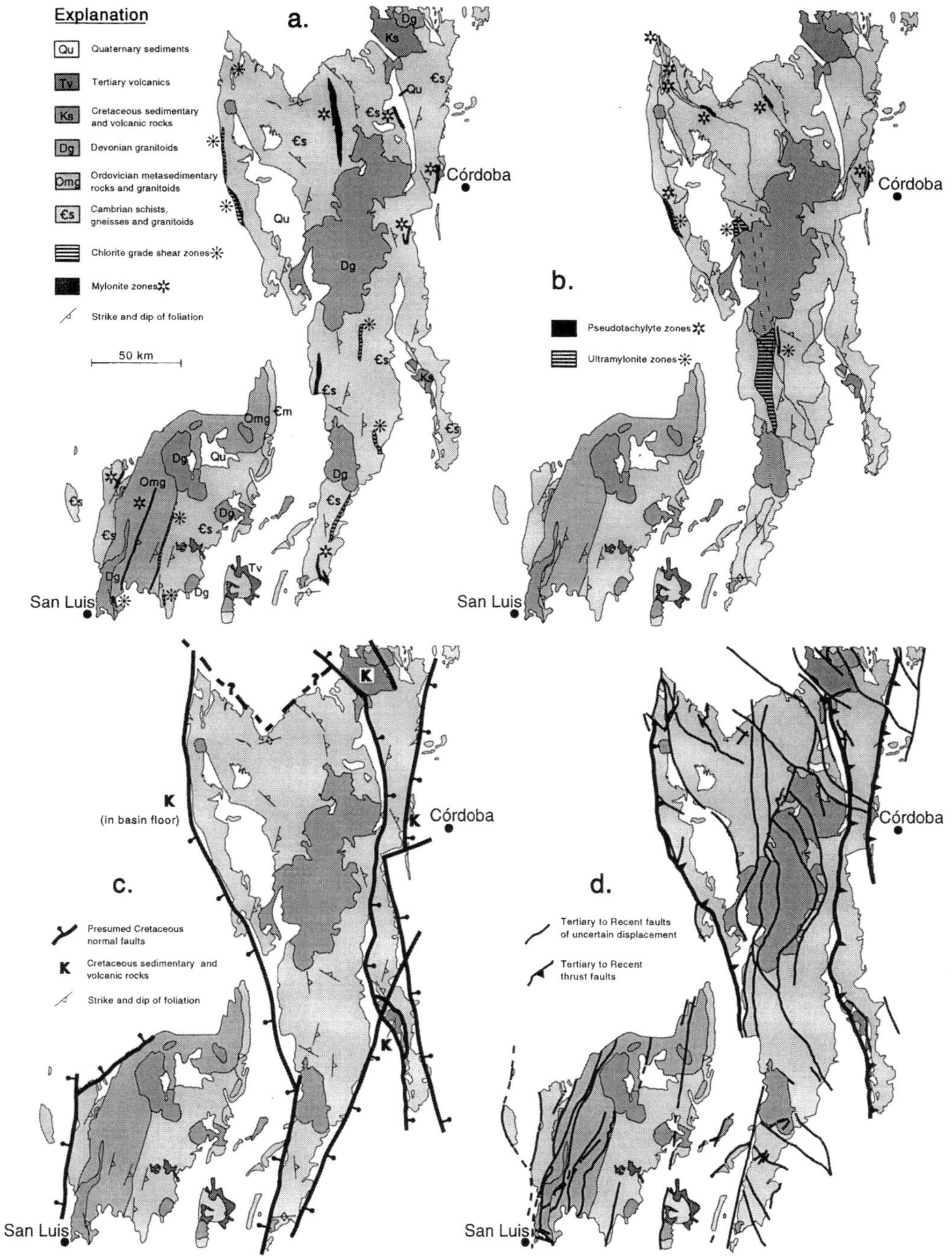

Fig. 5. Reactivation of early formed foliations in basement rocks of the Eastern Sierras Pampeanas. (**a**) Location of Palaeozoic mylonite zones and principal chlorite-grade shear zones. (**b**) Major pseudotachylyte localities, orientations of foliations in paragneisses, and principal ultramylonite zones. (**c**) Location of presumed Cretaceous basin-bounding normal faults (based on Schmidt *et al.* 1995). (**d**) Neogene reverse faults and subvertical fault zones of unknown displacement sense (from Lucero Michaut *et al.* 1995).

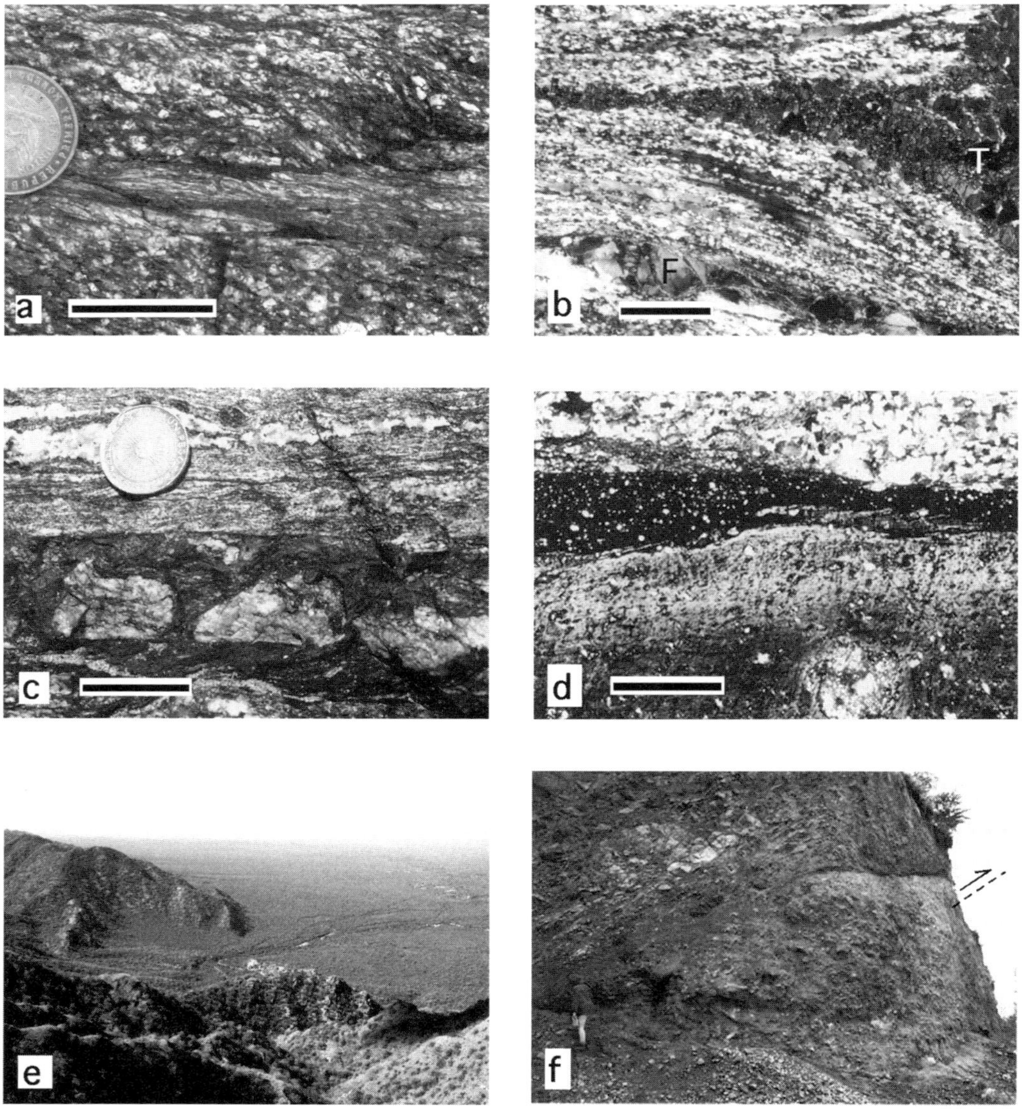

Fig. 6. (**a**) Narrow band of ultracataclasite in S-type granite migmatite, La Higuera, Sierras de Córdoba. (**b**) Photomicrograph of dynamically recrystallized quartz and brittle fracturing of feldspar (F) and tourmaline (T) in semi-brittle mylonitic pegmatite vein, La Higuera. Crossed Nicols. (**c**) pseudotachylyte vein parallel to foliation in cordierite-bearing gneiss, La Puerta, northern Sierras de Córdoba. (**d**) Photomicrograph of two-phase pseudotachylyte vein (dark) in hypersthene gneiss, La Calera. Lower half of image shows paler-coloured pseudotachylyte vein analysed for $^{40}Ar/^{39}Ar$ (Northrup *et al.* 1998b); feldspar inclusions (light) are large and relatively easy to avoid in this sample. Crossed Nicols. (**e**) Neogene range-bounding fault scarp on western margin of Sierras de Córdoba. View to west. (**f**) View to SW of Neogene west-directed thrust that places Palaeozoic gneiss over Tertiary conglomerate. Height of section *c*. 12 m. Left side of photograph is close to true dip; right side is apparent dip section. Scale bars for (**a**) and (**c**) represent 2.5 cm; for (b), 2 mm; for (**d**), 1 mm.

gressed zones clearly transect the higher-grade gneisses and are derived from them, but their age is uncertain. On the basis of their very low metamorphic grade we consider them to be younger than the mainly Ordovician, greenschist-grade deformation zones.

The subsequent Devonian granite plutons are generally circular in outcrop pattern (Fig. 2) and

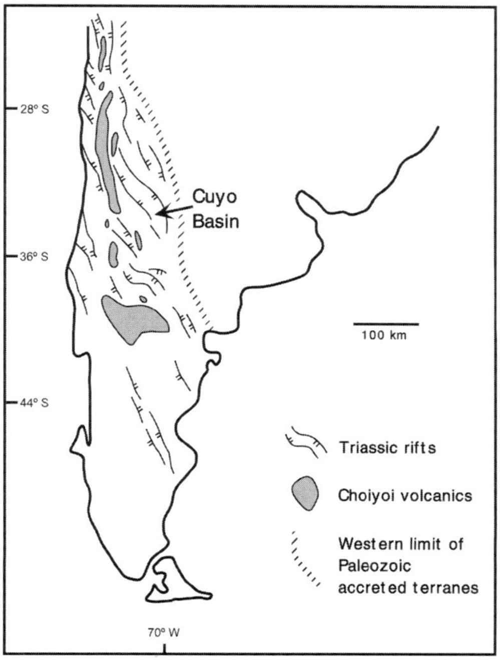

Fig. 7. Distribution of rift basins and location of Jurassic Choiyoi volcanic rocks in the Triassic. Modified from Ramos & Kay (1991).

show little evidence of syn- or post-emplacement strain. However, map-scale patterns of the late Devonian (368 ± 2 Ma) Achala Batholith in the Sierras de Córdoba (U–Pb zircon, Dorais *et al.* 1997) and especially the early Devonian (404 ± 5 Ma, U–Pb zircon, Camacho & Ireland 1997) La Escalerilla porphyrytic granite in the Sierra de San Luis (Fig. 2) are consistent with syntectonic emplacement in elongate transtensional structures. In addition, we have observed shear fabrics at the margins of aplite dykes within the La Escalerilla granite, indicating at least local Achalian ductile deformation of these rocks.

Brittle reactivation

Carboniferous to Cretaceous brittle reactivation

The Paganzo Basin (Fig. 1) developed over the docked Pampean, Precordilleran, and Chilenian terranes in Carboniferous and Permian times (Fernandez-Seveso & Tankard 1995). Outcrops of these continental sedimentary rocks are now relatively small and isolated in the Sierras de San Luis and Córdoba, although there are large outcrops of Paganzo rocks in La Rioja province, to the NW. According to Fernandez-Seveso & Tankard (1995), the Paganzo basin initiated as a series of isolated pull-apart basin depocentres, within a presumed right-lateral strike-slip fault system (see also Ramos 1988). Continued extension of isolated basins and consequent shortening of their linking strike-slip faults allowed them to amalgamate into the main Paganzo Basin (Fernandez-Seveso & Tankard 1995). The basin-bounding strike-slip faults are sub-parallel to, and may have been controlled in orientation by, the earlier basement structures. As an example, small outcrops of Paganzo sedimentary rocks SW of Tuclame (Fig. 2) in the northern Sierras de Córdoba, occur within a graben that is bounded by right-lateral strike-slip faults superimposed on mylonite and protomylonite zones in the underlying basement rocks.

The principal Triassic basins in this region of South America form an en echelon pattern (Fig. 7; Charrier 1979; Criado Roque *et al.* 1981; Uliana & Biddle 1988) and include the Beazley, Cuyo, Ischigualasto, and Marayes basins (Fig. 1), each of which is normal fault-bounded (Jordan & Ortiz 1987; Uliana & Biddle 1988). As with the Carboniferous–Permian Paganzo Basin, the Triassic basins appear to have been initiated as isolated depocentres in pull-apart basins and eventually amalgamated to form the larger basins (Uliana & Biddle 1988). Many of these basins are located above presumed boundaries of terranes accreted during Palaeozoic time (Ramos & Kay 1991). For example, the western margin of the Cuyo basin extends for over 700 km and is located on the boundary between the Precordillera and Chilenia; the Ischigualasto and Beazley basins are located on the border of the Sierras Pampeanas and Precordillera terranes (Ramos & Kay 1991). The map distribution of these basins and the granite rhyolites of the Choiyoi Formation (Fig. 7) is strongly suggestive of a sinistral strike-slip or transtensional tectonic environment that lasted from Carboniferous to Triassic time. The en echelon pull-apart basins in central and southern South America probably extended as far east as the present Sierras de Córdoba, where relict sections of the Cretaceous rocks are preserved (Figs. 2 and 5c). Sinistral transtensional movement contrasts with the apparent dextral motion displayed in the Mendoza region and documented by Ramos (1988) and Jordan & Ortiz (1987).

Tertiary brittle reactivation

Mid- to Late Tertiary to Recent fault reactivation of Cambrian to Cretaceous fault zones occurred on moderate- to high-angle faults that form a conjugate set, striking NW and NNE (Fig. 5d), and include the range-bounding reverse faults of the Sierras de Córdoba, Sierra de San Luis, and La Rioja. Recent west-directed thrust faulting can also be seen in the interior regions of the Sierras de Córdoba (Costa & Vita-Finzi 1996), well exemplified by a west-directed reverse fault near Santa Rosa that puts basement Cambrian paragneisses on Tertiary unconsolidated sediments (Fig. 6f).

Block rotation about a horizontal axis is a significant component of the Neogene faults, as evidenced by the 1000 m abrupt topographic change across the western edge of the Sierras de Córdoba (Fig. 6e), and similar topographic relief on the western border of the Sierra de San Luis. The eastern margins of these ranges gently descend and merge with the surrounding plains. This geometry is consistent with reversal of movement along initially listric, Cretaceous or older normal faults. The Sierra de San Luis and El Gigante blocks (Fig. 2) also appear to have undergone a component of dextral rotation about a vertical axis relative to the southern Sierras de Córdoba; all basement structures in the Sierra de San Luis strike to the NNE and those in El Gigante strike ENE, whereas those in the Sierras de Córdoba strike mainly NNW (Fig. 2). Tertiary volcanic rocks now crop out in the intervening valley (Fig. 2). The mechanism for block rotation about a vertical axis may be similar to that in the Mojave Desert of California (e.g. Dokka & Travis 1990), in which the steep, block-bounding brittle faults root into shallowly dipping ductile faults at depth.

Most of the Cretaceous basin sedimentary formations have undergone shortening during Tertiary time. Ramos & Kay (1991, fig. 3b) showed classic tulip-shaped flower structures in an interpreted seismic section of the Cuyo Basin. Allmendinger *et al.* (1983) illustrated extreme shortening by thrusting and reverse faulting during Tertiary time. Preliminary field observations in the Cretaceous section (with G. Rubay and P. Condat) in the Mendoza region revealed a minimum of 150% along-strike extension by minor structures on bedding planes that are shortened in the east–west direction by thrusting and thrust-related folding (Fig. 8).

Discussion

The theoretical basis for fault-zone reactivation

Spatial concentration of faults with very different periods of activity, as documented here, suggests that fault reactivation is a regional phenomenon of great importance over long periods of time. There is abundant evidence for the evolution of continental crust by cycles of terrane accretion, supercontinent formation, and supercontinent break-up. If the associated patterns of fault zones were random, then cratons would be smaller and more numerous, and stable blocks between orogenic belts would be cut by numerous randomly oriented deformation zones with a wide range of ages. The existence of orogenic belts that contain long histories of fault reactivation demonstrates that reuse of existing crustal-scale deformation zones is mechanically preferable to creation of new zones. Before discussing the field observations, we here review the theoretical basis for fault reactivation.

Anderson's law (Anderson 1942) limits the possible orientations of new surface faults (Fig. 9a). Observing that the surface of the Earth must be a plane of no shear stress (apart from negligible wind shear), Anderson concluded that two of the principal axes of the stress ellipsoid must always be contained in the surface and the third must therefore be oriented vertically. Combining this law with Mohr–Coulomb failure theory, which predicts that the maximum principal stress, σ_1, bisects the acute angle between conjugate fault planes (Fig. 9a), we are left with only three permissible types of fault: (1) high-angle normal faults; (2) low-angle thrusts; (3) vertical strike-slip faults. However, field observations demonstrate abundant occurrences of: (4) low-angle detachments; (5) high-angle reverse faults; (6) oblique slip faults (Fig. 9b). The stress states are compatible with ductile shear zones at depth, but not with high-level brittle faults.

Classic models of fault inversion (which we here term 'coaxial inversion') go part of the way toward explaining the range of faults that occur in nature. Coaxial models envisage a reverse of the sense of displacement on a fault surface with a 180° change in slip azimuth. The fact that faults can be reactivated despite orientations far from the optimal failure orientation given by Mohr–Coulomb theory, is explained in Fig. 10. Here, the Mohr circle represents the crustal stress state and every point on its perimeter represents the combination of normal and

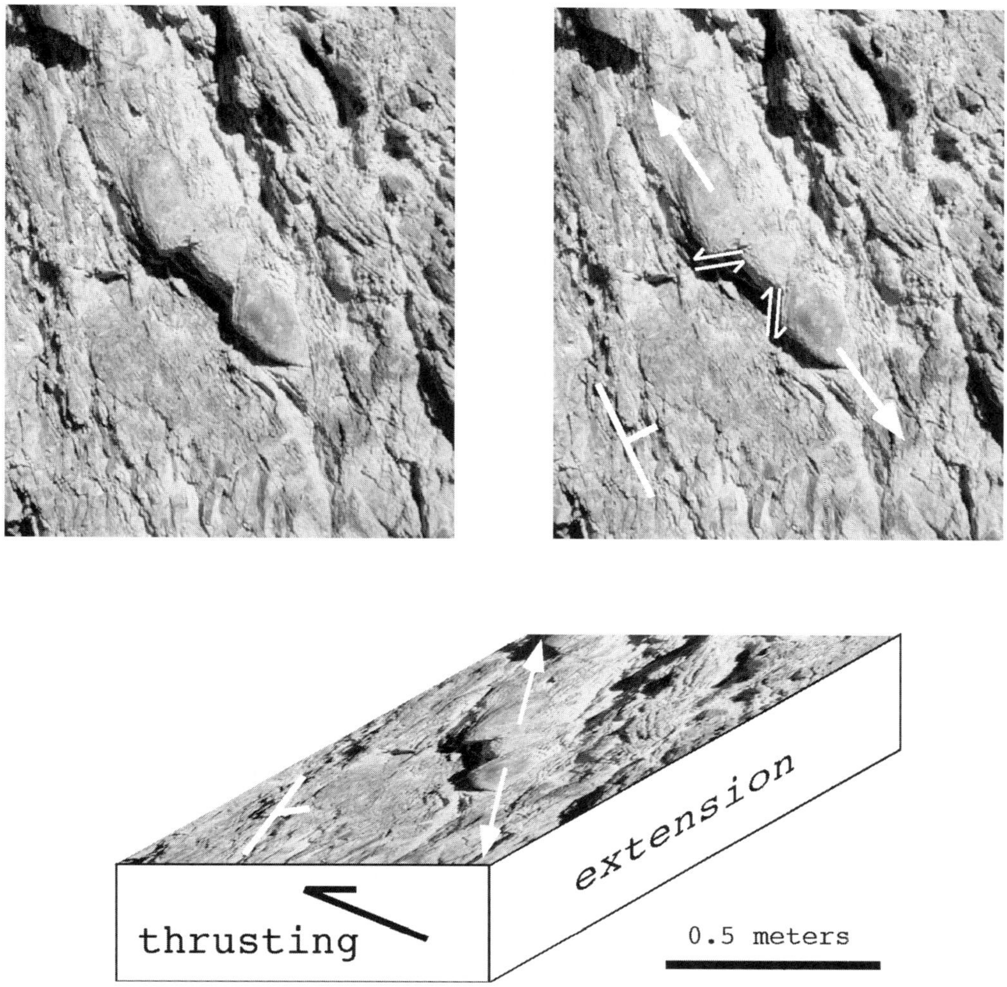

Fig. 8. Extension of bedding in a direction parallel to the strike of thrusts in Cretaceous sedimentary rocks of the Cuyo Basin, Mendoza Province, Argentina. (**a**) Conjugate normal faults extend bedding. (**b**) Annotated version of (**a**) showing bedding orientation. (**c**) Perspective version of (**b**) with orientation of thrusts.

shear stress on a given plane. The failure criterion for movement on pre-existing fracture planes is represented by the low-strength, straight Mohr envelope (a representation of Byerlee's Law) whereas the creation of new faults in pristine rock is governed by the Mohr–Coulomb failure criterion, represented by the portion of the parabolic envelope that touches the Mohr circle. Thus, for a given stress state, a pre-existing fault may become unstable before the Mohr circle reaches the pristine rock envelope (e.g. Handin 1969). If the stress state causes the circle to reach the latter envelope, a new fault will be generated in a more suitable, minimum energy orientation. Such faults are commonly called 'short cuts'.

Fault reactivation short cuts in the Sierras de Córdoba

Three orientations of short cuts, corresponding to normal, thrust, and strike-slip tectonism, are illustrated in Fig. 10. Occurrence of small faults in a short-cut orientation is itself evidence that reactivation of less suitably oriented faults was occurring before short-cut formation. Thus the occurrence of near-surface, high-angle reverse faults, in contradiction to the Mohr–Coulomb–Anderson theory, can be explained by reuse of early listric normal fault surfaces during later thrust movements (Fig. 10b). The occurrence of detachments may be aided, at least in part, by reuse of extant low-angle thrust surfaces during

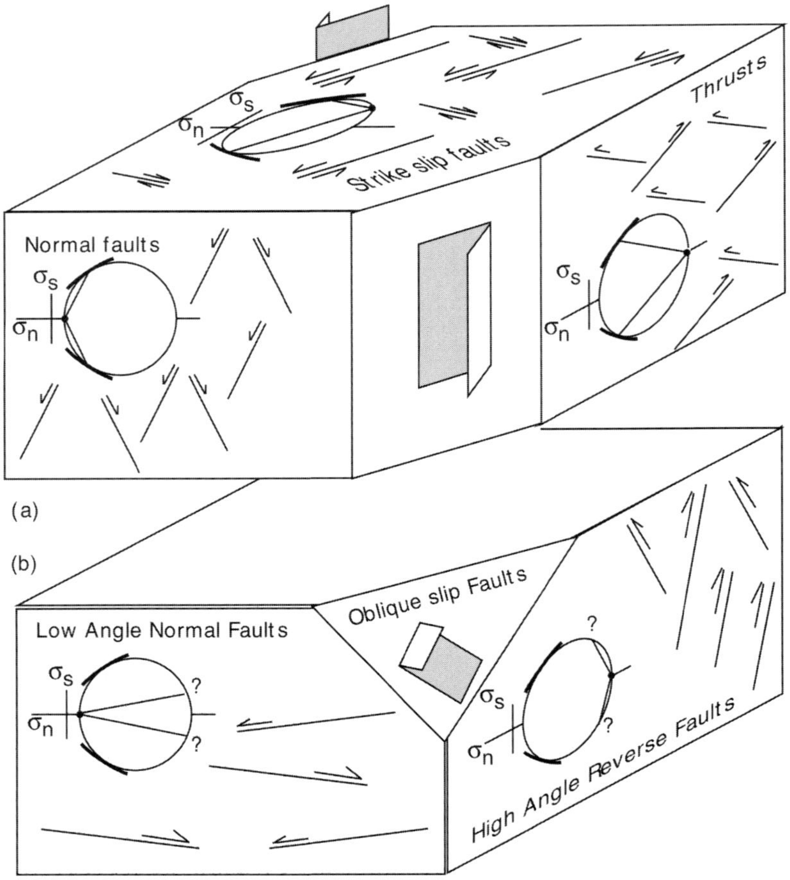

Fig. 9. Theoretical basis for reactivated fault orientation using Mohr–Coulomb–Anderson theory. (**a**) Orientations of new normal faults in intact rocks are given by joining Mohr circle's pole (•) to intersections of Mohr circle with Coulomb failure envelope for intact rock (bold black lines represent relevant segments of failure envelope). (**b**) Low-angle normal faults, oblique slip faults, and high-angle reverse faults form on planes with normal –shear stress combinations (question marks) that are not accounted for by the Mohr–Coulomb–Anderson theory.

later extension (Fig. 10a). Highly oblique, or even orthogonal, basin inversion may use the original bounding fault surface (Fig. 10c). The latter model may be likely where fault traces are of the same order of magnitude as their displacements.

An example of fault reactivation using short cuts is well illustrated on a small scale by the geology near La Calera, NW of Córdoba (Fig. 11), where there are close spatial and orientation relationships among early gneisses with high-temperature–high-strain zones, later mylonite and pseudotachylyte-bearing shear zones, Mesozoic normal faults, and Tertiary inversion faults. In this region, the predominant NNW-striking, steeply east-dipping structural grain was followed by each subsequent reactivation event. The ductile and brittle faults are close together and subparallel to each other, rather than exactly superimposed on each other, and their relative positions are consistent with the younger event in each case taking a short cut from the principal fault zone at depth. The main Mesozoic normal fault that bounds the basin into which the proximal Cretaceous sediments were deposited, stepped eastward of the pre-existing thrust-related mylonite and pseudotachylyte zone (Fig. 11b), which is consistent with a normal fault short cut during reactivation of an earlier thrust zone (see Fig. 10a). The Neogene reverse faults (Fig. 11c) can also be interpreted as short cuts, in this case from the

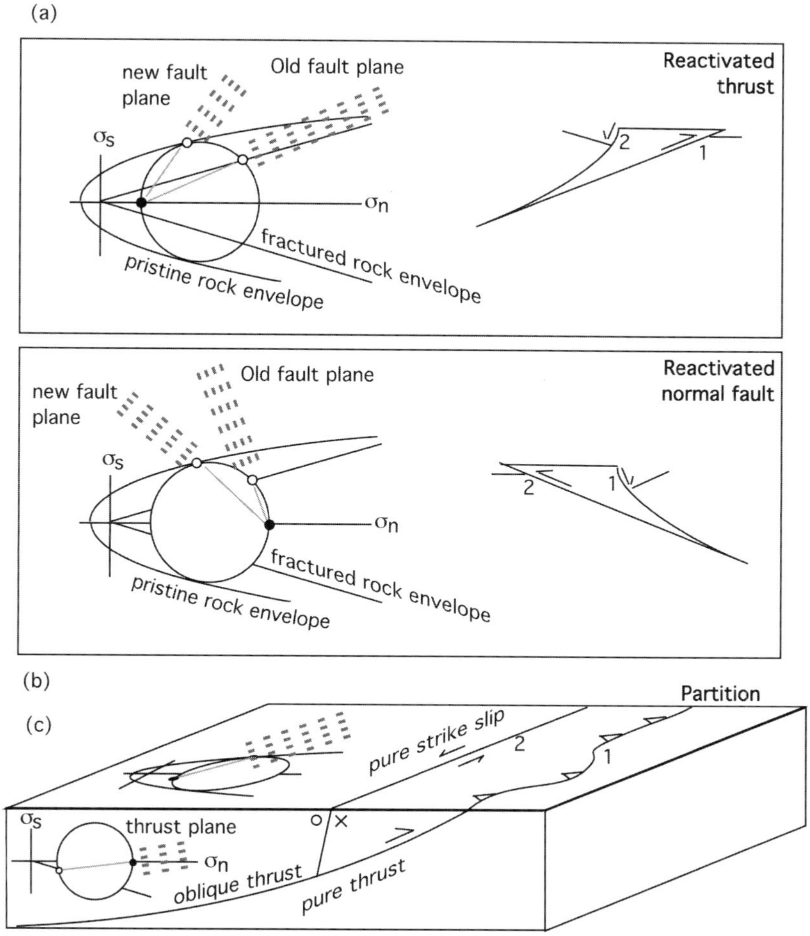

Fig. 10. Fault short cuts may indicate reactivation of earlier faults. Mohr constructions include parabolic failure envelope for intact rock and linear envelope for pre-fractured rock. •, Mohr circle poles; ○, failure stresses; dashed lines indicate failure plane orientations. (**a**) Initial thrust fault reactivated as normal fault. (**b**) Initial normal fault reactivated as thrust fault. (**c**) Oblique subduction may cause partitioning into pure thrust and strike-slip faults.

reactivation of the basin-bounding normal fault at depth (see Fig. 10b).

Net displacement and fault reversal

Other field evidence for fault reactivation includes reversal of slip on a single fault trace. A combination of growth sedimentation and slip reversal may lead to a null-slip point, which separates the part of a fault that shows net thrust displacement from the part that has experienced less reversal than the initial normal displacement (De Paor & Eisenstadt 1987). Such field data have rarely been reported (see, however, Powell 1987). What is commonly seen, both in the region discussed here and globally, is evidence for thrust displacement (slickensides, en echelon veins, minor folds, etc.) on a fault that places young, proximal basin sediments onto older, high-grade basement, such as illustrated in Fig. 11.

The hypothesis that displacement sense on a fault has undergone a complete reversal is strongly supported by a breach of the 'bow-and-arrow' rule of Elliott (1976) (Fig. 12). Elliott determined that thrust fault displacement was generally one-tenth the map length of the fault trace (Fig. 12a). When thrust faults extend several kilometres along strike, but show only minor displacement in the centre, it can be argued that much of the thrusting event was wasted on reversing the effects of an earlier

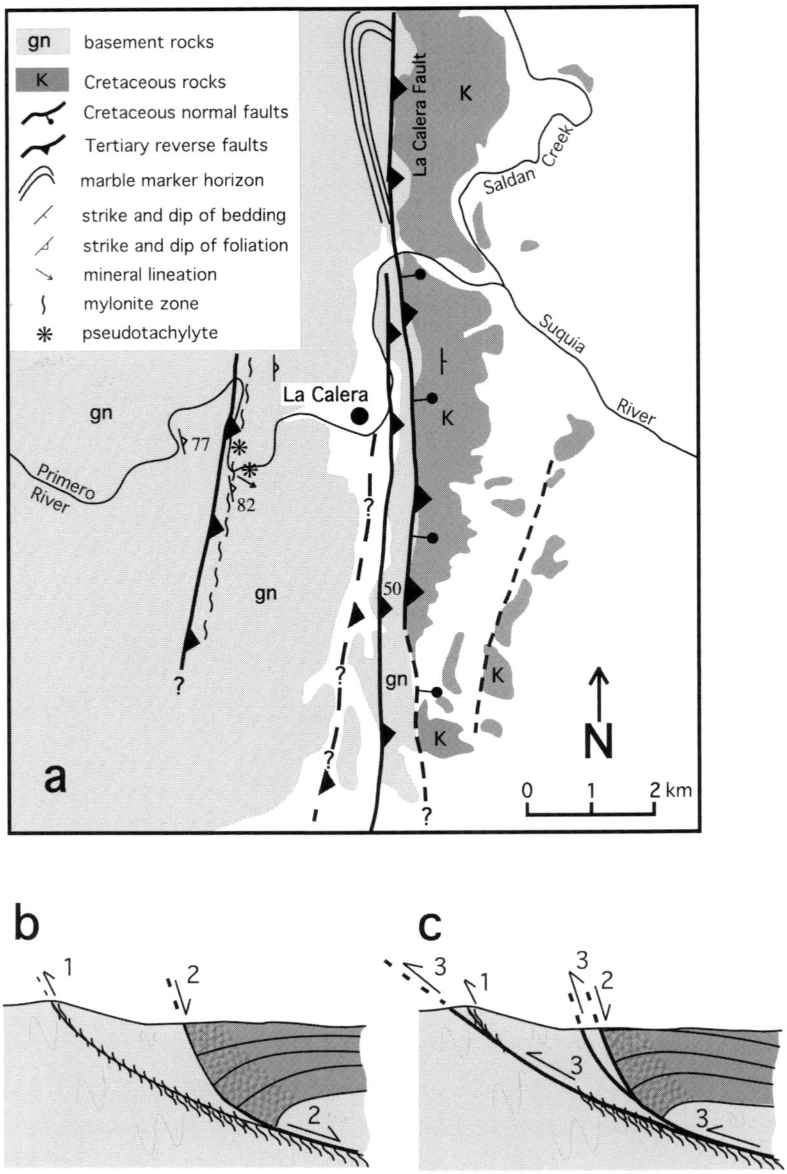

Fig. 11. (a) Simplified geology of the area near La Calera, Córdoba province, illustrating close proximity and orientation of early foliations, east-over-west mylonite zone, the Cretaceous basin boundary normal fault and Neogene reverse faults. (**b**) Schematic diagram of Cretaceous normal fault short cut (2) during reactivation of Cambrian thrust-related mylonite zone (1). (**c**) Tertiary reverse fault reactivation (3) of Cretaceous normal fault (2) produces west-stepping short cuts.

extension (Fig. 12b–f). The process is somewhat analogous to the way in which passive linear strain markers may undergo incremental shortening and then lengthen to become lines of no finite longitudinal strain, before finally entering the field of net extension. In a similar manner, fault displacement may start as extensional, become partially reversed, and pass through a stage of no net slip before finally producing a net thrust displacement. Thus the small net displacement on an inverted fault such as shown in Fig. 12f, may fail to account for the amount of lateral tip propagation that had developed by the time of peak absolute displacement.

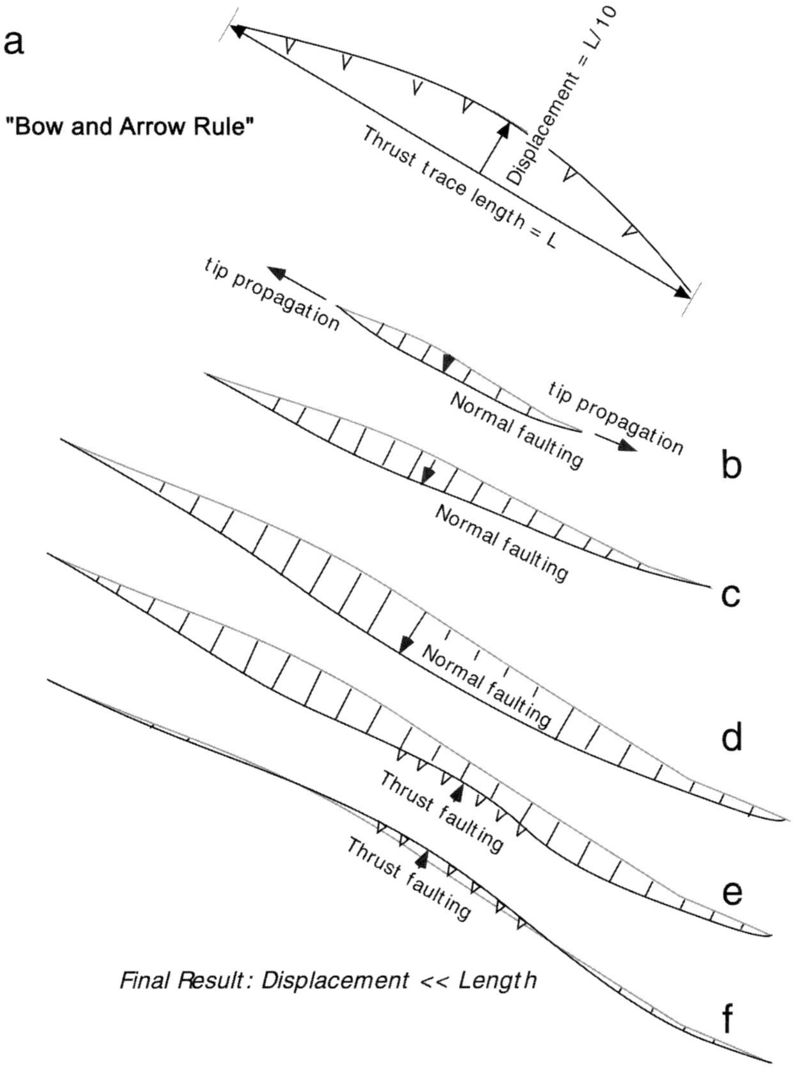

Fig. 12. (**a**) The 'bow-and-arrow rule' of Elliott (1976) explains thrust faults whose displacement is *c*. 10% fault trace length. (**b**) Early normal faulting allows tip propagation, which increases with extensional displacement (**c**) and (**d**). (**e**) Reversal of movement on same fault begins to reduce net displacement and can produce a small final thrust displacement to length ratio (**f**).

The 'bow-and-arrow rule' of Elliott (1976) explains thrust faults whose displacement is *c*. 10% fault trace length, but the mylonite zones and Neogene faults with reverse movement sense in Eastern Sierras Pampeanas have displacement far less than 10% of the fault length. Most of the mylonite zones in the Sierras de Córdoba are thin (a few metres or less) and have a rather small net displacement, as indicated by the lack of significant offset of metamorphic isograds, which is inconsistent with their previously mapped length of several hundred kilometres (e.g. Martino 1993) unless they are reactivated early normal or strike-slip faults. The absence of any kinematic evidence for early normal or strike-slip displacement within these shear zones may indicate that they are short-cut faults that root into major early strike-slip or extensional faults at depth. Alternatively, these deformation zones may be much shorter than has been previously inferred, or they may have been reactivated in a normal sense during one of the subsequent brittle events.

The presence of a small but significant number of thick, moderately dipping, retrogressed ultramylonite zones in the Sierras de Córdoba is consistent with the presence of major fault zones at depth that became reactivated during the several extensional and compressional episodes that this part of South America underwent. These zones have the same strike as the earlier greenschist- and higher-grade zones but have a more gentle easterly dip and may have originated as normal faults during the Permian–Triassic or Carboniferous transtensional events. In the absence of clear age constraints we cannot rule out that they may be even older and related to the Achalian event.

Basin inversion on reactivated faults

The typical small-scale, brittle, upper-crustal structure associated with bulk simple shear strain in a strike-slip orientation is a combination of normal and thrust faults. However, rocks do not deform homogeneously at shallow depths, and thus the simple shear model cannot accommodate significant displacement. Brittle rocks near the surface are more likely to undergo transtension as they are much weaker in tension than in compression. Therefore a period of transtension, which permits the opening of basins, begins first (Fig. 13a). It should be noted that the angle of tear faults to the regional shear is not necessarily 45°. As the basins continue to open, the active, strike-slip basin boundary faults become shorter, and this eventually results in stress concentrations in the region of the basin corners (Fig. 13b) as the component of driving force perpendicular to the transfer fault is spread over a smaller area of rigid wall rock. Although the basin is partially filled by sediments and there is ductile flow in the basement beneath, the crust is thinner and weaker within the basin. Continued strike slip on the overall zone will result in thrusting within the basins when the active branches of strike-slip transfer faults are minimal in length (Fig. 13c). The process can be likened to displacement on a surface with giant-scale frictional asperities.

After extensive transtensional or transpressional motion, pull-apart basins may undergo block rotation, so that basin-inverting thrust faults approach parallelism with the regional strike-slip direction whereas early basin-bounding normal faults become more oblique to the zone boundary. It should be noted that the inferred sinistral sense of shear used in this model is not critical: an equivalent model of orthogonal inversion would be viable in a regional zone of dextral strike-slip tectonics.

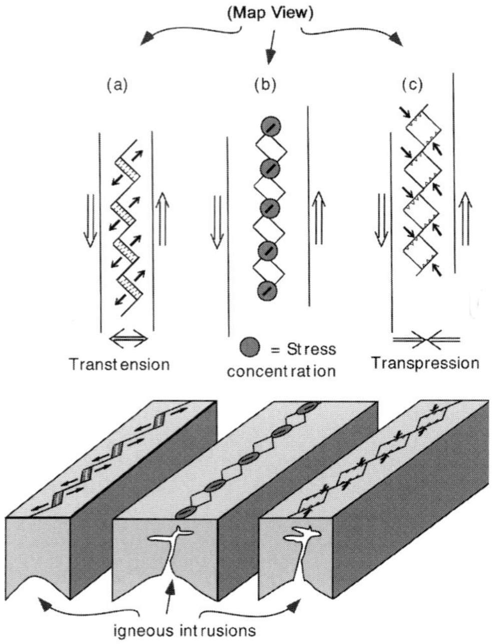

Fig. 13. Orthogonal inversion model for Tertiary inversion of Triassic–Cretaceous rift basins. (**a**) Formation of pull-apart basins in transtension. (**b**) Overlapping basin corners (shaded circles) are sites of stress concentration as active portions of strike-slip fault segments (bold lines inside shaded circles) are reduced in length. (**c**) Basins undergo inversion as transtensional zone changes to orthogonal transpression.

Mesozoic basin inversion in western Argentina

Previous workers considered the Triassic–Cretaceous basins of western Argentina to have formed as a consequence of propagation of the southern Atlantic extensional regime right across the South American continent (e.g. Schmidt *et al.* 1995), however, South America's Permo-Cretaceous basins did not form in the centre of a supercontinent, but rather within the sphere of influence of the continent's active Pacific margin. We therefore look to the subduction regime rather than the southern Atlantic opening event for the source of stress. An alternative model, proposed by Ramos & Kay (1991), is that the rift basins formed as a result of extensional collapse after the accretion of

Palaeozoic terranes to South America, with subsequent renewal of compressional tectonics in Tertiary time. A third possibility is that they are pull-apart basins with reversal of displacement sense on the dip-slip boundary faults, as illustrated in Fig. 13c, in which stress partitioning led to transtension followed by transpression. Thus we envisage the following scenario: (1) partition of subduction-related stresses into thrust and dextral strike-slip components during early Palaeozoic time; (2) termination of the thrust component by terrane docking in mid- to late-Palaeozoic time (e.g. the Precordillera and Chilenia) with continuation of the strike-slip component on terrane boundaries; (3) further partition of that strike-slip component at shallow, brittle crustal levels, leading to pull-apart basin formation in a dextral transtensional setting in Carboniferous to Cretaceous time; (4) a return to transpression as a prelude to the Tertiary Andean orogeny; (5) reactivation of the Mesozoic basin-bounding faults as reverse faults during Tertiary to Recent time. (Minor late, low-angle extensional faults thin the thrust wedge, but these dip to the east and are not reactivation faults).

Surface basin formation may have been accompanied by thermal pull-ups at depth (e.g. Fig. 13b), permitting granite–rhyolite volcanism (the Jurassic-aged Choiyoi volcanic rocks (see Fig. 7)). As illustrated in Fig. 7, the axis of volcanism is north–south in orientation but the long axes of most basins are NW–SE, in keeping with the sinistral transtensional model of Ramos & Kay (1991) for basin development.

A consequence of orthogonal basin inversion is that thrust sheets must rise along extra steep bounding faults related to the strike-slip portions of the initial pull-aparts. Consequently, we expect to see extremely tight to isoclinal 'buttress' folds near basin margins and thrust short cuts climbing through the wall-rock. Thus the Neogene faults that localized on earlier high-strain zones or foliations in the Sierras Pampeanas (e.g. Gordillo & Lencinas 1979; Kraemer & Martino 1993) may have accommodated much of the Cretaceous rift basins' inversion and facilitated the severe folding of the sedimentary rocks within the basins.

Conclusions

Mid-Cambrian high-temperature–low-pressure metamorphism transposed an ubiquitous pressure solution cleavage in early Cambrian accretionary prism sedimentary rocks into gneissic and schistose foliations that dipped steeply east, toward the palaeo-margin of Gondwana. The metamorphic event was short lived, but produced voluminous migmatites and S-type granites. Subsequent deformation, presumed to be during accretion of an Ordovician arc and/or the Precordillera terrane to the west, produced greenschist-grade mylonites, chlorite-grade shear bands, thick and retrogressed ultramylonites, and pseudotachylytes, all of which formed in zones parallel to the pre-existing ENE-dipping foliations. $^{40}Ar/^{39}Ar$ spot fusion data from two pseudotachylyte veins indicate that reactivation occurred during Silurian and/or early Devonian time, and may be related to the final accretion of the Precordillera or Chilenia terranes to the west.

Opening of Carboniferous to Triassic pull-apart basins, into which continental sediments were deposited, is consistent with oblique subduction of the palaeo-Pacific plate at that time. The bounding faults of these basins formed above the amalgamated terrane boundaries. Continued strike-slip deformation throughout Cretaceous time in response to oblique subduction caused transpressional opening of successor basins that merged into larger basins. Transpressional basin inversion occurred in Tertiary time. Evidence for reverse fault reactivation of Cretaceous normal faults is provided by the geometry of short-cut faults, as exemplified by those in the La Calera region. Theoretical considerations of fault geometry, displacement to length ratios, and short-cut faults, indicates that the majority of the Tertiary faults are probably reactivated Mesozoic faults that themselves were reactivated Palaeozoic structures. Thus the initial geometry of accretionary prism fabrics and assembled terranes has exercised control over the geometry and location of all subsequent deformation events.

We are indebted to the late R. Caminos for his generosity in sharing his knowledge of the geology of the Eastern Sierras Pampeanas. C.S. expresses thanks to R. Martino for initially introducing us to the area. Logistical support from SEGEMAR (Córdoba branch) and the University of San Luis is gratefully acknowledged. Discussions with C.J. Northrup at an early stage of the work helped clarify our understanding of the tectonics. This work was supported by NSF grants EAR-9628158 and EAR-9903166 to C.S. and EAR-9628285 to L.P.G.

References

ABBRUZZI, J., KAY, S.M. & BICKFORD, M.E. 1993. Implications for the nature of Precordilleran basement from Precambrian xenoliths in Miocene volcanic rocks, San Juan Province, Argentina. *XII Congreso Geológico Argentina, Actas*, **III**, 331–339.

ACHILLI, F. & 26 OTHERS 1997. *Mapa Geológico de la República Argentina*. Servicio Geológico Minero Argentino, Buenos Aires.

ALLMENDINGER, R.W., RAMOS, V.A., JORDAN, T.E., PALMA, M. & ISACKS, B.L. 1983. Paleogeography and Andean structural geometry, northwest Argentina. *Tectonics*, **2**, 1–16.

ANDERSON, E.M. 1942. *The Dynamics of Faulting*. Oliver and Boyd, Edinburgh, 183.

ASTINI, R.A. 1998. Stratigraphical evidence supporting the rifting, drifting, and collision of the Laurentian Precordillera terrane of western Argentina. *In*: PANKHURST, R.J. & RAPELA, D.W. (eds) *The Proto-Andean Margin of Gondwana*. Geological Society, London, Special Publications **142**, 11–34.

ASTINI, R.A., BENEDETTO, J.L. & VACCARI, N.E. 1995. The early Paleozoic evolution of the Argentine Precordillera as a Laurentian rifted, drifted, and collided terrane: a geodynamic model. *Geological Society of America Bulletin*, **107**, 253–273.

BOBYARCHICK, A.R. 1998. Evolving concepts in tectonics and the interpretive history of the Brevard Fault Zone. *Geological Society of America Abstracts with Programs*, **30**, 4.

CAMACHO, A. & IRELAND, T.R. 1997. *U–Pb geochronology: final report. Geoscientific mapping of the Sierras Pampeanas. Argentine–Australian Cooperative Project*. Australian Geological Survey Organization, Canberra, A.C.T.

CAMINOS, R. 1979. Cordillera Frontal. *In*: TURNER, J.C.M. (ed.) *Segundo Symposio Geologiá Regional, Argentina*. **1**, 397–454. Academia Nacional de Ciencias, Córdoba.

CAMINOS, R. 1988. *Mapa geológico estructural del area situada entre Villa de Soto y La Candelaria, Córdoba*. Servicio Geologico Argentino (SEGEMAR), Buenos Aires.

CERREDO, M.E. 1996. Metamorphic evolution of high grade metapelites of Sierra de Comechingones, Córdoba, Argentina. *XIII Congreso Geológico Argentino y III Congreso de Exploración de Hidrocarburos, Actas*, **V**, 531.

CHARRIER, R. 1979. El Triásico de Chile y regiones adyacentes de Argentina; una reconstrucción paleogeográfica y paleoclimática: Santiago. *Communicacion—Departmento de Geologia, Facultad de Ciencias Fisicas y Matematicas, Universidad de Chile*, **26**, 1–37.

COIRA, B.L., DAVIDSON, J.D., MPODOZIS, C. & RAMOS, V.A. 1982. Tectonic and magmatic evolution of the Andes of northern Argentina and Chile. *Earth-Science Reviews*, **18**, 303–332.

COSTA, C. & VITA-FINZI, C. 1996. Late Holocene intraplate faulting in the southeast Sierras Pampeanas, Argentina. *Geology*, **24**, 1127–1130.

CRIADO ROQUE, P., MOMBRÚ, C. & RAMOS, V.A. 1981. Estructura e interpretatión tectónica. *In*: YRIGOYEN, M. (ed.) *Geología y Recursos naturales de la Provincia de San Luis. 8th Congreso Geológico Argentino, Relatorio*. Asociación Geológica Argentina, Buenos Aires, 155–192.

DALLA SALDA, L.H., CINGOLANI, C.A. & VARELA, R. 1992. Early Paleozoic orogenic belt of the Andes in southeastern South America: result of Laurentia–Gondwana collision? *Geology*, **20**, 617–620.

DALLA SALDA, L.H., DALZIEL, I.W.D., CINGOLANI, C.A. & VARELA, R. 1992. Did the Taconic Appalachians continue into South America? *Geology*, **20**, 1059–1062.

DALZIEL, I.W.D., DALLA SALDA, L.H. & GAHAGAN, L.M. 1994. Paleozoic Laurentia–Gondwana interaction and the origin of the Appalachian–Andean mountain system. *Geological Society of America Bulletin*, **106**, 243–252.

DE PAOR, D.G. & EISENSTADT, G. 1987. Stratigraphic and structural consequences of fault reversal: an example from the Franklin Basin, Ellesmere Island. *Geology*, **15**, 948–949.

DICKERSON, P.W. & KELLER, M. 1998. The Argentine Precordillera: its odyssey from the Laurentian Ouachita margin towards the Sierras Pampeanas of Gondwana. *In*: PANKHURST, R.J. & RAPELA, C. (eds) *The Proto-Andean Margin of Gondwana*. Geological Society, London, Special Publications **142**, 159–179.

DOKKA, R.K. & TRAVIS, C.J. 1990. Late Cenozoic strike slip faulting in the Mojave Desert, California. *Tectonics*, **9**, 311–340.

DOLL, W.E., DOMORACKI, W.J., COSTAIN, J.K., ÇORUH, C., LUDMAN, A. & HOPECK, J.T. 1996. Seismic reflection evidence for the evolution of a transcurrent fault system: the Norumbega fault zone, Maine. *Geology*, **24**, 251–254.

DORAIS, M.J., LIRA, R., CHEN, Y. & TINGEY, D. 1997. Origin of biotite–apatite-rich enclaves, Achala Batholith, Argentina. *Contributions to Mineralogy and Petrology*, **130**, 31–46.

DURAND, F.R. 1996. La transición Precámbrico-Cámbrico en el sur de Sudamérica. *In*: BALDIS, B. & ACEÑOLAZA, F.G. (eds) *Early Paleozoic Evolution in NW Gondwana*. Universidad Nacional de Tucumán, Serie Correlación Geológica **12**, 195–205.

ELLIOTT, D. 1976. The motion of thrust sheets. *Journal of Geophysical Research*, **81**, 949–963.

FANTINI, R., GROMET, L.P., SIMPSON, C. & NORTHRUP, C.J. 1998. Timing of high-temperature metamorphism in the Sierras Pampeanas of Córdoba, Argentina: implications for Laurentia–Gondwana interactions. *X Congreso Latinoamericano de Geología y VI Nacional de Geología Económica, Actas*, **II**, 388–392.

FERNANDO-SEVESO, F. & TANKARD, A.J. 1995. Tectonics and stratigraphy of the late Paleozoic Paganzo basin of western Argentina and its regional implications. *In*: TANKARD, A.J., SUAREZ SORUCO, R. & WELSINK, H.J. (eds) *Petroleum Basins of South America*. Memoirs, American Association of Petroleum Geologists **62**, 285–301.

GORDILLO, C.E. 1984. Migmatitas cordieríticas de la Sierras de Córdoba: condiciones físicas de la migmatización. *Boletín de la Academia Nacional de Ciencias, Córdoba*, **68**, 1–40.

GORDILLO, C.E. & LENCINAS, A.N. 1979. Sierras Pampeanas de Cordoba y San Luis. *II Simposio de Geologia Regional Argentina, Academia Nacional de Ciencias*, **I**, 577–650.

GROMET, L.P. & SIMPSON, C. 1999. Age of the Paso del Carmen pluton and implications for the duration of the Pampean Orogeny, Sierras de Córdoba, Argentina. *XIV Congreso Geologico Argentino, Salta, Actas*, **I**, 149–151.

HANDIN, J. 1969. On the Coulomb–Mohr failure criterion. *Journal of Geophysical Research*, **74**, 5343–5348.

HANMER, S. & PASSCHIER, C. 1991. Shear-sense indicators; a review. Geological Survey of Canada, Paper **90-17**, 72.

HUFF, W.D., BERGSTRÖM, S.M., KOLATA, D.R., CINGOLANI, C.A. & ASTINI, R.A. 1998. Ordovician K-bentonites in the Argentine Precordillera: relations to Gondwana margin evolution. *In*: PANKHURST, R.J. & RAPELA, C.W. (eds) *The Proto-Andean Margin of Gondwana*. Geological Society of London, Special Publications **142**, 107–126.

JORDAN, T.E. & ALLMENDINGER, R.W. 1986. The Sierras Pampeanas of Argentina: a modern analogue of Rocky Mountain foreland deformation. *American Journal of Science*, **286**, 737–764.

JORDAN, T.E. & ORTIZ, A. 1987. Tiempo de generacion del petroleo en Mendoza Norte; time of petroleum genesis in Mendoza Norte. *Actas del Congreso Geologico Argentino*, **10**, 271–276.

KARLSTROM, K.E. & DANIEL, C.G. 1993. Restoration of Laramide right-lateral strike slip in northern New Mexico by using Proterozoic piercing points: tectonic implications from the Proterozoic to the Cenozoic. *Geology*, **21**, 1139–1142.

KRAEMER, P.E., ESCAYOLA, M.P. & MARTINO, R.D. 1995. Hipótesis sobre la evolución tectónica neoproterozoica de las Sierras Pampeanas de Córdoba (30°40′–32°40′). *Revista de la Asociación Geológica Argentina*, **50**, 47–59.

KROL, M.A. & SIMPSON, C. 1999. $^{40}Ar/^{39}Ar$ Cooling ages from micas in the eastern Sierras Pampeanas accretionary prism rocks. *Geological Society of America, Abstracts with Programs*, **31**, 114–115.

LIRA, R., MILLONE, H.A., KIRSCHBAUM, A.M. & MORENO, R.S. 1996. Calc-alkaline arc granitoid activity in the Sierra Norte–Ambargasta Ranges, central Argentina. *Journal of South American Earth Sciences*, **10**, 157–177.

LUCERO MICHAUT, N.H., GAMKOSIAN, A., JARSUN, B., ZAMORA, Y.E., SIGISMONDI, M., CAMINOS, R. & MIRO, R. 1995. *Mapa Geológica de la Province de Córdoba, República Argentina, 1:500 000*. Ministerio de Economía y Obras y Servicios Públicos, SEGEMAR, Buenos Aires.

LUDMAN, A., LANZIIROTTI, A., LUX, D. & CHUNZENG, W. 1999. Constraints on timing and displacement of multistage shearing in the Norumbega fault system, eastern Maine. *In*: LUDMAN, A. & WEST, D.P. JR (eds) *Norumbega Fault System of the Northern Appalachians*. Geological Society of America, Special Papers **331**, 179–194.

LYONS, P., SKIRROW, R.G. & STUART-SMITH, P.G. 1997. *Report on Geology and Metallogeny of the 'Sierras Septentrionales' de Córdoba: 1:250 000 map sheet, Province of Córdoba. Geoscientific Mapping of the Sierras Pampeanas. Argentine–Australian Co-operative Project*. Australian Geological Survey Organization, Canberra, A.C.T.

MARTINO, R.D. 1993. La faja de deformación Guamanes; petrografía, estructura interna y significado tectónico, Sierra Grande de Córdoba, Argentina. *Revista de la Asociación Geológica, Argentina*, **48**, 21–32.

MARTINO, R.D., LAW, R.D. & SIMPSON, C. 1993*a*. Evidence for orthogonal contractional orogeny in the Pampean Ranges of Córdoba, Central Argentina. *EOS Transactions, American Geophysical Union*, 16, **74**, 302.

MARTINO, R.D., SIMPSON, C. & LAW, R.D. 1993*b*. Taconic- (Ocloyic-) aged west-directed ductile thrusts in basement rocks of the Sierras Pampeanas, Argentina. *Geological Society of America, Abstracts with Programs*, **25**, 6, A-233.

MCDONOUGH, M., RAMOS, V.A., ISACHSEN, C.E., BOWRING, S.A. & VUJOVICH, G.I. 1993. Nuevas edades de circones del basamento de la Sierra de Pie de Palo, Sierras Pampeanas Occidentales de San Juan: sus implicancies para los modelos del supercontinente proterozoico de Rodinia. *XII Congreso Geológico Argentina, Actas*, **III**, 343–357.

MPODOZIS, C. & RAMOS, V.A. 1990. The Andes of central Chile and Argentina. *In*: ERICKSON, G.E., CAÑAS PINOCHET, M.T. & REINIMUD, J.A. (eds) *The Geology of he Andes and its Relation to Hydrocarbon and Mineral Resources*. Circum-Pacific Council for Energy and Mineral Resources, Earth Science Series **11**, 59–90.

NAESER, C.W., DUDDY, I.R., ELSTON, D.P., DUMITRU, T.A. & GREEN, P.F. 1989. Fission-track dating: ages for Cambrian strata and Laramide and post-Middle Eocene cooling events from the Crand Canyon, Arizona. *In*: ELSTON, D.P., BILLINGSLEY, G.H. & YOUNG, R.A. (eds) *Geology of the Grand Canyon, Northern Arizona (with Colorado River Guides)*. American Geophysical Union, Washington, DC, 139–144.

NORTHRUP, C.J., SIMPSON, C. & GROMET, L.P. 1998*a*. Early Paleozoic history of the Eastern Pampeanas, Argentina: development of a Cambrian arc and accretionary prism along the margin of Gondwana. *X Congreso Latinoamericano de Geología y VI Nacional de Geología Económica, Actas*, **II**, 400–403.

NORTHRUP, C.J., SIMPSON, C. & HODGES, K.V. 1998*b*. pseudotachylyte in fault zones of the Sierras De Córdoba, Argentina: petrogenesis and $^{40}Ar/^{39}Ar$ geochronology. *Geological Society of America, Abstracts with Programs*, **30**, A-325.

OMARINI, R.H. 1991. The Puncoviscana Mobile Belt. *In*: VIRAMONTE, J.G. (ed.) *Horizontal and Vertical Mass and Energy Transport in Andean Type Orogeny, Alfred Wegener Conference Post-Conference Excursion Guide*. University of Salta, Argentina, 1–6.

OTAMENDI, J.E. & RABBIA, O.M. 1996. Petrology of high-grade gneisses from Macizo Rio Santa Rosa: evidence of decompression in the Eastern Sierras Pampeanas. *XIII Congreso Geológico Argentino y III Congreso de Exploración de Hidrocarburos, Actas*, **V**, 527.

PANKHURST, R.J., RAPELA, C.W., SAAVEDRA, J., BALDO, E., DAHLQUIST, J. & PASCUA, I. 1996. Sierras de Los Llanos, Malanzan and Chepes: Ordovician I- and S-type granitic magmatism in the Famatinian orogen. *XIII Congreso Geológico Argentino y III Congreso de Exploración de Hidrocarburos, Actas*, **V**, 415.

PANKHURST, R.J., RAPELA, C.W., SAAVEDRA, J., BALDO, E., DAHLQUIST, J., PASCUA, I. & FANNING, C.M. 1998. The Famatinian magmatic arc in the central Sierras Pampeanas: an Early to Mid-Ordovician continental arc on the Gondwana margin. *In*: PANKHURST, R.J. & RAPELA, C.W. (eds) *The Proto-Andean Margin of Gondwana*. Geological Society, London, Special Publications **142**, 343–367.

PASSCHIER, C.W. & SIMPSON, C. 1986. Porphyroclast systems as kinematic indicators. *Journal of Structural Geology*, **8**, 831–843.

PÉREZ, M.B., RAPELA, C.W. & BALDO, E.G. 1996. Geología de los granitoides del sector septentrional de la Sierra Chica de Córdoba. *XIII Congreso Geológico Argentino y III Congreso de Exploración de Hidrocarburos, Actas*, **V**, 493–505.

PIETERS, P. & SKIRROW, R. 1997. *Informe Geológico y Metalogénico de las Sierras de Chepes, Ulapes, y Las Minas, Provincia de La Rioja. Argentine–Australian Co-operative Project. Unpublished report 26*. Australian Geological Survey Organization, Canberra, A.C.T.

POWELL, C.M. 1987. Inversion tectonics in southwest Dyfed. *Proceedings of the Geologists Association*, **98**, 193–203.

RAMOS, V.A. 1984. Chilenia; un terreno alóctono en la evolucíon paleozoica de los Andes Centrales. *9th Congreso Geológico Argentino, Actas*, **II**, 84–106.

RAMOS, V.A. 1988. The tectonics of the central Andes, 30°–33°S latitude. *In*: CLARK, S.P. JR, BURCHFIEL, B.C. & SUPPE, J. (eds) *Processes in Continental Lithospheric Deformation*. Geological Society of America, Special Papers **218**, 31–54.

RAMOS, V.A. & KAY, S.M. 1991. Triassic rifting and associated basalts in the Cuyo basin, central Argentina. *In*: HARMON, R.S. & RAPELA, C.W. (eds) *Andean Magmatism and its Tectonic Setting*. Geological Society of America, Special Papers **265**, 79–91.

RAMOS, V.A., DALLMEYER, R.D. & VUJOVICH, G. 1998. Time constraints on the Early Palaeozoic docking of the Precordillera, central Argentina. *In*: PANKHURST, R.J. & RAPELA, C.W. (eds) *The Proto-Andean Margin of Gondwana*. Geological Society, London, Special Publications **142**, 143–158.

RAMOS, V.A., JORDON, T.E., ALLMENDINGER, R.W., MPODOZIS, C., KAY, S.M., CORTÉS, J.M. & PALMA, M. 1986. Paleozoic terranes of the central Argentine–Chilean Andes. *Tectonics*, **5**, 855–880.

RAPELA, C.W., PANKHURST, R.J., CASQUET, C., BALDO, E., SAAVEDRA, J. & GALINDO, C. 1998*a*. Las colisiones continentales Pampeana y Famatiniana. *X Congreso Latinoamericano de Geología y VI Nacional de Geología Económica, Actas*, **II**, 404.

RAPELA, C.W., PANKHURST, R.J., CASQUET, C., BALDO, E., SAAVEDRA, J., GALINDO, C. & FANNING, C.M. 1998. The Pampean Orogeny of the southern proto-Andes: Cambrian continental collision in the Sierras de Córdoba. *In*: PANKHURST, R.J. & RAPELA, C.W. (eds) *The Proto-Andean Margin of Gondwana*. Geological Society, London, Special Publications **142**, 181–218.

RAPELA, C.W., PANKHURST, R.J., CASQUET, C., BALDO, E., SAAVEDRA, J. & GALINDO, C. 1998*c*. Early evolution of the Proto-Andean margin of South America. *Geology*, **26**, 707–710.

SCHMIDT, C.J., ASTINI, R.A., COSTA, C.H., GARDINI, C.E. & KRAEMER, P.E. 1995. Petroleum Basins of South America. *In*: TANKARD, A.J., SUÁREZ SORUCO, R. & WELSINK, H.J. (eds) *Memoirs, American Association of Petroleum Geologists*. **62**, 341–358.

SIMPSON, C. & DE PAOR, D.G. 1993. Strain and kinematic analysis in general shear zones. *Journal of Structural Geology*, **15**, 1–20.

SIMPSON, C. & NORTHRUP, C.J. 1998. Diffusional mass transfer in the formation of banded gneisses: examples from the Sierras Pampeanas of Central Argentina. *EOS Transactions, American Geophysical Union*, **79**, S350.

SIMPSON, C., LAW, R.D.W., NORTHRUP, C.J. & MARTINO, R.D. 1998*a*. Crustal shortening of the Cambrian arc and post-metamorphic shear zones in the Sierras Pampeanas. *X Congreso Latinoamericano de Geología y VI Nacional de Geología Económica, Actas*, **II**, 394–399.

SIMPSON, C., UZZELLE, G.H. & DE PAOR, D.G. 1998. Pressure solution transpaction as an accommodation mechanism for granitic intrusion. *EOS Transactions, American Geophysical Union*, **79**, S344.

SIMS, J., IRELAND, T. R., CAMACHO, A. & 5 OTHERS 1998. U–Pb, Th–Pb and Ar–Ar geochronology from the Southern Sierras Pampeanas, Argentina: implications for the Paleozoic tectonic evolution of the western Gondwana margin. *In*: PANKHURST, R.J. & RAPELA, D.W. (eds) *The Proto-Andean Margin of Gondwana*. Geological Society, London, Special Publications, **142**, 259–281.

SIMS, J., STUART-SMITH, P., LYONS, P. & SKIRROW, R. 1997. *Informe geológico y metalogénico de Las Sierras de San Luis y Comechingones, Provincias de San Luis y Córdoba. Argentine–Australian Cooperative Project*. Australian Geological Survey Organization, Town.

THOMAS, W.A. & ASTINI, R.A. 1996. The Argentine Precordillera: a traveller from the Ouachita embayment of North American Laurentia. *Science*, **273**, 752–757.

THOMAS, W.A. & SIMS, W.J. 1996. Contrasts in reactivation history of basement faults at the margins of the Uncompahgre uplift, Ancestral Rocky Mountains. *Geological Society of America, Abstracts with Programs*, **28**, 447.

TIKOFF, B. & TEYSSIER, C. 1994. Strain modelling of displacement field partitioning in transpressional orogens. *Journal of Structural Geology*, **16**, 1575–1588.

TOSELLI, A.J., TOSELLI, R.J.N., PELLITERO, E. & SAAVEDRA, J. 1993. El arco magmático granitico del Paleozoico inferior en el Sistema de Famatina, Argentina. *XII Congreso Geológico Argentino, Actas*, **IV**, 7–15.

ULIANA, M.A. & BIDDLE, K.T. 1988. Mesozoic–Cenozoic paleogeographic and geodynamic evolution of southern South America: São Paolo. *Revista Brasileira de Geosciencias*, **18**, 172–190.

VUJOVICH, G.I. & KAY, S. 1998. A Laurentian? Grenville-age oceanic arc/back-arc terrane in the Sierra de Pie de Palo, Western Sierras Pampeanas, Argentina. *In*: PANKHURST, R.J. & RAPELA, C.W. (eds) *The Proto-Andean Margin of Gondwana*. Geological Society, London, Special Publications **142**, 159–179.

WEST, D.P. JR. 1999. Timing of displacements along the Norumbega fault system, south–central and south–coastal Maine. *In*: LUDMAN, A. & WEST, D.P. JR (eds) *Norumbega Fault System of the Northern Appalachians*. Geological Society of America, Special Papers **331**, 167–178.

WILLNER, A.P. & MILLER, H. 1986. Structural division and evolution of the Lower Paleozoic Basement in the NW Argentine Andes. *Zentralblatt für Geologie und Palaontologie, Teil I, 9*, **10**, 1245–1255.

Rheological partitioning during multiple reactivation of the Palaeozoic Brevard Fault Zone, Southern Appalachians, USA

ROBERT D. HATCHER JR

Department of Geological Sciences, University of Tennessee, Knoxville, TN 37996-1410, USA (e-mail: bobmap@utk.edu)

Abstract: The Brevard Fault Zone is a linear, NE-trending, gently to moderately SE-dipping fault zone traceable some 750 km from Alabama to Virginia in the crystalline southern Appalachians. It ranges from 1 to 3 km wide and contains a mappable lithostratigraphy. The Brevard Fault Zone has been interpreted as a thrust, strike-slip fault (both dextral and sinistral), a suture and terrane boundary, and a fundamental crustal tectonic boundary. Deformation was partitioned in space and time, and motion was both strike-slip (dextral) and dip-slip (thrust). Early strike-slip and thrust movement was coupled to map-scale structures in the deep Inner Piedmont, late Palaeozoic dextral motion was confined to a zone of 1–3 km width, and the latest reactivation consisted of brittle thrusting confined to a zone of 100 m width. The fault zone is cut by undeformed NW-trending Mesozoic dolerite dykes. The Brevard Fault Zone is characterized by the presence of a prominent retrograde (chlorite–muscovite stable) S–C fabric that indicates dextral motion. This fabric is related to late Palaeozoic (Alleghanian) dextral reactivation of the fault zone, with an unknown displacement at a time when huge volumes of fluid were fluxed through the zone. The deformation overprints an earlier (Acadian) high-temperature (garnet–staurolite–kyanite) fabric that also yields a dextral motion sense, and involved a component of thrusting. This mid-Palaeozoic deformation was coupled with west-directed, near-metamorphic peak thrusting and flow from the deep Inner Piedmont (to the east) that was buttressed against the primordial Brevard Fault Zone so that the motion became SW directed, and plastic flow became constricted in this narrow 1–3 km zone. Both of these plastic deformations were overprinted by late Alleghanian NW-directed dip-slip brittle deformation confined to the NW side of the Brevard Fault Zone. This last deformation involved at least 10 km of displacement and was related to reactivation of this block of crust as part of the late Alleghanian, NW-directed Blue Ridge–Piedmont megathrust sheet, and formed out of sequence with respect to the megathrust sheet. The Brevard Fault Zone was clearly a zone of crustal weakness that had a suitable mechanical stratigraphy that imparted sufficient anisotropy to localize the initial Acadian fault(s). Early Alleghanian fluid fluxing weakened the already strongly anisotropic fault zone and probably focused ductile reactivation at a shallower crustal depth during the early Alleghanian event. Late Alleghanian reactivation occurred even shallower as an almost discrete boundary in the brittle regime. This is one of the few faults in the Appalachians to have undergone deformational partitioning to permit multiple reactivation during Palaeozoic time.

Faults that become reactivated must by their very nature have properties that render them mechanically weak with respect to the surrounding crust. Necessary conditions for reactivation include the following: (1) a fault must exist for reactivation to occur and suitable mechanical conditions (e.g. anisotropy and weak zone in the mechanical stratigraphy) must exist for the fault to have formed in the first place; (2) the fault zone must remain weak (and not be annealed) through time; (3) the fault must continue to concentrate stress for a variety of reasons; (4) both the fault and the stress field must be suitably oriented for reactivation to occur. Other criteria for, and phenomena associated with, reactivation were discussed by Holdsworth *et al.* (1997).

Large faults occur in all orogenic belts, but relatively few are reactivated several times over millions of years; fewer are reactivated in both strike- and dip-slip regimes. The Gander–Avalon boundary in the Newfoundland Appalachians has undergone a history analogous to the Brevard Fault Zone. It was initially a zone of sinistral transpression that was overprinted by dextral ductile deformation. Later brittle faulting

From: HOLDSWORTH, R.E., STRACHAN, R.A., MAGLOUGHLIN, J.F. & KNIPE, R.J. (eds) 2001. *The Nature and Tectonic Significance of Fault Zone Weakening*. Geological Society, London, Special Publications, **186**, 257–271. 0305-8719/01/$15.00

was concentrated along some of the earlier ductile faults (Holdsworth 1994). Holdsworth concluded that strain softening of the crust here permitted later concentration of the younger faults and shear zones along this boundary. The Outer Hebrides Fault Zone in Scotland is another extensively reactivated fault zone involving initial sinistral strike-slip movement with accompanying influx of hydrous fluids into the zone that produced extensive retrogression. This resulted in a zone of weakened crust that was reactivated again in extension (Imber *et al.* 1997). One of the best examples of inheritance and reactivation of earlier faults lies in the Late Triassic–Early Jurassic border faults of the Mesozoic basins of eastern North America. Wherever they are not covered by Coastal Plain sediments from the Carolinas to Newfoundland, these Mesozoic faults are localized on segments of older Palaeozoic faults that were reactivated during the new rifting cycle (precursor to opening of the Atlantic) (Manspeizer & Cousminer 1988).

The purpose of this paper is to discuss one of the largest faults in the Appalachians, the Brevard Fault Zone, as an example of a fault that has been reactivated throughout Palaeozoic time and presumably at different erosional depths so its rheological properties changed through time. The Brevard Fault formed in continental crust initially at great depth under ductile conditions (the viscous creep regime of Schmid & Handy (1991)), but was reactivated under ductile conditions near the brittle–ductile transition (the frictional–viscous transition of Schmid & Handy (1991)), and then again under the brittle conditions (the brittle–frictional flow regime of Schmid & Handy (1991)) of the upper crust.

The Brevard Fault Zone is a very linear NE-trending fault zone that is traceable some 750 km from Alabama to Virginia in the crystalline southern Appalachians (Fig. 1). It ranges from 1 to 3 km wide, dips gently to moderately toward the SE, and appears listric in seismic reflection profiles (e.g. Costain *et al.* 1989). This fault zone has been the subject of controversy for many years because it separates two major geological realms and has a complex history. It also is easily recognized and studied in the field, on topographic maps, and in satellite imagery. The attributes of the Brevard Fault Zone on which there is agreement include its extent from Alabama to Virginia, its linearity, presence of a polyphase history of ductile and

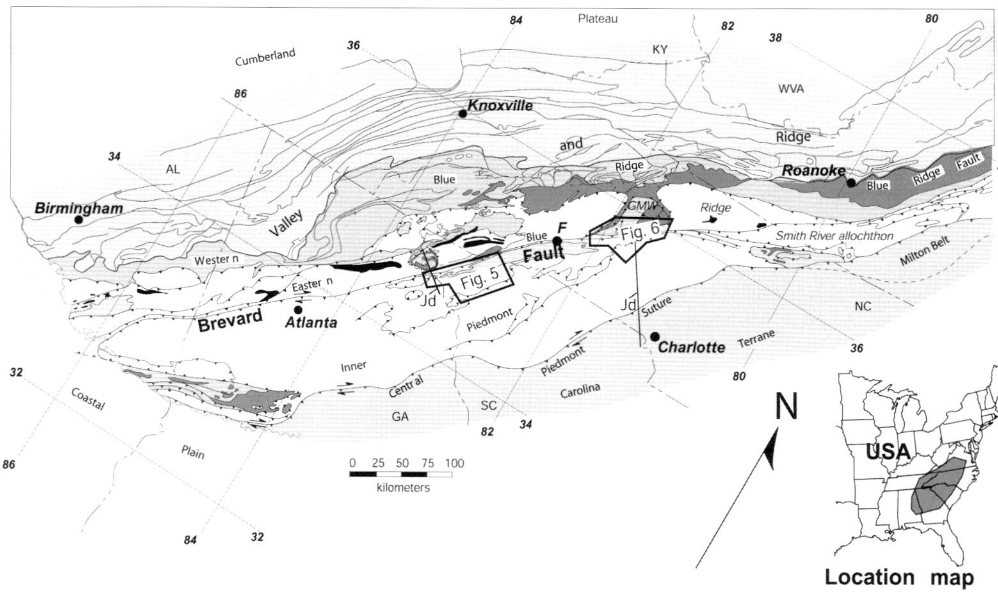

Fig. 1. Simplified tectonic map of the southern Appalachians showing the location of the Brevard Fault Zone. Boxes indicate locations of Figs. 5 and 6. GMW, Grandfather Mountain window; F, Fletcher, North Carolina; Jd, Jurassic dolerite dykes (continuous lines represent overlapping dyke segments). The foreland fold–thrust belt (Valley and Ridge and Cumberland Plateau) have a light screen; the rifted margin (the western Blue Ridge, Sauratown Mountains Window and Pine Mountain Window) has a medium screen, with 1.1 Ga Grenvillian basement inliers shown in a dark screen. The eastern Blue Ridge and Inner Piedmont are white, and the Carolina exotic terrane and Milton belt have a light–medium screen. Black areas are Palaeozoic plutons.

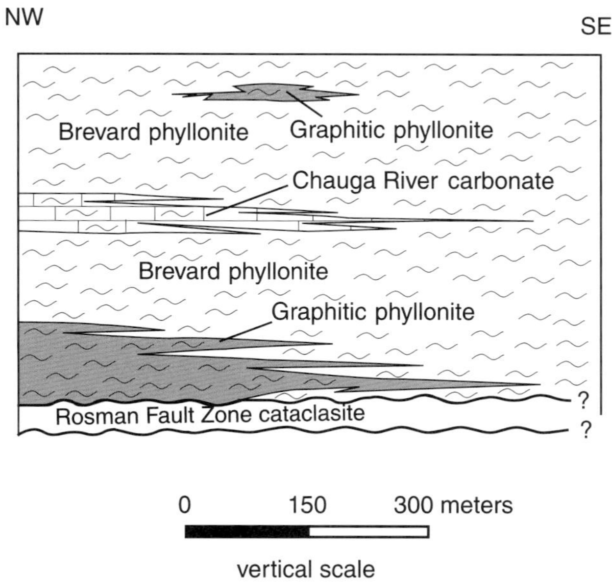

Fig. 2. Lithostratigraphic succession in the Brevard Fault Zone (Chauga River Formation). Thickness is unknown, but the present thickness is undoubtedly less than the original thickness. All of the rocks in this sequence are mylonitic. The asymmetry of the sequence and presence of both graphitic and carbonate rocks precludes generation of this sequence from tectonic processes alone. This unit is mappable continuously from northeastern Georgia through South Carolina into North Carolina, and probably continues farther NE and SW but cover of much of the Brevard Fault Zone by Quaternary stream deposits over much of its extent prevents recognition of the sequence elsewhere.

brittle motion resulting in retrogression of earlier prograde rock sequences, listric flattening of the fault zone toward the SE in seismic reflection profiles (Costain *et al.* 1989), a mappable lithostratigraphy for at least 200 km (Fig. 2), stratigraphically controlled faults that propagated into strongly anisotropic (Christensen & Szymanski 1988) graphitic and micaceous rocks, and exotic horses of platform (Cambro-Ordovician Knox Group) carbonate rocks in the fault zone along the northwestern brittle boundary with the Blue Ridge (called the Rosman Fault by Horton & Butler (1986)) and a horse of basement(?) rocks on its southeastern flank.

The Brevard Fault Zone is a complex structure that had an earlier (Acadian, Late Devonian–Early Carboniferous) history of dip- and strike-slip motion and a later (Alleghanian, Late Carboniferous–Early Permian) history of both strike- and dip-slip motion. It is a major structure within the Blue Ridge–Piedmont megathrust sheet (see Hatcher 1989; Hatcher & Hooper 1992). The Brevard Fault was first recognized by Keith (1907), who interpreted it as a syncline of lower-grade rocks flanked by higher-grade rocks. Since Keith's discovery, this structure has probably been interpreted in more different ways than any other structure in the world. In addition to a syncline, it has been interpreted as a thrust (Jonas 1932; Hatcher 1971, Hatcher, 1972), a dejective zone (King 1955), a large fault with a later history of strike-slip movement (King 1959), a sinistral strike-slip fault (Reed & Bryant 1964; Higgins 1966), an Alpine root zone (Burchfiel & Livingston 1967), a dextral strike-slip fault with a thrust component (Reed *et al.* 1970), a back-limb thrust (Hatcher 1971), a stratigraphically controlled thrust (Hatcher 1975, Hatcher, 1978), an Abscherung zone (Griffin 1971), a suture (Dewey & Kidd 1974), a transported suture (Rankin 1975), and a dextral fault. Edelman *et al.* (1987) recognized that the Brevard Fault Zone marks the known western edge of the Alleghanian strike-slip domain in the southern and central Appalachians (Hatcher *et al.* 1989). Roper & Justus (1973) outlined a number of additional alternative interpretations of the Brevard Fault Zone. Bobyarchick (1984) first confirmed the dextral movement history on the Brevard Fault by correctly interpreting the well-developed S–C fabric and other shear-sense indicators. Edelman *et al.* (1987); Bobyarchick *et al.* (1988) summarized the evidence for both dip-slip and strike-slip motion on the fault zone. Hatcher (1978; 1989) summarized the reasons the Brevard Fault Zone cannot be a suture, principally because the same sequence of paragneisses (Tallulah Falls–Ashe Formation) occurs on both sides of it. In addition, we now know that this sequence on both sides of the Brevard Fault Zone contains detrital zircons with the same Grenvillian provenance (Bream *et al.* 2000).

Timing of movement

Several lines of evidence provide clues to the timing of movement of the Brevard Fault Zone. Correlation of deformational events with prograde and retrograde metamorphic assemblages and rheological state of the rocks when deformation occurred provides some broad limits. Earliest deformation occurred near peak metamorphism. U–Pb monazite ages (Dennis & Wright 1997) and U–Pb ion probe ages of zircon rims (Bream *et al.* 2000) from the Inner Piedmont in the range of 360 Ma suggest that peak metamorphism occurred at this time (Late Devonian–Early Carboniferous, Acadian Orogeny). Odom & Fullagar (1973) also reported a 360 Ma Rb–Sr age from Henderson orthogneiss mylonite in the Brevard Fault Zone, and Sinha *et al.* (1988) determined a Rb–Sr age of 273 Ma as a time of 'fluid-enhanced deformation' of the Henderson orthogneiss accompanying retrograde metamorphism. The Henderson orthogneiss has a crystallization age of 509 Ma (Rb–Sr, Sinha *et al.* 1989) or 492 Ma (U–Pb ion probe, Vinson 1999). The 360 Ma age may be related to peak Acadian metamorphism, whereas the 273 Ma age probably correlates with Alleghanian reactivation under lower temperature–pressure conditions. The process of fluid-enhanced deformation described by Sinha *et al.* (1988) could have affected weakening of the fault zone so that it was reactivated several times throughout this event and focused later reactivation.

Post-Palaeozoic deformation on the Brevard Fault Zone is difficult to document, and it appears not to have been reactivated again after the Alleghanian events. Any post-Palaeozoic movement, however, had to have occurred before intrusion of NW-striking Mesozoic dolerite dykes (Jurassic, Ragland *et al.* 1983) that cross the Brevard Fault Zone in at least two locales separated by more than 150 km (Fig. 1). These dykes are undeformed and cross the fault zone without any displacement. Nearly east–west coeval Mesozoic faults containing siliceous cataclasite also appear to cross the Brevard Fault Zone in the Carolinas (Garihan *et al.* 1988; Dennison *et al.* 1997). These structures did not reactivate the Brevard Fault Zone possibly because the stress field was not suitably oriented, or because the stress field was tensional.

Early history

The Brevard Fault Zone had an early movement history that was coeval with formation of many of the penetrative structures (foliations, mineral lineations) in the eastern Blue Ridge and Inner Piedmont. This history produced at least three sets of superimposed ductile folds (two sets are not coaxial), and at least two of these events transposed the rocks sufficiently to produce regionally penetrative foliations (see Hopson & Hatcher 1988). Both Taconian (Mid-Ordovician to Early Silurian) and Acadian foliations are present in the eastern to central Blue Ridge. These events have been recognized using U–Pb age dates on foliated plutons, such as the 467 Ma Whiteside Granite, the 455 Ma Persimmon Creek tonalite, and the 375 Ma Rabun pluton in the eastern Blue Ridge (Miller *et al.* 2000) and the 464 Ma Dysartsville granodiorite (Bream *et al.* 2000) in the Inner Piedmont. Relict (Taconian?) foliations present in boudins and intrafolial folds are wrapped by the dominant (Acadian) foliation in the easternmost Blue Ridge and Inner Piedmont. These earliest structures have been transposed by the Alleghanian overprint in the Brevard Fault Zone, and are visible only as relict high-grade assemblages preserved as aligned inclusion trails in unretrograded garnets (see photographs in the study by Roper & Dunn (1973)).

Evidence for the early (Acadian?) history of the Brevard Fault Zone consists primarily of ties to the high-temperature structures (formed under middle- to upper-amphibolite facies conditions) in the Inner Piedmont. Brevard phyllonite occurs SE of the Brevard Fault Zone in a lower-grade part of the Inner Piedmont called the Chaiga belt (Hatcher 1972) as part of the Poor Mountain Formation (Hatcher 1969). There it consists of a lower- to middle-amphibolite facies assemblage and does not contain the strong Alleghanian retrograde overprint characteristic of the Brevard Fault Zone. Moreover, the quartz slip system in the rocks of this part of the Inner Piedmont is typical of high-temperature ductile deformation processes (Davis 1993). Movement of thrust sheets in the Inner Piedmont appears to be tracked by the prominent, high-temperature mineral stretching lineation (composed of quartz, feldspar, micas) that occurs throughout the Inner Piedmont and becomes transposed by younger (early Alleghanian) C and L fabrics in the Brevard Fault Zone (Davis *et al.* 1991; Hatcher 1998). Davis *et al.* (1991); Davis (1993) first suggested that this lineation tracks the ductile flow of the plastic Type F (fold-related) thrust sheets (Hatcher & Hooper 1992) that dominate here. The primordial Brevard Fault Zone in this model served as a buttress to deflect westward-moving thrust sheets in the central part of the Inner Piedmont toward the SW along the western margin (Fig. 3). The buttressing effect is indicated by the strong coaxial NE–SW mineral lineation within and immediately SE of the Brevard Fault Zone and less strong alignment farther east. The net result would have been dextral flow with respect to the Blue Ridge to the NW, and sinistral flow with respect to the more internal (central and eastern) parts of the Inner Piedmont. This phenomenon is very similar to that described by Jones *et al.* (1997); Dewey *et al.* (1998) as lateral extrusion (or escape) in transpression zones. In the present example, the Brevard Fault Zone would have buttressed the obliquely converging Inner Piedmont thrust sheets and caused them to extrude (escape) toward the SW. At least two of their three boundary conditions, wrench simple shear and pure shear in *xy*, can be identified in the mesofabrics present in the Brevard Fault Zone. Tikoff & Tessier (1994) might also have considered this to be a form of strain partitioning of contractional and strike-slip components in an obliquely converging strain field.

Alleghanian Late Carboniferous–Early Permian history

The Carboniferous–Permian Alleghanian Orogeny was the culminating event in the formation of the Appalachian orogen. The classic foreland fold–thrust belt that makes up the external part of the southern and central Appalachians is a product of the collision of Africa with Laurentia driving the Blue Ridge–Piedmont megathrust sheet forelandward and pushing the foreland fold–thrust belt snowplough-like in front of it.

The Brevard Fault Zone was reactivated from its primordial beginnings in the Acadian Orogeny as the northwesternmost boundary of a domain of dextral strike-slip faults that formed

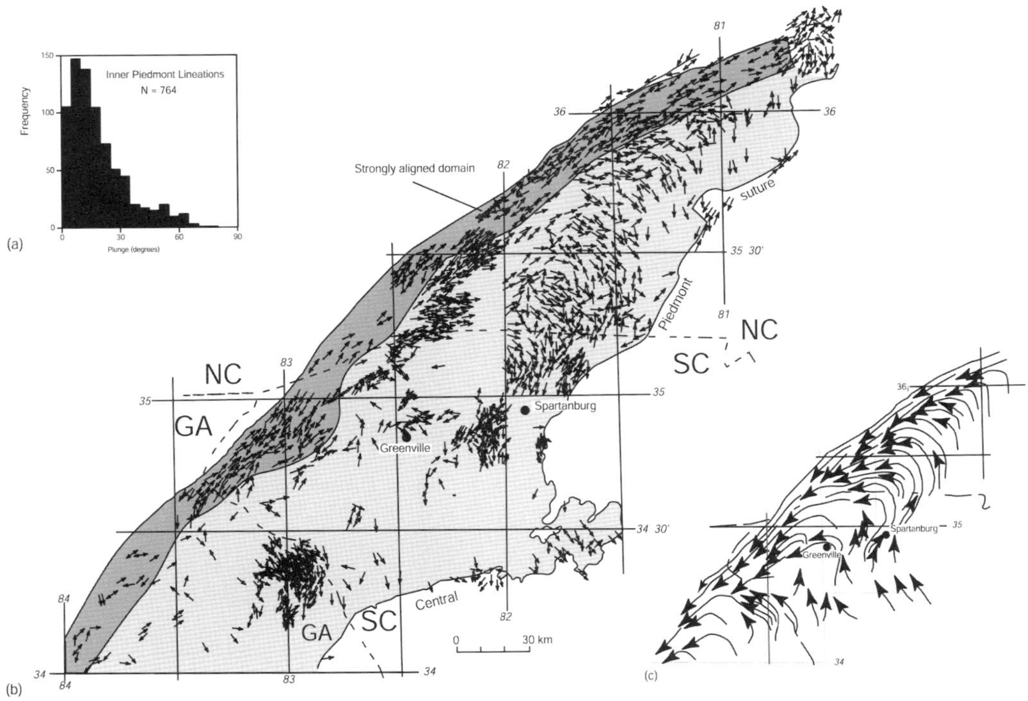

Fig. 3. Mineral stretching lineation pattern in the Inner Piedmont and interpretation. It should be noted that the lineations become closely aligned along the Brevard Fault Zone, suggesting constrictional flow as thrust sheets were buttressed against the fault zone and deflected toward the SW. (**a**) Histogram of plunges of lineations indicating an overall shallow plunge. (**b**) Lineation map. (**c**) Interpreted flow pattern.

during the initial stages of transpressional collision in the central to eastern parts of the mountain chain (Hatcher 1989). Sufficient uplift of the mountain chain had occurred by this time and enough fluid was present that high-temperature assemblages were strongly retrograded to lower greenschist facies (quartz–chlorite–muscovite) during the formation of a strong ductile S–C fabric in the 1–3 km wide fault zone (Fig. 4). Sinha *et al.* (1988) demonstrated that enormous amounts of fluid had moved through the fault zone during formation of the mylonitic rocks associated with this event, resulting in extensive metasomatism and migration of a number of mobile elements in the Ordovician Henderson orthogneiss. The prominent imbricate structure in the Brevard Fault Zone formed at this time (Figs 5 and 6). Initially, this map pattern suggests thrust faulting, but the pattern can similarly be explained by dextral strike-slip motion. Part of the complexity of the map pattern is derived from the crosscutting relationships between earlier thrusts cut and displaced by dextral faults that belong to this event. Reed & Bryant (1964), using map and quartz *c*-axis data, suggested that the prominent, strongly NE-trending, subhorizontal mineral lineation present here was produced by strike-slip motion.

Late Alleghanian history

The late Alleghanian history of the Brevard Fault Zone is probably related to the final emplacement history of the Blue Ridge–Piedmont thrust sheet and thus to the deformation of the foreland fold–thrust belt. This event is manifested as a cataclasite zone of 100 m thickness (Fig. 7) along the western margin of the primary fault zone. This zone of cataclasite and associated faulting, called the Rosman Fault, is indicative of shallower burial and low-temperature deformation. The Rosman Fault locally contains horses of platform carbonate rocks that were probably derived from the footwall beneath the megathrust sheet. The location in South Carolina of one of these horses is shown in Fig. 5. Direct evidence of low-temperature deformation (probably below 350 °C) is derived from several generations of stylolitic surfaces present in the carbonate horses, although the earliest of the selvages is recrystallized to chlorite. Additional

Fig. 4. Photomicrograph of Brevard phyllonite from South Carolina. The strong S–C fabric and dextral shear sense in the rock should be noted, along with the annealed quartz vein and garnets that have not been retrograded, indicating upper-greenschist to lowermost-amphibolite facies conditions of deformation. Graphite and chlorite are also present in the mica fish. Width of field is 11.1 mm.

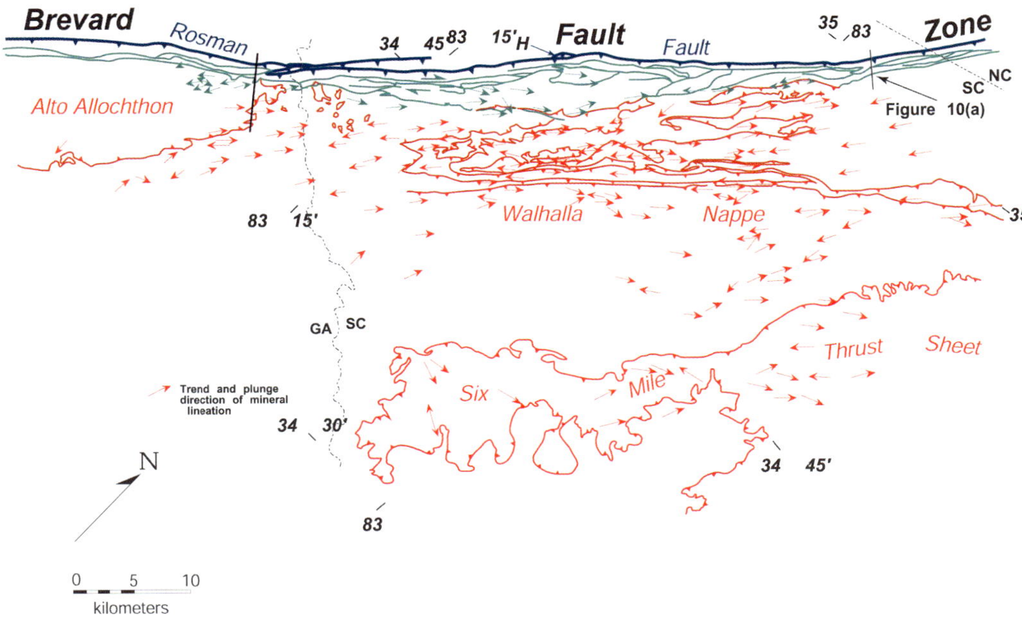

Fig. 5. Faults and orientations of mineral stretching lineations in the western Inner Piedmont in northwestern South Carolina, and nearby Georgia and North Carolina. Red indicates early structures associated with the Acadian (Late Devonian–Early Carboniferous) high-temperature ductile thrusting in the Inner Piedmont. Red arrows show orientation of mineral stretching lineations related to the early deformation event. Green indicates faults associated with the early Alleghanian (probably late Early Carboniferous) dextral strike-slip phase of deformation (ductile). Blue indicates Late Alleghanian (Late Carboniferous–Early Permian) brittle faulting along the boundary between the Brevard Fault Zone and the Blue Ridge. H, Carbonate horse; Jd, Jurassic dolerite dykes (undeformed). Location shown in Fig. 1.

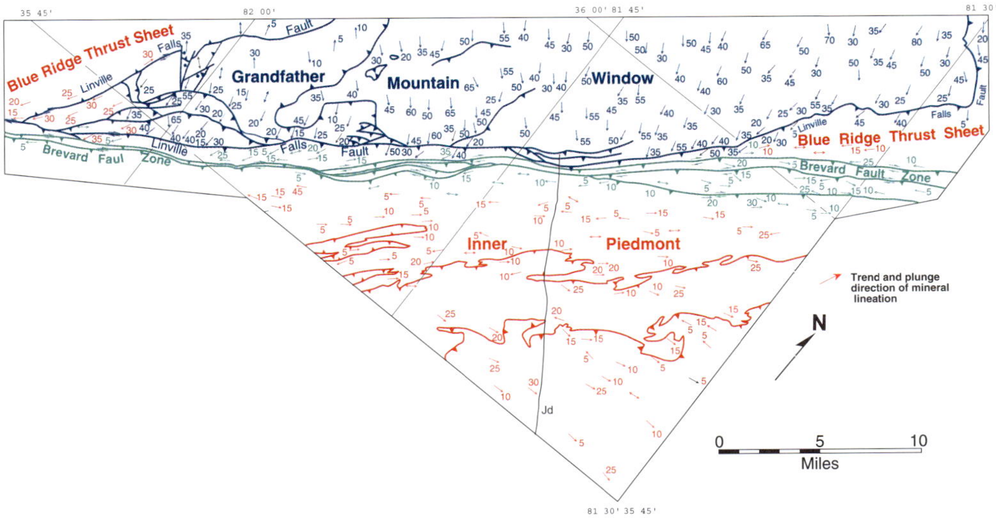

Fig. 6. Faults and orientations of mineral stretching lineations from the southern part of the Grandfather Mountain window across the Brevard Fault Zone and into the Inner Piedmont. Early, high-temperature lineations are overprinted by similarly oriented early Alleghanian mineral lineations that are characterized by a retrograde mineral assemblage of chlorite and muscovite. Red indicates faults and lineations related to the early Acadian deformation event. Green indicates faults and lineations related to the early Alleghanian dextral deformation. Blue indicates faults and lineations related to the late Alleghanian deformation along the northwestern edge of the Brevard Fault Zone and to the emplacement of the Blue Ridge Thrust Sheet. It should be noted that the Jurassic dolerite dykes (Jd, undeformed) crosscut all of the thrust and strike-slip faults in this diagram. Location shown in Fig. 1. Modified from Reed & Bryant (1964).

evidence for low-temperature deformation is derived from prominent grain-size reduction and down-dip-oriented slickensided surfaces observed in some of the larger exposures of the Rosman Fault, with steps indicating hanging-wall up (thrust) motion. Horton & Butler (1986); Bobyarchick *et al.* (1988) concluded that the motion on the Rosman Fault is dip-slip, with the hanging wall up.

The Rosman Fault probably formed as an out-of-sequence (back-limb) thrust, possibly as the frontal (Blue Ridge) part of the Blue Ridge–Piedmont thrust sheet slowed or partially locked. The Rosman Fault propagated into the suitably oriented western flank of the existing Brevard Fault Zone, possibly rotating more gently dipping earlier faults into a more moderate dip (Fig. 7). Minimum dip-slip displacement of the Rosman Fault is 10 km, based on retro-deformation of the existing fault geometry (Hatcher 1971, 1972), and more recently observed with the aid of seismic reflection profiles in South Carolina (Costain *et al.* 1989).

Exotic carbonate horses are one of the primary lines of evidence for dip-slip thrust motion along the Rosman Fault, the northwesternmost fault in the Brevard Fault Zone. They are composed of dolomite and limestone and minor amounts of black chloritic phyllonite and schist (Fig. 8). The carbonate components are chemically most similar to rocks of the Upper Cambrian–Lower Ordovician Knox Group in the Valley and Ridge foreland fold–thrust belt (Hatcher *et al.* 1973). The most likely source of the platform carbonate rocks is the footwall beneath the Blue Ridge–Piedmont megathrust sheet, suggesting that the horses were emplaced as the Brevard Fault Zone acted as an out-of-sequence thrust. The outcrop pattern of repetition of units where the Brevard Fault Zone is sufficiently well exposed and can be mapped in detail also suggests dip-slip (thrust) displacement (Fig. 5). Slickenlines that indicate dip-slip motion occur in a large exposure of the Rosman Fault in a quarry at Fletcher, North Carolina (Fig. 9). The Rosman Fault is also the structure in the Brevard Fault Zone that produces the strong reflector(s) on seismic reflection profiles (see Costain *et al.* 1989; Hatcher *et al.* 1989, plate 8). Additional discussion of timing relations of the Rosman and other Brevard faults may be found in the studies by Hatcher (1978, 1989), Horton & Butler (1986), and Hatcher *et al.* (1989).

Discussion

The Brevard Fault Zone obviously meets several of the criteria required for reactivation of faults outlined at the beginning of this paper, but we do not know the degree to which multiple reactivation occurred within any of the events affecting this fault zone. This scenario of an early ductile event at elevated temperatures and pressures followed by later reactivation in upper-crustal conditions that produce cataclasis is typical for faults that undergo reactivation (Holdsworth *et al.* 2001). Other major faults, such as the Alpine Fault in New Zealand, where Reed (1964) recognized Late Jurassic to Early Cretaceous mylonite superimposed by retrograde mylonite and then Quaternary cataclasite that has been thrust over glacial moraine deposits, have experienced recurrent movement and fluid fluxing through the fault zone similar to that in the Brevard Fault Zone. The western boundary of the Alpine Fault Zone is likewise considered a major thrust boundary (Reed 1964). The Alpine Fault was also probably weakened by fluid being pumped through the fault zone and then it subsequently remained suitably oriented to be reactivated several times in both the shallow ductile and brittle regimes.

Detachment fault zones (crustal listric normal faults formed by lithospheric extension, e.g. Wernicke 1985) in the Basin and Range province commonly contain a plastic, mylonitic lower part and a brittle upper part. The plastic lower portion is commonly best developed in the footwall rocks, whereas the brittle upper part is best developed at the contact with the upper plate (Coney 1980; Davis 1980), an inverted relationship compared with compressional faults such as the Brevard Fault.

The earlier (Acadian) movement events on the Brevard Fault Zone occurred well below the brittle–ductile transition at maximum temperatures ranging from 535 to 670 °C and pressures in the range of 3–5 kbar, near minimum melting conditions for these rocks (Davis 1993). Movement on the Brevard Fault Zone would probably have been accomplished by steady-state plastic flow as the Inner Piedmont block moved southwestward alongside the eastern Blue Ridge.

The block of continental crust that contains the Brevard Fault would have been exhumed over periods of millions of years (into Late Carboniferous and Early Permian time) until the present erosion level was just below the brittle–ductile transition. This also corresponded to the time when the initial transpressional collision of Laurentia with Gondwana occurred and the major strike-slip system developed in the Appalachian internides (Mosher 1983; Hatcher *et al.* 1989). The Brevard Fault was reactivated at this time under greenschist facies conditions as the northwesternmost boundary of the early Alleghanian strike-slip domain in the southern Appalachians (Hatcher 1989). The displacement on the various Brevard faults that moved at this time has not been determined because displacement markers have not been recognized across the fault zone. Reed *et al.* (1970) estimated the strike-slip displacement to be *c.* 200 km: of the same order of magnitude as the dip-slip displacement of the Blue Ridge–Piedmont megathrust sheet. Detailed geological mapping and structural studies along several segments and in the best exposed segment across South Carolina have neither disputed this figure nor provided a more precise estimate for the displacement.

The Rosman Fault formed at a time when the crust had been exhumed and unloaded sufficiently to produce frictional–brittle behaviour at a depth corresponding to the present erosion level. This probably is related to the time (the Early Permian period) when Laurentia and Gondwana were in head-on collision and the Blue Ridge–Piedmont megathrust sheet had formed, was moving northwestward, and was pushing the foreland fold–thrust belt ahead of it (Fig. 8). It is likely that the more frontal parts of the megathrust sheet slowed their forward motion so that some of the excess energy was utilized in reactivation of the mechanically weak northwestern margin of the Brevard Fault Zone as a brittle to semibrittle backlimb or out-of-sequence thrust. Abundant mesoscopic and map-scale brittle thrusts exist here, but chlorite remains a stable phase in the Rosman Fault Zone. The present erosion level could have, at that time, been right at the ductile–brittle transition, and a more rapid strain rate along the fault zone may have forced its behaviour into the frictional–brittle realm, whereas slower rates may have permitted plastic behaviour to prevail. The exotic carbonate horses (Figs. 8 and 9) along the Rosman Fault coupled with the down-dip slickenfibres in the fault zone and extensive brittle behaviour suggest that the Rosman Fault represents an end member of partitioning in time and rheological behaviour from the other Brevard faults.

Formation of the composite Brevard Fault Zone involved deformation in the deep crust, in the shallower crust still below the ductile–brittle transition, and in the upper crust probably within or very close to the ductile–brittle transition (Fig. 10). The change in thickness of each of the fault zones related to each separate event is at least an order of magnitude from the thickness of the Acadian fault zone (>10 km) to the early Alleghanian fault zone (1–3 km) to the late Alleghanian brittle fault zone (100 m). Time and exhumation of the crust effectively superimposed the zones on each other at the same present erosion level. Undoubtedly, each separate event involved movement throughout the crust, and each would have produced a section not unlike the ideal fault zone described by Sibson (1977), with the exception of thinning of the fault zone toward the more brittle components and toward the surface. Handy (1990) suggested that the deformation in contractional faults migrates through time toward the footwall. The behaviour of the Brevard Fault Zone during various stages of reactivation may typify this behaviour.

Conclusions

(1) The Brevard Fault Zone originally formed in a strongly anisotropic mechanical stratigraphy that was part of the southern Appalachian Acadian (Late Devonian–Early Carboniferous) crystalline core. Initial faults propagated under middle-amphibolite facies conditions into this suitably oriented stratigraphy in response to the buttressing affect of westward advancing thrust sheets in the deeper Inner Piedmont. West-directed thrusting buttressed against NE–SW-oriented anisotropy that produced extrusion or escape as thrusting was deflected into SW-directed dextral strike slip. This event was driven by the collision and docking of the Carolina Exotic Terrane with Laurentia.

(2) The mylonitic products of Acadian deformation enhanced the anisotropy in the Brevard Fault Zone so that plastic reactivation occurred under greenschist facies conditions as the orogen was erosionally unroofed by the beginning of the Alleghanian (Late Carboniferous–Permian) Orogeny. Enormous volumes of fluid were fluxed through the Brevard Fault Zone at this time, probably weakening it and localizing several episodes of renewed dextral movement. At this time, the Brevard Fault Zone formed the westward boundary of a domain of dextral strike-slip deformation that was active in the central and eastern parts of the orogen. The early Alleghanian events mark the onset of transpressional collision between Laurentia and Gondwana.

(3) Continued uplift of the mountain chain occurred as a result of the collision already under way. As Laurentia and Gondwana rotated toward each other, early transpression gave way to head-on collision and the main phase of the Alleghanian Orogeny (in Early Permian time).

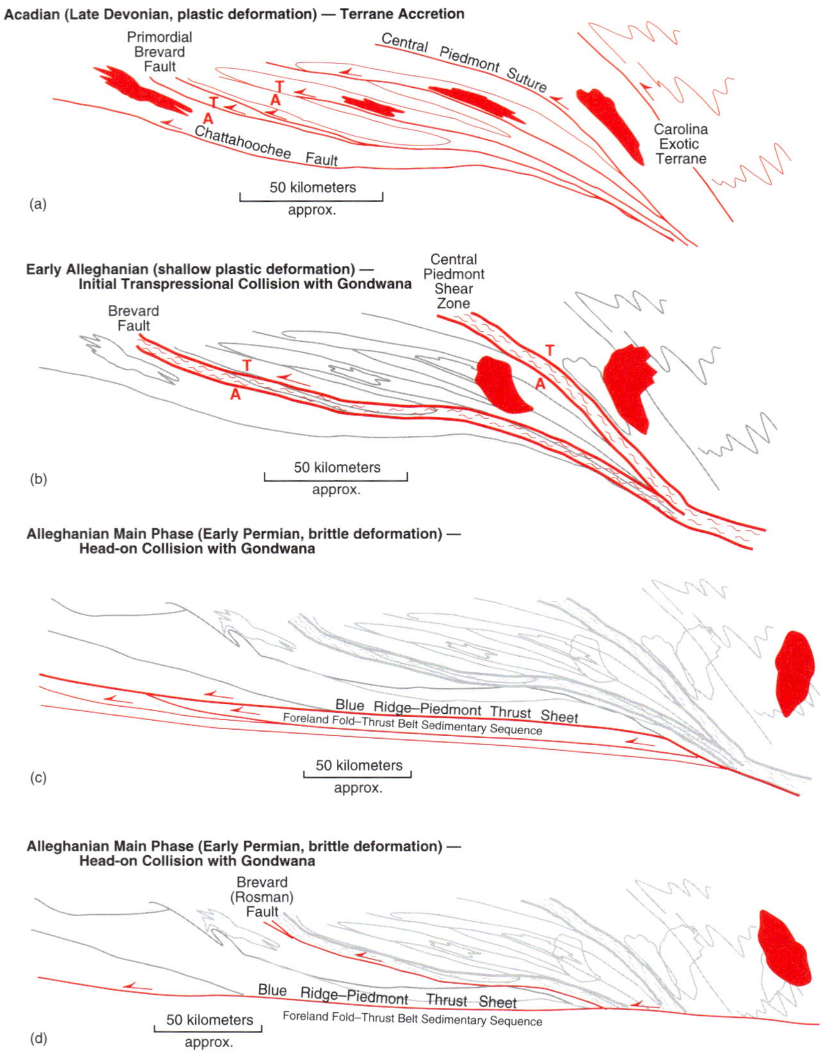

Fig. 7. Sequential development of the Brevard Fault Zone through mid- and late Palaeozoic time. Structures and other boundaries active at a particular time are indicated by red lines; inactive features are indicated by grey lines. Plutons actively being intruded are shown as red filled areas; cooled plutons are indicated by grey outline shapes. T indicates relative motion toward observer; A indicates relative motion away from observer. (**a**) Acadian (*c.* 360 Ma) plastic deformation, metamorphism, and plutonism in the Appalachian Inner Piedmont and formation of the primordial Brevard Fault Zone. This event is thought by the author to be related to docking of the Carolina Exotic Terrane with Laurentia. (**b**) Early Alleghanian (*c.* 325 Ma) renewed dextral movement on and fluid fluxing through both the Brevard Fault Zone and the Central Piedmont Suture, producing retrograde (greenschist facies) metamorphism and some plutonism. This renewed movement and weakening of these faults was driven by the initial transpressive collision of Laurentia and Gondwana. (**c**) Main phase of the Alleghanian Orogeny; initiation of the Blue Ridge–Piedmont megathrust sheet and propagation of the sheet into the sedimentary rocks of the Palaeozoic Laurentian margin. This phase of deformation, metamorphism, and plutonism is related to head-on collision of Laurentia with Gondwana. (**d**) Continued Alleghanian main-phase deformation (possibly near the end of the orogeny) and cratonward movement of the Blue Ridge–Piedmont thrust sheet. At this time, the frontal parts of the sheet may have become locked and the hinterlandward parts continued to move forward, propagating renewed movement into the already weak Brevard Fault Zone and producing the brittle Rosman Fault along its northwestern margin.

Fig. 8. Sample from the carbonate horse located in Fig. 5. Light material is dolomite with some calcite and silica (recrystallized chert?); dark material is chlorite on stylolitic selvages and a small amount of graphite. Deformation in the carbonate is penetrative and ductile superimposed by brittle deformation that produced faults and stylolites. Coin is a US dime (10 cent piece, 18 mm diameter).

Fig. 9. Exposure of carbonate horse along the SE-dipping Rosman Fault in a quarry near Fletcher, North Carolina. The dip here is *c*. 45° and the mineral lineation plunges directly down the dip. Steps are present on the fault surface, indicating that the last motion was hanging-wall up. View is toward the NE. (Note truck for scale.)

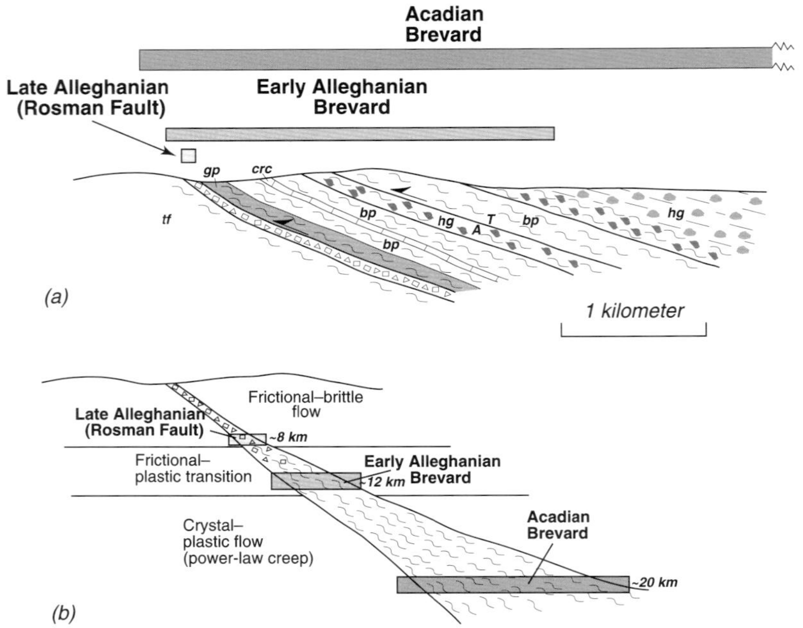

Fig. 10. (**a**) Simplified cross-section through the Brevard Fault Zone in South Carolina (location shown in Fig. 1). Shaded horizontal bars correspond to the extent of deformation at different times and to the depths in the crust shown in (**b**). tf, Tallulah Falls Formation (undivided in the Blue Ridge footwall). Chauga River Formation units (also see Fig. 2): gp, graphitic phyllonite (muscovite–chlorite–graphite pelite); bp, Brevard phyllonite (muscovite–chlorite pelite); crc, carbonate (calcite–quartz–muscovite marble). hg, Henderson orthogneiss (mylonitic overprint indicated by black augen-shaped objects). T indicates relative motion toward observer; A indicates relative motion away from observer. (**b**) Relationship of the different episodes of movement on the Brevard Fault Zone to an ideal crustal fault model (e.g. Sibson 1977). The boxes indicate the locations of the events in the different behavioural regimes. It should be noted that the thickness of each deformational regime in the actual fault zone changes by more than an order of magnitude from the oldest (>10 km thick) to the intermediate (1–3 km thick) to the youngest (50–100 m thick).

The Blue Ridge–Piedmont megathrust sheet in the southern and central Appalachians is the main product of this collision. During emplacement of the Blue Ridge–Piedmont sheet, the frontal parts of the sheet may have slowed and the northwestern edge of the Brevard Fault Zone was reactivated again, in the brittle regime, forming the out-of-sequence Rosman Fault. This fault was localized in a zone of cataclasite only 100 m thick.

(4) Mesozoic (Late Triassic–Early Jurassic) tectonism in eastern North America involved formation of a series of rift basins, NW–SE- and north–south-oriented dolerite dykes, and east–west- and north–south-trending normal and sinistral strike-slip faults defined by zones of siliceous cataclasite as the present Atlantic Ocean began to open. Both the dykes and faults crosscut the Brevard Fault Zone, thus limiting its movement to Palaeozoic time. Either the Brevard Fault Zone was not suitably oriented in the early Mesozoic stress field or the tensional stress field was not sufficient to reactivate the fault zone.

Support for this research was provided by US National Science Foundation grants GA-1409, EAR-8206949, and EAR-8417894; and Nuclear Regulatory Commission Contract FIN B10538. Reviews by B. Holdsworth, I. Alsop, and A. Dennis resulted in considerable improvement of the manuscript, and are very much appreciated. I remain culpable for any and all misrepresentations of fact or interpretation.

References

BOBYARCHICK, A.R. 1984. A late Paleozoic component of strike-slip in the Brevard zone, southern Appalachians. *Geological Society of America, Abstracts with Programs*, **19**, 126.

BOBYARCHICK, A.R., EDELMAN, S.H. & HORTON, J.W. JR 1988. The role of dextral strike-slip in the displacement history of the Brevard zone.

In: SECOR, D.T. (ed.) *Southeastern Geological Excursions*. Geological Society of America Southeastern Section Field Trip Guidebook. South Carolina Geological Survey, Columbia, 53–154.

BREAM, B.R., HATCHER, R.D. JR, MILLER, C.F. & FULLAGAR, P.D. 2000. Paragneiss geochemistry and preliminary ion microprobe geochronology of detrital zircons from the southern Appalachian crystalline core. *Geological Society of America, Abstracts with Programs*, **32**, A31.

BURCHFIEL, B.C. & LIVINGSTON, J.L. 1967. Brevard zone compared to Alpine root zones. *American Journal of Science*, **265**, 241–256.

COSTAIN, J.K., HATCHER, R.D. JR & ÇORUH, C. 1989. Appalachian ultradeep core hole (ADCOH) site investigation seismic lines and geologic interpretation. *In*: HATCHER, R.D. JR, THOMAS, W.A. & VIELE, G.W. (eds) *The Appalachian–Ouachita Orogen in the United States*. The Geology of North America, **F-2**. Geological Society of America, Boulder, CO, Plate 8.

CHRISTENSEN, N.I. & SZYMANSKI, D.L. 1988. Origin of reflections from the Brevard Fault Zone. *Journal of Geophysical Research*, **93**, 1087–1102.

CONEY, P.J. 1980. Cordilleran metamorphic core complexes: an overview. *In*: CRITTENDEN, M.D. Jr, CONEY, P.J. & DAVIS, G.H. (eds) *Cordilleran Metamorphic Core Complexes*. Geological Society of America, Memoirs, **153**, 7–34.

DAVIS, G.H. 1980. Structural characteristics of metamorphic core complexes, southern Arizona. *In*: CRITTENDEN, M.D. JR, CONEY, P.J. & DAVIS, G.H. (eds) *Cordilleran Metamorphic Core Complexes*. Geological Society of America, Memoirs, **153**, 35–78.

DAVIS, T. L. 1993. Stratigraphy, structure, and metamorphism of a crystalline thrust terrane, western Inner Piedmont, North Carolina. Ph.D. thesis, University of Tennessee, Knoxville.

DAVIS, T.L., HATCHER, R.D. JR, LIU, A. & TABOR, J.R. 1991. Southern Appalachian western Inner Piedmont: a progressive crustal-scale shear zone. *Geological Society of America, Abstracts with Programs*, **23**, A138.

DENNIS, A.J. & WRIGHT, J.E. JR. 1997. Middle and late Paleozoic monazite U–Pb ages. *Inner Piedmont, South Carolina. Geological Society of America, Abstracts with Programs*, **29**, 12.

DENNISON, J.M., STEWART, K.G., ROTH, E.A., ALCOTT, A.E., LOWE, A.L. & VLAHOVIC, G. 1997. Post-orogenic linear fracture systems in the Great Smoky–Blue Ridge region of North Carolina, Tennessee, and Georgia. *Geological Society of America, Abstracts with Programs*, **29**, 13.

DEWEY, J.F. & KIDD, W.S.F. 1974. Continental collisions in the Appalachian–Caledonide orogenic belt: variations related to complete and incomplete suturing. *Geology*, **2**, 543–546.

DEWEY, J.F., HOLDSWORTH, R.E. & STRACHAN, R.A. 1998. Transpression and transtension. *In*: HOLDSWORTH, R.E., STRACHAN, R.A. & DEWEY, J.F. (eds) *Continental Transpressional and Transtensional Tectonics*. Geological Society, London, Special Publications, **135**, 1–14.

EDELMAN, S.H., LIU, A. & HATCHER, R.D. JR. 1987. Brevard zone in South Carolina and adjacent areas: an Alleghanian orogen-scale dextral shear zone reactivated as a thrust fault. *Journal of Geology*, **95**, 793–806.

GARIHAN, J.M., RANSON, W.A., PREDDY, M. & HALLMAN, T.D. 1988. Brittle faults, lineaments, and cataclastic rocks in the Slater, Zirconia, and part of the Saluda 7 1/2-minute quadrangles, northern Greenville County, SC, and adjacent Henderson and Polk Counties, NC. *In*: SECOR, D.T. JR (ed.) *Southeastern Geological Excursions*. Geological Society of America Southeastern Section Field Trip Guidebook. South Carolina Geological Survey, Columbia, 266–349.

GRIFFIN, V.S. JR. 1971. Stockwork tectonics in the Appalachian Piedmont of South Carolina. *Geologische Rundschau*, **60**, 868–886.

HANDY, M.R. 1990. The exhumation of cross-sections of the continental crust: structure, kinematics and rheology. *In*: SALISBURY, M.H. & FOUNTAIN, D.M. (eds) *Exposed Cross-Sections of the Continental Crust*. Kluwer, Dordrecht, 485–507.

HATCHER, R.D. JR. 1969. Stratigraphy, petrology and structure of the low rank belt and part of the Blue Ridge of northwesternmost South Carolina. *South Carolina Division of Geology Geologic Notes*, **13**, 105–141.

HATCHER, R.D. JR. 1971. Stratigraphic, petrologic, and structural evidence favoring a thrust solution to the Brevard problem. *American Journal of Science*, **270**, 177–202.

HATCHER, R.D. JR. 1972. Developmental model for the southern Appalachians. *Geological Society of America Bulletin*, **83**, 2735–2760.

HATCHER, R.D. JR. 1975. The second Penrose field conference: the Brevard zone. *Geology*, **3**, 149–152.

HATCHER, R.D. JR. 1978. Tectonics of the western Piedmont and Blue Ridge: review and speculation. *American Journal of Science*, **278**, 276–304.

HATCHER, R.D. JR. 1989. Tectonic synthesis of the U.S. Appalachians. *In*: HATCHER, R.D. JR, THOMAS, W.A., VIELE, G.W. (eds) *The Appalachian–Ouachita Orogen in the United States*. The Geology of North America, **F-2**. Geological Society of America, Boulder, CO, 511–535.

HATCHER, R.D. JR. 1998. Structure of the Appalachian Inner Piedmont. *Geological Society of America Abstracts with Programs*, **30**, 17.

HATCHER, R.D. JR & HOOPER, R.J. 1992. Evolution of crystalline thrust sheets in the internal parts of mountain chains. *In*: MCCLAY, K.R. (ed.) *Thrust Tectonics*. Chapman & Hall, London, 217–234.

HATCHER, R.D. JR, PRICE, V. JR & SNIPES, D.S. 1973. Analysis of chemical and paleotemperature data from selected carbonate rocks of the southern Appalachians. *Southeastern Geology*, **15**, 55–70.

HATCHER, R.D. JR, THOMAS, W.A., GEISER, P.A., SNOKE, A.W., MOSHER, S. & WILTSCHKO, D.V. 1989. Alleghanian orogen. *In*: HATCHER, R.D. JR, THOMAS, W.A. & VIELE, G.W. (eds) *The Appalachian–Ouachita Orogen in the United States*. The Geology of North America, **F-2**. Geological Society of America, Boulder, CO, 233–318.

HIGGINS, M.W. 1966. The Geology of the Brevard Lineament near Atlanta, Georgia. *Georgia Department of Mines, Mining, and Geology Bulletin*, 77.

HOLDSWORTH, R.E. 1994. Structural evolution of the Gander–Avalon terane boundary: a reactivated transpression zone in the NE Newfoundland Appalachians. *Journal of the Geological Society, London*, **151**, 629–646.

HOLDSWORTH, R.E., BUTLER, C.A. & ROBERTS, A.M. 1997. The recognition of reactivation during continental deformation. *Journal of the Geological Society, London*, **154**, 73–78.

HOLDSWORTH, R.E., STEWARD, M. & IMBER, J. 2001. The structure and rheological evolution of reactivated continental fault zones. *In*: MILLER, J.A., BUICK, I.S., HAND, M. & HOLDSWORTH, R.E. (eds) *Continental Reworking and Reactivation*. Geological Society, London, Special Publications, **184**, 115–137.

HOPSON, J.L. & HATCHER, R.D. JR. 1988. Structural and stratigraphic setting of the Alto allochthon, northeast Georgia. *Geological Society of America Bulletin*, **100**, 339–350.

HORTON, J.W. JR & BUTLER, J.R. 1986. The Brevard fault zone at Rosman, Transylvania County, North Carolina. *In*: NEATHERY, T.L. (ed.) *Southeastern Section of the Geological Society of America Centennial Field Guide, 6*. Geological Society of America, Boulder, CO, 251–256.

IMBER, J., HOLDSWORTH, R.E., BUTLER, C.A. & LLOYD, G.E. 1997. Fault-zone weakening processes along the reactivated Outer Hebrides Fault Zone, Scotland. *Journal of the Geological Society, London*, **154**, 105–109.

JONAS, A.I. 1932. Structure of the metamorphic belt of the southern Appalachians. *American Journal of Science, Series 5*, **24**, 228–243.

JONES, R.R., HOLDSWORTH, R.E. & BAILEY, W. 1997. Lateral extrusion in transpression zones: the importance of boundary conditions. *Journal of Structural Geology*, **19**, 1201–1217.

KEITH, A. 1907. *Description of the Roan Mountain Quadrangle (Tennessee–North Carolina)*. US Geological Survey Geologic Atlas Folio, **151**.

KING, P.B. 1955. A geologic cross section across the southern Appalachians, an outline of the geology in the segment in Tennessee, North Carolina, and South Carolina. *In*: RUSSELL, R.J. (ed.) *Guides to Southeastern Geology*. Geological Society of America Annual Meeting Guidebook, 332–373.

KING, P.B. 1959. *The Evolution of North America*. Princeton University Press, Princeton, NJ.

MANSPEIZER, W. & COUSMINER, H.L. 1988. Late Triassic–Early Jurassic synrift basins of the U.S. Atlantic margin. *In*: SHERIDAN, R.E. & GROW, J.A. (eds) *The Geology of North America, The Atlantic Continental Margin, U.S.*. The Geology of North America, **I-2**. Geological Society of America, Boulder, CO, 197–216.

MILLER, C.F., HATCHER, R.D. JR, AYERS, J.C., COATH, C.D. & HARRISON, T.M. 2000. Age and zircon inheritance of eastern Blue Ridge plutons, southwestern North Carolina and Northeastern Georgia, with implications for magma history and evolution of the southern Appalachian orogen. *American Journal of Science*, **300**, 142–172.

MOSHER, S. 1983. Kinematic history of the Narragansett basin, Massachusetts and Rhode Island: constraints on late Paleozoic plate reconstruction. *Tectonics*, **2**, 327–344.

ODOM, A.L. & FULLAGAR, P.D. 1973. Geochronologic and tectonic relationships between the Inner Piedmont, Brevard zone, and Blue Ridge belts. *American Journal of Science*, **273A**, 133–149.

RAGLAND, P.C., HATCHER, R.D. JR & WHITTINGTON, D. 1983. Juxtaposed Mesozoic diabase dike sets from the Carolinas: a preliminary assessment. *Geology*, **11**, 394–399.

RANKIN, D.W. 1975. The continental margin of eastern North America in the southern Appalachians: the opening and closing of the proto-Atlantic Ocean. *American Journal of Science*, **275**, 298–336.

REED, J.C. JR & BRYANT, B. 1964. Evidence for strike-slip faulting along the Brevard zone in North Carolina. *Geological Society of America Bulletin*, **75**, 1177–1196.

REED, J.C. JR, BRYANT, B. & MYERS, W.B. 1970. The Brevard zone: a reinterpretation. *In*: FISHER, G.W., PETTIJOHN, F.J., REED, J.C. JR & WEAVER, K.N. (eds) *Studies of Appalachian Geology: Central and Southern*. Wiley–Interscience, New York, 261–269.

REED, J.J. 1964. Mylonites, cataclasites, and associated rocks along the Alpine fault, South Island, New Zealand. *New Zealand Journal of Geology and Geophysics*, **7**, 645–684.

ROPER, P.J. & DUNN, D.E. 1973. Superposed deformation and polymetamorphism, Brevard zone, South Carolina. *Geological Society of America Bulletin*, **84**, 3373–3386.

ROPER, P.J. & JUSTUS, P.S. 1973. Polytectonic evolution of the Brevard zone. *American Journal of Science*, **273**, 105–132.

SCHMID, S.M. & HANDY, M.R. 1991. Towards a genetic classification of fault rocks: geological usage and tectonophysical implications. *In*: MULLER, D.W., MCKENZIE, J.A. & WEISSERT, H. (eds) *Controversies in Modern Geology*. Academic Press, San Diego, CA, 339–361.

SIBSON, R.H. 1977. Fault rocks and fault mechanisms. *Journal of the Geological Society, London*, **133**, 191–213.

SINHA, A.K., HEWITT, D.A. & RIMSTIDT, J.D. 1988. Metamorphic petrology and strontium isotope geochemistry associated with the development of mylonites: an example from the Brevard Fault Zone, North Carolina. *American Journal of Science*, **288**, 115–147.

SINHA, A.K., HUND, E.A. & HOGAN, J.P. 1989. Paleozoic accretionary history of the North American plate margin (central and southern Appalachians): constraints from the age, origin, and distribution of granitic rocks. *In*: HILLHOUSE, H.W. (ed.) *Deep Structure and Post-kinematics of Accreted Terranes*. Geophysical Monograph, American Geophysical Union, **50**, and International Union of Geodesy and Geophysics, **5**, 219–238.

TIKOFF, B. & TESSIER, T. 1994. Strain modeling of displacement-field partitioning in transpressional orogens. *Journal of Structural Geology*, **16**, 1575–1588.

VINSON, S. 1999. Ion probe geochronology of granitoid gneisses of the Inner Piedmont, North Carolina and South Carolina. M.S. thesis, Vanderbilt University, Nashville, TN.

WERNICKE, B. 1985. Uniform-sense normal simple shear of the continental lithosphere. *Canadian Journal of Earth Science*, **22**, 108–125.

Repeated reactivation in the Apennine–Maghrebide system, Italy: a possible example of fault-zone weakening?

ENRICO TAVARNELLI[1], FRANCESCO ANTONIO DECANDIA[2], PIETRO RENDA[2], MARIANO TRAMUTOLI[3], ERWAN GUEGUEN[4] & MAURO ALBERTI[1]

[1]*Dipartimento di Scienze della Terra, Università di Siena, Siena, Italy*

[2]*Istituto di Geologia e Geodesia, Università di Palermo, Palermo, Italy*

[3]*Amministrazione Regionale della Basilicata, Potenza, Italy*

[4]*ITIS, Consiglio Nazionale delle Ricerche, Matera, Italy*

Abstract: Italy owes its complex geological structure to a double switch in tectonic regime, which involved the opening of the Tethys Ocean during Early Mesozoic time, its closure leading to development of the Apennine–Maghrebide fold-and-thrust belt during the Eocene–Recent interval, and the post-orogenic opening of the Tyrrhenian Sea since Miocene time. This history of tectonic inversion is partly preserved within two major fault zones, the Valnerina Line, in the central Apennines, and the Gratteri–Mount Mufara Line, in central–northern Sicily, which were repeatedly reactivated with different kinematic characters. The relatively long life of these structures indicates that strain was localized along anisotropies inherited from early deformation episodes. However, the progressive widening of both fault zones through time may result from strain-hardening fault-rock behaviour during subsequent deformations, thus suggesting that fault reactivation does not imply fault-zone weakening as is often assumed.

Extensive research in the field of thrust tectonics over the last three decades has shown that the deformation style of foreland fold-and-thrust belts is generally thin skinned, i.e. dominated by shallow structures propagating across undeformed sedimentary cover rocks (Bally *et al.* 1966; Boyer & Elliott 1982). As a result, many geologists tend to assume that most orogenic belts are thin skinned in style (e.g. see discussion by Coward (1983)). In more recent years, however, the increasing recognition of tectonic inversion processes and basement fault reactivation in thrust belts has challenged the indiscriminate use of thin-skinned thrust models (Williams *et al.* 1989; Holdsworth *et al.* 1997). In fact, inversion tectonic interpretations often assume that fault reactivation has occurred, but the hard evidence to support this process is rarely provided (e.g. see discussion by Roberts & Holdsworth (1999)).

Fault reactivation is generally regarded as intimately associated with an intrinsic strain-weakening behaviour of faults and shear zones, but this assumption is often incorrect (Holdsworth *et al.* 2001). Examples of faults that experienced repeated reactivation are relatively rare; yet documentation of long-lived faults is important because it can provide a scenario for the study of how rheological fault-rock properties evolve through time, leading to either strain-weakening or strain-hardening behaviour. In this paper we describe two examples of fault zones that were repeatedly reactivated with different kinematic characters during successive deformation episodes. Our examples are derived from the Apennine–Maghrebide fold-and-thrust belt of Italy.

The Apennine–Maghrebide system

The Apennine–Maghrebide system of Italy (Fig. 1) is an arcuate fold-and-thrust belt made up of tectonic slices that were detached from the African (e.g. Hyblean) and Adriatic (e.g. Apulian) continental plates during Tertiary and Quaternary time (Boccaletti & Guazzone 1974; Boccaletti *et al.* 1980; Robertson & Grasso 1995; Gueguen *et al.* 1998). The piled tectonic slices mainly consist of sedimentary sequences detached from the underlying basement (Bally *et al.* 1986), and record extension related to the opening and evolution of the Mesozoic Tethys Ocean (Bernoulli 1967; Colacicchi *et al.* 1970; D'Argenio & Alvarez 1980). Extensional deformation episodes, with development of important synsedimentary normal faults, were active

From: HOLDSWORTH, R.E., STRACHAN, R.A., MAGLOUGHLIN, J.F. & KNIPE, R.J. (eds) 2001. *The Nature and Tectonic Significance of Fault Zone Weakening*. Geological Society, London, Special Publications, **186**, 273–286. 0305-8719/01/$15.00

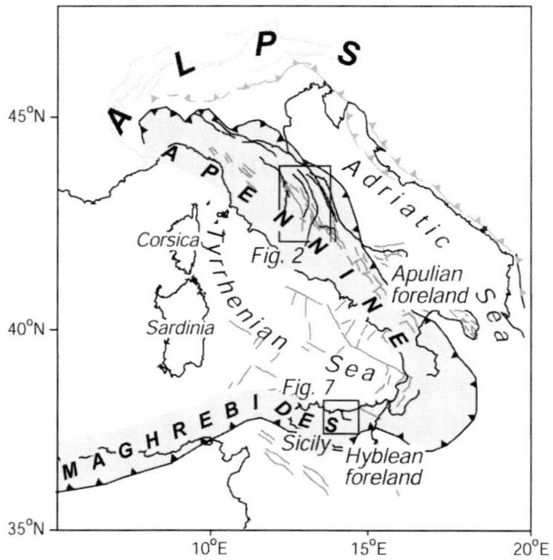

Fig. 1. Tectonic sketch map of the Apennine–Maghrebide system, with locations of the structures described in this study.

during the Triassic (Martini *et al.* 1986), Jurassic (Colacicchi *et al.* 1970; Centamore *et al.* 1971; Alvarez 1990) and Cretaceous–Paleogene time intervals (Decandia 1982; Montanari *et al.* 1989). The earliest normal faults dismembered a Hercynian basement and an overlying wide tidal platform of mid-Triassic age (Zappaterra 1994) into differently subsiding blocks, producing Mesozoic–Cenozoic pelagic basins (e.g. the Umbria–Marche Basin of central Italy; the Imerese Basin of Sicily) and coeval survivor platforms (e.g. the Latium–Abruzzi Platform of central Italy; the Panormide Platform of Sicily). From Eocene time onwards, the sedimentary sequences were affected by contractional deformations, with development of regionally important thrusts and related folds (Boccaletti *et al.* 1980). Contractional deformation largely occurred under non-metamorphic conditions, migrated progressively from the present peri-Tyrrhenian areas eastwards and southwards, and is probably still active in the vicinities of the Apulian and Hyblean foreland sectors (Bianco *et al.* 1998; Gueguen & Tomasi 1999). Late Tertiary contraction was followed by the onset of post-orogenic extensional deformations, related to the opening and evolution of the Tyrrhenian Sea (Carmignani *et al.* 1994). Although locally very important, post-orogenic extension did not significantly modify the architecture of pre-existing structures to be described below, and therefore its effects are not considered in this paper.

There is increasing evidence that the structures inherited from early extensional episodes of Triassic, Jurassic and Cretaceous–Paleogene age locally influenced the geometry and location of regionally important thrusts during the Late Tertiary evolution of the Apennine–Maghrebide fold-and-thrust belt (Tavarnelli 1996, 1999). The best documented example is the so-called Ancona–Anzio Line, a major arcuate fault zone in the central Apennines (Fig. 2a) that separates the basin-derived Umbria–Marche tectonic units to the west from the platform-derived Latium–Abruzzi units to the east (Calamita & Deiana 1988; Lavecchia *et al.* 1988). This fault zone experienced extensional, strike-slip and contractional deformations during Jurassic, Miocene and Pliocene times, respectively (Castellarin *et al.* 1982). In the following sections, we present additional evidence for repeated reactivation along two regionally important fault zones of the Apennine–Maghrebide system: the Valnerina Line, in the Umbria–Marche region of the central Apennines, and the Gratteri–Mount Mufara Line, in the Madonie Mountains of central–northern Sicily (Fig. 1).

The Valnerina Line

The Valnerina Line is a major, SSW–NNE-trending transpressional feature that extends for *c.* 50 km from Terni in the south to Camerino in the north (Fig. 2a). Its present expression consists of alternating north–south-trending thrusts and folds, that are linked by SW–NE-trending, right-lateral strike-slip faults of Late Messinian–Early Pliocene age (Fig. 2b and c; Decandia 1982; Alberti 1998). These contractional and strike-slip structures occur within a deformation zone of 3 km width, and affect a carbonate sequence that was deposited in the Umbria–Marche Basin, whose local stratigraphy consists of a massive shelf limestone of Early Liassic age, overlain by well-bedded pelagic limestones, marls and calcarenites of Mid-Lias–Early Miocene age.

The current position of the Valnerina Line approximately follows that of an older, major stratigraphic boundary, along which the Scaglia Rossa Formation of Late Cretaceous–Eocene age displays significant facies and thickness variations (Fig. 2b; Renz 1936; Decandia 1982; Montanari *et al.* 1989). Decandia (1982) first suggested that this stratigraphic boundary represents evidence for a major synsedimentary fault, active during the Late Cretaceous–Eocene interval, an interpretation further supported by Lavecchia (1985) and Alberti (1998). The results of our integrated stratigraphic and

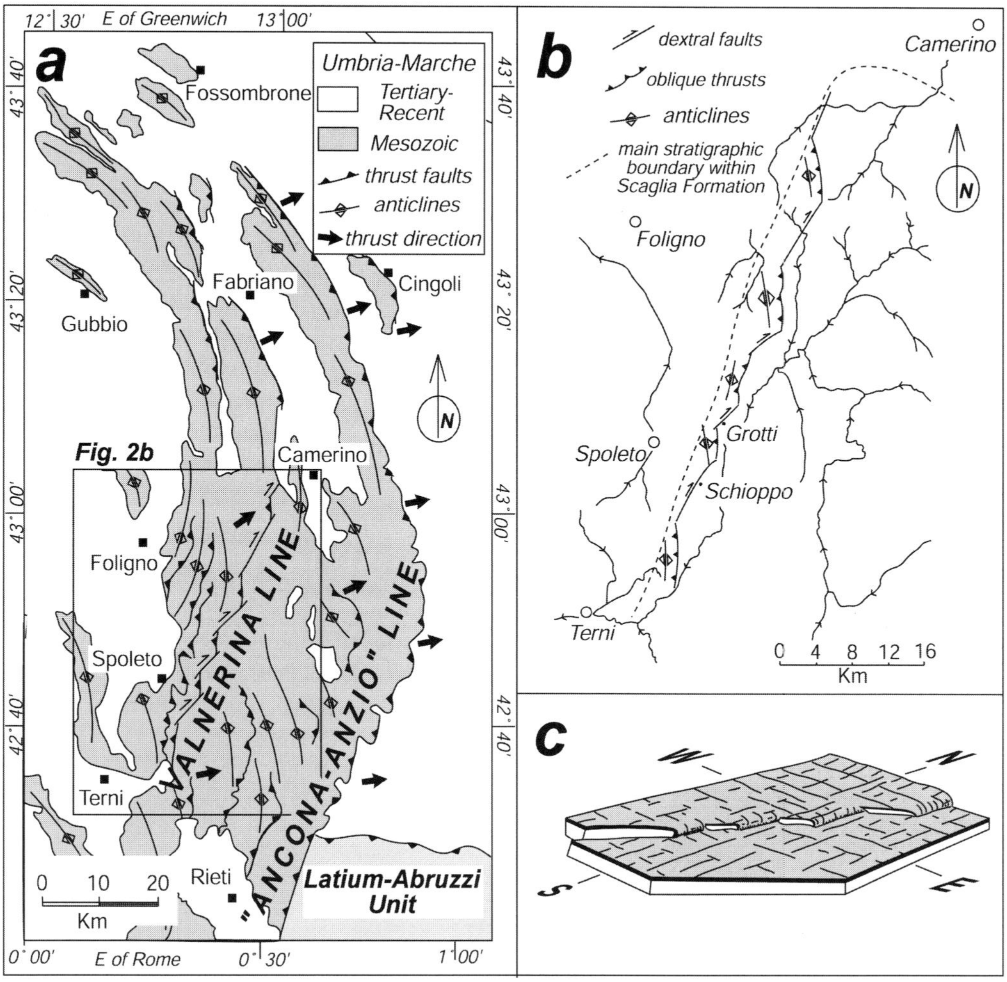

Fig. 2. (**a**) Sketch map of the Umbria–Marche Apennines, with location of the 'Ancona–Anzio' and Valnerina lines. Post-orogenic normal faults are not shown. (**b**) Main tectonic features of the Valnerina Line; its parallelism to the main stratigraphic boundary within the Upper Cretaceous–Eocene Scaglia Rossa Formation should be noted (modified after Decandia 1982). (**c**) Highly simplified 3D representation of the Valnerina Line.

structural analyses are broadly consistent with this interpretation, and indicate an even more complex tectonic history for the Valnerina Line, with repeated reactivations during the Late Cretaceous–Eocene, Oligocene–Late Miocene, and Late Miocene–Early Pliocene time intervals.

Late Cretaceous–Eocene interval

The Scaglia Rossa Formation of Late Cretaceous–Eocene age displays significant thickness and facies variations on both sides of the Valnerina Line. A condensed, 80 m sequence of well-bedded pelagic limestones and marls on the western side passes into a 850 m composite sequence of alternating pelagic limestones and calcareous turbidites, with abundant slumping, on the eastern side (Fig. 3a). Palaeocurrent measurements at the base of turbidite beds indicate a southeasterly provenance for the calcareous turbidites, whose composition is consistent with a shelf origin, suggesting that the main source was probably the Latium–Abruzzi Platform, adjacent to the Umbria–Marche Basin.

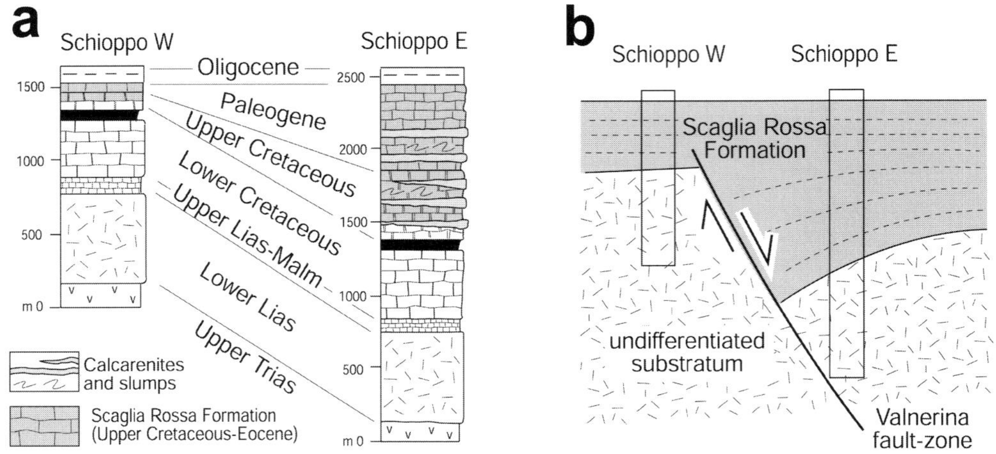

Fig. 3. (**a**) Stratigraphic logs of the Umbria–Marche sequence across the Valnerina Line near Schioppo. (**b**) Proposed model to account for the stratigraphic variations within the Scaglia Rossa Formation.

These stratigraphic variations reflect extensive tectonism during deposition of the Scaglia Rossa Formation (Colacicchi & Baldanza 1986; Montanari *et al.* 1989). Normal faults of Late Cretaceous–Paleogene age are recognized throughout the Umbria–Marche region (Winter & Tapponnier 1991), and are particularly abundant in the vicinity of the Valnerina Line. Some of these faults produce offsets of several hundred metres, and are often truncated by regionally important thrusts (Tavarnelli 1996, 1999). The outcrop-scale evidence for extensional synsedimentary deformation consists of minor structures, such as a pervasive bedding-parallel stylolite fabric associated with bedding-normal en echelon dilatant calcite veins (Fig. 4a). These structures are indicative of an overall extensional stress field. Bedding-parallel stylolites and bedding-normal extension veins are often associated with minor growth normal faults (Fig. 4b), that have often experienced deflection and folding as a consequence of the superimposed Late Tertiary contraction (Fig. 4c).

Because of the severe overprint of later deformations, it is not possible to document the early kinematics of the Valnerina fault zone. However, the systematic association of stratigraphic variations with normal faults of different scales within the Scaglia Formation on both sides of the Valnerina Line suggests an extensional character for this structure during the Late Cretaceous–Eocene time interval (Fig. 3b), in general agreement with previous interpretations by Decandia (1982), Lavecchia (1985) and Alberti (1998).

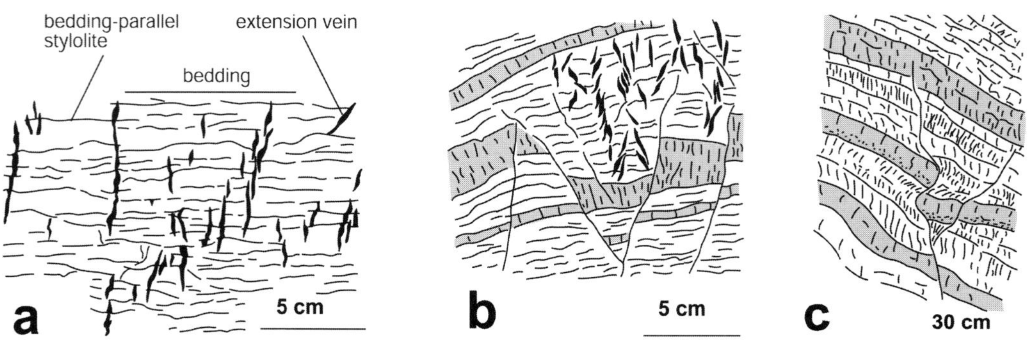

Fig. 4. Mesoscopic structures within the Scaglia Rossa Formation in the vicinity of the Valnerina Line. (**a**) Coeval bedding-parallel stylolites and bedding-normal veins. (**b**) Normal growth fault associated with bedding-parallel stylolites and bedding-normal veins. (**c**) Minor normal fault array deformed during regional contraction.

Oligocene–Mid-Miocene interval

The extensional structures within the Scaglia Rossa Formation along the Valnerina Line are overprinted by distinct sets of later deformation fabrics, which also affect younger stratigraphic units. The earliest recognized fabric consists of a carbonate-rich fault gouge of 1–2 m width; SW–NE-trending, steeply dipping to subvertical fault planes of centimetre to decimetre scale are locally distinguished within the fault gouge. A careful analysis of mechanical slickenlines and fibres on stepped calcite veins reveals a consistent left-lateral strike-slip sense of shear (Fig. 5a). We interpret these shear-sense indicators as relicts of an episode of left-lateral strike-slip deformation along the Valnerina fault zone. The overprinting relationships among minor fabrics indicate that left-lateral movements occurred after deposition of the Scaglia Rossa Formation of Late Cretaceous–Eocene age, and before the onset of transpressional deformation during the Late Miocene–Early Pliocene interval. Therefore, left-lateral strike-slip movement along the Valnerina Line is tentatively referred to the Oligocene–Mid-Miocene time interval.

Late Miocene–Early Pliocene interval

The SW–NE-trending faults of the Valnerina Line display abundant evidence for right-lateral, transpressional deformation (Figs. 5b and 6a–d), that overprints pre-existing left-lateral strike-slip structures. Along the Schioppo and Grotti faults (see Fig. 2b for location) a steep-to-vertical SSW–NNE-trending pressure-solution foliation associated with SW–NE-trending shear planes defines an asymmetric fabric. Mechanical striations and calcite fibres on the shear planes indicate a consistent oblique dextral sense of shear (Fig. 6b). These faults link en echelon, north–south- to NNW–SSE-trending, west- to WSW-dipping thrust faults and related ramp anticlines (Fig. 2b), whose deformation is accommodated by zones of highly sheared

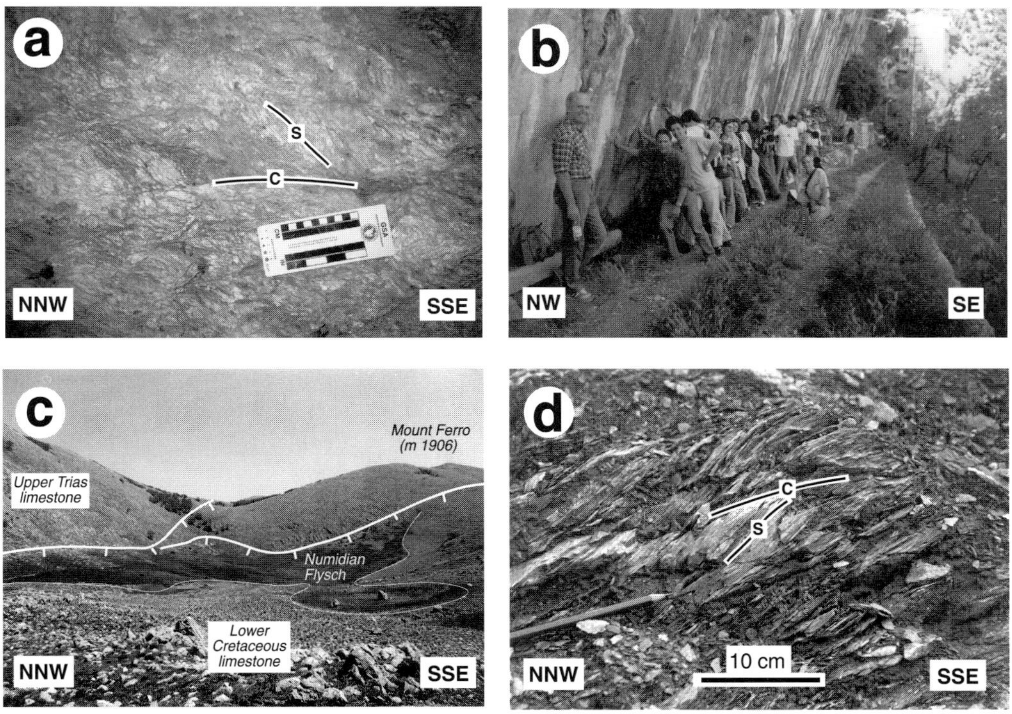

Fig. 5. Structural features of the Valnerina and Gratteri–Mount Mufara lines, and associated structures. (**a**) Fault gouge within the Scaglia Rossa Formation along the Valnerina Line, indicating early sinistral kinematics. (**b**) The Schioppo fault, a strand of the Valnerina Line. (**c**) Synsedimentary normal fault array at Piano Battaglia, west of the Gratteri–Mount Mufara Line, affecting platform-derived carbonates and the overlying Numidian Flysch. (**d**) Mesoscopic fabrics along the Mount Mufara–Mount Ferro thrust, indicating a top-to-the-SSE sense of shear. S, pressure-solution cleavage; C, shear surfaces.

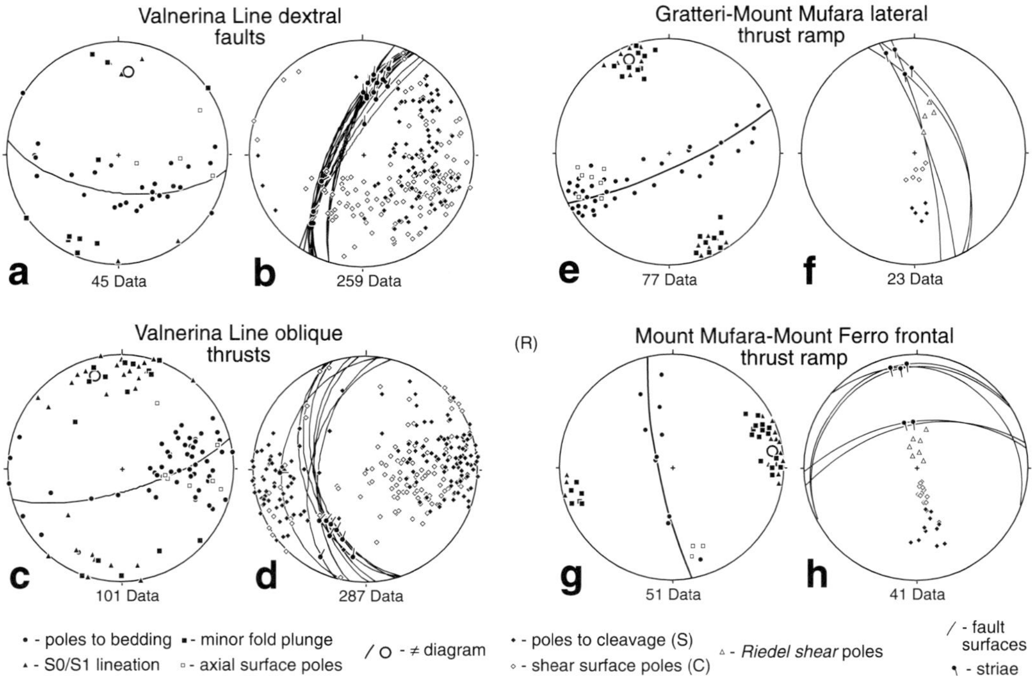

Fig. 6. Orientation of structures along the Valnerina (**a–d**) and Gratteri–Mount Mufara (**e–h**) lines (equal-area projection, lower hemisphere). (**a**) Mean plunge of minor dextral fault-related folds: 12°/197. Mean plunge of the host major fold, inferred from the distribution of poles to bedding (π diagram): 28°/008. (**b**) Mean slip direction inferred from striae along dextral faults: 16°/012. (**c**) Mean plunge of minor oblique thrust-related folds: 16°/016. Mean plunge of host major fold, inferred from the distribution of poles to bedding (π diagram): 21°/348. (**d**) Mean slip direction inferred from striae along oblique thrusts: 26°/211. (**e**) Mean plunge of minor folds: 6°/341. Mean plunge of lateral ramp anticline, inferred from the distribution of poles to bedding (π diagram): 9°/337. (**f**) Mean slip direction inferred from striae along the lateral thrust ramp: 13°/349. (**g**) Mean plunge of minor folds: 10°/073. Mean plunge of frontal ramp anticline, inferred from the distribution of poles to bedding (π diagram): 6°/079. (**h**) Mean slip direction inferred from striae along the frontal thrust ramp: 24°/348.

rocks of 2–10 m thickness. The penetrative shearing fabric within these deformation zones consists of a steep, WSW-dipping pressure-solution cleavage and shallow WSW-dipping shear surfaces that are roughly parallel to the deformation zone boundaries. Steps in fibrous calcite veins along the shear surfaces indicate displacement of the hanging walls towards the NE (Fig. 6d): the related thrust segments are therefore defined as oblique ramps, whose kinematics is broadly consistent with that inferred for the adjacent right-lateral transpressional faults (Fig. 6b).

The available regional stratigraphic constraints (Decandia & Giannini 1977; Deiana & Pialli 1996) indicate that these kinematically linked transpressional deformations developed during the Late Miocene–Early Pliocene interval.

The Gratteri–Mount Mufara Line

The Gratteri–Mount Mufara Line, in the Madonie Mountains of central–northern Sicily, is a major, NNW–SSE-trending fault zone that extends for >20 km from Gratteri, to the north, to the southern flank of Mount Mufara, to the south (Fig. 7a). This fault zone has the features of a thrust ramp along which the platform-derived Panormide unit to the east overrode the basin-derived Imerese unit to the west during the Miocene–Pliocene time interval (Lentini & Vezzani 1974; Catalano & D'Argenio 1978; Renda *et al.* 1999, and references therein). Inte-

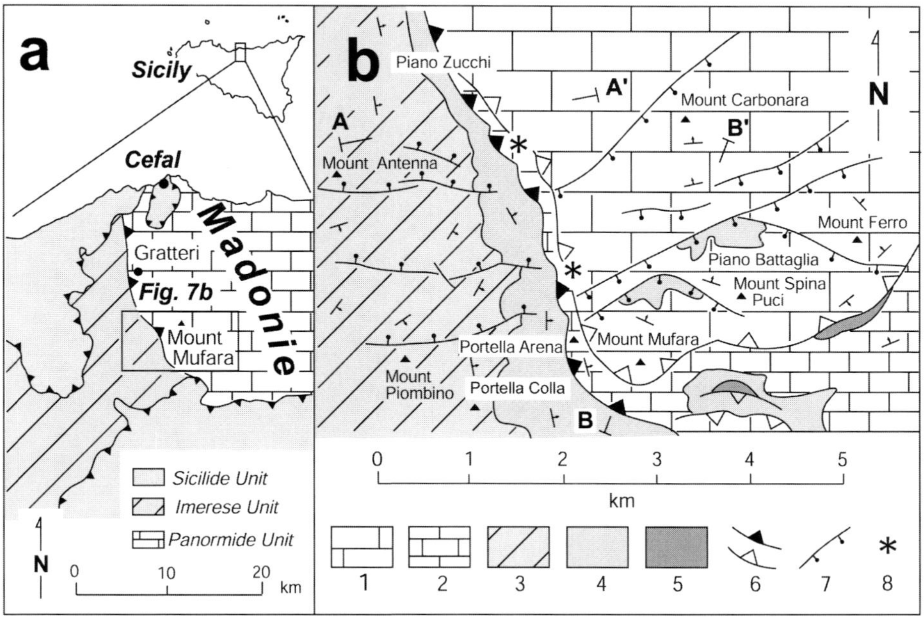

Fig. 7. (**a**) Sketch map of the Madonie Mountains, with location of the Gratteri–Mount Mufara Line. (**b**) Simplified structural map of the Mount Mufara area. 1, Deposits of the Panormide Platform (Upper Trias–Lower Cretaceous sediments); 2, platform–basin transitional facies (mainly Upper Trias sediments); 3, deposits of the Imerese Basin (Upper Trias–Lower Oligocene sediments); 4, Numidian Flysch (Upper Oligocene–Middle Miocene sediments); 5, Trubi Fm (Lower Pliocene sediments); 6, thrust faults; 7, pre-thrusting normal faults of possible Oligocene–Miocene age; 8, location of the relicts of platform–basin bypass margins.

grated stratigraphic and structural analyses carried out at its southern termination (Fig. 7b) allow for reconstruction of earlier movements along the Gratteri–Mount Mufara Line, and indicate that it was active during the Late Triassic–Oligocene, Oligocene–Miocene and Late Miocene–Pliocene time intervals.

Late Triassic–Oligocene interval

The stratigraphic units of central–northern Sicily display significant facies and thickness variations (Fig. 8; Catalano & D'Argenio 1978; Renda *et al.* 1999). Well-bedded pelagic marls, limestones and radiolarites of Late Triassic–Cretaceous age were deposited within the Imerese Basin, whereas coeval massive and nodular limestones and dolostones were deposited onto the adjacent Panormide Platform (Fig. 8a). The platform–basin transition zone was characterized by deposition of dolostone breccias. Although intense deformations have modified the original contact between the platform and basin domains, it is locally possible to observe well-exposed relics of a platform–basin bypass margin along the Gratteri–Mount Mufara fault zone (Fig. 7b), with erosional grooves on the Triassic platform marginal escarpment filled with younger sediments. At some localities, the escarpment preserves a fault zone of 2 m width with slickenlines that display a dip-slip extensional component, suggesting that it behaved as a normal fault. It is tempting to relate this deformation to the episode of Neogene–Quaternary post-orogenic extension; however, the Cretaceous age of sediments sealing the normal fault escarpment unequivocally indicates that this extensional deformation occurred during the Late Triassic–Cretaceous time interval (Fig. 8a and b). Further differences in thickness and facies of the overlying carbonatic sequences suggest that the normal fault was probably reactivated during the Cretaceous–Oligocene time interval (Fig. 8c).

Oligocene–Miocene interval

The sedimentary sequences of central–northern Sicily record an abrupt change in composition, from carbonatic to siliciclastic; this change occurred during Oligocene time, when both the Imerese and Panormide domains became the site of deposition of continental pelites and quartzites of African provenance (i.e. the so-

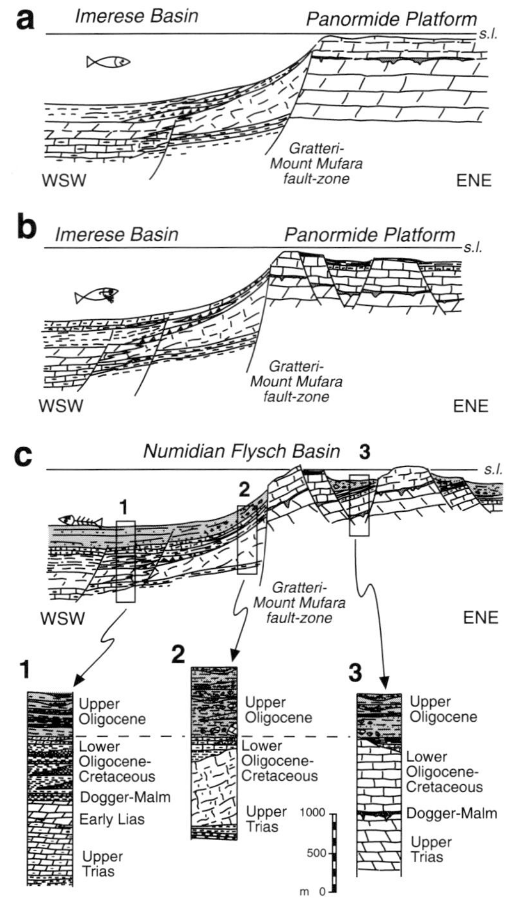

Fig. 8. Stratigraphic evolution of the central Madonie Mountains, inferred from analysis of the relationships among coeval basin and platform sediments. The stratigraphic logs (1, 2, 3) show the sequences of the Imerese, transitional, and platform sequences reconstructed at Mount Antenna, Piano Zucchi and Mount Spina Puci (location shown in Fig. 7b). (**a**) Late Trias–Early Cretaceous time. (**b**) Late Cretaceous–Early Oligocene time. (**c**) Late Oligocene–Mid-Miocene time. (Note the role of the Gratteri–Mount Mufara fault zone as a major synsedimentary boundary.)

called Numidian Flysch; Pescatore *et al.* 1987). Similarly to the underlying pelagic and platform sequences, the Numidian Flysch displays variations in thickness and facies, which reflect the activity of synsedimentary normal faults (Fig. 5c). These structures produce offsets up to 1000 m, and control the distribution of decametre-scale platform-derived olistoliths (Fig. 8c), particularly abundant in the vicinities of the platform–basin transition zone.

The trend of these important extensional structures is east–west, i.e. normal to that of the Gratteri–Mount Mufara fault zone, across which their dip direction changes (Fig. 7b). These relationships suggest that during the Oligocene–Miocene interval both east–west- and north–south-trending faults behaved as a kinematically coherent extensional fault system, where the Gratteri–Mount Mufara Line behaved as a transfer boundary, linking south-dipping normal faults, to the east, with coeval north-dipping normal faults, to the west.

Late Miocene–Pliocene interval

From Mid-Miocene time onwards, the Triassic–Early Miocene sedimentary sequences of the Madonie Mountains experienced contractional deformations, with development of macroscopic thrust ramps and related folds of different orientations, ranging from NNW–SSE (i.e. the Gratteri–Mount Mufara Line) to WSW–ENE (i.e. the Mount Mufara–Mount Ferro thrust segment: Figs. 6e–h and 7b). Thrusting and folding were accompanied by development of related mesoscopic structures, such as reverse faults, shear fabrics and minor parasitic folds, whose orientation is consistent with that of the larger, host anticlines and synclines (Fig. 6f and g).

Thrust kinematics is best inferred from mechanical slickensides and calcite fibres measured on sheared rocks adjacent to thrust ramps within a 5–10 m deformation zone (Fig. 5d). The NNW–SSE-trending thrust segments are characterized by dominantly right-lateral strike-slip movements (Fig. 6f), thus being kinematically defined as lateral ramps, whereas the WSW–ENE-trending segments display reverse-slip indicators (Fig. 6h), thus being best defined as frontal ramps (Fig. 9).

The youngest sediments in the footwall of the Gratteri–Mount Mufara lateral thrust ramp belong to the Numidian Flysch, an observation consistent with regional stratigraphic data (Catalano & D'Argenio 1978) according to which the onset of contractional deformation in this part of Sicily can be referred to the Oligocene–Miocene interval. However, Lower Pliocene marls belonging to the Trubi Formation are perched by the Mount Mufara–Mount Ferro thrust segment (Fig. 7b), suggesting that contractional deformations continued at least to Early Pliocene time (see also Abate *et al.* 1991).

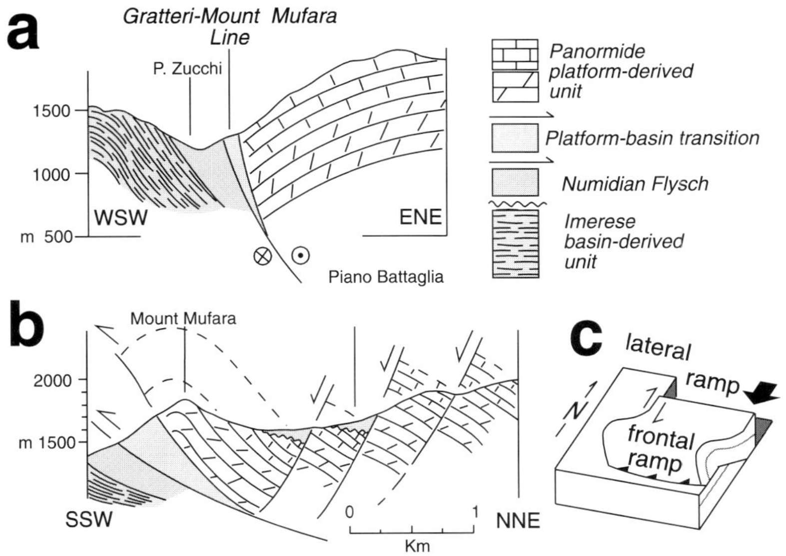

Fig. 9. Geometrical features of the Madonie Mountains. (**a**) Cross-section across Piano Zucchi (trace AA′ of Fig. 7b). (**b**) Cross-section across Mount Mufara and Piano Battaglia (trace BB′ of Fig. 7b). (**c**) Schematic representation of the sections (**a**) and (**b**), outlining their kinematic role as the lateral and frontal ramps of a linked thrust system.

Tectonic significance of fault reactivation

The data presented in the previous sections show that the Valnerina and the Gratteri–Mount Mufara fault zones experienced repeated reactivation with different kinematic characters during time spans that cover much of the history of the Apennine–Maghrebide system, from the opening of the Tethys Ocean to its closure and consequent construction of the fold-and-thrust belt (Fig. 10). This observation poses the question of the tectonic significance of fault-reactivation processes in the framework of the Mesozoic–Tertiary evolution of the belt. On the basis of the partial overlap between the various episodes of activation and reactivation of both fault zones (Fig. 10), we can recognize three sequentially younger stages, I, II and III, that correspond to the Late Triassic–Eocene, the Oligocene–Early Miocene, and the Late Miocene–Pliocene time intervals (Fig. 10).

Stage I

The Mesozoic activity of the Valnerina and Gratteri–Mount Mufara fault zones (a′ and a″ of Fig. 10) is characterized by extensional movements, inferred either from solely stratigraphic (along the Valnerina Line) or kinematic data (along the Gratteri–Mount Mufara Line). These elements, in combination, are broadly consistent with the proposed evolution of the Mesozoic Tethys Ocean (e.g. see D'Argenio & Alvarez 1980), and provide further support for the long-recognized extensional character of Triassic, Jurassic and Cretaceous–Paleogene deformations along its southern margin (Decandia 1982; Martini *et al.* 1986; Alvarez 1990).

Stage II

The reactivation of the Gratteri–Mount Mufara fault zone as the transfer segment of a linked extensional system during the Oligocene–Miocene interval (b″ of Fig. 10) poses the problem of its significance in the framework of the proposed tectonic evolution of northern Sicily. Regional stratigraphic data (Catalano & D'Argenio 1982; Pescatore *et al.* 1987) indicate that the earliest manifestations of contractional deformation occurred during Late Oligocene time, i.e. the time of deposition of the Numidian Flysch. The development of coeval synsedimentary extensional structures, as seen in the Madonie Mountains, seems at odds with the onset of an overall contractional regime. Recent research on synorogenic deposition from belt–foredeep–foreland systems may provide a key for the interpretation of this paradox. Numerous studies (Sinclair 1997; Lemieux 1999) show

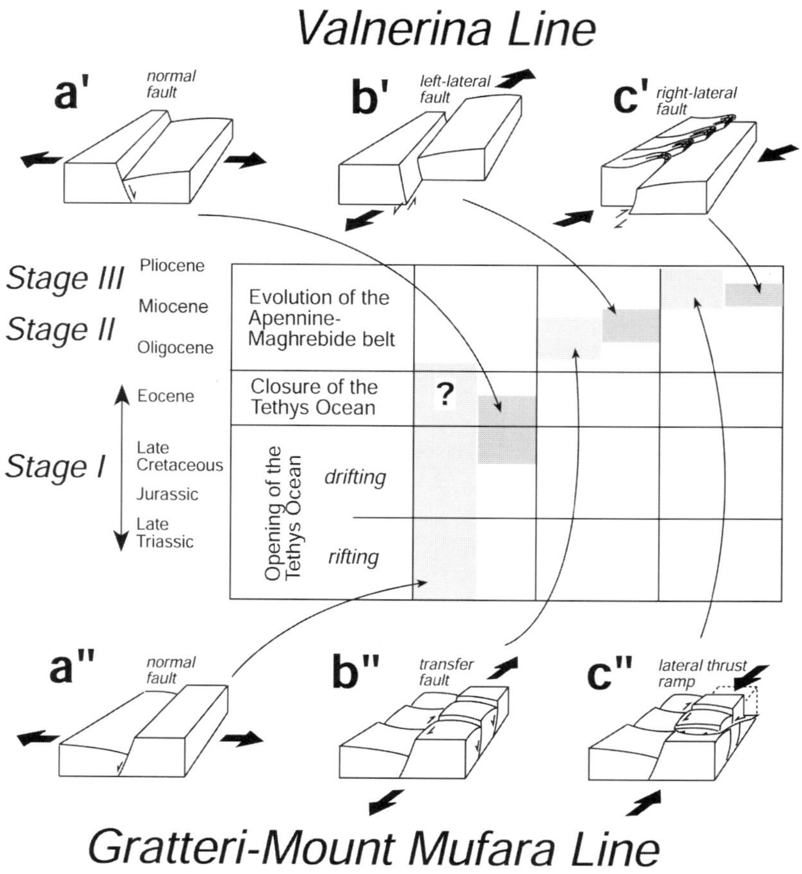

Fig. 10. Proposed reactivation history for the Valnerina (a′–c′) and Gratteri–Mount Mufara (a″–c″) lines, in the framework of the evolution of the Apennine–Maghrebide system. Stages I–III are also shown.

that normal faults are common within foredeep basins, and especially abundant at the foredeep–foreland transition zones, where they may be connected by transfer faults to isolate uplifted peripheral bulges. Once formed, these extensional structures are usually truncated, or reactivated, by advancing thrust faults (Scisciani *et al.* 2001). By analogy, we propose that the extensional fault system of the Madonie Mountains, with east–west-trending normal faults linked by the reactivated Gratteri–Mount Mufara transfer fault, represents the flexural response of the Hyblean foreland plate to the load produced by the advancing thrust pile during the Oligocene–Miocene evolution of northern Sicily.

Stage III

The history of reactivation of the Valnerina and Gratteri–Mount Mufara lines during Late Miocene and Pliocene times can be tentatively related to the Neogene geodynamic framework of the Apennine–Maghrebide system (b′, c′ and c″ of Fig. 10). The present shape of the fold-and-thrust belt results from its segmentation into two main arcs, i.e. the northern Apennines and the southern Apennines, that migrated eastwards and southwards at different rates at different times (Fig. 11; Malinverno & Ryan 1986; Royden *et al.* 1987; Gueguen *et al.* 1998). The Ancona–Anzio Line is considered by most workers (e.g. see Castellarin *et al.* 1982, and references therein) to represent the junction between the northern and southern Apennines. Numerous palaeogeographical reconstructions (e.g. Mantovani *et al.* 1997; Gueguen *et al.* 1998; Gueguen & Tomasi 1999) show that during Tortonian time the southern arc migrated faster than the northern arc, thus producing sinistral wrench along the Ancona–Anzio Line (Fig. 11a; Royden *et al.* 1987). From Late Mes-

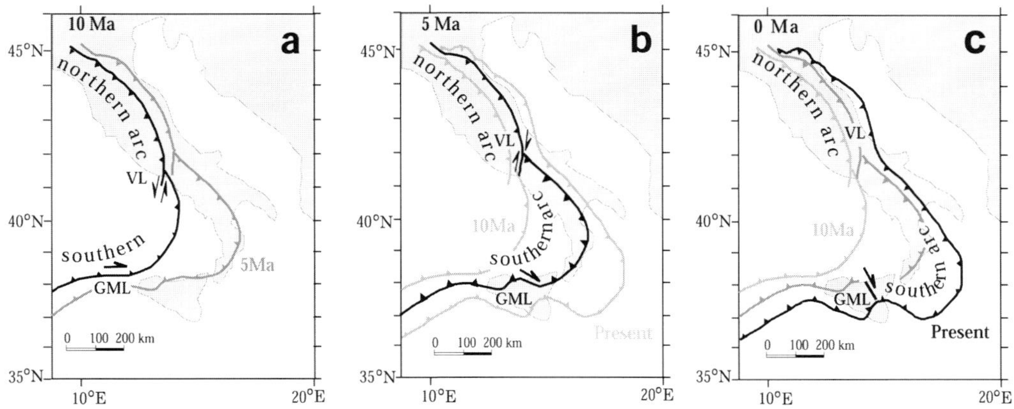

Fig. 11. Geodynamic evolution of the northern and southern arcs of the Apennine–Maghrebide system (modified after Gueguen *et al.* 1998), showing the Neogene activity along the Valnerina (VL) and Gratteri–Mount Mufara (GM) lines. (Note the reversal in the overall shear sense, from sinistral to dextral, along the Valnerina Line at *c.* 5 Ma, and the significant clockwise rotations experienced by the Gratteri–Mount Mufara Line.)

sinian time onwards the northern arc migrated faster that the southern arc, causing an inversion in the relative sense of shear, from sinistral to dextral, along the Ancona–Anzio Line (Fig. 11b; Castellarin *et al.* 1982). Our data from the Valnerina fault zone provide further, independent evidence supporting an abrupt reversal in the arc migration pattern since Late Messinian time, separating an early episode of left-lateral strike-slip deformation (b′ of Fig. 10) from a late episode of dextral transpression (c′ of Fig. 10). Our data also suggest that the junction between the northern and the southern Apenninic arcs could correspond to the deformation zone between the Valnerina Line and the Ancona–Anzio Line, rather than be restricted to the latter tectonic feature, as previously proposed by Castellarin *et al.* (1982).

The palaeogeographical reconstructions of Fig. 11 also outline the role of the Gratteri–Mount Mufara Line during the Neogene evolution of the Apennine–Maghrebide system in Sicily. Regional stratigraphic considerations indicate that the Panormide unit was emplaced onto the Imerese unit during Late Miocene time (Catalano & D'Argenio 1982) along a thrust system of which the Gratteri–Mount Mufara Line represents a lateral ramp (c″ of Fig. 10; see also Fig. 11a). Continued thrusting caused clockwise rotations of the Panormide and Imerese units (Channell *et al.* 1980; Grasso *et al.* 1987; Oldow *et al.* 1990), and reactivation of the Gratteri–Mount Mufara Line after Early Pliocene time (Fig. 11b and c).

Discussion and conclusions

The results of our analyses along the Valnerina and Gratteri–Mount Mufara lines have implications for the tectonic evolution of the Apennine–Maghrebide fold-and-thrust belt. The deformation style of this orogenic system has long been considered to be thin skinned, i.e. dominated by thrusts and related folds that affect only the sedimentary cover units (see Bally *et al.* 1986, and references therein). However, recent seismic profiles across the belt have questioned the validity of this assumption, by showing that conspicuous basement slices are also locally involved within the structures that affect the overlying sedimentary cover (Barchi *et al.* 1998; Decandia *et al.* 1998). Our data indicate that Mesozoic fault zones cutting through the pre-Triassic basement and younger cover rocks were repeatedly reactivated during the Tertiary evolution of the fold-and-thrust belt, thus supporting the view that deep basement faults locally controlled the distribution and geometry of contractional structures within the overlying sedimentary cover.

The documented history of repeated reactivation along pre-existing faults of the Apennine–Maghrebide system has more general implications for investigation of the processes that operate during the development and evolution of fault rocks in orogenic belts. Multiple fault reactivation through time is often considered as indicative of fault-zone softening or weakening. However, this assumption is not always correct, as deformation is partitioned into faults and shear zones, and this localization occurs in both strain-softening and strain-hardening regimes (e.g. see discussion by Holdsworth *et al.* (2001)). In general, materials deforming by mechanisms that involve dilatancy (i.e. frictional flow, hydrofracture veining, etc.) lead to strain-hardening behaviour, whereas materials deforming by non-dilational mechanisms (e.g. crystal plasticity) lead to strain-induced weakening. In particular, some fault reactivation events in most geological environments, particularly at upper-crustal levels, involve dilatancy and may lead to the onset of strain-hardening behaviour (Holdsworth *et al.* 2001). Our data from the Valnerina and Gratteri–Mount Mufara lines indicate that strain was repeatedly localized along, or close to pre-existing faults. Multiple reactivation occurred under dominantly brittle conditions, leading to progressive fault-zone widening, from a few metres to a few tens of metres. Slip along reactivated faults occurred largely at the expense of previously undeformed fault wall rocks, thus suggesting that these were more favourable sites for promoting faulting than the pre-existing fault surfaces themselves. This inference supports the view that, although strains were repeatedly localized along pre-existing faults, the deformation processes that operated during reactivation ultimately led to a strain-hardening behaviour of previously formed fault rocks, favouring deformation of the surrounding wall rocks and their involvement within a progressively widening fault zone. Therefore, we believe that the history documented from the Apennine–Maghrebide system represents an example of how fault reactivation may not always cause fault-zone weakening, as is often assumed.

We are grateful to B. Holdsworth, R. Strachan and J. K. Blom for very helpful advice, suggestions and discussion on fault-weakening processes, which led to a significantly improved paper. Constructive reviews by M. Watkeys and A. M. Michetti are also gratefully acknowledged. This work has been financially supported by MURST (40% funded to F.A. Decandia).

References

ABATE, B., DI STEFANO, E., INCANDELA, A. & RENDA, P. 1991. Evidenze di una fase tettonica pliocenica nelle Madonie occidentali (Sicilia centro-settentrionale). *Memorie della Società Geologica Italiana*, **47**, 225–234.

ALBERTI, M. 1998. Ruolo cinematico e dinamico di lineamenti sinsedimentari mesozoic durante la tettogenesi appenninica: Linea della Valnerina, Umbria. Ph.D. thesis, University of Siena.

ALVAREZ, W. 1990. Pattern of extensional faulting in pelagic carbonates of the Umbria–Marche Apennines of central Italy. *Geology*, **18**, 407–410.

BALLY, A.W., BURBI, L., COOPER, C. & GHELARDONI, R. 1986. Balanced sections and seismic reflection profiles across the Central Apennines. *Memorie della Società Geologica Italiana*, **35**, 237–310.

BALLY, A.W., GORDY, P.L. & STEWART, G.A. 1966. Structure, seismic data, and orogenic evolution of the Southern Canadian Rocky Mountains. *Canadian Petroleum Geologists Bulletin*, **14**, 337–381.

BARCHI, M., MINELLI, G. & PIALLI, G.P. 1998. The CROP 03 profile: a synthesis of results on deep structures of the Northern Apennines. *Memorie della Società Geologica Italiana*, **52**, 383–400.

BERNOULLI, D. 1967. Probleme der Sedimentation im Jura Westgriechenlands und des zentralen Apennin. *Verhandlungen der Naturforschenden Gesellschaft in Basel*, **78**, 35–54.

BIANCO, G., DEVOTI, R., FERMI, M., LUCERI, V., RUTIGLIANO, P. & SCIARRETTA, C. 1998. A contribution in the estimation of tectonic motion in crucial areas: the CGS96 SLR solution. *Tectonophysics*, **294**, 225–236.

BOCCALETTI, M. & GUAZZONE, G. 1974. Remnant arcs and marginal basins in the Cenozoic development of the Mediterranean. *Nature*, **252**, 18–21.

BOCCALETTI, M., COLI, M., DECANDIA, F.A., GIANNINI, E. & LAZZAROTTO, A. 1980. Evoluzione dell'Appennino settentrionale secondo un nuovo modello strutturale. *Memorie della Società Geologica Italiana*, **21**, 359–373.

BOYER, S.E. & ELLIOTT, D. 1982. Thrust systems. *AAPG Bulletin*, **66**, 239–267.

CALAMITA, F. & DEIANA, G. 1988. The arcuate shape of the Umbria–Marche–Sabina Apennines (Central Italy). *Tectonophysics*, **146**, 139–147.

CARMIGNANI, L., DECANDIA, F.A., FANTOZZI, P.L., LAZZAROTTO, A., LIOTTA, D. & MECCHERI, M. 1994. Tertiary extensional tectonics in Tuscany (Northern Apennines, Italy). *Tectonophysics*, **238**, 295–315.

CASTELLARIN, A., COLACICCHI, R., PRATURLON, A. & CANTELLI, C. 1982. The Jurassic–lower Pliocene history of the Ancona–Anzio line (Central Italy). *Memorie della Società Geologica Italiana*, **24**, 325–336.

CATALANO, R. & D'ARGENIO, B. 1978. An essay of palinspastic restoration across the Western Sicily. *Geologica Romana*, **17**, 145–159.

CATALANO, R. & D'ARGENIO, B. 1982. Schema geologico della Sicilia. *In*: CATALANO, R. & D'ARGENIO, B. (eds) *Guida alla Geologia della Sicilia Occidentale*. Società Geologica Italiana, Guide Geologiche Regionali **24**, 9–42.

CENTAMORE, E., CHIOCCHINI, M., DEIANA, G., MICARELLI, A. & PIERUCCINI, U. 1971. Contributo alla conoscenza del Giurassico dell'Appennino umbro-marchigiano. *Studi Geologici Camerti*, **1**, 7–89.

CHANNELL, J.E.T., CATALANO, R. & D'ARGENIO, B. 1980. Palaeomagnetism and deformation of the Mesozoic continental margin in Sicily. *Tectonophysics*, **58**, 391–407.

COLACICCHI, R. & BALDANZA, A. 1986. Carbonate turbidites in a Mesozoic pelagic basin: Scaglia Formation, Apennines—comparison with siliclastic depositional models. *Sedimentary Geology*, **48**, 81–105.

COLACICCHI, R., PASSERI, L. & PIALLI, G. 1970. Nuovi dati sul Giurese Umbro-Marchigiano ed ipotesi per un suo inquadramento regionale. *Memorie della Società Geologica Italiana*, **9**, 839–874.

COWARD, M.P. 1983. Thrust tectonics, thin skinned or thick skinned, and the continuation of thrusts to deep in the crust. *Journal of Structural Geology*, **5**, 113–123.

D'ARGENIO, B. & ALVAREZ, W. 1980. Stratigraphic evidence for crustal thickness changes on the southern Tethyan margin during the Alpine cycle. *Geological Society of America Bulletin*, **91**, 2558–2587.

DECANDIA, F.A. 1982. Geologia dei Monti di Spoleto (Prov. di Perugia). *Bollettino della Società Geologica Italiana*, **101**, 291–315.

DECANDIA, F.A. & GIANNINI, E. 1977. Studi geologici nell'Appennino umbro-marchigiano. 1—Evidenze di due fasi tettoniche terziarie a SE di Spoleto (Prov. di Perugia). *Bollettino della Società Geologica Italiana*, **96**, 713–722.

DECANDIA, F.A., LAZZAROTTO, A., LIOTTA, D., CERNOBORI, L. & NICOLICH, R. 1998. The CROP 03 traverse: insights on post-collisional evolution of the northern Apennines. *Memorie della Società Geologica Italiana*, **52**, 427–439.

DEIANA, G. & PIALLI, G.P. 1996. Le province strutturali dell'Appennino Umbro-Marchigiano. *Memorie della Società Geologica Italiana*, **51**, 27–36.

GRASSO, M., MANZONI, M. & QUINTILI, L. 1987. Misure paleomagnetiche sui Trubi infrapliocenici della Sicilia orientale: possibili implicazioni stratigrafiche e strutturali. *Memorie della Società Geologica Italiana*, **38**, 459–474.

GUEGUEN, E. & TOMASI, P. 1999. *Geodynamic Interpretation of the Central Mediterranean Combining Geological and VLBI Data. Proceedings of the 13th Working Meeting on European VLBI, Viechtach (Germany)*. Bundesamt für Kartographie und Geodäsie, Fundamentalstation Wettzell, 198–205.

GUEGUEN, E., DOGLIONI, C. & FERNANDEZ, M. 1998. On the post-25 Ma geodynamic evolution of the Western Mediterranean. *Tectonophysics*, **298**, 259–269.

HOLDSWORTH, R.E., BUTLER, C.A. & ROBERTS, A.M. 1997. The recognition of reactivation during continental deformation. *Journal of the Geological Society, London*, **154**, 73–78.

HOLDSWORTH, R.E., STEWART, M., IMBER, J. & STRACHAN, R.A. 2001. The structure and rheological evolution of reactivated continental fault zones: a review and case study. *In*: MILLER, J.A., BUICK, I.S., HOLDSWORTH, R.E. & HAND, M. (eds) *Continental Reactivation and Reworking*. Geological Society, London, Special Publications **134**, 115–137.

LAVECCHIA, G. 1985. Il sovrascorrimento dei Monti Sibillini: analisi cinematica e strutturale. *Bollettino della Società Geologica Italiana*, **104**, 161–194.

LAVECCHIA, G., MINELLI, G. & PIALLI, G. 1988. The Umbria-Marche arcuate fold belt (Italy). *Tectonophysics*, **146**, 125–137.

LEMIEUX, S. 1999. Seismic reflection expression and tectonic significance of Late Cretaceous extensional faulting of the Western Canada Sedimentary Basin in southern Alberta. *Canadian Petroleum Geology Bulletin*, **47**, 375–390.

LENTINI, F. & VEZZANI, L. *Carta Geologica delle Madonie (Sicilia centro-settentrionale). Scala 1:50 000*. Litografia Artistica Cartografica, Florence.

MALINVERNO, A. & RYAN, W.B.F. 1986. Extension in the Tyrrhenian Sea and shortening in the Apennines as a result of arc migration driven by sinking of the lithosphere. *Tectonics*, **5**, 227–245.

MARTINI, I.P., RAU, A. & TONGIORGI, M. 1986. Syntectonic sedimentation in a Middle Triassic rift, Northern Apennines, Italy. *Sedimentary Geology*, **47**, 191–219.

MANTOVANI, E., ALBARELLO, D., BABBUCCI, D. & TAMBURELLI, C. 1997. Recent/Present tectonic processes in the Italian region and their relation with seismic and volcanic activity. *Annales Tectonicae*, **11**, 27–57.

MONTANARI, A., CHAN, L.S. & ALVAREZ, W. 1989. Synsedimentary tectonics in the Late Cretaceous-Early Tertiary pelagic basin of the Northern Apennines. *In*: CREVELLO, P., WILSON, J.L., SARG, R. & REED, F. (eds) *Controls on Carbonate Platform and Basin Development*. Society of Economic Paleontologists and Mineralogists, Special Publications **44**, 379–399.

OLDOW, J.S., CHANNELL, J.E.T., CATALANO, R. & D'ARGENIO, B. 1990. Contemporaneous thrusting and large-scale rotations in the Western Sicilian fold and thrust belt. *Tectonics*, **9**, 661–681.

PESCATORE, T., RENDA, P. & TRAMUTOLI, M. 1987. Facies ed evoluzione sedimentaria del Bacino Numidico nelle madonie Occidentali (Sicilia). *Memorie della Società Geologica Italiana*, **38**, 297–316.

RENDA, P., TAVARNELLI, E. & TRAMUTOLI, M. 1999. La distensione tetidea ed il suo controllo sulle strutture compressive del sistema appenninico-maghrebide: l'esempio dei Monti delle Madonie (Sicilia centro-settentrionale). *Bollettino delle Società Geologica Italiana*, **118**, 179–190.

RENZ, O. 1936. Stratigraphische un mikropaleontologische Untersuchung der Scaglis (Obere Kreide

Tertiar) im zentralen Apennin. *Eclogae Geologicae Helvetiae*, **29**, 11–35.

ROBERTS, A.M. & HOLDSWORTH, R.E. 1999. Linking onshore and offshore structures: Mesozoic extension in the Scottish Highlands. *Journal of the Geological Society, London*, **156**, 1061–1064.

ROBERTSON, A.H.F. & GRASSO, M. 1995. Overview of the Late Tertiary–recent tectonic and palaeoenvironmental development of the Mediterranean region. *Terra Nova*, **7**, 114–127.

ROYDEN, L., PATACCA, E. & SCANDONE, P. 1987. Segmentation and configuration of subducted lithosphere in Italy: an important control on thrust-belt and foredeep–basin evolution. *Geology*, **15**, 714–717.

SCISCIANI, V., CALAMITA, F., TAVARNELLI, E., RUSCIADELLI, G., ORI, G.G. & PALTRINIERI, W. 2001. Foreland-dipping normal faults in the inner edges of syn-orogenic basins: a case from the Central Apennines, Italy. *Tectonophysics*, **330**, 211–224.

SINCLAIR, H.D. 1997. Tectonostratigraphic model for underfilled peripheral foreland basins: an Alpine perspective. *Geological Society of America Bulletin*, **109**, 324–346.

TAVARNELLI, E. 1996. The effects of pre-existing normal faults on thrust ramp development: an example from the Northern Apennines, Italy. *Geologische Rundschau*, **85**, 363–371.

TAVARNELLI, E. 1999. Normal faults in thrust sheets: pre-orogenic extension, post-orogenic extension, or both? *Journal of Structural Geology*, **21**, 1011–1018.

WILLIAMS, G.D., POWELL, C.M. & COOPER, M.A. 1989. Geometry and kinematics of inversion tectonics. *In*: COOPER, M.A. & WILLIAMS, G.D. (eds) *Inversion Tectonics*. Geological Society, London, Special Publications **44**, 3–15.

WINTER, T. & TAPPONNIER, P. 1991. Extension majeure post-Jurassique et ante-Miocène dans le centre de l'Italie: données microtectoniques. *Bulletin de la Societé Geologique de France*, **162**, 1095–1108.

ZAPPATERRA, E. 1994. Source-rock distribution model of the Periadriatic region. *AAPG Bulletin*, **78**, 333–354.

Weak zones in Precambrian Sweden

C.J. TALBOT

Hans Ramberg Tectonic Laboratory, Department of Earth Sciences, Uppsala University, SE-752 36 Uppsala, Sweden (e-mail: Christopher.Talbot@geo.uu.se)

Abstract: Examples of weak fault zones in and bordering Precambrian Sweden are reviewed and then analysed in terms of the factors that rendered them weak. The criterion taken here for weak zones is evidence of post-glacial uplift having reactivated old shear zones that are still active now. Strain analysis of the Singö shear zone demonstrates that it was already weak while it was deep and ductile between 1.86 and 1.6 Ga. Thus the orientations of strain ellipsoids indicate pure rather than simple shear across shear strands in which the dissolution of quartz and feldspar indicate high synshear fluid pressures. The characteristics of weak strands of the Baltic–Bothnian shear zones and the Senja–Bothnia shear zones are long histories of superimposed ductile, semi-ductile and brittle structures, indicating repeated reactivations with different kinematics. The shear zones that weakened are those that reactivated in several episodes during their history. Repeated reactivation renders major shear zones in the 'brittle' upper crust 'ductile' by rounding major angularities along subvertical and subhorizontal zones. This generates continuous seams of cataclasites in which pore fluids can pressurize so that internal strains can remain ductile and almost aseismic by frictional sliding or flow. This review ends by discussing how multiple reactivations weaken major faults and how reactivation remains focused on particular zones.

This work is based on the premise that deformation zones in old continental crust are weak if they have long histories of activity and are currently active without much seismicity.

Like other peneplained collages of continental Precambrian crust, the crystalline basement of Fennoscandia (Fig. 1) consists of terranes constructed in different environments being sutured together along zones of intense deformation (Gaal & Gorbatschev 1987; Stephens *et al.* 1997). Each of the individual terranes of Sweden accumulated marginal and internal shear zones as continental convergence passed from vertical to lateral escape (or vice versa) with more or less gravitational collapse of successive mountain chains and their roots. The orientations and kinematics of each subsequent generation of shear zones reflect changes in their boundary conditions and stress fields. The style and mineralogies of their fabrics and structures in relevant portions record the metamorphic grades during repeated reactivations.

The Precambrian rocks of Sweden (2.45–0.6 Ga) accumulated fractures every time (at least five) oceans opened or closed nearby (Munier & Talbot 1993). By about the end of the Caledonian orogeny (0.4 Ga), the rock masses constituting Precambrian Sweden had become fracture saturated. This means that blocks jostled along old fractures rather than generating new fractures when the boundary conditions changed throughout the remaining Phanerozoic time (Munier & Talbot 1993).

Most of Sweden is now a thick (40–50 km) cool (<70 mW m^2) Precambrian shield or craton, which has traditionally been considered to have been essentially stable for the last 1 Ga (Husebye *et al.* 1986). However, weak zones in this crustal unit are particularly obvious because they are the old zones of repeated shear that have been active during post(Pleistocene) glacial uplift, which is still occurring at rates of up to 0.9 mm a^{-1} in the north (Ekman 1996).

Several lines of evidence in combination localize the major weak zones that flaw the Swedish lithosphere. Thus, zones of oxidation pick out aeromagnetic lineaments of low magnetic intensity (Henkel 1991) and some major shear zones occur along linear gravity anomalies (e.g. Bergman & Sjöström 1994). Similarly, known irregularities in the contours of post-glacial uplift in Fennoscandia (Ekman 1996) can be attributed to uplift at different rates on either side of active faults indicated by geomorphology, structural geology and seismic monitoring (Talbot & Slunga 1989). Many such faults are indicated by breaks in slope of uplift rate along profiles produced by the first two precision

From: HOLDSWORTH, R.E., STRACHAN, R.A., MAGLOUGHLIN, J.F. & KNIPE, R.J. (eds) 2001. *The Nature and Tectonic Significance of Fault Zone Weakening*. Geological Society, London, Special Publications, **186**, 287–304. 0305-8719/01/$15.00

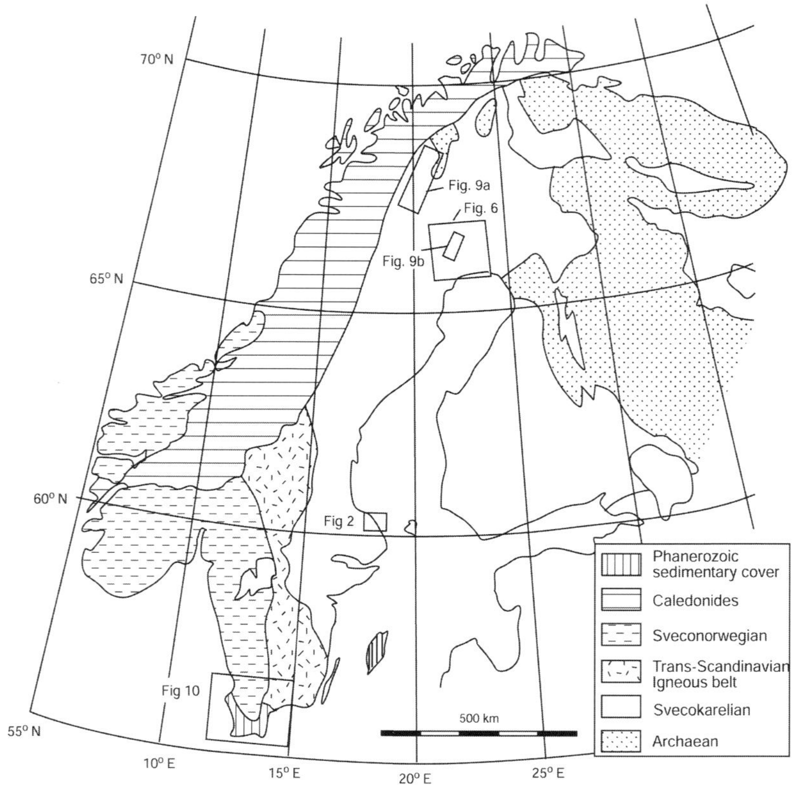

Fig. 1. Simplified geological map of Fennoscandinavia with outlines of areas discussed.

surveys of Sweden. However, the evidence for a reverse dip-slip component of displacement is statistically sound along only one of these faults, and that at a rate that averaged *c.* 1 mm a^{-1} between 1892 and 1960 (Talbot 1992).

This work starts with a summary of two lines of evidence that demonstrate that one of the weak zones in central Sweden was already weak when it sheared in a ductile state at depth. The structural histories of two suites of major deformation zones in the north of Sweden are then outlined, and the lessons reinforced by studies of the End Glacial faults in Lapland. The major earthquakes 9–8 ka ago along a particular category of not particularly weak zones in Lapland gave way to the current low seismicity of Fennoscandia (Arvidsson 1996). The current activity along the weak zones is being measured by repeated global positioning system (GPS) surveys (e.g. Pan *et al.* 1999).

The discussion opens by contrasting the characteristics of short-lived shear zones with long-lived weak zones in Sweden. It is argued that weak zones are weakened by their repeated reactivation so that they become capable of 'ductile' shear right through the lithosphere. Considering how reactivations weaken the lithosphere leads on to the much more significant question of how reactivations remain focused on the same zone.

Singö deformation zone

The comparatively well-known ductile strains in part of the Singö deformation zone in east–central Sweden (Fig. 2) give an instructive glimpse of the early stages of one of the weak zones in Sweden.

The Singö gneiss zone trends WNW along 70 km of the east coast and is probably a major terrane boundary within Svecokarelian (Palaeoproterozoic) crust, which is *c.* 46 km thick to the SW and *c.* 48 km thick to the NE (BABEL working group 1993). The amphibolite-facies gneisses of the Singö zone are *c.* 5 km wide onshore but may reach a width of 50 km offshore to the NNE (Eriksson & Henkel 1988). Before 1.83 Ga, transpression of the Singö

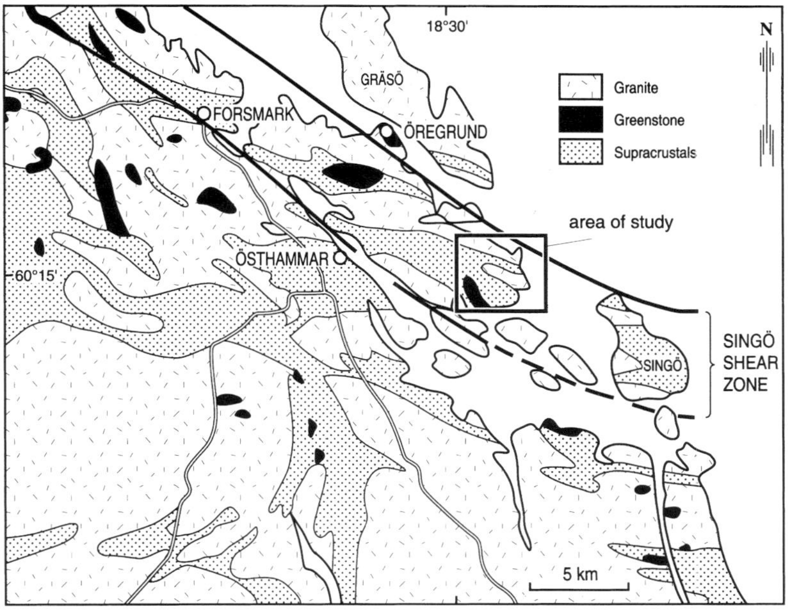

Fig. 2. The Singö deformation zone in east–central Sweden and the district analysed.

gneiss zone smeared clockwise the regional map pattern of polygonal supracrustal belts (*c.* 2.0–1.9 Ga) intruded by granitoid plutons (1.8–1.7 Ga) (Stålhös 1984, 1991) along a WNW-trending (*c.* 1.8 Ga) swarm of mafic dykes (Welin *et al.* 1980; Wickman *et al.* 1983; Welin & Stålhös 1986). These dykes are decimetres wide and metres apart in the gneiss zone but, in subvertical mylonites about 30 m wide that anastomose along the gneiss zone, have been thinned to widths and spacings of millimetres (Talbot & Sokoutis 1995).

The dykes and associated mafic enclaves in their granitoid country rock zone deformed as incompetent inclusions within the Singö zone. The dykes extended uniformly or developed mullions on scales of decimetres, depending on their orientations. By distinguishing the orientations of planar elements of dykes that elongated (uniformly) from those that shortened (and developed fold mullions), Talbot & Sokoutis (1995) constrained the surface of no elongation for 32 ellipsoids ($X \geqslant Y \geqslant Z$) for 25 localities with volumes of about 2 m × 2 m × 10 m. Strains in the 25 localities integrate within a district ellipsoid with a volume of about 3 km × 3 km × 0.1 km (Fig. 3).

Their main findings were as follows.

(1) Tie-lines between ellipsoids constructed using up to three different types of strain markers at the same localities (dykes, inclusions and quartzo-feldspathic veins), demonstrated that the linear strain field is also the strain path ($k = 0.70$ on a logarithmic Flinn plot; Fig. 3). This strain path reveals that locality ellipsoids lost a uniaxial volume of *c.* 3% per 10% total shortening along *Z*. This volume loss was attributed to the uniaxial dissolution of quartz and microcline from the granitoids along their axis of shortening. A small proportion of this volume was reprecipitated as quartzo-feldspathic veins locally within some of the mafic dykes.

(2) Early right-lateral escape represented by gneissose fabrics (*c.* 1.86–1.83 Ga) is interpreted (Fig. 4) to have given way to later vertical escape along dip-flow mylonites (*c.* 1.7–1.6 Ga). The gneiss lenses bound by anastomosing mylonites may have been the cores of Pennine-type nappes extruded out of their (Singö) root zone.

(3). The *X*-axes for locality ellipsoids rotate from subhorizontal in Singö gneisses with natural octahedral unit strains of $\bar{\varepsilon}_s \approx 1.2$, through 60° in mylonites ($2 < \bar{\varepsilon}_s < 3$) to become vertical in ultramylonites ($\bar{\varepsilon}_s > 3$), where they parallel the ubiquitous subvertical mineral shape and orientation lineation in all rock types (Fig. 5).

(4) Most relevant to the concept of weak ductile shear zones is the finding that there are surprisingly few signs of rotation within the Singö shear zone, and those few are all around unexpected axes. Instead of *X* and *Z* having rotated about *Y*, as usually expected of simple shear,

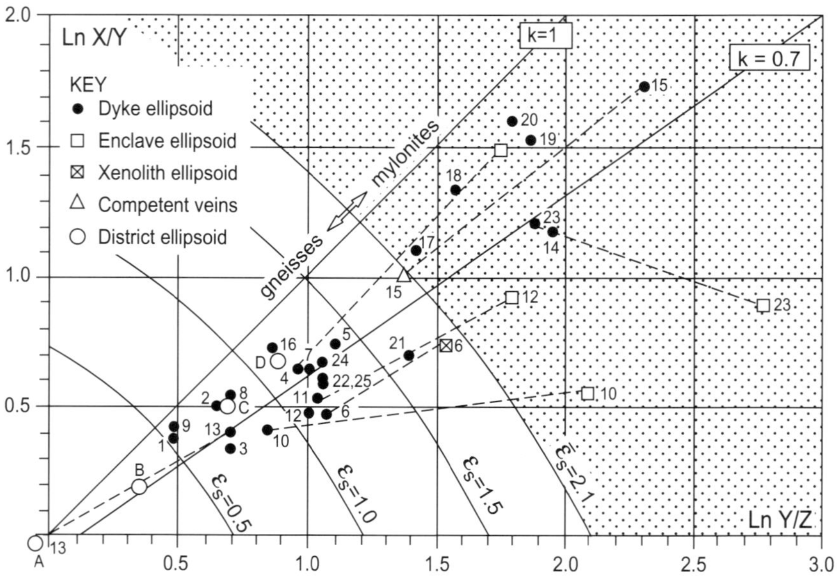

Fig. 3. The 32 ellipsoids from 25 localities in the Singö shear zones lie in a linear deformation field on a logarithmic Flinn plot. Tie-lines between ellipsoids constructed using different markers at the same localities indicate that the deformation field was also the linear deformation path ($k = 0.7$), which indicates 3% volume loss for every 19% total shortening along *Z* (from Talbot & Sokoutis 1995).

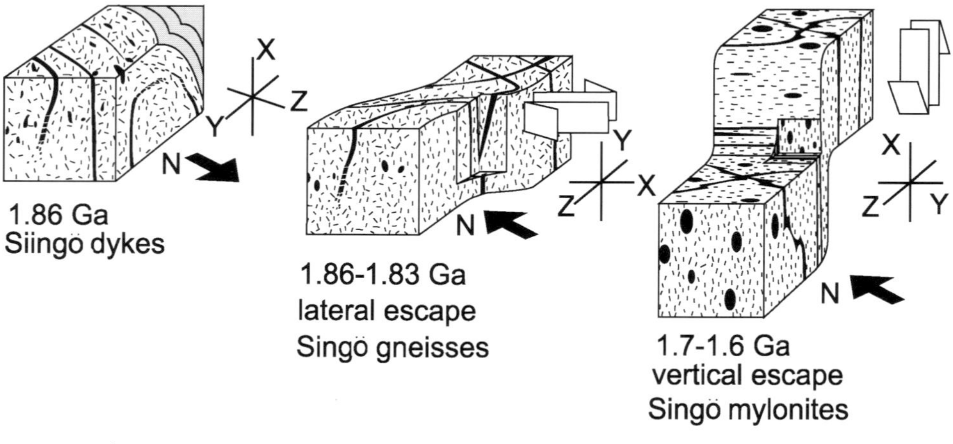

Fig. 4. Block diagrams illustrating the main three stages in the development of the Singö shear zone. The different orientations of blocks relative to present north should be noted (see Talbot & Sokoutis (1995) for recovery of this history on a radial Hsu plot).

the *Z*-axes are essentially constant whatever the strain, and are everywhere orthogonal rather than oblique to the length and dip of the Singö zone and its internal planar fabrics (Fig. 5). As a result, the foliation is planar and parallel ($\pm$ *c.* 2°) to the *XY*-planes of enclaves and all 32 strain ellipsoids. The foliation is also axial planar to mullions in the mafic sheets, and to buckles in the quartzo-feldspathic veins and granitoid screens within the mafic sheets. These relationships indicate that the Singö mylonites involved pure irrotational shear rather than simple rotational shear.

Talbot & Sokoutis (1995) compared this situation with the first-recognized weak fault, the current San Andreas fault, which had also been attributed to pure rather than simple shear (Evans & Wong 1992; but see Scholtz 2000). The principal *in situ* stresses in the much younger rocks near that active plate boundary

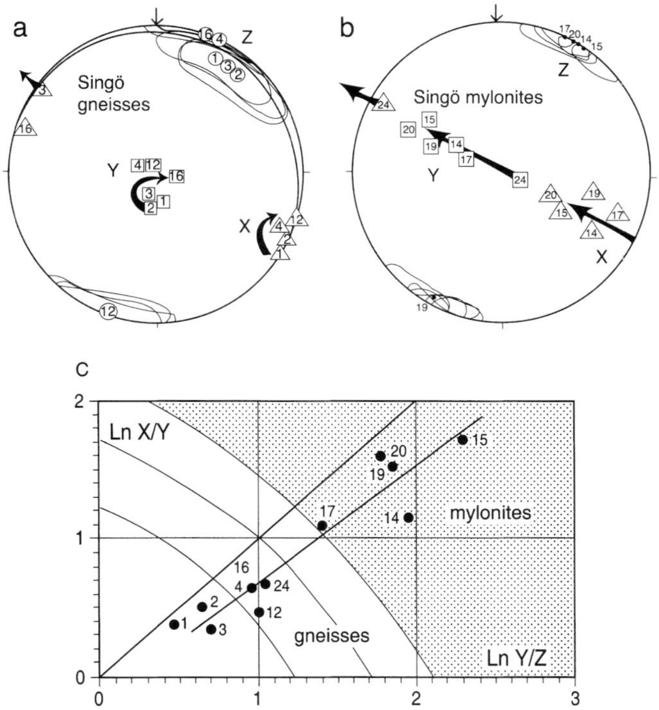

Fig. 5. The maximum (*X*) and intermediate (*Y*) principal axes of strain rotate about the minimum axis (*Z*) when traced from Singö gneisses (**a**) into Singö mylonites (**b**), indicating pure rather than simple shear. Numbers on lower-hemisphere equal area projections indicate locality ellipsoids with shapes plotted on the logarithmic Flinn plot (**c**) (simplified from Talbot & Sokoutis 1995).

refract to become locally orthogonal to the San Andreas fault. This geometry was attributed to high pore-fluid pressures further weakening already weak fault gouges in the shallow crust (Evans & Wong 1992). Talbot & Sokoutis (1995) speculated that the principal stresses in regions adjoining the Singö zone also refracted to become orthogonal to the Singö zone as deep gneisses strained to mylonites subject to high pressures in the pore fluids responsible for their volume loss.

As appears to be general for weak zones throughout Sweden, the Singö mylonite zones were later overprinted by centimetre-thick seams of ultramylonites and then millimetre-thick seams of palaeoseismic fault melts (now devitrified). During their rise through the ductile–brittle transition in greenschist facies between 1.6 and 1.5 Ga, the Singö mylonites, ultramylonites and frictional melts were veined by epidote. Whereas the penetrative mineral lineation developed in the ductile gneisses and mylonites had been subvertical, the slickenlines in the many generations of minerals infilling superimposed brittle faults indicate different kinematics during numerous subsequent brittle reactivations. Vertical offsets of the sub-Cambrian peneplain that are locally veneered by Cambrian sandstones in the Singö zone demonstrate that reactivations continued into Phanerozoic time. A few small earthquakes have registered modern activity along the Singö shear zone (Slunga 1991).

Baltic–Bothnian and Senja–Bothnia shear zones

Figure 6 shows an aeromagnetic map and an interpretation of the irregular strands of major weak zones in nine 50 km × 50 km map squares covering part of northern Sweden. The weak zones are indicated by lineaments of low magnetic intensity up to 200 m wide and hundreds of kilometres long attributed to oxidation of magnetic minerals (Henkel 1989, 1991).

The rocks of this region are mainly of Palaeoproterozoic age and consist of small volumes of supracrustal rocks among huge volumes of

Fig. 6. (**a**) Caption opposite.

coarse-grained granitoid sheets and plutons. The shear zones with N–NNE trends in the eastern map squares are strands of the *c.* 45 km wide Baltic–Bothnian shear zone (BBSZ). These strands are about 200 m wide and dip about 80° west along the border between Sweden and Finland (Fig. 6). Berthelsen & Marker (1986), who named this zone, extrapolated it for <1000 km southward down the length of the Baltic Sea. Similar N–NNE-trending strands *c.* 50 km further west have no name. The other main weak zones in Fig. 6 trend NW and are narrower (*c.* 50 m) and dip steeply SW in outcrop (Talbot *et al.* 1989). These are strands of the *c.* 100 km wide Senja–Bothnia shear zone (SBSZ), which extends over 350 km from the Atlantic to the Bay of Bothnia and may be a continuation of a similar zone in Finland offset (since 1.7 Ga) across the BBSZ (Henkel 1991).

Supracrustal rocks in the BBSZ belong to the twice migmatized *c.* 2.45 Ga Korpilombolo Group, whereas those of the *c.* 2.05 Ga Raneå Group to the west record only the younger migmatization (see Skiöld (1982) for references to dating). The north- to NE-trending penetrative foliation imposed by the first migmatization on the Korpilombolo Group accounts for the conformable and penetrative aeromagnetic grain

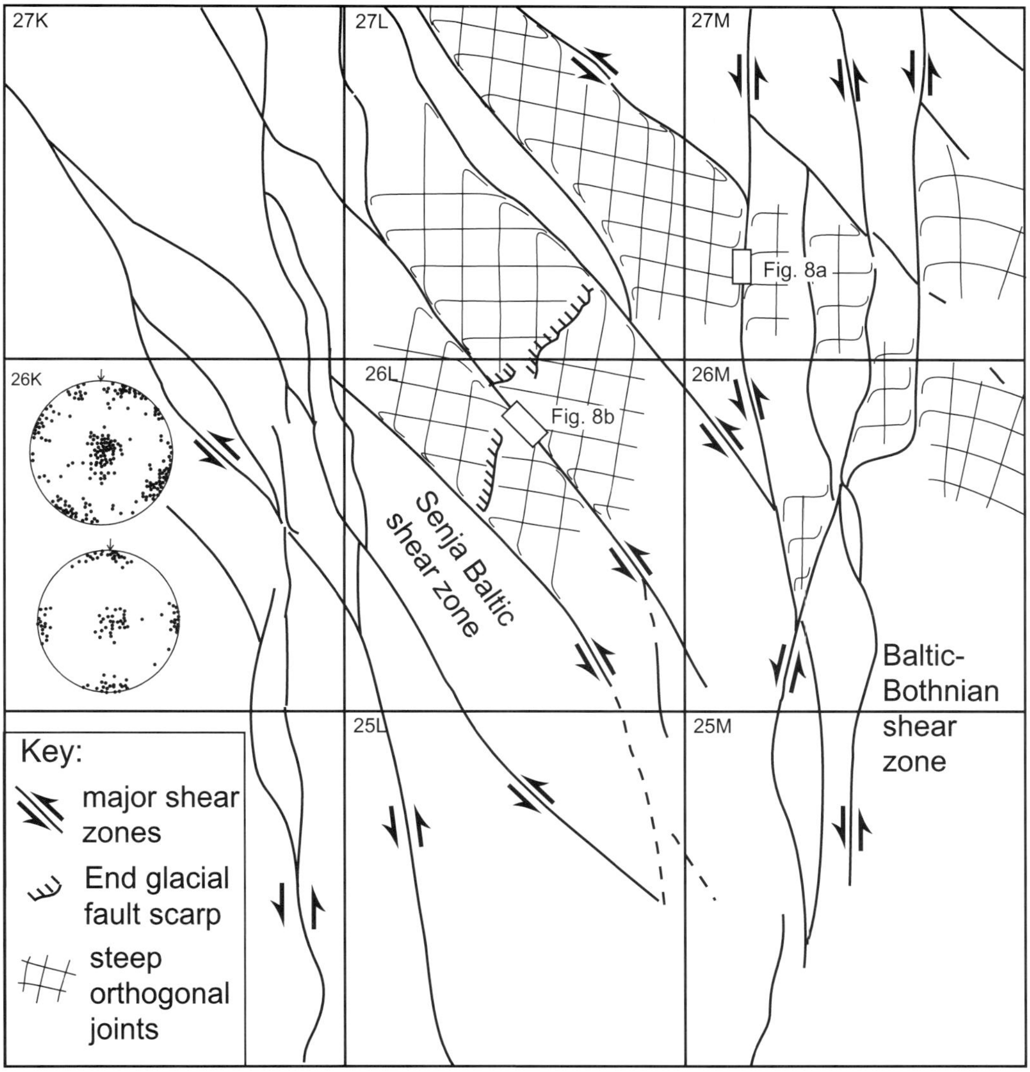

Fig. 6. (**a**) Aeromagnetic anomaly map (total field minus reference field) of part of northern Sweden indicated in Fig. 1 (Swedish Geological Survey Flygmagnetiska Karta 24J SV-28N NO, scale 1: 500 000, 2000). (**b**) Interpretation of (**a**) with nine 50 km × 50 km map squares labelled (simplified from Talbot *et al.* 1989). Lower hemisphere equal-area stereograms on left represent orientations of fractures in the Senja–Baltic shear zone (top) and the Baltic–Bothnian shear zone (bottom).

seen between the granitoids and the various individual shear strands (Fig. 6a). This foliation is subparallel to layered rocks intruded by the oldest grey Haparanda granitoids (diorite to granodiorite) and uniform red medium-grained hornblendic Skröven granites during the *c.* 1.9–1.85 Ga Svecokarelian orogeny (Fig. 7).

The syn-Svecokarelian Haparanda and Skröven granitoids are injected by sheets of three essentially contemporaneous facies of the 1.8–1.7 Ga Lina granitoids. The emplacement of the Lina granitoids west of the BBSZ probably accounts for the second pervasive synkinematic regional migmatization at 1.8–1.7 Ga (Skiöld 1982). Quartzo-feldspathic veins of the second migmatites are deformed in the supracrustal gneisses to the west of the BBSZ but undeformed where they have been superimposed on the oldest mylonites of the BBSZ.

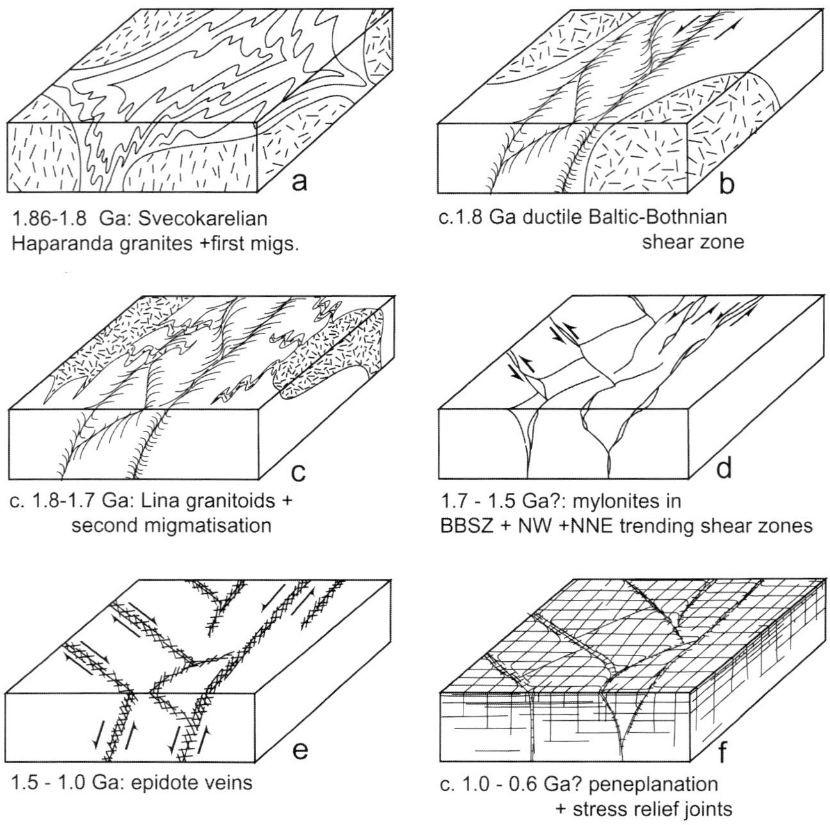

Fig. 7. Simplified tectonic history of the region shown in Fig. 6 (modified after Talbot *et al.* 1989).

Ductile shear zones

The BBSZ therefore began to develop late in, or after, the Svecokarelian orogeny (Fig. 7a) and before the emplacement of the Lina granitoids (Fig. 7c). Most strands of the NW-trending SBSZ shear zones developed later than the north to NNE zones of the BBSZ and after the Lina granitoids (Fig. 7d). However, all major shear zones in the region were reactivated many times after they first developed; seismicity (Slunga 1991; Arvidsson 1996) and unpublished preliminary GPS readings indicate that some components are still active.

The oldest deformation fabric in the region is a steep north- to NE-trending penetrative foliation with a subvertical lineation that is axial planar to kilometre-scale upright folds in the supracrustal rocks. Haparanda granitoids of uniform grain size are gneissose because of aligned hornblende aggregates and ellipsoidal quartz and feldspar grains with axial ratios of about 3:2:1. Previous workers supposed the Lina granitoids to postdate all regional deformations. However, although they foliated less readily than their surroundings, all those south of the northern third of map sheet 27L (Fig. 6b) display, faintly but clearly, the steep north to NE foliation of the region (Talbot *et al.* 1989). Furthermore, some of the Lina aplites and pegmatites less than *c.* 1 m thick are clearly pinched along, or folded about, the same steep foliation as that superimposed on their country rocks (Fig. 7c).

Although of different ages, all the north- and NW-trending deformation zones studied on the ground (Fig. 6b) appear to have shared similar general histories (Talbot *et al.* 1989). All began as zones of penetrative ductile shear in amphibolite metamorphic facies. However, whereas displacements along north to NNE strands of the BBSZ were sinistral throughout their history, displacements along NW strands of the SBSZ were usually sinistral but were dextral oblique during part of their passage through greenschist facies (Fig. 7e).

All the major shear zones in the region narrowed during successive reactivations through mylonites (± pseudotachylytes) in greenschist facies before broadening to later vein systems and still younger brittle fracture zones (Fig. 7f). All shear zones are marked on the surface by low topographic lineaments partially infilled by glacial deposits, post-glacial marine sands, organic mires, or lakes. Many of these long linear depressions are flanked by fault- or fault-line scarps hundreds of metres long in bedrock.

Foliations

Although the central 50–200 m of major shear zones is nowhere exposed, the regional foliation intensifies and swings toward them in nearby outcrops. Most protoliths diminish in grain size toward shear zone centre-lines so that gneisses become schists or protomylonites where they are subparallel to the shear zone. Foliated metamafic sheets display mullions with axial surfaces parallel to the shear zone. A new penetrative foliation is common in all major shear zones. In schistose rocks, this synkinematic foliation is planar and parallel to the major shear and forms a compound foliation together with the rotated regional foliation. In gneisses, the new synkinematic foliation tends to form spaced crenulations superimposed on the regional foliation (S–C fabrics). In both cases, any hornblende surviving in the regional foliation was retrogressed to biotite with the exsolution of magnetite and, very locally, synkinematic garnets exist. Some of the major strands are relatively simple symmetrical shear zones but most are compound shear zones with various degrees of complexity (e.g. Fig. 7b).

Ductile–brittle transition

Some strands of mylonite are completely annealed. Some of the oldest quartz–epidote veins were deformed by some of the youngest mylonites. Thin (<2 mm) and long (>5 m) planar veins of black recrystallized pseudotachylyte and micro-cataclasites occur in a few exposures; most are superimposed on mylonites or the earliest quartz–epidote veins but some occur in unfoliated aplites. These complex cross-cutting relationships of brittle and ductile structures suggest that aplites could act as brittle asperities in major deformation zones that were undergoing episodes of shear that alternated from penetrative ductile shear (responsible for the mylonites) through semiplastic (veins) to seismic (pseudotachylytes), possibly because of changes in strain rate in the greenschist facies.

Vein and fracture systems

Repeated semi-ductile and brittle reactivations of the early ductile shear zones led to vein systems with complex multiple mineral infills. In natural exposures, these preserve successive generations of only relatively insoluble minerals such as quartz, quartz + epidote, and epidote ± chlorite ± Fe oxides. Such veins range from a few millimetres to decimetres in width and from a few centimetres to many metres in length. Some parallel the zone but most are en echelon at various angles to the older ductile fabrics on which they were superimposed. Internal slip surface display strike-slip slickenslines for most of their history but oblique or dip-slip normal fault slickenlines or fibres in greenschist infills. Longer (and younger?) fractures infilled by weaker and more soluble minerals are missing from natural outcrops but display various combinations of laumontite, prehnite, calcite ± sulphides in road cuts, drill cores and trenches (Bäckblom & Stanfors 1989).

The general loss of soluble mineral infills from natural exposures means that comparatively few veins or fractures can be dated, even in relative terms. However, it is clear from matching infills with fracture orientations that new fractures developed with different kinematics during many reactivations. Fracture spacing is of the order of a metre on the margins to decimetres in outcrops nearest the unexposed centre-line.

Faults

Only short segments of three definite faults have been found exposed in the region. All appear to have acted as metre-scale transpressive flower structures formed in greenschist facies. One thrust westward out of a north–south strike-slip fault duplex (at Solkoberget, Fig. 6a). Another thrusts graphic granite seamed by illite southeastward out of a NW-trending strike-slip fault over metabasic rocks extensively altered and veined by epidote and chlorite with a block volume of <1 cm^3 (at Vitberget, Fig. 6b). The third is a metre-scale transpressional imbricate strike-slip fault zone in a NW-trending shear zone.

Superimposition of orthogonal joints

Högbom (1925); Ödman (1957) reported 'cubic' joint patterns to be ubiquitous in all outcrops of Lina granites in the region (Fig. 7c). Talbot *et al.* (1989) found that one or more of these three orthogonal joint sets are represented in every exposure of every rock type visited throughout the region.

A stereogram of poles to joints measured everywhere in the region would obscure a general simplicity; three orthogonal joint sets (with different orientations) are usual in the vast majority of individual outcrops. Orthogonal sets of joints are usually attributed to stress relief (Engelder 1993, pp. 218–227). In Svecokarelian Sweden they presumably began developing as the rocks approached exposure in association with sub-Jotnian (*c.* 1.7 Ga) and/or sub-Cambrian (*c.* 0.6 Ga) peneplanation of the craton, long after the major shear zones were already in existence.

In the major rock blocks between weak zones throughout this region (indeed, throughout Svecokarelian Sweden), the only obvious brittle structures consist of one subhorizontal joint set and two subvertical joint sets that strike north to NNE and west to WNW (Fig. 6b). These define orthogonal joint blocks that can exceed *c.* 30 m^3 at low levels in the relief but generally decrease in volume toward free surfaces, particularly toward hilltops.

The oldest set of orthogonal joints at any particular locality parallels the dominant foliation, the next parallels the closest free surface (or its initial planar envelope), and the third is essentially orthogonal to the other two. Because the regional foliations are subvertical, subhorizontal joints are ubiquitous. Differences between patterns of stress relief joints in the major shear zones, the intervening rock masses and their mutual transition zones, imply that the planar mechanical anisotropy imparted by the foliation influenced early stress-relief more than the orientation of the nearest free surface.

In significant areas between the major shear zones, early, widely spaced 'subhorizontal' joints can dip as much as 20° to remain perpendicular to the foliation, whereas closer (younger?) joints are orthogonal to the free surface rather than to the foliation. This effect, which implies that the early significance of the foliation diminished with local block volume, is particularly clear near the summits of hills, which exfoliate like onions. The mechanical influence of the foliation is even more dramatic in and beside major shear zones. Where the regional foliation swings into a major shear zone, the two subvertical orthogonal joint sets swing in concert to remain parallel and perpendicular to the curving foliation (Figs 6b and 8). This occurs within *c.* 100 m of NW-trending strands of the SBSZ and rather further from north–south strands of the BBSZ. At first sight it appears that the joints have been sheared but it seems more likely that the younger (1–0.6-Ga) joints merely exploited a strong mechanical anisotropy that was already distorted.

This local refraction of orthogonal fractures into each major shear zone reduces their block volume, commonly from >1 m^3 to <1 cm^3 over the 100 m approaching the outcrop closest to their centre-lines. The later addition of orthorhombic fractures to the older shear-induced fractures presumably further weakened the surficial levels of every major shear zone.

Breccias and gouges

Apart from a small outcrop of orange fault gouge with clasts <1 cm temporarily exposed in a dammed lake, a zone of no exposure, typically 50–200 m wide, is ubiquitous along the centre-line of all the major shear zones visited on the ground (Fig. 6b). The soils and lakes along these ice-scoured topographic lineaments obscure the thick oxidized and saturated fault breccias and gouges indicated by aerial and ground magnetic surveys, seismic refraction and electromagnetic traverses (Henkel 1989) and drilling and trenching (Bäckblom & Stanfors 1989).

End Glacial faults

The fault with the most northeasterly trend in the region just described is the Lansjärv End Glacial Fault (shown barbed in Fig. 6b). This is one of six or so End Glacial faults (EGFs), which together have scarps lengths exceeding 500 km in Lapland (Lagerbäck 1990). The eskers and ice-moulded moraines that defined the ground surface when the last ice melted were displaced by up to 10–15 m (locally 25 m), usually by steep reverse dip-slip, cliff-like fault scarps and wrinkle ridges along these EGFs. Each scarp appears to have formed in a single event as the last ice of the Pleistocene ice ages melted 8–9 ka ago. Forty or so major (5×10^5 to $10^6 m^3$) contemporaneous landslides (involving toppling and sliding in rock falls and circular slides or liquefied flows in sandy moraines) disturbed slopes that are now stable above the highest marine shoreline. Soil slopes were presumably saturated when the landslides moved down slopes of between 10° and 3° around obstructions that are no longer present. Kujansuu (1964) was the first to interpret these invisible obstructions as the last remnants of the ice and to recognize that the generation of each scarp must have been accompanied by a major earthquake, an interpretation since documented by convincing palaeoseismic studies based on trenching (Olesen 1988; Lagerbäck 1990).

Comparison of plots of surface-breaking fault lengths v. maximum displacement of Lapland EGFs with plots of length v. magnitude of measured associated earthquakes, suggests that the EGFs had magnitudes of about eight, much larger than any earthquakes in the same region in historical times (Talbot 1986; Arvidsson 1996).

Bedrock outcrops are rare along the EGFs of Lapland. However, every outcrop makes it clear that the EGFs exploited old structures with long histories of repeated reactivation with different kinematics (Talbot 1986; Talbot *et al.* 1989). Different scarps of EGFs (Olesen 1988; Lagerbäck 1990; Dehls *et al.* 2000) join segments of old structures that range in age from at least Svecokarelian (1.7 Ga) to Caledonian (0.42 Ga). Two of the best known EGFs in Sweden (the 80 km Lansjärv and the 160 km Pairvie faults) have the geometries of transpressive flower structures striking generally NNE; pop-ups extrude from depths of <1 km along local segments (Fig. 9).

Structural contours combined with ground studies of the Lansjärv EGF reveal that the gneissose foliation that generally dips 70° to the NW or SE locally intensifies as it decreases to dips between 30° and 10° along the bases of zones up to 50 m thick. Many of the brittle thrusts that reactivated at End Glacial times were along the bases of old ductile (mylonitic) precursors. Veins and fracture zones are superimposed on most, but not all such reactivated shear zones that have staircase trajectories with listric imbricate ramps above fault flats.

Not all workers believe the shallow imbricate thrust flakes shown in Fig. 9 (e.g. Muir-Wood 1989, 2000) because a borehole drilled to test one such imbricate fault system found so many open fractures that nobody agreed on whether the EGF had been intersected or not (Bäckblom & Stanfors 1989). However, it was agreed that all the open fractures were old (and coated by chlorite, Fe oxides, etc.) and that not a single new fracture below surficial levels could be attributed to syn- or post-glacial activity (Bäckblom & Stanfors 1989).

It is noticeable that the transpressive reverse EGFs of Lapland reactivated old structures that trend mainly NE–SW with a spacing of *c.* 100 km (thicker than the 48 km crust involved). The forces that generated the EGFs appear to have had an axis of maximum horizontal compression sufficiently close to that expected of

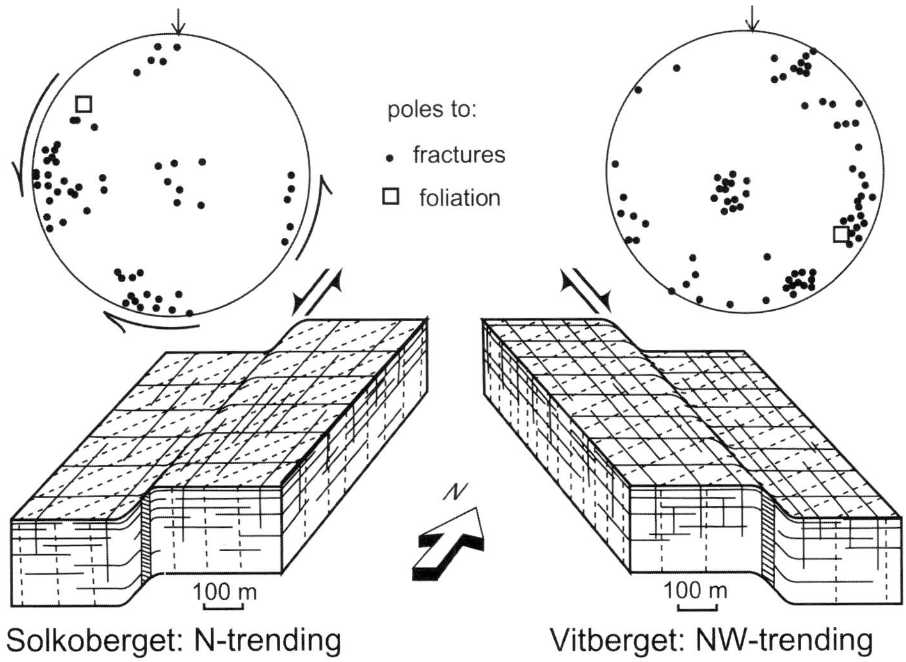

Fig. 8. Lower-hemisphere equal area projections and block diagrams of fractures and foliation in outcrops beside representative strands of the north–south-trending BBSZ and the NW-trending SBSZ (from Talbot *et al.* 1989).

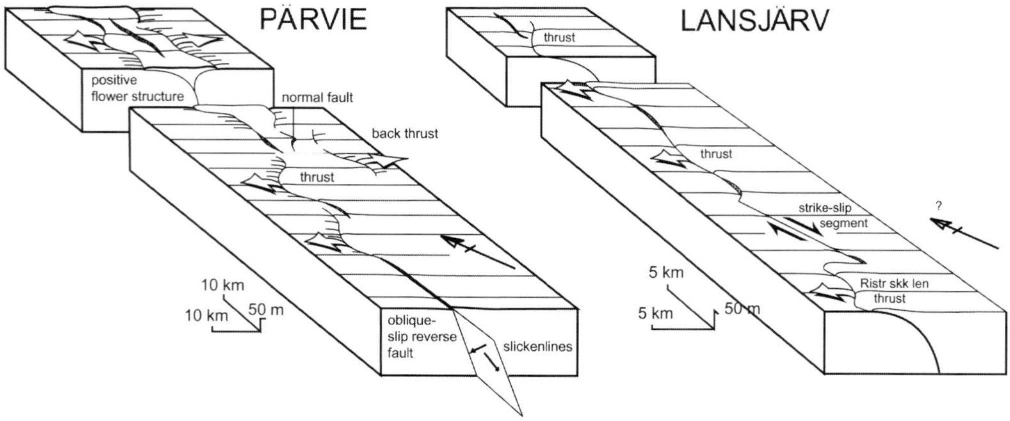

Fig. 9. Schematic block diagrams of parts of the two (of about six) End Glacial Faults located in Fig. 1 (simplified from Talbot 1986). Only the southernmost 85 km of the 160 km Pärvie fault are shown with form lines about 5 km apart along strike. About 25 km of the 60 km Lansjärv fault are shown with form lines about 2 km apart along strike.

plate tectonic forces of the last 34 Ma (Talbot & Slunga 1989) not to have exploited significant lengths of the weak strands of the nearby BBSZ or the SBSZ (Fig. 6b). Some of the major NW-trending shear zones obviously acted as linked strike-slip transfer faults where NE–SW segments of the EGFs are en echelon. Nevertheless, however likely such linkages were to have propagated away from the EGFs, no contemporary strike-slip displacements have been demonstrated in the glacial scenery without a significant dip-slip component making them obvious.

Although the EGFs may still be seismically active (Arvidsson 1996) they do not appear to be as seismically active as weak zones with other orientations (Talbot & Slunga 1989). Furthermore, the large range of ages of the shear zones that were reactivated in different segments of the EGFs suggest that these were not as long lived as the nearby, much wider strands of the BBSZ and SBSZ. The same impression is given by the (<1 m thick) breccias (with oblique-slip epidote and chlorite slickenfibres and lines) and decimetre thick Fe-oxide-rich gouges with dip-slip slickenlines revealed in trenches across the Lansjärv fault. It took major earthquakes to move the EGFs in special conditions (Talbot 1999a) because they are not as weak as the other examples described here.

GPS strains

Since 1993, Sweden has had the SWEPOS array of 21 permanent global positioning system (GPS) stations, which are measured continuously (Scherneck *et al.* 1998; Milne *et al.* 2001). The accumulating vertical motions confirm the long-held picture of post-glacial uplift. To a first approximation, the (more accurate) horizontal strains emerging from this permanent GPS network can be attributed to post-glacial unloading of an unflawed lithosphere (Johnston *et al.* 1998).

However, the Swedish lithosphere is flawed by major weak zones (Fjeldskaar *et al.* 2000). Local networks of GPS stations have been established across examples of such active fault zones with low seismicity under the FENTEC programme (Talbot 1992). These stations have been occupied for a few days at a time at intervals of a few years since 1989. The only FENTEC GPS network for which results have been published so far straddles the WNW–ESE-trending Sorgenfrei–Tornquist zone in southern Sweden (Pan *et al.* 1999).

This zone is the boundary between the 42–45 km thick Precambrian (>1 Ga) gneisses of the Baltic shield to the north from thinner (*c.* 30 km) Phanerozoic Europe to the south (Tryggvason *et al.* 1998). The structurally complex Sorgenfrei–Tornquist zone can be traced from south of Norway to the Black Sea and is seismically visible to a depth of 600 km (Husebye *et al.* 1986). Initiated <1 Ga ago, this zone formed as a Neoproterozoic continental margin that was sutured in Caledonian times and has had a complex history of displacements ever since (Zeigler 1988).

Talbot & Slunga (1989) used fault-plane solutions for 200 earthquakes in southern Sweden between 1980 and 1984 to identify the Sorgenfrei–Tornquist zone as a strike-slip subplate boundary in line with the Iceland transform in

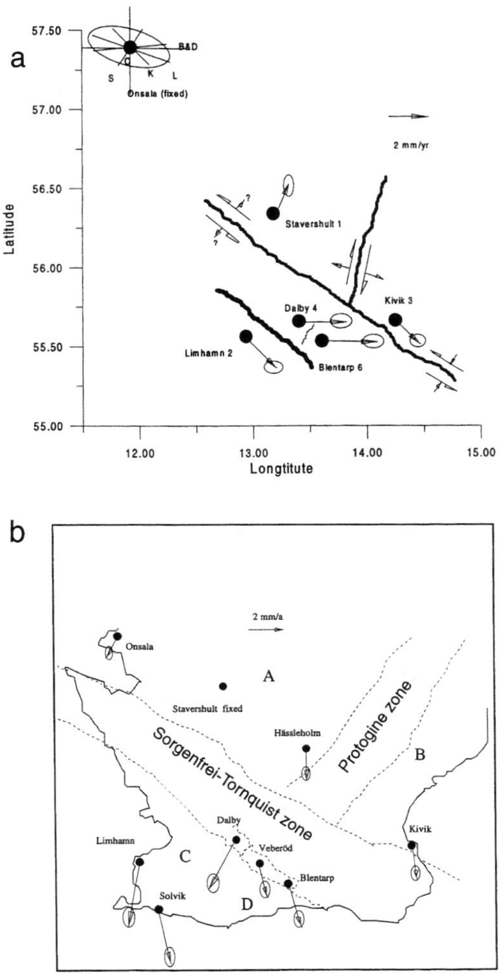

Fig. 10. Two interpretations of displacements of FENTEC GPS stations straddling the Sorgenfrei–Tornquist zone in Skåne, southern Sweden (see Fig. 1 for location of region). (**a**) Scaled vectors for displacements, from data for 1989–1996, relative to a fixed site at Onsala. The strain ellipse best fitting vectors representing horizontal displacements of the stations moving from 1989 to 1996 are also shown centred on Onsala (from Pan *et al.* 1999). Irregular lines of various thickness are inferred faults. (**b**) Scaled vectors for displacements, from data for 1992–1998, relative to a fixed Stavershult (from Pan *et al.* 2001).

the North Atlantic for the last 38 Ma. This model predicts oblique left lateral strike-slip shearing along the Sorgenfrei–Tornquist line in Skåne that is expected to be transpressive along most of its length but a negative transtensional flower structure exists on a crustal scale in NW Skåne. This picture is consistent with the principal axis of compression of the *in situ* stress field in Skåne being more WNW–ESE than the NW axis general in most of Europe (Stephansson 1989).

The vectors representing the horizontal displacements of different suites of FENTEC GPS stations are shown in Fig. 10a and b relative to different fixed stations. If the permanent GPS station at Onsala is taken as stationary, displacement vectors calculated from data for 1989–1996 indicate 4.0 $\pm$ 0.9 mm a^{-1} of slightly transpressive left-lateral strike-slip motion along the Sorgenfrei–Tornquist zone (Fig. 10a). Many geodynamic studies take Onsala to represent 'Sweden' or even 'the Baltic shield'. It is therefore sobering to find that Onsala may be on a shear pod shearing off the rest of Sweden (Talbot & Slunga 1989). The FENTEC station at Stavershult may therefore be more typical of Sweden than Onsala. As a result, GPS readings from 1992 to 1998 are referred to a fixed Stavershult station in Fig. 10b (Pan *et al.* 2001). This shows that Onsala moved *c.* 1.5 mm a^{-1} to the SSW, which agrees well with the finding of Argus *et al.* (1999), who measured a displacement rate of Onsala to the SW, away from the centre of post-glacial uplift, of 1.3 $\pm$ 0.8 mm a^{-1} between 1976 and 1997. Rather than emphasizing transpression along the Sorgenfrei–Tornquist zone (as in Fig. 10a), data from 1992 to 1996 (Fig. 10b) suggest normal faulting with a minor strike-slip component across the Sorgenfrei–Tornquist zone at a rate of *c.* 6.2 mm a^{-1}. This is in agreement with one of the two models suggested by a recent reinterpretation of a major deep seismic profile across the Tornquist zone (Tryggvason *et al.* 1998, fig. 5b).

Provisional unpublished data from other FENTEC GPS networks indicate that similar rapid strains can be expected for other fault zones measured using networks of 15 stations in central Sweden and 12 stations along a 130 km traverse NE across the BBSZ and SBSZ in the north (Fig. 5b). All the GPS networks were designed to monitor general strains, not locate individual active faults. Whether the measured strains occur along the known major faults or other faults between the stations cannot be distinguished. Nevertheless, whichever faults are active, those in Fig. 10 must be weak because no significant seismic activity was registered along any of them during the GPS monitoring.

Summary

This work has confirmed the findings of previous studies of shear zones in other parts of

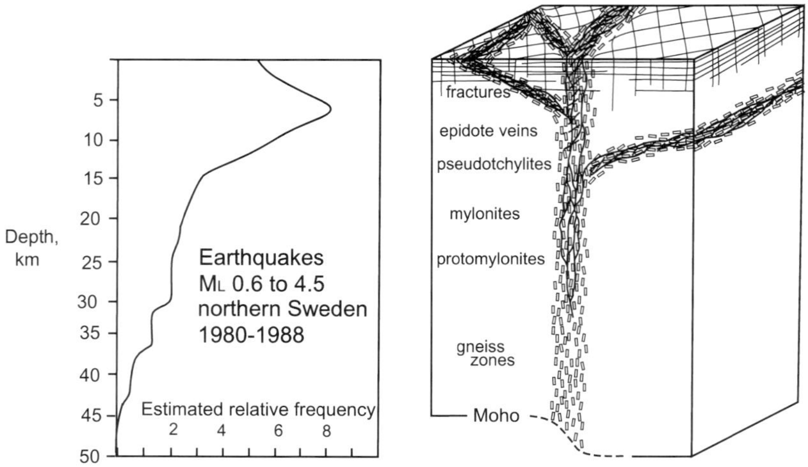

Fig. 11. All the various ductile and brittle structures generated at different levels in Swedish weak zones have accumulated throughout their long histories and are superimposed in their surface expressions.

Sweden (e.g. Talbot & Heeroma 1989; Munier & Talbot 1993; Bergman & Sjöström 1994). These showed that foliations inherited from crustal construction were penetrative on regional scales until strain began to localize in gneiss zones a few kilometres wide. As the rocks now exposed rose toward the ductile–brittle transition at greenschist metamorphic facies, parts of many such gneiss zones narrowed through mylonites tens of metres wide to seismic faults only millimetres wide. As each rock mass cooled through the ductile–brittle transition, it cratonized, and the upper crust embrittled and stress-relieved to a surface being peneplained. Subsequent reactivations jostled the craton along some of the formerly ductile shear zones, which widened to zones of brittle fracture zones seamed by cataclasites (Munier & Talbot 1993).

If the Eurasian plate were rigid, a linear displacement gradient could be expected between its pole of spreading in Siberia and its equator of spreading in the Mediterranean. However, Talbot & Slunga (1989) argued that two major weak zones (complicated by major shear pods) cross Fennoscandia, and suggested that these weak zones act as boundaries to corridors of different rates of rotation of the Eurasian plate around its pole of rotation in Siberia. The low displacement rates expected of a single rigid plate are partitioned to differential strains of a few mm a^{-1} across the boundaries to these corridors. The very high ratio of total to seismic strain in Sweden (Slunga 1991 suggests that strains in weak zones are close to stable right through the lithosphere.

Discussion

Characteristics of weak zones in Precambrian Sweden

Swedish weak zones appear to conform to Sibson's (1977) famous schematic profile of a crustal-scale shear zone with only one modification. Sibson showed broad gneiss zones at depth narrowing upward through mylonites to seismic melts among the lower levels of brittle fracture zones that widen upward to the surface. However, Swedish weak zones share an additional simple, common, and obvious characteristic. As well as being active at different levels now, all the various ductile and brittle structures and fabrics generated at different levels in Swedish weak zones have accumulated through their long histories to be superimposed in their surface expressions (Fig. 11).

Continental weak zones (as distinct from plate boundaries) can be classified on the basis of the scale of the intervening blocks that jostle along them. Blocks of continental lithosphere hundreds of kilometres wide (sub-plates) jostle along first-order weak zones. Blocks of continental crust tens of kilometres wide jostle along second-order weak zones. Blocks of brittle upper crust up to *c.* 10 km wide jostle along third-order weak zones. Narrower blocks of cover sequences jostle along boundaries of fourth order. Any fifth- and higher-order shear zones are not weak in the sense adopted here.

Weak zones in the Precambrian rock masses of Sweden are distinctive for their lengths, spacings, shapes, orientations, internal properties and locations. First- and second-order weak zones appear to be long (1000 km), wide ($\geqslant$50 m), widely spaced ($\geqslant$20 km), subvertical or subhorizontal, smoothly sinuous, and to involve seams of permeable cataclasites that necessarily continue downward to deep ductile extensions located along major heterogeneities in the continental crust. Furthermore, and probably most important, they have long histories with complicated kinematics. These characteristics are considered here to be the symptoms of repeated reactivations.

Survival of short-lived shear zones

Not all ductile shear zones that develop in deep warm rock subsequently reactivate. Before discussing how repeated reactivations weaken long-lived shear zones and how successive reactivations remain focused on the same zones for so long, it is worth contrasting weak shear zones with those that did not reactivate.

Some of the factors that contribute to lowering the strength of particular weak zones can be reversed by subsequent strengthening effects. Thus, hydrated serpentitious or clay-rich shear zones can dehydrate, evaporites can dissolve or migrate elsewhere, or (more relevant to Sweden) fluid overpressures can dissipate. Similarly, fine-grained mylonites or brittle fractures can be distorted or completely annealed by heating (by burial or nearby plutons) and porous cataclasites can be cemented. However, these special conditions will be ignored here so as to focus on the general case, which does not involve strength.

Throughout Sweden, large numbers of ductile shear zones survive without further distortion in rock blocks bound by younger shear zones that did reactivate. Similarly, brittle shear zones can be preserved at any stage in their development (from one set of Riedel shears to two, three or more sets of fractures). In general, short-lived shear zones, whether ductile, brittle or anywhere in between, are characteristically short, narrow, closely spaced, and of any orientation. Most important, such minor shear zones appear to have had comparatively short lives with simple kinematics.

Short-lived shear zones presumably cease shearing when the stress fields and boundary conditions responsible for them change so that their orientations become inappropriate for continued activity in the new regional stress fields. Minor shear zones survive by inactivating as regional strains progressively localize to contemporaneous shear zones that lengthen.

Weakening of ductile shear zones

So how do shear zones become weak? It has already been argued that the first example discussed here was sufficiently weakened by fluid pressures high enough to refract the regional stress field so that the Singö shear zone underwent pure rather than simple shear. The early ductile stages of the Singö zone can be taken as a model for what is happening at depth beneath the San Andreas fault zone now. However, there is a more general argument pointing to situations in which ductile shear zones can weaken.

Focusing on the simplest macroscopic aspects of ductile shear zones, Talbot (1999*b*) argued that shear zones in rocks localize to degrees solely dependent on the stress sensitivity of the strain rate of the rocks when they sheared. Displacement gradients across natural ductile shear zones of any scale (many from Swedish rocks) fit, at least locally, one or other of the theoretical displacement curves for strain-rate weakening pseudoplastics sheared in boundary shear zones (Talbot 1999*b*). This suggests that ductile shear localizes along planar zones because silicate rocks have a particular simple material property. They deform as strain-rate softening power-law pseudoplastics. This means that their rate of steady longitudinal strain, $\bar{\varepsilon}_s$, relates to the stress difference, σ ($=\sigma_1-\sigma_3$), by $\bar{\varepsilon}_s = A\sigma^n$. Here n is the stress sensitivity of the strain rate of the rock as it sheared. Displacement gradients across natural shear zones of any scale fit the theoretical displacement gradients for strain-rate weakening pseudoplastics ($n = 1$) counterflowing in paired boundary shears retarded along mutual no-slip boundaries (Talbot 1999b). Just as brittle rock masses can deform by developing spontaneous new planar boundaries called fractures (with or without reactivating old fractures), so pseudoplastic rock masses can deform by developing new planar boundaries called ductile shear zones (with or without reactivation of old ductile shear zones).

The displacement gradients across natural shear zones therefore provide a simple measure of the stress sensitivity of their strain rate represented by n, the power-law exponent of the stress when they sheared. Although there appears to be no theoretical necessity for it, empirical studies suggest that n values of rocks relate to temperature. This is because the fits imply that the rocks deformed as strain-rate softening pseudoplastics with n being low ($n < 2$) in migmatites shearing in upper amphibolite facies, intermediate ($n = 3$–7) in rocks shearing at lower amphibolite facies, and increasing rapidly ($n = 29$) in rocks shearing in greenschist facies.

The general implication of ductile silicate rocks shearing as strain-rate softening pseudoplastics is that, if temperatures fall during continuous ductile shear, the n value of the rock mass increases and shear is increasingly localized to the counterflow boundary, which could evolve into a counterslip fault. This is in agreement with the common observation that zones of superimposed ductile shear generally narrow.

How brittle reactivations weaken

The major weak faults of Sweden are essentially aseismic right through even the thick cool lithosphere of Fennoscandia. Ductile shear can be expected at depths where aseismicity indicates that even feldspar becomes ductile, but what about the so-called brittle rocks above?

Brittle reactivations with slightly different kinematics break jogs, round angular boundaries of blocks on large scales and smooth them on smaller scales. Just as important, multiple reactivations of brittle fracture zones generate en echelon dilational shear zones, which bound pods that can rotate and break to potentially permeable breccias and gouges. Weak fracture zones seamed by cataclasites act as conduits for ground water that can weaken them still further by chemical processes and/or overpressures.

In effect, fracture zones in the 'brittle crust' become weak when they develop overpressured clay-rich cataclasites that shear by 'ductile' frictional sliding or flow (Estrin & Bréchet 1996).

'Ductile' shear right through the crust

Having argued that both the ductile and the brittle components of long-lived crustal-scale shear zones are weak because they deform by essentially ductile processes right through the crust, we must consider how or whether ductile shear zones are maintained through or across the ductile–'brittle' transition.

Surely the most important reactivations are those that maintain the identity of weak zones by maintaining ductile shear through the ductile–'brittle' transition. These are helped by reactivations focusing ever-narrowing ductile weak zones to the same narrow zones that will widen during the formation of successive generations of semi-brittle veins and brittle fractures in ever-widening fracture zones. It may be difficult to visualize ductile shear zones propagating upward, or frictional sliding or flowing cataclasites propagating downward, but outcrops with ductile shear zones, pseudotachylytes and epidote veins superimposed in any order are not uncommon in greenschist facies.

So how do existing shear zones reactivate?

The crucial question is not how repeated reactivations weaken long-lived shear zones but how reactivations remain focused on the same zones for so long. The characteristics of weak zones offer clues.

Their orientations should remain (or become) appropriate for continued activity in regional stress fields with as wide a range of orientations as possible (from slight rotations to complete swaps of the principal stresses). As demonstrated by structural inversions being so common, subvertical and subhorizontal zones of regional extent are likely to remain generally capable of displacement during both subhorizontal shortening and extension.

Weak zones are likely to be spaced on the length scale (thickness or width) of the jostling unit. Crustal-scale zones of marked planar anisotropy between blocks of different rock types and histories often began as crustal-scale heterogeneities, such as terrane boundaries (like the Sorgenfrei–Tornquist zone) and can remain weak if they continue to reactivate.

Conclusions

Studies of Swedish weak zones suggest that they were already weak when they initiated as ductile shear zones at deep crustal levels, because most silicate rocks strain-rate soften and cool over long periods so that they become increasingly strain-rate sensitive (Talbot 1999*b*, 2001). Detailed strain analysis of ductile mylonite zones along one weak zone demonstrates pure rather than simple shear. This suggests that the regional stress fields refracted to become orthogonal to this weak zone, in which volume loss suggests that high fluid pressures can weaken even deep ductile shear zones.

Shear zones at other crustal levels (even in the upper crust) are also weak when cataclasites along them are capable of aseismic frictional flow or sliding. This happens where repeated reactivation rounds on large scales and smooths on small scales and generates narrow seams of fine-grained gouge that can be further weakened by fluids that can both overpressure and alter rocks to clays.

Most of the fieldwork behind this review was funded by the Swedish Natural Science Foundation or SKB, the company responsible for handling Sweden's nuclear fuels and waste. FENTEC GPS studies were funded between 1989 and 1999 by the Swedish Natural Science Foundation. Thanks are due to K. Hult of the Swedish Geological Survey for providing Fig. 5a, and to H. Fossen and T. Andersen for their constructive reviews.

References

ARGUS, D.F., PETIER, W.R. & WATKINS, M.M. 1999. Glacial isostatic adjustment observed by Very Long Baseline Interferometry and Satellite Laser Ranging geodesy. *Journal of Geophysical Research*, B12, **104**, 29077–29093.

ARVIDSSON, R. 1996. Fennoscandian earthquakes: whole crustal rupturing related to postglacial rebound. *Science*, **274**, 744–746.

BABEL working group 1993. Integrated seismic studies of the Bothnian shield using data in the Gulf of Bothnia region. *Geophysical Journal International*, **112**, 305–324.

BÄCKBLOM, G. & STANFORS, R. (eds) 1989. *Interdisciplinary Study of Post-glacial Faulting in the Lansjärv Area, Northern Sweden*. Progress Report, **89-31**. Swedish Nuclear Fuels and Waste Management Company, Stockholm.

BERGMAN, S. & SJÖSTRÖM, H. *The Storsjön–Edsbyn deformation zone, central Sweden*. Research Report. Geological Survey of Sweden, Uppsala.

BERTHELSEN, A. & MARKER, M. 1986. 1.9–1.8 Ga old strike-slip megashears in the Baltic shield and their plate tectonic implications. *Tectonophysics*, **128**, 163–181.

DEHLS, J.F., OLESEN, O., OLSEN, L. & BLIKRA, L.H. 2000. Neotectonic faulting in northern Norway; the Stuoragurra and Nordmanvikdalen postglacial faults. *Quaternary Science Reviews*, **19**, 1447–1460.

EKMAN, M. 1996. A consistent map of the postglacial uplift of Fennoscandia. *Terra Nova*, **8**, 158–165.

ENGELDER, T. 1993. *Stress Regimes in the Lithosphere*. Princeton University Press, Princeton, NJ.

ERIKSSON, L. & HENKEL, H. 1988. *Storstrukturella zoner in Forsmarksområdet*. Geological Survey of Sweden, Uppsala, Geofysiskrapport, **BRAP 88402**.

ESTRIN, Y. & BRÉCHET, Y. 1996. On a model of frictional sliding. *Pageoph*, **147**, 754–762.

EVANS, B. & WONG, T.-F. (eds) 1992. *Fault Mechanics and Transport Properties of Rocks*. Academic Press, New York.

FJELDSKAAR, W., LINDHOLM, C., DEHLS, J.F. & FJELDSKAAR, I. 2000. Postglacial uplift, neotectonics and seismicity in Fennoscandia. *Quaternary Science Reviews*, **19**, 1413–1422.

GAAL, G. & GORBATSCHEV, R. 1987. An outline of the Precambrian evolution of the Baltic Shield. *Precambrian Research*, **35**, 15–52.

HENKEL, H. 1989. Tectonic studies in the Lansjärv region. *In*: BÄCKBLOM, G. & STANFORS, R. (eds) *Interdisciplinary Study of Post-glacial Faulting in the Lansjärv Area, Northern Sweden*. Progress Report 89-31. Swedish Nuclear Fuels and Waste Management Company, Stockholm, 2:1–2:22.

HENKEL, H. 1991. Magnetic crustal structures in northern Fennoscandia. *Tectonophysics*, **192**, 57–79.

HÖGBOM, I. 1925. Der berg Luppio. *Bulletin of Geological Institutions of Uppsala*, **XIX**, 223–240.

HUSEBYE, E.S., HOVLAND, J., CHRISTOFFERSON, A., ÅSTRÖM, K., SLUNGA, R. & LUND, C.-E. 1986. Tomographic mapping of the lithosphere and asthenosphere beneath southern Scandinavia and adjacent areas. *Tectonophysics*, **128**, 229–250.

JOHNSTON, P., WU, P. & LAMBERT, K. 1998. Dependence of horizontal stress magnitude on load dimensions in glacial rebound models. *Geophysical Journal International*, **132**, 41–60.

KUJANSUU, R. 1964. Nuorista siirroksistta Lapissa. (Summary in English: Recent faults in Lapland). *Geologi*, **16**, 30–36.

LAGERBÄCK, R. 1990. Late Quaternary faulting and paleoseismicity in northern Fennoscandia with particular reference to the Lansjärv area, northern Sweden. *Geologiska Föreningens Förhandlingar, Stockholm*, **112**, 333–355.

MILNE, G.A., DAVIS, J.L., MITROVICA, J.X., SCHERNECK, H.-G., JOHANSSON, J.M., VERMEER, M. & KOIVULA, H. 2001. Space-geodetic constraints on glacial isostatic adjustments in Fennoscandia. *Science*, **291**, 2381–2385.

MUIR-WOOD, R. 1989. Extraordinary deglaciation reverse faulting in northern Fennoscandia. *In*: GREGERSEN, S. & BASHAM, P.W. (eds) *Earthquakes at North-Atlantic Passive Margins: Neotectonics and Postglacial Rebound*. Kluwer, Dordrecht, 141–173.

MUIR-WOOD, R. 2000. Deglaciation seismotectonics: a principal influence on intraplate seismogenesis at high latitudes. *Quaternary Science Reviews*, **19**, 1399–1411.

MUNIER, R. & TALBOT, C.J. 1993. Segmentation, fragmentation and jostling of cratonic basement in and near Äspö, southeast Sweden. *Tectonics*, **12**, 713–727.

ÖDMAN, O.H. 1957. *Beskrivning till berggrundskarta över urberget in Norbottens Län. Series Ca No. 41*. Geological Survey of Sweden, Uppsala.

OLESEN, O. 1988. The Stuoragurra fault, evidence for neotectonics in the Precambrian of Finnmark, northern Norway. *Norsk Geologisk Tidsskrift*, **68**, 107–118.

PAN, M., SJÖBERG, L.E. & TALBOT, C.J. 2001. Crustal movements in Skåne, Sweden, between 1992 and 1998 as observed by GPS. *Journal of Geodynamics*, in press.

PAN, M., SJÖBERG, L.E., TALBOT, C.J. & ASENJO, E. 1999. GPS measurements of crustal deformation in Skåne, Sweden, between 1989 and 1996. *Geologiska Föreningens Förhandlingar, Stockholm*, **121**, 67–72.

SCHERNECK, H.G., JOHANSSON, J.M., MITROVICA, J.X. & DAVIS, J.L. 1998. The BIFROST project: GPS determined 3-D displacement rates in Fennoscandia from 800 days of continuous observations in the SWEPOS network. *Tectonophysics*, **294**, 305–321.

SCHOLTZ, C.H. 2000. Evidence of a strong San Andreas fault. *Geology*, **28**, 163–166.

SIBSON, R.H. 1977. Fault rocks and fault mechanisms. *Journal of the Geological Society, London*, **133**, 191–213.

SKIÖLD, T. 1982. Radiometric ages of plutonic and hypabyssal rocks from the Vittangi–Karesuando area, N. Sweden. *Geologiska Föreningens Förhandlingar, Stockholm*, **103**, 317–329.

SLUNGA, R.S. 1991. The Baltic shield earthquakes. *Tectonophysics*, **189**, 323–331.

STÅLHÖS, G. 1984. Svecokarelian folding and interfering macrostructures in eastern Central Sweden. *In*: KRÖNER, A. & GREILING, R. (eds) *Precambrian Structures Illustrated*. Schweizerbart'sche, Stuttgart, 369–379.

STÅLHÖS, G. 1991. *Beskrivning till berggrundskartorna Östhammar NV, NO, SV, SO. (1:50 000 bedrock maps and description with English summary).*. Series Af, 161, 166, 169 and 172. Swedish Geological Survey, Uppsala.

STEPHANSSON, Ö. 1989. Stress measurements and modelling of crustal rock mechanics in Fennoscandinavia. *In*: GREGERSEN, S. & BASHAM, P.W. (eds) *Earthquakes at North-Atlantic Passive Margins: Neotectonics and Postglacial Rebound*. Kluwer, Dordrecht, 213–229.

STEPHENS, M.B., WHALGREN, C.H. & WEIHED, P. 1997. Sweden. *In*: MOORES, E.M. & FAIRBRIDGE, R.W. (eds) *Encyclopedia of European and Asian Geology*. Chapman and Hall, London, 690–704.

TALBOT, C.J. 1986. *A Preliminary Structural Analysis of the Patterns of Post-glacial Faults in Northern Sweden*. Progress Report, **86-20**. Swedish Nuclear Fuels and Waste Management Company, Stockholm.

TALBOT, C.J. 1992. GPS networks to measure active strains in Sweden. *Geologiska Föreningens Förhandlingar, Stockholm*, **114**, 378–380.

TALBOT, C.J. 1999*a*. Ice ages and nuclear waste isolation. *Engineering Geology*, **52**, 177–192.

TALBOT, C.J. 1999*b*. Ductile shear zone as counterflow boundaries in power law fluids. *Journal of Structural Geology*, **21**, 1535–1551.

TALBOT, C.J. 2001. Ductile shear zone as counterflow boundaries in pseudoplastic fluids: Reply. *Journal of Structural Geology*, **23**, 157–159.

TALBOT, C.J. & HEEROMA, P. 1989. Cover/basement relationships in the SW Swedish gneisses near Varberg. *Geologiska Föreningens Förhandlingar, Stockholm*, **111**, 105–119.

TALBOT, C.J. & SLUNGA, R. 1989. Patterns of active shear in Fennoscandinavia. *In*: GREGERSEN, S., BASHAM, P.W. (eds) *Earthquakes at North-Atlantic Passive Margins: Neotectonics and Postglacial Rebound*. Kluwer, Dordrecht, 441–466.

TALBOT, C.J. & SOKOUTIS, D. 1995. Mylonitisation and volume loss in the Singö gneiss zone, E Sweden. *Journal of Structural Geology*, **17**, 927–948.

TALBOT, C.J., RIAD, L. & MUNIER, R. 1989. Reactivation of Proterozoic shear zones. *In*: BÖCKBLOM, G. & STANFORS, R. (eds) *Interdisciplinary Study of Post-glacial Faulting in the Lansjärv Area, Northern Sweden*. Technical Report, **89-31**. Swedish Nuclear Fuels and Waste Management Company, Stockholm, 4:1–4:17.

TRYGGVASON, A., LUND, C.-E. & FRIBERG, M. 1998. A two-dimensional seismic velocity model across the transition zone between the Baltic shield and the North German Basin—the EUGENO profile 1 revisited. *Tectonophysics*, **290**, 47–58.

WELIN, E. & STÅLHÖS, G. 1986. Maximum age of the synmetamorphic Svecokarelian fold phases in south central Sweden. *Geologiska Föreningens Förhandlingar, Stockholm*, **108**, 31–34.

WELIN, E., KÄHR, A.-M. & LUNDEGÅRDH, P.H. 1980. Rb–Sr isotope systematics at amphibolite facies conditions, Uppsala region, eastern Sweden. *Precambrian Research*, **13**, 87–101.

WICKMAN, F.E., ÅBERG, G. & LEVI, B. 1983. Rb–Sr dating of alteration events in granitoids. *Contributions to Mineralogy and Petrology*, **83**, 358–362.

ZEIGLER, P.A. 1988. Evolution of the Arctic–North Atlantic and the western Tethys. *AAPG Memoir*, **43**.

The role of fault zones and melts as agents of weakening, hardening and differentiation of the continental crust: a synthesis

M.R. HANDY[1], A. MULCH[1,2], M. ROSENAU[1,3] & C.L. ROSENBERG[1]

[1]*Freie Universität Berlin, Geowissenschaften, Malteserstrasse 74–100, Haus B, Geologie, D-12249 Berlin, Germany (e-mail: mhandy@zedat.fu-berlin.de)*

[2]*Present address: Université de Lausanne, Institut de Minéralogie et Pétrographie, BFSH 2, CH-1015 Lausanne, Switzerland*

[3]*Present address: Geoforschungszentrum Potsdam, Haus C, D-14473 Potsdam, Germany*

Abstract: The rheology of crustal fault zones containing melts is governed primarily by two strain-dependent mechanical discontinuities: (1) a strength minimum parallel to mylonitic foliation just below the active brittle–viscous (b–v) transition; (2) the anatectic front, which marks the upper depth limit of anatectic flow. The mode of syntectonic melt segregation in fault zones is determined by the scale of strain localization and melt-space connectivity, to an extent dependent on strain, strain rate and melt fraction in the rock. Melt drains from the mylonitic wall rock into dilatant shear surfaces, which propagate sporadically as veins. Anatectic flow at natural strain rates therefore involves melt-assisted creep punctuated by melt-induced veining. On the crustal scale, dilatant shear surfaces and vein networks serve as conduits for the rapid, buoyancy-driven ascent of transiently overpressured melt from melt-source rocks at or just below the anatectic front to sinks higher in the crust. Strength estimates for natural rocks that experienced anatectic flow indicate that melts weaken the continental crust, particularly in depth intervals where they spread laterally beneath low-permeability layers or along active shear zones with a pronounced mylonitic foliation. However, acute weakening associated with strength drops of more than an order of magnitude occurs only during short periods (10^3–10^5 a) of crustal-scale veining. Cooling and crystallization at the end of these veining episodes is fast and hardens the crust to strengths at least as great as, and in some cases greater than, its pre-melting strength. Repeated melt-induced weakening then hardening of fault zones may be linked to other orogenic processes that occur episodically (shifting centres of clastic sedimentation and volcanism) and has implications for stress transmission across orogenic wedges and magmatic arcs.

Molten rock originates and accumulates at structurally controlled sites within the continental crust in a wide range of tectonic settings, as documented in a growing literature on melt-related tectonics (for examples and references, see Fig. 1). Melting is a harbinger of nascent plate boundaries, as well as a process at active and formerly active, divergent, convergent and transcurrent boundaries (Prichard *et al.* 1993 and papers therein). Although there is broad consensus that melting is spatially and temporally related to crustal-scale fault zones (Hollister & Crawford 1986; Speer *et al.* 1994; Vigneresse 1995) on at least some scales of observation (Paterson & Schmidt 1999), the causes and processes underlying this close relationship remain controversial.

Debate has centred on the question of whether melting causes strain localization or vica versa. Melts either deform rock by forcing their way and creating space (Hutton 1988*a*; Paterson *et al.* 1989; De Saint Blaquat *et al.* 1998), or fill space opened by incompatible rock deformation on both the local (Collins & Sawyer 1996) and regional scales (e.g. Hutton 1988a). This chicken-or-egg-first dilemma has been apparently resolved by the insight that melting and deformation are linked in a positive feedback loop (Brown & Solar 1998) such that melt-enhanced strain localization leads to further melt segregation (Brown 1994) and crustal differentiation. We will have more to say about this below, but point out that the main problem underlying debates on melt transport and emplacement is our limited knowledge of the deformation mechanisms and rheology of partially melted rock at high strains.

From: HOLDSWORTH, R.E., STRACHAN, R.A., MAGLOUGHLIN, J.F. & KNIPE, R.J. (eds) 2001. *The Nature and Tectonic Significance of Fault Zone Weakening*. Geological Society, London, Special Publications, **186**, 305–332. 0305-8719/01/$15.00

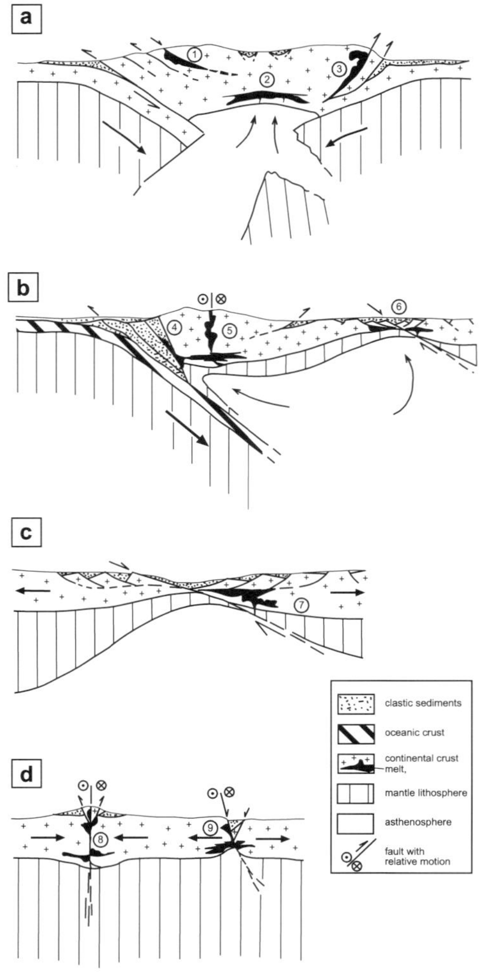

Fig. 1. Phanerozoic tectonic settings of fault-related melt transport and emplacement. Circled letters correspond to settings and references numbered sequentially below. (**a**) Collisional orogens: (1) granitic rocks in footwall of synorogenic extensional faults, e.g. Tertiary leucogranites beneath North Himalayan Normal Fault (Burg *et al.* 1984; England & Molnar 1993); (2) melts beneath high intramontane plateaux, e.g. Tibetan plateau north of the Himalaya (Nelson *et al.* 1996), Altiplano–Puna plateau in the Central Andes (Scheuber & Reutter 1992; Allmendinger *et al.* 1997); (3) intrusive rocks along late-orogenic backthrusts bounding the retro-wedge of orogens, e.g. Tertiary intrusive rocks along the Periadriatic (Insubric) Line bordering the Central Alps (Dietrich 1976; Rosenberg *et al.* 1995; von Blanckenburg & Davies 1995; Davidson *et al.* 1996), Late Cretaceous–Early Tertiary tonalite along Valdez Creek Shear Zone bordering the MacLaren Glacier metamorphic belt, south–central Alaska (Hollister & Crawford 1986; Davidson *et al.* 1992). (**b**) Magmatic arcs: (4) forearc intrusive rocks, e.g. Tertiary S-type granites in forearc, Eastern Gulf of Alaska (Barker *et al.* 1992); (5) transpressional strike-slip systems between forearcs and back-arcs, e.g. Great Sumatran and Mentawi Faults (De Saint Blaquat *et al.* 1998); (6) footwall of uniform-sense, back-arc extensional shear zones, e.g. Tertiary granitic rocks beneath low-angle normal faults, northern Tyrrhenian Sea (e.g. Daniel & Jolivet 1995), Aegean Sea (Lister & Baldwin 1993). (**c**) Continental rifts: (7) granites in footwall of low-angle normal faults, e.g. Cordilleran metamorphic core complexes (Crittenden *et al.* 1980), basic melts beneath Tertiary to Recent Basin-and-Range and Rio Grande rift systems (Olsen *et al.* 1987), Tertiary East African rift system (Bailey 1992). (**d**) Continental transforms: (8) calc-alkaline intrusive rocks within transpressive shear zones, e.g. Cadomian North Armorican Shear Zone (D'Lemos *et al.* 1992), Late Palaeozoic South Armorican Shear Zone (Gapais 1989; Vigneresse 1995), Devonian Central Main Belt shear zone, northern Appalachians (Brown & Solar 1998, Brown & Solar 1999), Ox Mountains igneous complex (McCaffrey 1992); (9) calc-alkaline intrusive rocks within transtensional shear zones, e.g. Early Permian Cossato–Mergozzo–Brissago Line in Ivrea–Verbano Zone, Southern European Alps (Handy & Streit 1999, and this paper), gabbroic melts beneath Salton Trough, Gulf of California (Lachenbruch *et al.* 1985), Main Donegal Granite, NW Ireland (Hutton 1982), granites along transtensional splay of Great Glen Fault, Scotland (Hutton 1988b).

The fact that some melt-bearing shear zones accommodated large displacements (greater than tens of kilometres) suggested long ago that melts may enhance strain localization and facilitate rapid exhumation of the continental crust (e.g. Hollister & Crawford 1986). This inference is apparently consistent with experiments showing that silicic melt viscosities are many orders of magnitude lower than those of crystalline rocks undergoing solid-state, viscous flow by dislocation creep at comparable strain rates (Shaw 1980; Cruden 1990; Talbot 1999, and references therein). Some experimentalists have predicted that as little as 5 vol. % melt can reduce the strength of continental crust by 50% (e.g. Dell'Angelo & Tullis 1988), whereas others have argued that the crust is weakened by several orders of magnitude once a critical melt percentage (10–30 vol. %, Arzi 1978) is attained. Yet deformation induces melt segregation, and partial loss of melt can cause rocks to harden (Renner *et al.* 2000), particularly if strain rates are high ($\geqslant 10^{-8}$ s^{-1}). Also, strain localization does not necessarily coincide with weakening; rocks can harden during localization if the deformation involves a component of dilation (Hobbs *et al.* 1990) as, for example, during melting or syntectonic metamorphic reactions (Brodie & Rutter 1985).

Thus, a number of vexing questions remain that we will address in this paper: Does melting always involve crustal weakening, as is conventionally thought? How do rock structure and rheology affect melt segregation, transport and emplacement on different time and length scales? What are the driving forces and rate-controlling steps during syntectonic melt segregation and transport? Do these forces and rates differ at different levels of the crust? What is the role of melt in the long-term, high-strain rheology of the continental crust?

This paper synthesizes field and experimental studies of deformed rocks and rock analogues, to characterize the structure and rheology of fault zones in partially melted continental crust. In the next section of this paper, we describe the strain- and depth-dependent mechanical transitions in the crust that govern high-strain crustal rheology. We then propose a conceptual model of melt-assisted creep punctuated by melt-induced fracturing and melt flow along dilatant shear surfaces. A fracture analysis of such melt-filled shear surfaces in anatectic rock allows us to place upper limits on its strength at the time of flow and fracturing. We then cast our model of anatectic flow in the broader context of strain-dependent changes in melt permeability. In particular, we propose two strain-dependent melt segregation thresholds that separate three modes of syntectonic melt flow, each of which is related to strain localization and dilation on a different length scale. We then apply the concepts in the previous sections to answer the questions posed above. In so doing, we refer to two exhumed fault zones that offer a rare glimpse of rheological transitions and syntectonic magmatic structures. We conclude with a model for the transient thermomechanical properties of melt-bearing fault zones and mention some of the implications of this model for the dynamics of basins, orogens and magmatic arcs.

Structure and rheology of crustal-scale fault zones containing melt

Depth-dependent changes in fault zones

The structure and rheology of fault rocks change with depth, as shown in the profile in Fig. 2a through a prototypical fault zone. Unlike most generic models, this fault zone contains melt at its base. Cataclasites in the brittle upper crust grade downward to mylonites undergoing viscous, solid-state flow (Sibson 1977, 1986; Scholz 1990), which in turn yield to partially melted gneisses and segregated melts undergoing anatectic flow at depth. This tripartite fault structure is generally valid for an anomalously high geothermal gradient along the shear zone ($\geqslant 30\,^{\circ}\mathrm{C\ km^{-1}}$), but to avoid overinterpretation we have not marked lithostatic pressure and temperature units on the vertical axis. Also, only a half-width section of the fault zone is shown in Fig. 2a, to emphasize our neglect of across-strike variations in fault-zone structure, particularly variations related to thermal disequilibrium during the juxtaposition of hot and cold crust across the fault zone (e.g. Handy 1990).

There is broad consensus that the strength of unmelted continental crust generally increases downward with increasing lithostatic pressure, P_l, in the brittle regime, but decreases with rising temperature, T, and/or decreasing strain rate, $\dot{\gamma}$ in the solid-state viscous flow regime (Sibson 1977, 1986; Brace & Kohlstedt 1980). Since its inception, however, this rather simple view has been modified to include transient mechanical behaviour related to rate-dependent frictional sliding (Scholz 1990, 1998, and references therein), depth-dependent increases in pore-fluid pressure in the brittle regime (Streit 1997), and polymineralic frictional–viscous flow in the solid-state, viscous regime (Handy *et al.* 1999b).

The shape of the strength v. depth curve in Fig. 2b reflects the modifications of Streit (1997) and Handy *et al.* (1999b), respectively, for the brittle and solid-state viscous flow regimes, but also includes two first-order structural–mechanistic transitions within the crust: (1) the transitional interval from brittle deformation to solid-state viscous flow (interval between dashed lines labelled b and v in Fig. 2 and subsequent figures); (2) the transition from solid-state viscous (mylonitic) flow to anatectic flow. Both transitions, particularly their mechanical characteristics, directly affect the transport and emplacement of melts in the crust.

The brittle-to-viscous transition

The brittle-to-viscous (b–v) transition is a depth interval characterized in nature by very localized deformation (Fig. 2a) and mutually overprinting relationships between cataclasites and mylonites (Handy 1998). Rock strength within this interval is expected initially to fluctuate as a result of the alternation of cataclasis and thermally activated creep mechanisms. At high strains, however, strain weakening within this depth interval is inferred to effect a decrease in strength to less than that of the over- and underlying crust

(Handy 1989), and the b–v transition narrows and rises to somewhat shallower levels (Fig. 2b; Schmid & Handy 1991). This strain weakening is attributed to switches from cataclasis (just above the initial b–v transition) or grain-size insensitive (dislocation) creep (just below the initial b–v transition) to grain-size sensitive creep (diffusion creep, diffusion- or reaction-accommodated granular flow) in fine-grained, polymineralic aggregates (Handy 1989). Rocks with small grain sizes and a grain-boundary fluid have lower laboratory strengths than coarser-grained rocks undergoing either cataclastic flow (Rutter & White 1979) or grain-size insensitive dislocation creep at comparable strain rates and temperatures (Schmid *et al.* 1977; Brodie & Rutter 1987).

Our model of a mobile, strain-dependent strength minimum just below the b–v transition modifies the widely held view of a stationary, strain-independent strength maximum at the b–v transition (e.g. Brace & Kohlstedt 1980; Sibson 1986; Scholz 1990). This view is based on the extrapolation of steady-state laboratory flow laws for rocks undergoing frictional sliding on smooth surfaces (Byerlee 1978) and thermally activated, grain-size insensitive (dislocation) creep (e.g. Weertman 1968). The rocks (usually quartzite and anorthosite, respectively, for the upper and lower crust) are tacitly assumed to have a strain-invariant microstructure, and to deform at a constant strain rate and geothermal gradient. Therefore, a strength maximum at the b–v transition as predicted with extrapolated laboratory flow laws probably exists in nature only at the onset of deformation (dashed lines in Fig. 2b) before progressive grain-size reduction weakens the rocks. As shown below, a strain-dependent strength minimum just below the ultimate b–v transition has important implications for the transport and emplacement of melts within the crust.

The transition to anatectic flow

The transition from solid-state viscous (mylonitic) flow to anatectic flow (Fig. 2a) obviously coincides with the onset of crustal melting, which varies from about 650 °C for hydrous granite (Johannes & Holtz 1996) to 750–970 °C for water-absent dehydration melting of micas in pelitic rocks at lithostatic pressures of 200–500 MPa (Singh & Johannes 1996, and references therein). Anatectic flow (Fig. 2a) involves the simultaneous operation of several deformation mechanisms in the presence of a melt. In naturally deformed, anatectic rocks, these mechanisms include dislocation creep (Rosenberg & Riller 2000), diffusion-assisted granular flow (Nicolas & Ildefonse 1996), melt-induced cataclasis (Paquet *et al.* 1981; Bouchez *et al.* 1992) and veining (Nicolas & Jackson 1982; Davidson *et al.* 1994). Melt-induced veining is analogous to hydrofracturing in rocks and soils (e.g. Shaw 1980; Wickham 1987) because melt viscosity is sufficiently low for the melt to behave as a pore fluid that counteracts normal stress on potential fracture surfaces such as grain boundaries (Law of Effective Stress or Terzaghi's Law). We use the term vein to refer generally to dilatant, melt-filled fractures that are concordant or discordant (in the sense of sills and dykes, respectively) to the structural anisotropy in the adjacent wall rock. Melt-induced veining is favoured by near- to supralithostatic melt

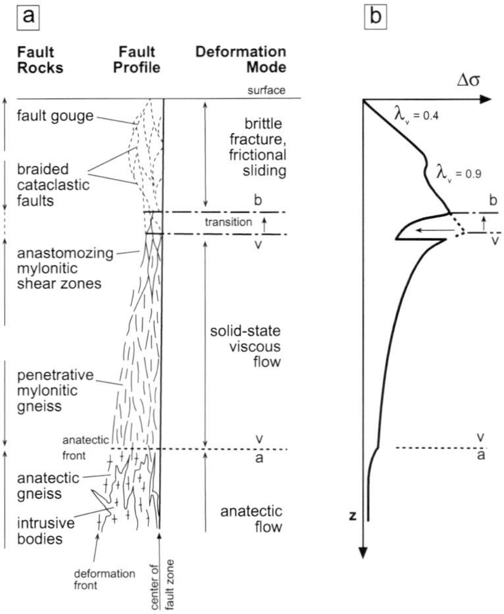

Fig. 2. Idealized section through an active, melt-bearing fault zone in continental crust at a geothermal gradient of about 30 °C km^{-1} parallel to the shear zone. (**a**) Profile showing the relative width of the fault zone from its centre to its margin (active deformation front) as a function of depth. Three deformation modes (brittle, solid-state viscous, anatectic) are separated by two major rheological transitions: the brittle-to-viscous (b–v) transition (dashed lines) and the anatectic front (dotted line) at the transition from solid-state viscous creep (marked v) to anatectic flow (marked a). (**b**) Schematic strength v. depth curve for the fault zone in (**a**). Strength curve in the brittle domain from Streit (1997) includes depth-dependent change in the pore-fluid factor, λ_v ($\lambda_v = P_{fluid}/P_{lithostatic}$) from 0.4 to 0.9, corresponding to hydrostatic and near-lithostatic pore-fluid pressures, respectively.

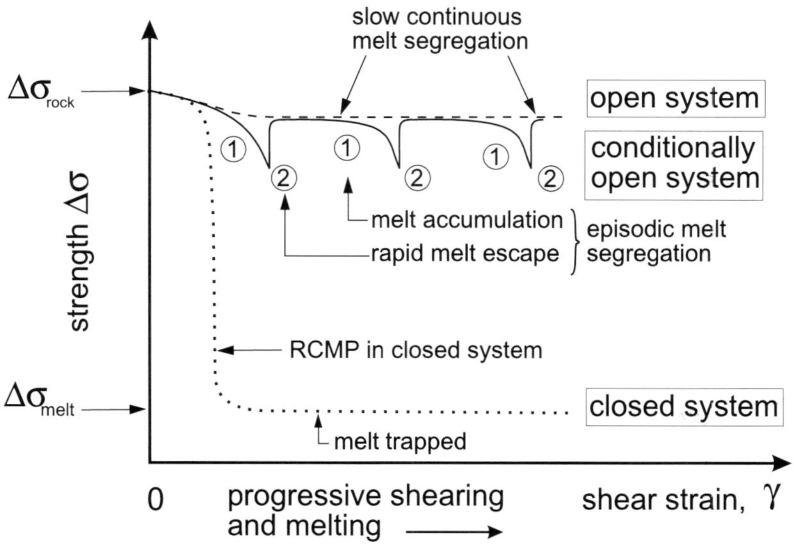

Fig. 3. Strength v. shear strain curves for three types of anatectic rock system: closed system, inspired by Arzi (1978); Van der Molen & Paterson (1979); open system, inferred from Dell'Angelo & Tullis (1988); conditionally open system, from Rosenberg & Handy (2000) and this paper. Stages numbered 1 and 2 along this curve are, respectively, periods of melt accumulation and melt escape discussed in the text.

pressures, P_m. The limit of melt-induced veining within the crust can occur on either side of the anatectic front, depending on the strain rate, melt fraction and melt pressure (see discussion below).

Deforming anatectic rocks are therefore more than just a two-phase mixture of solid and liquid; they represent an amalgam of mechanical phases with contrasting rheologies. Clearly, to characterize the bulk rheology of such a composite material with a single flow law for one, dominant deformation mechanism over the entire range of temperatures, strain rates, effective pressures, melt contents and compositions would be very unrealistic indeed. On the other hand, some structural and mechanistic simplifications are necessary to predict the bulk rheology of anatectic rocks. To model successfully the rheology of anatectic rocks is to understand the relationship between melt distribution and strain localization on the granular and supragranular scales.

Melt segregation and strain localization in closed and open systems

When attempting to understand melt segregation in a deforming rock, it is important to distinguish between melt flow in 'closed' and 'open' systems. A system is taken here to be a composite volume of deforming rock and melt. In closed (undrained) systems, melt remains within the rock on the time scale of deformation, whereas in open (drained) systems melt can flow out of, or into, the rock quickly on the same time scale. Real anatectic rocks are hybrids of these two end-member systems, but we will first consider these ideal cases before returning to nature and its complexities.

In a closed system undergoing melting, melt initially collects in the interstices of grains and forms weak pockets within a strong, load-bearing framework. Deformation takes place at constant volume and the melt pressure (P_m) is approximately equal to the mean stress in the rock. Coaxial deformation experiments at high strain rates indicate that the strength of closed systems drops drastically when the melt makes up a critical percentage or fraction of the rock (curve for closed system in Fig. 3). This so-called 'rheologically critical melt percentage' or RCMP is generally believed to coincide with the breakdown of a load-bearing framework of solid grains to a structure comprising interconnected layers of melt that accommodate almost all the strain of the partially melted rock (Renner *et al.* 2000; Rosenberg 2001). Experimentally derived values for the RCMP range from melt fractions, ϕ_m (melt volume/total volume of rock + melt) of 0.1–0.3 (Arzi 1978) to 0.4–0.6 (Lejeune & Richet 1995). At higher melt fractions and strains, the rheology of the aggregate is dominated by that of the melt. It is important to realize that the strength drop at the RCMP occurs at a critical strain (*c.* 2% coaxial strain according to Van der Molen & Paterson (1979)) and therefore can be defined only in dynamic systems.

Unfortunately, the RCMP is often equated with a magma segregation threshold at a fixed melt fraction (Sawyer 1994; Vigneresse *et al.* 1996), above which the magma (melt + phenocrysts) is thought to become mobile and ascend from the melt source region (e.g. Wickham 1987). This is somewhat misleading, as there is strong reason to believe that the RCMP depends not only on the melt fraction, but also on the bulk strain, the bulk strain rate (Miller *et al.* 1988), and the shapes (Arzi 1978; Van der Molen & Paterson 1979) and rheologies of the constituent phases. Sawyer (1994) and Vigneresse *et al.* (1996) pointed out that melt can segregate from host rocks at melt fractions below the RCMP. In fact, the deformation experiments of Rosenberg & Handy (2001) show that syntectonic melt segregation occurs at melt fractions as small as 0.01. Both the interconnectedness of melt and the melt segregation rate increase with bulk strain as strain localizes on the granular scale (Rosenberg & Handy 2000). We therefore argue that melt segregation thresholds (henceforth abbreviated as MST) in partially melted rocks are more appropriately defined in terms of a critical shear strain for localization in the presence of melt at specified melt fraction, bulk strain rate and effective pressure (effective pressure (P_e) = lithostatic pressure (P_l) − melt pressure (P_m)).

In contrast to closed systems, open systems retain only a small amount of melt (much less than the RCMP, say $\phi_m < 0.10$) at any time during deformation, because it is assumed that the melt production rate only slightly and locally exceeds the melt segregation rate. Open systems can undergo substantial volume change through loss (or gain) of melt, and the melt pressure may deviate significantly from the mean stress in the rock. In this context, it is not important *how* the melt leaves the system but *that* it does this. Irrespective of the melt extraction mechanism(s), what is important is that most siliceous melts wet the grain boundaries of the solid grains under static (e.g. Laporte *et al.* 1997) as well as dynamic conditions (Jin *et al.* 1994; Rosenberg & Riller 2000). The melt is able to form tubules along grain boundaries and triple junctions for the rapid diffusion of melted species between grains. In this role, the melt does not accommodate the bulk strain directly, but allows diffusive mass transfer to occur on the granular scale and to accommodate the change in shape of neighbouring grains (Paterson 2001). The melt therefore enhances diffusional or granular flow of the solid grains. Deformation experiments on partially melted, hydrous granite (3–5% melt, Dell'Angelo & Tullis 1988) and melt-bearing peridotite (2–10% basalt melt, Cooper & Kohlstedt 1986; Hirth & Kohlstedt 1995a, 1995b) show that melt distributed along grain boundaries weakens aggregates undergoing dynamic recrystallization and enhances grain-size sensitive creep in very fine-grained aggregates. In the case of melt-bearing peridotite (Cooper & Kohlstedt 1986), the creep rate increased by factors of 2–5 compared with melt-free experiments that involve the same load. In partially melted granites, especially quartz-rich varieties, creep enhancement is expected to be greater because of the smaller wetting angles for quartz–granitic melt interfaces (22–23° in experiments cited by Laporte *et al.* (1997); 27° in naturally deformed granite measured by Rosenberg & Riller (2000)) compared with olivine–basaltic melt interfaces (30–40°, Cooper & Kohlstedt 1986). The strength of an open-system anatectic rock is therefore less than that of a melt-free host rock undergoing solid-state viscous creep, but significantly greater than that of a closed system or of the melt alone. This is reflected in Fig. 3, where the strength curve for the open system decreases at the onset of syntectonic melting to a constant level somewhat less than the strength of the unmelted rock.

Strain localization and episodic melt segregation in nature

Of course, no natural system is ideally open or closed on all length and time scales of deformation. The naturally and experimentally deformed rocks and rock analogues in Figs. 4 and 5 suggest that real anatectic rocks behave as conditionally open systems characterized by alternating periods of open- and closed-system behaviour.

The leucosomes in the anatectic gneisses in Fig. 4 occupy dilatant shear surfaces within the rock: the main foliation (S surfaces, Fig. 4a and b), the shear foliation (C surfaces, Fig. 4a) and a shear fracture (C′ surface, Fig. 4b). The leucosomes along the C surfaces in Fig. 4a are clearly derived from the intervening rock with the arcuate S surfaces because the C-parallel leucosomes have the same composition as the S-parallel leucosomes, the S- and C-parallel leucosomes do not truncate each other, and there are no

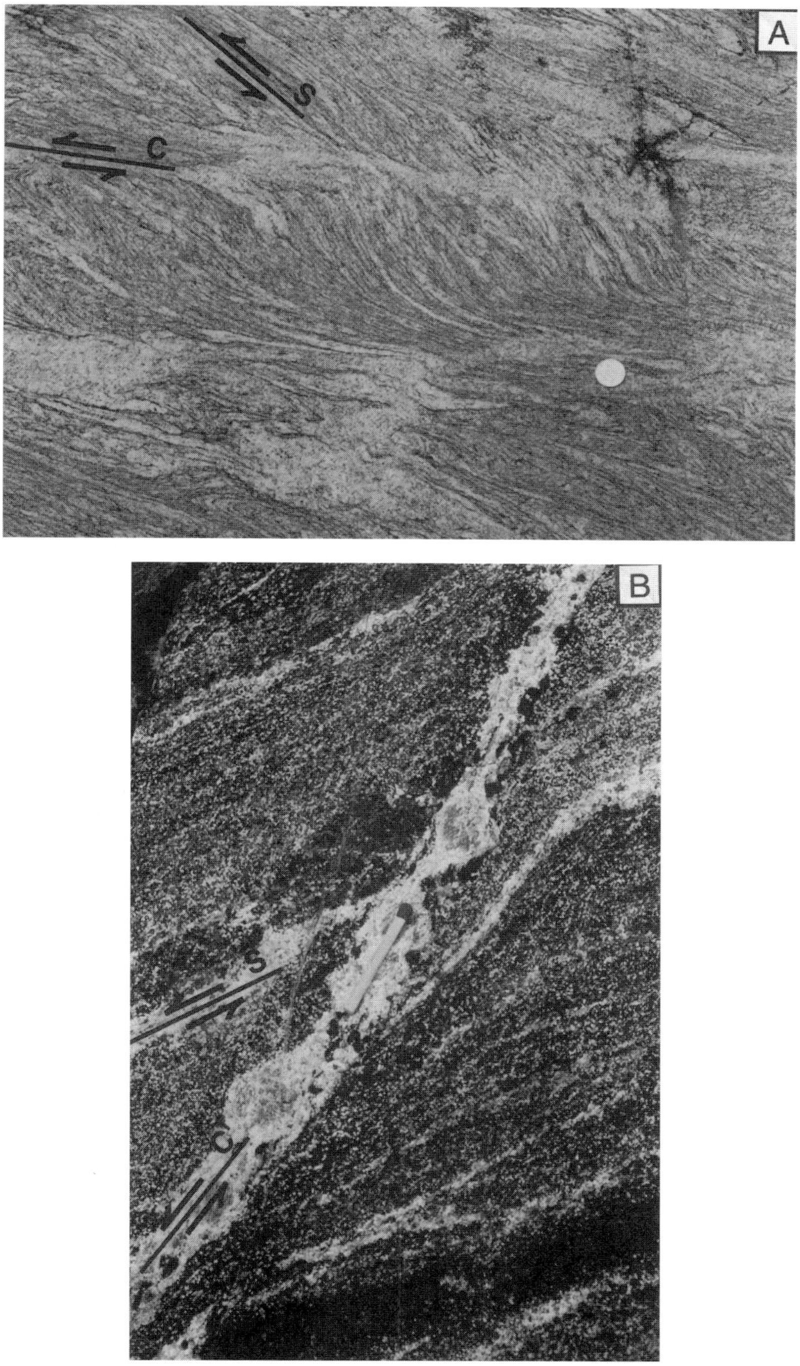

Fig. 4. Natural examples of localized deformation in the presence of melt. (**a**) Granitic leucosome occupies S and C surfaces of granitic biotite gneiss from the Southern (Insubric) Steep Belt in the Central Alps. Italian 500 Lire coin is 2.5 cm wide. (**b**) Tonalite intruded along S foliation and dilatant shear surface (C′ surface) in a garnet–pyroxene–amphibole mafic gneiss, Kapuskasing Uplift, Superior Province, Canada. Match is 4 cm long.

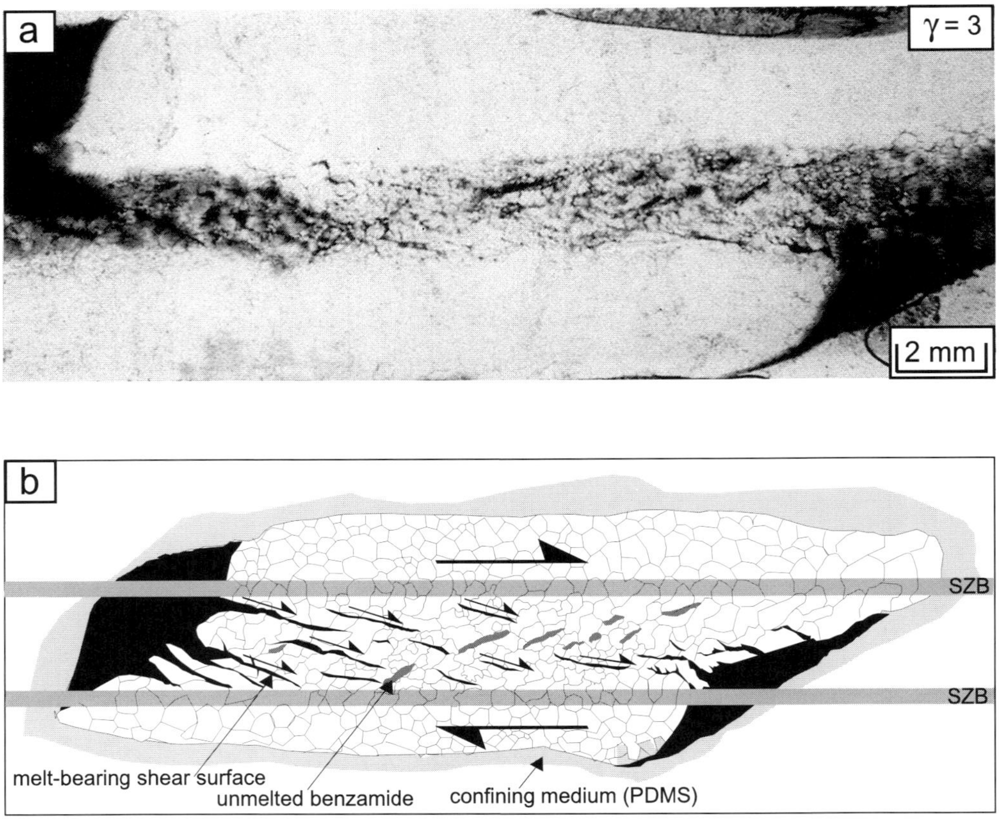

Fig. 5. Syntectonic melt segregation in a partly drained granite analogue (norcamphor–benzamide) from Rosenberg & Handy (2000). (**a**) Deformed sample viewed under plane-polarized light, perpendicular to shearing plane and parallel to shearing direction. (Note melt (back areas) and benzamide grains (grey patches) within the dynamically recrystallized norcamphor aggregate.) (**b**) Line drawing of grain aggregate in the same sample showing melt (black) along dilatant C′ shear surfaces and in sink regions at diametrically opposite ends of the shear zone. Confining medium is polydimethyl-siloxane (PDMS). SZB, shear-zone boundary.

other potential source rocks for the leucosomes in the area. The melt is therefore interpreted to have drained from source regions between the S surfaces and to have flowed into and along the dilatant C surfaces. The rock evidently remained in a partially melted state for the duration of deformation. The melt-filled C′ surface in Fig. 4b opened as an extensional shear fracture, as inferred from the displaced, angular edges of wall rock bordering the leucosome. This extensional shear geometry is diagnostic of supra-lithostatic melt pressure (i.e. $P_m > P_l$) and very low differential stress at the time of fracturing (Handy & Streit 1999).

Partially drained syntectonic melting experiments performed on a two-solid-phase organic aggregate (norcamphor and benzamide) directly under the optical microscope (Fig. 5) offer insight into the processes underlying melt segregation in the natural examples above. The experiments (Rosenberg & Handy 2000) were designed to simulate melt segregation during syntectonic melting of a granite and contained no more than $\phi_m = 0.1–0.15$ during the deformation. During the first increments of strain ($\gamma \leqslant 0.17$), melt that appeared along grain boundaries and interstices linked up along intergranular fractures (not shown in Fig. 5). These fractures eventually coalesced to form extensional shear surfaces within the dynamically recrystallizing grain aggregate (Fig. 5). Small amounts of melt also occupied elongate pockets in bridge structures parallel to S and linking the dilatant shear surfaces (Fig. 5). At $\gamma \geqslant 1$, the dilatant, melt-filled surfaces became conduits for the flow of melt down a local melt-pressure gradient to voids opening at diametrically opposite ends of the sample (Fig. 5). The melt fraction in the actively deforming part of the sample (between the frosted grips in Fig. 5) remained roughly constant at about 0.1–0.15 in experiments to higher shear strains ($\gamma = 3$). This range of ϕ_m values therefore represents a dynamic equilibrium at an MST between the rates of melt production and melt segregation. In similar experiments performed on an undrained system with 0.1–0.15 water fraction (instead of melt fraction) in pure polycrystalline norcamphor (Bauer *et al.* 2000*a*), the displacement rate parallel to dilatant, fluid-filled shear surfaces fluctuated with bulk shear strain; the displacement accelerated whenever the shear surfaces coalesced, but slowed during subcritical propagation and lengthening of individual shear surfaces.

The picture that emerges from the natural and experimental observations above is one of locally episodic melt accumulation and escape (Allibone & Norris 1992; Handy & Rosenberg 1999). Natural anatectic rocks are inferred to be conditionally open systems that are characterized at the granular to outcrop scales by alternating periods of mostly closed and mostly open behaviour. The hypothetical mechanical evolution of such an anatectite is shown in Fig. 3, with the two stages labelled 1 and 2 making up one of several cycles of melt segregation.

Melting rock initially behaves as a closed system (stage 1) and weakens slightly as the melt accumulates and the solid aggregate undergoes melt-assisted creep. At a critical strain that depends on a host of variables (e.g. strain rate, melt fraction, effective pressure, wetting angle, grain size), the melt nucleates rheological instabilities (dilatant shear surfaces) that localize strain and further weaken the system. At this point, the dilatant shear surfaces serve as temporary, local sinks for melt that is squeezed and/or drawn out of the deforming source areas between the shear surfaces. As the shear surfaces lengthen and interconnect, however, they function less as melt reservoirs and more as pathways for the melt to escape. The rock has attained the MST and now behaves like an open system (stage 2). Softening as a result of strain localization on shear surfaces is now more than balanced by hardening associated with the net loss of melt to a larger melt sink in an adjacent volume of rock. The rate of melt loss, and hence also the hardening rate, is highest if the dilatant shear surfaces interconnect suddenly to form a flow network or backbone (e.g. Rutter 1997). The cycle ends when a batch of melt leaves the deforming rock.

We refer to systems with this kind of cyclical behaviour as 'conditionally open' because their openess is primarily contingent on having bulk strain rates that are sufficiently high for the melt in the creeping aggregate to build up pressure and induce hydrofracturing. We expect that higher strain rates and/or lower temperatures favour lower critical strains to the MST and hence also shorter wavelength cycles compared with those shown in Fig. 3, whereas lower strain rates and/or higher temperature would have the opposite effect. At much lower strain rates, the melt would be squeezed out of the system fast enough to pre-empt the formation of dilatant instabilities and the system would become unconditionally open in the sense described in the previous section (see Dell'Angelo & Tullis (1988) for an experimental example of this behaviour).

Upper limits on the strength of rocks undergoing anatectic flow and melt-induced veining

The rheology of conditionally open anatectic rock systems is transient and therefore impossible to characterize in terms of a steady-state flow law. Extrapolating experimental (Rutter & Neumann 1995) or theoretical (Paterson 1995, 2001; Rutter 1997) flow laws for melt-bearing aggregates undergoing dislocation creep or granular flow yields an estimate of anatectic rock strength in an open system (Fig. 3). For various reasons, however, extrapolating such flow laws is problematic (Paterson 1987) and at best yields only order-of-magnitude strength estimates.

Limits on anatectic rock strength during melt-induced veining can be obtained from the geometry of leucosomes in the partially melted rocks pictured in Fig. 4 and depicted schematically in Fig. 6a and b. End-member stress states for these rocks are shown in Mohr diagrams in Fig. 6c and d. In these diagrams, rock strength or differential stress corresponds to the diameter of the Mohr circle and is constrained in the following way. As discussed in the previous section, the structures in Fig. 4 indicate that melt-filled fractures oriented both parallel and oblique to the S foliation opened during anatectic flow. This requires the Mohr circle to touch or cross two failure envelopes for fracturing oblique and parallel to the pre-existing S foliation (respectively labelled 'tr' and '||S' in Fig. 6) and a yield envelope for anatectic flow on this foliation (labelled 'flow on S' in Fig. 6). More specifically, the circle must be small enough to cross or touch the 'tr' envelope somewhere between this envelope's intersections with the abscissa (at $\sigma_n' = -T$) and the ordinate (at $\tau = 2T$) to account for the extensional shear geometry of

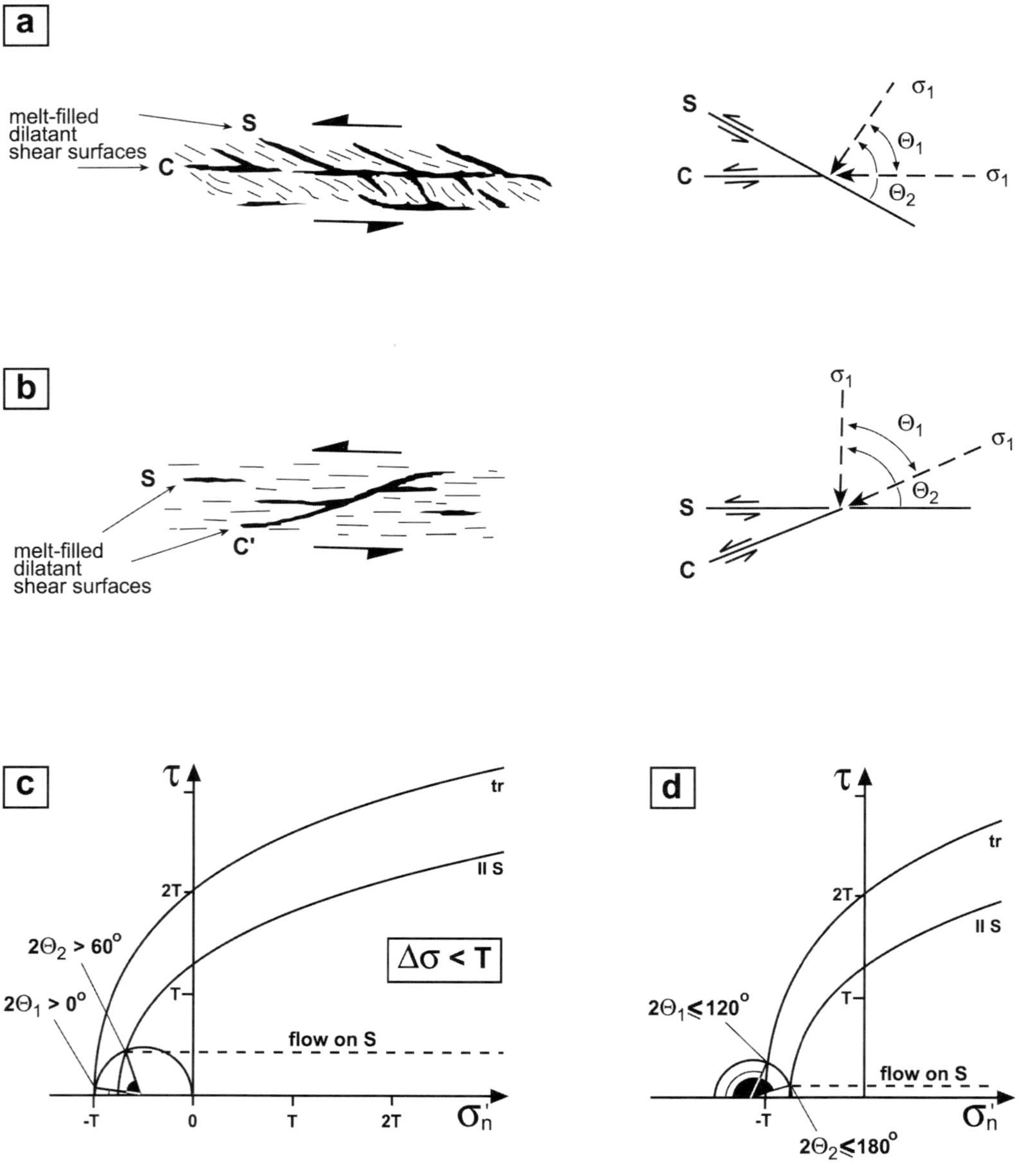

Fig. 6. Method for determining the strength limits of anatectic gneiss during coeval melt-assisted creep and melt-induced fracturing. (**a**) S and C foliations (dashes) in Fig. 4a partly filled with leucosome (black). (**b**) S and C′ surfaces in Fig. 4b filled with leucosome (black). (**c**) Mohr diagram with circle for the maximum differential stress assuming θ_1 was near 0°. θ_1 and θ_2 are fracture angles between the greatest principal stress direction, σ_1, and, respectively, the S and C′ surfaces. (**d**) Mohr diagram for unconstrained stress state (arbitrary Mohr circle) assuming θ_1 was near 60°. Horizontal and vertical axes of the Mohr diagrams are marked in units of tensile stress, T. Failure envelopes represent strength of mylonite in planes parallel to the pre-existing S foliation (||S) and in all other directions (tr, through the rock). Horizontal envelope (dashed line) is for anatectic flow on the S foliation.

the C and C′ surfaces oblique to the pre-existing S foliation (Fig. 6). Because all melt-filled fractures accommodated a component of shear parallel to their boundaries, we also know that $\sigma_n' \leqslant 0$ and $\tau > 0$ on these surfaces.

The tr failure envelope in the Mohr diagrams of Fig. 6 was constructed by using the modified Griffith criterion for failure of flawed materials (Secor 1965). The shape and position of the parallel-S failure envelopes with respect to the tr failure envelope reflect the lower cohesion and tensile strength of rocks parallel to pre-existing foliation surfaces, as documented in numerous experimental (Paterson 1978) and field studies (e.g. Cosgrove 1997). Melt-assisted creep is assumed to be non-dilational or nearly so, and is therefore represented by a horizontal visco-plastic yield envelope that is valid for the temperature, strain rate and average grain size in the wall rock. Both the viscoplastic yield envelope and the parallel-S failure envelope are valid only for flow and fracturing, respectively, along the pre-existing S foliation at an unfavourable angle (θ_2 in Fig. 6) to the σ_1 direction (Sibson 1985).

The measured angles between the oblique dilatant shear surfaces and the S foliation in Fig. 4a and b indicate that the fracture angles, θ_1, between the greatest principal stress axis, σ_1, and the melt-filled surfaces (C surface in Fig. 4a, C′ surface in Fig. 4b) must have been in the range of $0° < \theta_1 \leqslant 57°$ (Fig. 6a) and $0° < \theta_1 \leqslant 60°$ (Fig. 6b). In both cases, the minimum fracture angle near 0° and the presence of melt-bearing S surfaces that accommodated flow require that rock strength did not exceed the tensile strength of intact rock, T, as shown in Fig. 6c. On the other hand, the maximum possible fracture angles in both examples are near 60°, indicating that $\sigma_1' > -T > \sigma_3'$ (Fig. 6d). This information alone is insufficient to place a lower limit on rock strength at the time of veining. The Mohr circle in Fig. 6d is only one of a family of circles of variable diameter at $\theta_1 = 60°$. Unfortunately, we found no appropriate strain markers with which to reconstruct the strain ellipsoid and kinematic vorticity (Passchier & Urai 1988), and narrow the range of possible fracture angles.

The Mohr diagrams in Fig. 6 allow us to place a crude upper limit on anatectic rock strength. In a review of the rock mechanics literature, Etheridge (1983) determined that most metamorphic rocks have a tensile strength of 10 MPa or less, and in the presence of a metamorphic fluid, a value of 5 MPa is more realistic. We can therefore limit the maximum differential stress in the anatectic gneisses of Fig. 4 to somewhere in the range of 5–10 MPa. Strengths of <5 MPa are certainly conceivable; Rutter (1997, p. 106) speculated that partially melted granitic rocks have strengths of 1 MPa or less, a value apparently obtained from the undrained experiments of Van der Molen & Paterson (1979) on granite with 20% melt.

The upper limit on anatectic rock strength of 5–10 MPa is consistent with the maximum strength estimate by Handy & Streit (1999) of 10–20 MPa for anatectic, lower-crustal rocks near concordant mafic veins. Both estimates fall within the lower end of a broad range of strength estimates obtained from experimental flow laws (Kohlstedt *et al.* 1995) for quartz and feldspar aggregates undergoing dislocation creep and extrapolated to the conditions of deformation in Fig. 4. These results provide independent confirmation of the prediction that granitic crust undergoing coeval melt-assisted creep and melt-induced veining is weaker than crust undergoing solid-state viscous flow by dislocation creep.

Another interesting implication of our low strength estimates is that the tectonic overpressure (taken here to be $\sigma_m' - P_e$, the difference of the effective mean stress and effective pressure, assuming that $P_e = \sigma_3'$) is constrained to be only half of the differential stress, i.e. $\leqslant$2–5 MPa for the range of tensile strengths above. We will return to this point below in a discussion of melt driving forces, but turn next to the mechanisms underlying syntectonic melt segregation.

Modes of syntectonic melt flow

The evolution of melt permeability with strain

Episodic melt segregation in deforming anatectic rocks can be considered in terms of a time- and strain-dependent sequence of melt segregation modes, each of which is linked to strain localization on different scales and at different strain rates, melt fractions and effective pressures. These melt segregation modes and their respective microstructures are shown schematically in a diagram of permeability v. shear strain in Fig. 7. This diagram applies to a hypothetical rock that is homogeneous on the granular scale at the onset of deformation and melting. Permeability of melt is considered subparallel to the shear-zone boundary and is therefore controlled primarily by the connectivity of melt-filled space in this plane.

The three fields on the left side of Fig. 7 correspond to distinct modes of syntectonic melt flow in open and conditionally open rocks with melt fractions less than *c.* 0.30: (1) melt percolation along wetted or partly wetted grain boundaries, commonly referred to as porous flow (e.g. Brown *et al.* 1999); (2) melt transport within discrete supragranular, dilatant shear surfaces or bands (e.g. Gapais 1989; Rey *et al.* 1992; Vauchez & Egydio da Silva 1992), referred to here as shear surface flow; (3) channel flow through a network of interconnected, dilatant shear surfaces or veins (Clemens &

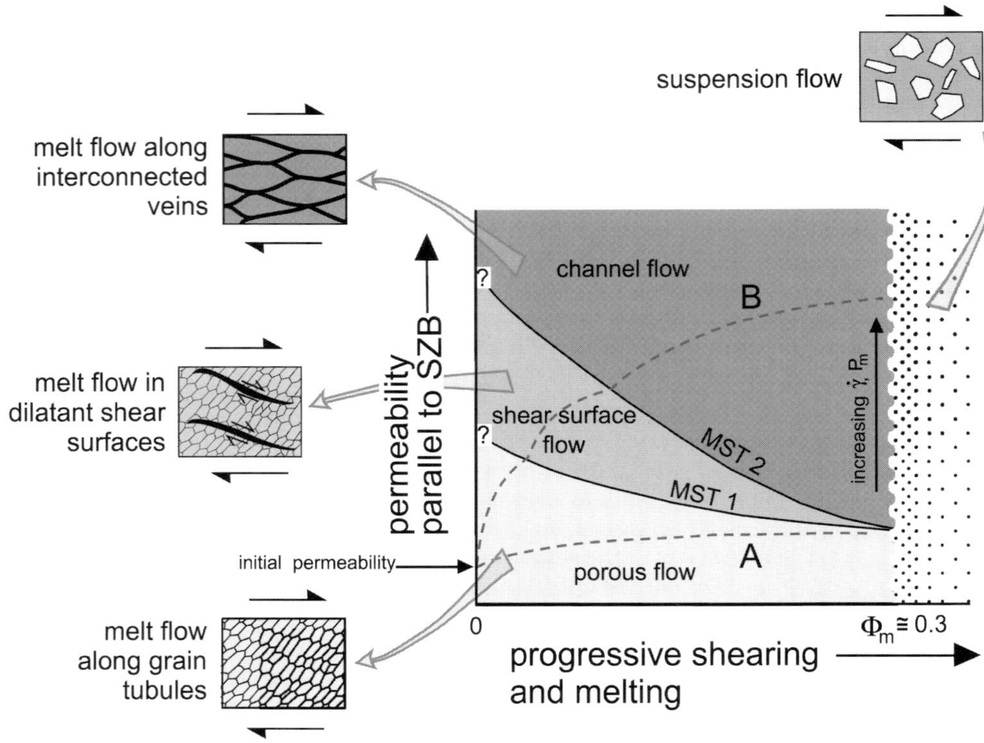

Fig. 7. Permeability of melt subparallel to the shear-zone boundary (SZB) v. shear strain for different modes of syntectonic melt flow. Insets depict microstructures corresponding to these modes, with melt-filled areas (black) set against a matrix of grains undergoing melt-assisted creep. Continuous curves represent melt segregation thresholds (MST). Dashed curves depict hypothetical evolution of permeability for progressive melting at low (A) and high (B) shear strain rates. Vertical wavy line marks transition to suspension flow (stippled field) at melt fractions, ϕ_m, of *c*. 0.3.

Mawer 1992; Rutter & Neumann 1995). The two continuous curves delimiting these fields represent MSTs that are discussed below. Two dashed curves labelled A and B show how permeability of melt evolves with structure during melting at ideal conditions of constant bulk shear strain rate. The strain rate of curve B is greater than for curve A. A fourth field on the right side of Fig. 7 for suspension flow at melt fractions greater than *c*. 0.30 pertains to closed systems in which melt-space is connected in all directions. This field is therefore not relevant to the discussion below. The shapes of the curves and fields in this diagram are speculative, as they are based partly on experimental data and partly on theoretical models of fluid flow in porous media. Nevertheless, the diagram is useful for discussing the possible mechanisms and rate dependences underlying the evolution of melt-filled space during deformation.

At low strain rates and melt fractions (curve A in Fig. 7), the entire system is expected to deform ductilely by one or more melt-assisted creep mechanisms (dislocation creep, granular flow, e.g. Dell'Angelo & Tullis 1988; Hirth & Kohlstedt 1995a, 1995b). In the absence of an externally imposed melt head (analogous to a hydraulic head), the melt flows because the deforming grains in the aggregate impart motion to the intragranular melt (Rutter 1997). Porous flow is slow compared with shear surface or channel flow because it is induced by changing pore size during deformation of the matrix grains. Melt connectivity and hence also permeability are expected to increase at a decreasing rate with shear strain as a result of the development of a steady-state microstructure, as indicated by the convex-upward shape of curve A in Fig. 7. Assuming that the grain boundaries remain wetted with melt throughout deformation, the overall increase in permeability is not just due to related increases in melt fraction and porosity, but may also be due to the formation of steady-state foliations subparallel to the shearing plane (Lamoureux *et al.* 1999), or to limited intergranular fracturing during dynamic recrystallization (Rosenberg & Riller 2000) and/or diffusion-accommodated granular flow. Irrespective of the creep mechanism, however, the strain rate is slow enough that the melt is never pressured sufficiently for it to induce pervasive hydrofracturing on the supragranular scale. Rather, the melt is squeezed out as fast as or faster than the rate at which it is produced, as discussed by Dell'Angelo & Tullis (1988).

At higher strain rates (curve B in Fig. 7), the melt pressure is sufficiently high to induce intergranular hydrofracturing at the lower MST (MST 1) in Fig. 7. This threshold marks the transition from melt transport by porous flow within a ductilely deforming aggregate to melt transport primarily within dilatant shear surfaces, as previously discussed for Fig. 4. With further strain, the melt-induced fractures lengthen, localizing the strain while draining melt by porous flow from the intervening rock. The rock is conditionally open, as described in the previous section. Curve B is inferred to flatten with strain if the melt-filled dilatant shear surfaces attain a steady-state aspect ratio and spacing. A similar relationship between effective porosity and strain was observed in deformation experiments conducted at near-lithostatic fluid pressures (Fischer & Paterson 1992; Zhang *et al.* 1994) and has been linked to the formation of dilatant fluid-filled shear surfaces in dynamically recrystallizing norcamphor, a quartz analogue material (Bauer *et al.* 2000a). If dilatant shear surfaces contain overpressured melt, they can propagate like zippered veins (e.g. Rubin 1998), opening and closing in the direction of propagation along the mechanically weak foliation subparallel to the shear-zone boundary, as observed in the experiments of Bauer *et al.* (2000b).

With further shearing and melting, melt-space connectivity and permeability increase rapidly with strain as the rock crosses the second MST. There, the dilatant shear surfaces interconnect to form a network of veins subparallel to the shear-zone boundary. Curve B can flatten (Fig. 7) if either the veins that channel the melt or the vein network itself stops widening. This steady-state width represents a dynamic equilibrium in both the rheological and fluid-dynamic senses: it manifests a balance of softening and hardening processes (e.g. Means 1995) as well as a balance of processes creating and destroying interconnected, melt-filled space. Effective porosity during channel flow is estimated to vary within the range of 10–25 vol. % in metasediments undergoing anatexis, based on the assumption that minimum porosity is equal to the leucosome volume (Brown *et al.* 1999, and references therein).

Network or channel flow facilitates the rapid transport of melt down melt-pressure gradients (Rutter 1997; Brown *et al.* 1998). Examples of melt networks include leucosome channels in stromatic migmatites (e.g. Brown *et al.* 1999) and mafic vein arrays in anatectic gneisses (e.g. Escher *et al.* 1976; Hanmer *et al.* 1997; Handy & Streit 1999). These melt-bearing vein networks are analogous to fluid-flow backbones along anastomosing shear zones formed at or just below the b–v transition (Cox 1999, and references therein). Networks of fluid flow and melt flow have in common the persistence of transient, near-lithostatic to supra-lithostatic pore-fluid pressures (P_f in metamorphic rocks, P_m in anatectic rocks) during deformation (e.g. Etheridge *et al.* 1983; Handy & Streit 1999). High melt pressures are necessary to prevent vein networks from collapsing at the high lithostatic pressures and temperatures of the deep crust. Such high melt pressures are maintained only for short times because melt can escape much more quickly along dilatant shear surfaces and vein networks than can enter these fast-drain channels by porous intergranular flow. Rutter & Mecklenburgh (2001) have estimated that melt permeability in channel flow is up to 10^6 times that in porous intergranular flow. This considerable contrast in melt fluxes for porous flow and channel flow accounts for the episodicity of melt segregation in anatectic rocks which, like those described in the previous section, were conditionally open.

Clearly, curve B in Fig. 7 only approximates the evolution of melt permeability in naturally deformed anatectic rocks. The actual evolution is probably much more irregular than either of the idealized trajectories in Fig. 7. Melt-space connectivity (and therefore permeability) fluctuates locally with changes in strain rate and melt pressure so that trajectories for small volumes of rock are expected to cross MSTs more than once. The MSTs themselves would shift their position in Fig. 7 if melt viscosity (dependent on temperature and volatile content) or shear regime (coaxial v. noncoaxial) were to vary during deformation.

The dynamic nature of MSTs

The two MSTs in Fig. 7 pertain to rocks melting during deformation and are therefore meaningless at static (i.e. stress-free) conditions. In this sense, they are similar to the MSTs of Sawyer (1994), but differ fundamentally from the melting and crystallization thresholds proposed by Vigneresse *et al.* (1996). The latter melting thresholds depend exclusively on melt content (liquid percolation and melt escape, respectively, at 8 and 20–25 vol. % melt) and are defined on the basis of bond-percolation theory in materials with a static melt distribution. The same is true of their two thresholds in crystallizing systems (rigid-percolation and particle-locking thresholds at 55 and 25–28 vol. % melt), except that these are based on site-percolation theory for static crystal mushes.

Although we agree with Vigneresse *et al.* (1996) that the pre-deformational structures of melting and crystallizing systems at rest differ, we predict that deformation has two modifying effects on any statically defined segregation thresholds: (1) deformation would cause structures in melting and crystallizing systems to converge towards a similar if not identical steady-state structure; (2) it would lower the amount of melt at such thresholds to an extent dependent primarily on strain rate, melt pressure, shape, size, and rheologies of constituent phases, as argued above. The first effect pertains if the rates of melting or crystallization are comparable with each other and if both are lower than the rate at which the microstructure adjusts to changes in melt content. The ubiquity of melt-filled shear surfaces in both melting (e.g. Rey *et al.* 1992) and crystallizing (e.g. Gapais 1989; McCaffrey 1994; Smith 1997) magmatic rocks with a similar range of melt fractions suggests that these conditions indeed pertain in nature.

A conclusive assessment of MSTs in melting and crystallizing rocks must await the outcome of experiments designed to test the effects of deformation on such thresholds. If our analysis is correct, however, then the dynamic MSTs we defined above in Fig. 7 are equivalent in both melting and crystallizing systems.

Melt segregation, transport and emplacement in crustal-scale shear zones: a discussion with examples

To understand how the modes of syntectonic melt flow in Fig. 7 operate on different time and length scales, we turn to some case studies of igneous activity in crustal-scale fault zones. Together with the concepts developed in the previous sections, these examples provide answers to some of the questions posed in the introduction.

How do structure and rheology affect melt segregation and crustal differentiation?

The Ivrea crustal cross-section in the southern European Alps is a good example of structurally controlled crustal differentiation. This exhumed cross-section of continental crust contains the remnants of an Early Permian transtensional fault zone and related magmatic rocks (Fig. 8). The composite section in Fig. 8 depicts this fault zone at its inception in Early Permian time. Steep, oblique-slip faults bounding elongate, asymmetrical basins in the upper crust (Collio basin, Schönborn & Schumacher 1994; Cassinis *et al.* 1995) extended downwards to moderately inclined mylonites of the CMB (Cossato–Mergozzo–Brissago) Line in the intermediate to lower crust (Handy *et al.* 1999a). These mylonites contain syntectonic, gabbro–dioritic to granitic intrusive bodies with a broad range of calc-alkaline compositions (Pinarelli *et al.* 1988; Mulch *et al.* 1999). These intrusive rocks were consanguineous with Early Permian mafic and ultramafic melts in the lower crust (Mafic Complex in Fig. 8; Rivalenti *et al.* 1984; Sinigoi *et al.* 1994) and similarly aged granitic plutons in the upper crust (Baveno granites in Fig. 8; Boriani *et al.* 1992).

The oblique-slip fault in the upper crust bounds these granitic plutons, the shallowest of which intruded at a depth of *c.* 5 km (Köppel 1974) along the steeply dipping, pre-Permian schistosity of the wall rock (Fig. 8). These plutons have hydrothermal contact aureoles and locally discordant contacts (Balk 1924), but did not induce partial melting or mylonitization of the wall rock. The limit of melt-induced veining is extended around the perimeter of the plutons, well above the Early Permian b–v transition (Fig. 8). This transition is not exposed in the field as a result of overprinting and tectonic excision during later Early Mesozoic and Tertiary faulting. However, the heat advected by the granitic plutons along the fault is expected to have raised the b–v transition in their vicinity, as shown in Fig. 8.

The mylonitic shear zones in the lower crust, including those of the CMB Line (Fig. 8), attenuated and partly exhumed the intermediate to lower crust during Early Permian transtension

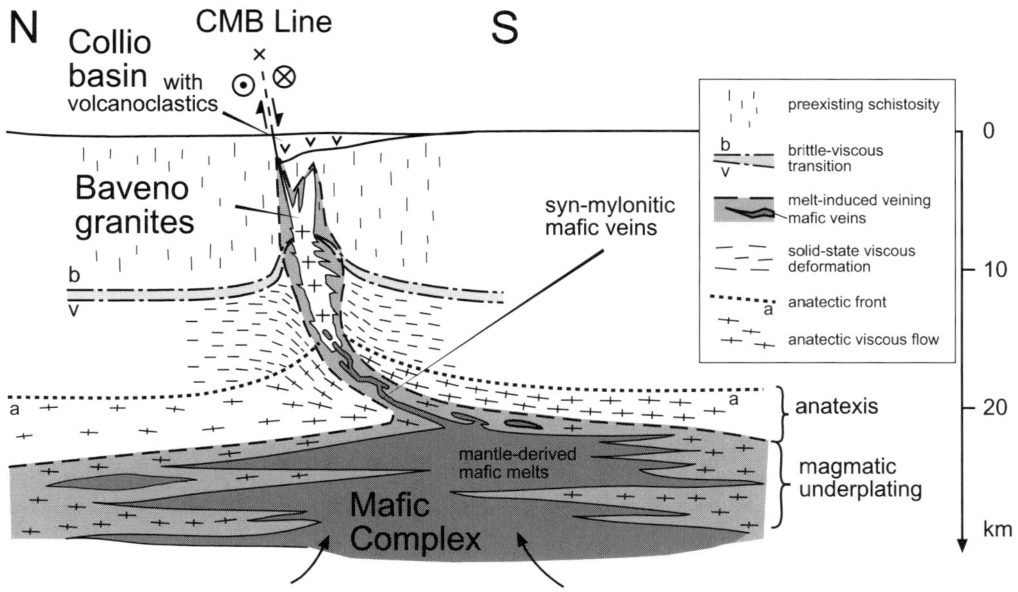

Fig. 8. Composite profile of the Ivrea crustal cross-section in Early Permian time showing the CMB Line (Cossato–Mergozzo–Brissago) and related Early Permian features. Curves for b–v transition (dashed–dotted), anatectic front (dotted) and limit of melt-induced veining (dashed) are same as in Fig. 2.

(Handy & Zingg 1991; Rutter *et al.* 1993; Quick *et al.* 1994). Mylonitization was associated with veining of mafic melts and anatexis of the metasedimentary wall rocks near these veins, both along the CMB Line (Handy & Streit 1999) and along the rim of the Mafic Complex (Snoke *et al.* 1999). The anatectic front therefore extended upwards within the crust only as far as the syntectonic mafic veins of the CMB Line, and downwards to just above the mantle-derived melts that ponded in the lower crust (Fig. 8).

The syn-mylonitic mafic veins of the CMB Line are very elongate (aspect ratios of >1000:1) and link up locally to form a vein network along the mylonitic foliation (Fig. 8). This vein geometry is obviously governed by the mechanical anisotropy of the mylonitic foliation and is consistent with low crustal strength ($\leqslant$10–20 MPa) and very high melt pressure ($\leqslant$600–700 MPa) during mafic veining (Handy & Streit 1999). The vein network channelled large amounts of mantle-derived melt from the lower to the intermediate crust, where these melts crystallized under amphibolite-facies conditions. There, anatectic melts from the mylonites near the mafic veins back-veined the latter. They fractured the partly consolidated mafic melt (angular mafic fragments in Fig. 9) and induced mingling and mixing of anatectic and mafic melts (lobate structures in Fig. 9). The products of this mixing are the aforementioned hybrid melts along the CMB Line whose broad range of main- and trace-element geochemistries spans the compositional gap between coeval gabbroic and granitic rocks, respectively, in the lower and upper crust (Mulch *et al.* 1999).

The CMB Line therefore played two important roles in differentiating the continental crust. First, it served as a melt conduit for the upward veining and channelled flow of transiently overpressured mantle-derived melts. In so doing, it advected sufficient heat to partially melt and thermally soften metasedimentary wall rocks in the lower crust. Second, it served as a melt reactor system, allowing veined mantle-derived melts to mingle and mix with anatectic melts derived from the adjacent mylonitic wall rock. The resulting hybrid melts then used the same shear zone to ascend to emplacement sites at and above the b–v transition.

The CMB fault zone functioned like a conditionally open system on the crustal scale in that it accommodated sporadic fluxes of melt into and out of the granite source region within mylonites at the base of the intermediate crust and in the lower crust. Melt veining and mixing within the active shear zone must have happened very quickly, judging from the short crystallization times predicted by thermal modelling of the hybrid veins along the CMB Line (seconds to years, Handy & Streit 1999). Thus,

Fig. 9. Concordant hybrid vein along the CMB Line containing fragments of partly consolidated mafic melt within a matrix of felsic melt. The mafic rock comprises hornblende–plagioclase with minor amounts of retrograde biotite and has a gabbro–dioritic geochemistry. The felsic rock is granitic and is derived from anatectic melt in the mylonitized metasedimentary wall rock. (Note fractures in mafic rock filled with felsic rock (lower right side) and lobate structures at their interface (across centre) indicative of mingling and assimilation.) Italian 100 Lire coin is 2.7 cm wide.

individual magmatic pulses along the shear zone were short lived compared with the total duration of Early Permian magmatism in the Ivrea crustal cross-section (265–290 Ma, radiometric ages compiled by Handy *et al.* 1999*a*).

The case study above suggests that rapid melt segregation and transport on the crustal scale are sited along shear zones because there the rocks are sufficiently anisotropic and weak, and melt pressures high enough for the rock to cross one or both of the MSTs in Fig. 7. Porous granular flow appears to be restricted to chemically and structurally more homogeneous rocks, for example, in the asthenosphere beneath slow-spreading mid-ocean ridges (Kelemen *et al.* 1997) or possibly in parts of the Archaean lower to middle crust that underwent partial convective overturn above huge volumes of underplated mafic melt (e.g. Pilbara Craton, Collins *et al.* 1998). Otherwise, inherited lithological heterogeneities tend to localize strain and focus melt flow.

How does rheology affect melt transport and emplacement?

Insight into this question comes from the Insubric Line and Bergell pluton, which form part of the late-orogenic transpressive Periadriatic fault system sited at the retro-wedge of the Tertiary Alpine orogen (Fig. 10). The reconstructed section in Fig. 10a shows that the steep Insubric mylonites contain the concordant feeder dyke of the Bergell pluton and extend down to where they tapped mantle-derived melts ponded at the base of previously subducted continental crust (Schmid *et al.* 1996). The Periadriatic Line and associated Tertiary plutons are interpreted to have formed in response to continuing collision above a detaching lithospheric slab (Dietrich 1976; von Blanckenburg & Davies 1995).

Davidson *et al.* (1996) proposed a two-stage emplacement history for the Bergell pluton (Fig. 10b and c). In the first stage, melt transport occurred within the steep, concordant feeder vein and emplacement outside the Insubric mylonite belt into space created by the northward movement of two Penninic nappes (the Tambo and Suretta nappes; Schmid *et al.* 1990) beneath the Late Cretaceous Tethyan suture. By the time of the Mid-Tertiary Bergell intrusion, the rocks just below this suture were the site of an east-dipping normal fault (Turba normal fault in Fig. 10c; Nievergelt *et al.* 1996). This normal fault was an active b–v transition that juxtaposed cold (<250 °C), brittle Austroalpine units in its footwall against warm, mylonitic Penninic units in its exhuming footwall (350–430 °C; Ferreiro Mählmann 1996). The steep Insubric mylonites were also an

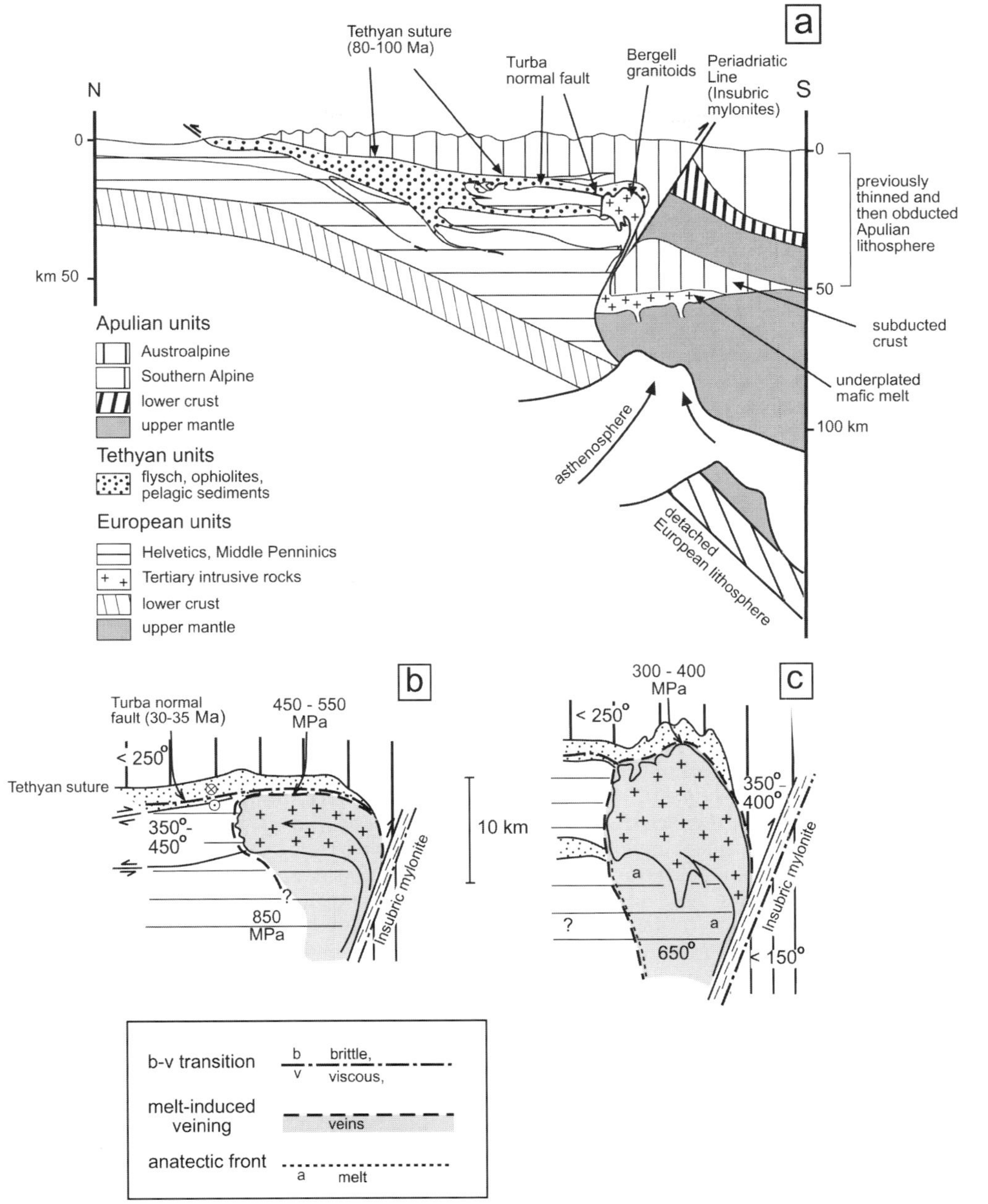

Fig. 10. (**a**) Cross-section of Tertiary Alpine Orogen (Central Alps) including the Bergell pluton and its elongate feeder vein along the Insubric Line (from Schmid *et al.* 1996, fig. 8e); (**b**) and (**c**) two-stage emplacement of Bergell pluton showing b–v transitions, anatectic front and limit of melt-induced veining (modified from Davidson *et al.* 1996, fig. 10). Pressure and temperature estimates from Davidson *et al.* (1996); Ferreiro Mählmann (1996); Trommsdorff & Connolly (1996).

active b–v transition at the time of intrusion, as they juxtaposed hot Tertiary metamorphic rocks of the Penninic basement against cold (<150 °C) pre-Alpine metamorphic rocks of the southern Alpine basement. In the second stage, continued melt transport within the feeder vein and pluton growth (ballooning) were coeval with lateral, north–south shortening (folding) and east–west stretching of the by then partly crystallized pluton. Emplacement during this second stage involved melt-induced veining and stoping along the upper sides of the pluton, and anatectic flow and folding above, along and beneath the pluton's base (Rosenberg *et al.* 1995). Although the wall rocks along the structurally higher margins of the pluton underwent contact metamorphism (Trommsdorff & Nievergelt 1983, and references therein), the temperatures and fluid activities during intrusion of these rocks were evidently not sufficiently high to induce partial melting and anatectic flow. A dashed line in Fig. 10b and c indicating the limit of melt-induced veining is therefore drawn around the pluton margin and extended downward to include the upper amphibolite facies anatectic gneisses (Fig. 4a) beneath the pluton. This dashed line truncates the Turba mylonites in Fig. 10c, in accordance with the observation that the Turba mylonites are locally cut by the Bergell granites (Nievergelt *et al.* 1996).

The evolution above reveals several important aspects of rheological control on melt ascent. First, the Bergell pluton initially intruded mylonitic crust that was bounded by two active b–v transitions: the subhorizontal Turba normal fault above and the steep Periadriatic mylonite belt to the south (Fig. 10b). Thus, melt-induced veining occurred well above the anatectic front and affected active mylonitic rocks just below and to the north of these transitions. The mylonitic zones were weak parallel to their foliations, as described above. The Insubric mylonites channelled the melt upwards within a long, concordant feeder vein, whereas the Turba mylonites arrested its further ascent. Together with the sink created by the northward movement of the Penninic nappes, these mylonites were responsible for the fist-like geometry of the pluton. Second, for large-scale similar folds to develop at the base of the Bergell pluton during the second stage (Fig. 10c) the partly crystallized melts within the pluton must have acquired a strength comparable with that of the anatectic country rocks beneath it (Davidson *et al.* 1996). Third, melt-induced veining during the second stage extended slightly above the defunct b–v transition at the Turba mylonite zone, indicating that ancient b–v transitions do not form barriers to intrusions.

We emphasize that mylonite below a b–v transition is an effective barrier to melt propagation only if two conditions are met: (1) the mylonitic foliation must obviously be oriented at high angles to the direction of magma ascent; (2) it must be active in order for the flow stress to be sufficiently low for the pressurized melt to induce extensional shear fracturing parallel to this foliation. This notion contrasts with other explanations that invoke the viscous strength maximum at (rather than a strength minimum just below) the b–v transition as a hindrance to melt ascent (e.g. Kriens & Wernicke 1990). Once melt ponds below the b–v transition, the thermal anomaly engendered by the melt accentuates the already existing strength minimum parallel to the foliation. Alternatively, Lister & Baldwin (1993) proposed that the b–v transition along the mylonitic front or metamorphic carapace of metamorphic core complexes was induced by the steep thermal gradients in the roofs of syntectonic sills (site 7 in Fig. 1). Only if the melt pressure increases to supralithostatic values can melt fracture through an active b–v transition. This probably happens regularly for magmas to reach the surface, as in the case of the Early Permian volcanism in the southern Alps (Fig. 8).

What are the rate-controlling processes during syntectonic melt segregation and transport? Do the rates of these processes vary with crustal level?

The pattern of curves for the anatectic front and the limit of melt-induced veining in Figs 8 and 10 has interesting implications for the relative rates of veining, heating and cooling in fault zones. In both the CMB and Insubric examples, the limit of melt-induced veining was frozen above the anatectic front. Because the anatectic rocks below this front supplied melt to the propagating veins above, anatexis and veining are linked processes that presumably cease at about the same time at the end of deformation. The distribution of curves in Figs. 8 and 10 is therefore an indication that melt-induced veining was fast compared with heating and melting of the wall rocks away from the veins in the fault zone. This is consistent with model studies predicting that overpressured, melt-filled veins propagate very quickly. Because mafic melts are one to three orders of magnitude less viscous than hydrous granitic melts (Clemens & Petford 1999), they propagate at much higher velocities (metres to kilometres per second, Handy & Streit 1999) than the latter (millimetres to centimetres per second, Clemens & Petford 1999). Either of these ranges is much greater than the characteristic thermal diffusion rate in basement rocks (0.1–1 mm s^{-1}). Estimates for time to melting and melt segregation range from months to years (Sawyer 1991) or even hundreds to thousands of years (Huppert & Sparks 1988). Thus, the limit of melt-induced veining in prograde fault zones will always ascend rapidly ahead of a slower, broadening anatectic front that marks the onset of melting and anatectic flow at depth. Given the different rates of vein propagation for mafic and felsic melts, one can even distinguish limits of veining for these two melt compositions. In the case of the CMB Line, field observations indicate that the upper limit of mafic melt-induced veining corresponds to the anatectic front (Fig. 8).

As mentioned above, the rate-limiting step in conditionally open systems such as the CMB Line is likely to be anatexis and/or the porous flow of melt in the source rock to dilatant sites (shear surfaces or veins). The rate of melting depends primarily on temperature, whereas the rate of porous flow (i.e. the melt flux) is a complex function of the melt viscosity, the melt pressure gradient in the direction of flow, and the permeability of the grain aggregate. All of these factors in porous flow are themselves functions of other factors: the melt viscosity varies with temperature and volatile content (Clemens & Petford 1999), the pressure gradient depends on the dilational rate of the veins and/or on the strain rate and creep mechanism(s) in the aggregate (Rutter 1997; Paterson 2001), and the permeability is a function of the porosity and tortuosity of the aggregate (Brown *et al.* 1999). Because both melting rates and porous flow rates are strongly temperature dependent, the rate at which deforming source rocks supply melt to dilatant shear surfaces or veins increases with increasing heat flow and strain rate in the fault zone.

A striking feature of rocks from numerous ancient anatectic fault zones is that they preserve fresh magmatic and solid-state deformational microstructures, even from deep crustal levels. This suggests that temperature and stress drops at the end of deformation were sufficiently fast to prevent structures from re-equilibrating later with ambient retrograde conditions. The time to crystallization of melt in veins is short compared with the total cooling time to the temperature of the wall rock (Paterson & Tobisch 1992). Davidson *et al.* (1992) calculated that even a 1 km thick tonalite vein at 700 °C in a wall rock at a temperature of 500 °C crystallizes after only 90 ka. The degree of preservation of magmatic structures generally increases upwards within fault zones, primarily as a result of increased rates of cooling and stress decay at progressively shallower levels of the crust. Knipe (1989) estimated that differential stress must decrease at a rate of at least 1 MPa per 10^3-10^4 a to preserve dislocation creep microstructures in quartz formed at temperatures of 300–400 °C and 50 MPa differential stress. Higher temperatures require faster stress drops to preserve these solid-state microstructures. This range of stress decay rates ($10^{-3}-10^{-4}$ MPa a^{-1}) is therefore a lower limit on the stress decay rates necessary to prevent magmatic microstructures formed at much higher temperatures from equilibrating with stress under subsolidus conditions.

The constraints above indicate that although there are order of magnitude differences in the rates of processes active during melt segregation and transport, none of these processes are slow. They operate on time scales that are remarkably short (10^3-10^5 a) compared with the duration of orogenesis or rifting (10^6-10^7 a).

What are the driving forces of melt ascent and emplacement?

The preservation of magmatic structures within vein networks along the CMB Line (Fig. 8) and within the long, narrow Bergell feeder vein along the Insubric Line (Fig. 10a) indicates that veins remained melt filled along their length on the same time scale as required for the melt to cool below the solidus. The vein geometry further suggests that the melts were overpressured, thus preventing the veins from collapsing (Berger *et al.* 1996; Handy & Streit 1999).

Several driving forces that have been proposed for syntectonic melt ascent within the crust, such as tectonic overpressure (Hutton 1998; De Saint Blaquat *et al.* 1998) or local dilation and compression along uneven fault planes (D'Lemos *et al.* 1992), cannot explain the melt distribution in our examples for the following reasons: (1) tectonic overpressure is a viable driving force for ascending melts only in compressional or transpressional settings (e.g. magmatic arcs, De Saint Blaquat *et al.* 1998), not in transtensional faults such as the CMB Line; (2) no irregular, dilating fault surfaces were seen on a large scale that could have squeezed or sucked the melt up or down the CMB or Insubric fault zones; (3) the low strengths estimated for the veined anatectic rocks from the CMB Line and the examples in Fig. 4 suggest that tectonic overpressure in partially melted intermediate to lower crust does not exceed 2–5 MPa (recall discussion above). These values are two orders of magnitude less than values cited elsewhere for tectonic overpressure (e.g. 500 MPa, Bott & Kuznir 1984, in De Saint Blaquat *et al.* 1998) and are also significantly lower than stresses at depth associated with the density contrast between melt and wall rock (Hogan *et al.* 1998). Rutter & Mecklenburgh (2001) have postulated that tectonically induced differential stress gradients (*c.* 1 MPa m^{-1}) exceed the gravitationally induced gradient (*c.* 0.003 MPa m^{-1}) only over relatively small distances (millimetres to metres) between fast-drainage conduits (veins, dilatant shear surfaces). Therefore, deformation cannot squeeze melt over greater distances, where gravity acting on density contrasts between melt and wall rock must be the dominant force driving vein propagation and melt ascent.

The dominance of gravity as the dominant driving force explains why the direction of melt flow within many fault zones, including the CMB and Insubric Lines, is independent of the orientation of the strain ellipsoid as inferred from mineral stretching lineations and leucosome rods in anatectic and mylonitic wall rocks. Although melt flow parallel to the main stretching axis is certainly possible (as documented in the study by Brown & Solar (1998) of melt flow along prolate-strain structures), it is not the rule in most shear zones.

The aspect ratio of mafic veins along the CMB Line is much less than that of the kilometre-long, millimetre-wide dykes predicted theoretically by Rubin (1998) with fracture and fluid mechanics. We suspect that the unrealistic vein dimensions obtained in the models of Rubin (1998) partly reflect the assumption that channelled melt flow in veins is perfectly laminar. They may also indicate that the flow of melt from its source in a dynamically recrystallizing aggregate into the vein is more efficient than assumed in his models. Neverthess, the basic conclusion of Rubin (1998) that veins (his dykes) can drain large amounts of melt into compliant plutons is consistent with our own conclusions for plutons beneath the b–v transition.

What is the long-term, high-strain rheology of the continental crust? Does melting always involve crustal weakening?

Crustal rheology during a complete magmatic cycle can be understood in terms of three stages of fault zone development, shown in Fig. 11: (1) a juvenile stage, with incipient melting at depth; (2) a climax stage, at the peak of magmatic activity; (3) a mature stage, just after the cessation of magmatism. Each of these stages has a characteristic distribution of melt, fault rocks and rheological transitions, depicted in the left-hand column of Fig. 11. The middle and right-hand columns show the corresponding transient geotherms and strength profiles. The temperature and strength curves in Fig. 11 were drawn by hand rather than calculated, to avoid pitfalls associated with thermal modelling of complex structures or with extrapolating laboratory flow laws to natural conditions. Nevertheless, they reflect the basic laws of thermal conduction and advection as applied to fault rocks with frictional and viscous rheologies. Fiducial temperature values at the b–v transition and anatectic front have been discussed above for Fig. 2.

Incipient heating during the juvenile stage (Fig. 11a) clearly weakens the base of fault zones. The primary cause of weakening is the enhancement of thermally activated dislocation creep (glide-plus-climb) in quartz and feldspar of the wall rocks. We note that incipient melting of wall rocks does not necessarily lead to weakening of the residue; partial melting can lower the water fugacity in unmelted rock and therefore inhibit hydrolytic weakening (Karato 1986). However, local hardening associated with this effect is probably masked once the melt is sufficiently abundant to induce diffusion-accommodated granular flow and thus weaken the rock.

Although mafic melts have low viscosities (log η of 1–3 Pa s according to Clemens & Petford (1999)), veins containing mafic melt are not expected to accommodate much bulk strain and affect the fault rheology significantly unless they form networks that remain open on time scales approaching that of a magmatic cycle (10^3–10^5 a; see below). The example of the CMB Line indicates that mafic veins propagate sporadically and crystallize very quickly (seconds to months, Handy & Streit 1999). Therefore, even if they form networks, mafic veins will actually harden rather than weaken the fault zone in the long term.

Increased strain rate as a result of weakening at the base of fault zones is expected to load the upper crust to an extent dependent on the mechanical coupling at the b–v transition. A high degree of coupling, favoured by low strains, and uniform crustal composition and structure, will obviously load the intermediate and upper crust rapidly. The corresponding increase in strain rates at these crustal levels can be expected to depress the b–v transition within the crust. Mechanical coupling is reduced, however, where inherited structural or compositional heterogeneities nucleate instabilities that localize strain.

During the climax stage (Fig. 11b), massive heat advection as a result of the intrusion of hot basic melts at the base of the fault zone drives the anatectic front to shallower depths and weakens all levels of the fault zone substantially. Strengths of 5–10 MPa derived in this paper (Fig. 6) represent an upper limit on the average strength of rocks undergoing anatectic flow at intermediate to deep levels of the fault zone at this stage. Higher in the intermediate crust, granitic feeder veins that connect plutons with the anatectic melt-source rocks at depth (Fig. 11b) can accommodate large displacements if they are sufficiently thick and the melt pressure remains high enough for them to remain open for a substantial time. The Bergell feeder vein in Fig. 10 or the kilometre-thick Valdez Creek tonalite within the MacLaren Glacier metamorphic belt in Alaska (Davidson *et al.* 1992) are good examples of such syntectonic veins. Davidson *et al.* (1992) estimated that the Valdez Creek tonalite accommodated a displacement of at least 10 km within the 90 ka to crystallization.

As shown in the examples of the Insubric and CMB Lines, the strength minimum just below the b–v transition is also driven upward and hinders ascent of granitic melts into the upper crust. The ponding of granitic melts just below this rheological transition engenders a local thermal bulge, which further reduces rock strength and accentuates the role of this level as an important intracrustal detachment horizon. Detachment will also occur at any depth interval where anatectic rocks are capped by stronger, less permeable rock layers (not shown in Fig. 11). There, anatectic rocks behave like closed systems and weaken drastically upon reaching an RCMP (recall Fig. 3).

Mature fault zones (Fig. 11c) generally harden as melts within them crystallize and the geotherm subsides. Not all parts of the fault

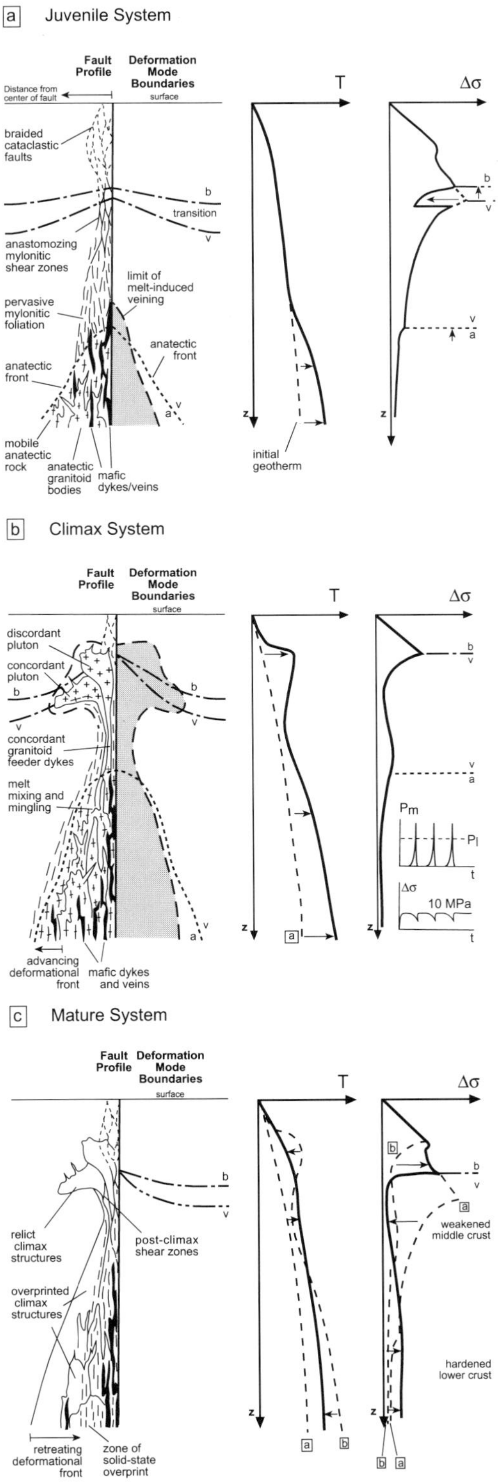

Fig. 11. Synthetic sections through a fault zone undergoing one cycle of syntectonic magmatism: (**a**) juvenile stage; (**b**) climax stage; (**c**) mature stage. Symbols in left column: dashed–dotted curve, brittle-to-viscous (b–v) transition; dotted curve, anatectic front at the transition from solid-state viscous creep (marked v) to anatectic flow (marked a); dashed curve, limit of melt-induced veining (stippled pattern). Other symbols in middle and right columns: bold continuous curves, temperature profiles (*T* v. *z*) and strength profiles ($\Delta\sigma$?v. *z*); bold dashed curves, temperature and strength profiles during previous stages labelled with boxed a and b; arrows indicate movement of curves since the previous stage(s). Inset diagrams in (**b**) show small, episodic variations in melt pressure, P_m, and strength, $\Delta\sigma$, as a function of time, *t*, and strain in the lower continental crust.

zone necessarily harden immediately or at the same time, however. Downward movement of the decaying thermal bulge around plutons emplaced just below the b–v transition can sustain or even accentuate weakening at intermediate levels of the fault zone whereas the over- and underlying levels cool and harden (Fig. 11c).

Fault zones or parts thereof can eventually attain strengths greater than their pre-melting strength if the melts that crystallize comprise minerals with higher solid-state creep strengths than those of the adjacent wall rock. For example, some of the crystallized mafic veins along the CMB Line (Fig. 8) are boudinaged (Handy & Streit 1999, fig. 2) and therefore represented a second, harder phase within the quartzo-feldpathic gneiss during mylonitization. The composite strength of this aggregate was greater than that of the gneiss alone, although in this case the gneisses still dominated the bulk rheology because of their interconnectedness parallel to the shear-zone boundary (Handy 1994). The large volumes of Carboniferous and Permian mafic and ultramafic intrusive rocks in the Mafic Complex (Fig. 8) probably hardened the lower crust, as indicated by the fact that most solid-state deformation during subsequent Early Mesozoic rifting was accommodated in hydrous, quartz-rich amphibolite-facies metasediments at the base of the intermediate crust rather than the crystallized intrusive rocks of the lower-crustal Mafic Complex (Handy & Zingg 1991). This observation is inconsistent with the widely held assumption that the lower continental crust is weak (e.g. McKenzie *et al.* 2000) and corroborates studies indicating that the lower continental crust beneath orogens (e.g. the Central Alps, Schmid *et al.* 1996) and non-volcanic rifted margins (Handy 1989; 1990) was as strong as or stronger than the intermediate crust and subcontinental upper mantle (Maggi *et al.* 2000).

The considerations above suggest that syntectonic melting induces cyclical weakening then hardening of the continental crust. Individual cycles coincide with melt ascent along many dilatant shear surfaces or vein networks through the crust. These episodes are relatively short lived (10^3–10^5 a) compared with the total duration of shearing at plate boundaries, which can last for as long as 10^6–10^7 a. Consequently, hardening of the crust or parts thereof may be the most important long-term mechanical effect of syntectonic melting.

Summary and conclusions

Continental fault zones in partially melted continental crust assume three roles: (1) they are agents of weakening and hardening at different levels of the crust; (2) they are conduits for the rapid movement of veined melts from melt-source regions at or below the anatectic front to melt sinks higher in the crust; (3) they serve as melt reactors in the intermediate to lower crust where mantle- and crustal-derived melts mix, mingle and fractionate to form hybrid melts. In the last capacity, they facilitate differentiation of the continental crust.

These varied roles played by fault zones are closely tied to two strain-dependent rheological discontinuities whose evolution is related to the large-scale distribution of melt: (1) a strength minimum parallel to subhorizontal mylonitic foliation in a depth interval just below the b–v transition hinders (but does not always prevent) the ascent of melts into the upper crust. Melt intrudes preferentially within this depth interval and, by advecting heat, accentuates the already existing strength minimum just below the b–v transition. However, ancient fossil b–v transitions in exhumed crust are not associated with strength minima and therefore do not hinder the ascent of melt. (2) A decrease in strength coincides with the upper depth limit of partial melting (the anatectic front). At this front, melt pressures are equal to and locally even exceed the lithostatic pressure.

The rheology of anatectic rock containing pressurized melt is determined by both the scale of strain localization and the pore connectivity, to an extent dependent on strain, strain rate and melt fraction in the rock. Below a critical shear strain, strain is homogeneous and melt is squeezed from the rock during melt-assisted dislocation creep or melt diffusion-accommodated granular creep. Beyond a critical shear strain, however, a first MST is crossed when strain localizes along dilatant shear surfaces. These surfaces draw melt from the adjacent deforming wall rock and propagate sporadically as veins. At higher strain rates and/or melt fractions, a further MST is crossed when the melt-filled shear surfaces interconnect to form vein networks. In most naturally deformed basement rocks, therefore, anatectic flow at natural strain rates in naturally deformed rocks involves a combination of melt-assisted dislocation creep or melt diffusion-accommodated granular creep, punctuated locally by melt-induced veining. Analysis of the melt-filled fracture geometries in such rocks yields upper strength limits in the range of 5–10 MPa. Thus, this type of anatectic crust is weaker than quartzo-feldspathic rock undergoing solid-state dislocation creep at comparable temperatures and strain rates.

Dilatant shear surfaces and interconnected veins form conduits for the rapid ascent of transiently overpressured melt from melt-source rocks below the anatectic front to sinks higher in the crust. Melt ascent along such crustal scale vein networks is short lived (10^3-10^5 a), as a result of the high flux of melt within and along the veins compared with that of melt flowing into the veins. Channelled melt flow along these networks is driven primarily by gravity acting on the contrasting densities of the melt and wall rock. The direction of melt flow is therefore not related to the orientation of the incremental strain ellipsoid in most mylonitic shear zones.

The high-strain rheology of the continental crust is predicted to be episodic: long-term creep of the crust is interrupted by ephemeral weakening events associated with the propagation and interconnection of melt-filled veins on the crustal scale. Following such events, the crust hardens to pre-intrusive levels as melts crystallize. Large volumes of crystallized mafic melt in the lower crust can even increase the strength of the crust to above pre-intrusive levels.

Episodic weakening then hardening of melt-bearing fault zones in the continental crust has broad implications for other, fault-related orogenic processes. For example, melts may be at least partly responsible for triggering rapid, in some cases even repeated, exhumation of basement rock ('tectonic surges' of Hollister & Crawford 1986) along fault zones in the retro-wedges of orogens. This may apply to the central Alps, where high Oligo-Miocene exhumation rates just north of the Insubric Line have been otherwise attributed to climatically controlled drainage and erosion (Schlunegger 1999). Episodic melt-induced fault slip may also be responsible for repeated, sudden shifts in sedimentary depocentres and volcanic fields along the margins of fault-bounded basins (e.g. Tertiary Basin-and-Range: Cemen *et al.* 1985; Link *et al.* 1985; Rio Grande Rift: Olsen *et al.* 1987; Early Permian Saar–Nahe Basin, Germany: Schäfer & Korsch 1998). The notion that acute weakening of the crust is ephemeral on the time scale of faulting along plate boundaries further suggests that the dynamic problem of transmitting shear stresses across supposedly weak magmatic arcs may not be as intractable as generally believed.

The concepts presented above are largely qualitative and require quantification if they are to be tested against other causes of fault-related weakening or strengthening. This is the gauntlet thrown to numerical modellers!

This paper benefited from the insightful reviews of K. McCaffrey and E. H. Rutter, as well as the critical comments of A. Berger and M. Brown. We thank E. H. Rutter and J.-L. Vigneresse for generously sending us their manuscripts in review. Our work was supported by the German Science Foundation (DFG) in the form of grants Ha 2403/2 (SPP 'Orogene Prozesse...') and RO 2177/1 (SPP 'Genese und Transport von Silikatschmelzen').

References

ALLIBONE, A.H. & NORRIS, R.J. 1992. Segregation of leucogranite microplutons during syn-anatectic deformation: an example from the Taylor Valley, Antarctica. *Journal of Metamorphic Geology*, **10**, 589–600.

ALLMENDINGER, R.W., JORDAN, T.E., KAY, S.M. & ISACKS, B.L. 1997. The evolution of the Altiplano–Puna Plateau of the Central Andes. *Annual Review of Earth and Planetary Sciences*, **25**, 139–174.

ARZI, A.A. 1978. Critical phenomena in the rheology of partially melted rocks. *Tectonophysics*, **44**, 173–184.

BAILEY, D.K. 1992. Episodic alkaline igneous activity across Africa: implications for the causes of continental break-up. *In*: STOREY, B.C., ALABASTER, T. & PANKHURST, R.J. (eds) *Magmatism and the Causes of Continental Break-up*. Geological Society, London, Special Publications, **68**, 91–98.

BALK, R. 1924. Zur Tektonik der Granitmassiv von Baveno und Orta in Oberitalien. *Geologische Rundschau*, **15**, 110–122.

BARKER, F., FARMER, G.L., AYUSO, R.A., PLAFKER, G. & LULL, J.S. 1992. The 50 Ma granodiorite of the Eastern Gulf of Alaska: melting in an accretionary prism in the forearc. *Journal of Geophysical Research*, B5, **97**, 6757–6778.

BAUER, P., PALM, S. & HANDY, M.R. 2000*a*. Strain localization and fluid pathways in mylonite: inferences from in-situ deformation of a water-bearing quartz analogue (norcamphor). *Tectonophysics*, **320**, 141–165.

BAUER, P., ROSENBERG, C. & HANDY, M.R. 2000*b*. Deformation experiments on rock-analogue materials at controlled confining pressure: a new approach for studying strain localization at the brittle to viscous transition. *Journal of Structural Geology*, **22**, 281–289.

BERGER, A., ROSENBERG, C. & SCHMID, S.M. 1996. Ascent, emplacement and exhumation of the Bergell Pluton within the Southern Steep Belt of the Central Alps. *Schweizerische Mineralogische und Petrographische Mitteilungen*, **76**, 357–382.

BORIANI, A., CAIRONI, V., GIOBBI ORIGONI, E. & VANNUCCI, R. 1992. The Permian intrusive rocks of Serie dei Laghi (Western Southern Alps). *Acta Vulcanologica*, **2**, 73–86.

BOTT, M.P. & KUZNIR, N.J. 1984. Origin of tectonic stress in the lithosphere. *Tectonophysics*, **105**, 1–14.

BOUCHEZ, J.L., DELAS, C., GLEIZES, G. & NÉDÉLEC, A., CUNEY, M. 1992. Submagmatic fractures in granites. *Geology*, **20**, 35–38.

BRACE, W.F. & KOHLSTEDT, D.L. 1980. Limits on lithospheric stress imposed by laboratory experiments. *Journal of Geophysical Research*, B11, **85**, 6248–6252.

BRODIE, K.H. & RUTTER, E.H. 1985. On the relationship between deformation and metamorphism with special reference to the behavior of basic rocks. *In*: THOMPSON, A.B. & RUBIE, D. (eds) *Kinematics, Textures and Deformation: Advances in Physical Geochemistry*. Springer, Berlin, 138–179.

BRODIE, K.H. & RUTTER, E.H. 1987. The role of transiently fine grained reaction products in syntectonic metamorphism. *Canadian Journal of Earth Sciences*, **24**, 556–564.

BROWN, M. 1994. The generation, segregation, ascent and emplacement of granite magma: the migmatite to crustally-derived granite connection in thickened orogens. *Earth-Science Reviews*, **36**, 83–130.

BROWN, M. & SOLAR, G.S. 1998. Shear zones and melts: positive feedback in orogenic belts. *Journal of Structural Geology*, **20**, 211–227.

BROWN, M. & SOLAR, G.S. 1999. The mechanism of ascent and emplacement of granite magma during transpression: a syntectonic granite paradigm. *Tectonophysics*, **312**, 1–34.

BROWN, M.A., BROWN, M., CARLSON, W.D. & DENISON, C. 1999. Topology of syntectonic melt-flow networks in the deep crust: inferences from three-dimensional images of leucosome geometry in migmatites. *American Mineralogist*, 11–12, **84**, 1793–1818.

BROWN, S., CAPRIHAN, A. & HARDY, R. 1998. Experimental observation of fluid flow channels in a single fracture. *Journal of Geophysical Research*, B3, **103**, 5125–5132.

BURG, J.P., BRUNEL, M., GAPAIS, D., CHEN, G.M. & LIU, G.H. 1984. Deformation of leucogranites of the crystalline Main Central Thrust Sheet in southern Tibet (China). *Journal of Structural Geology*, 5, **6**, 535–542.

BYERLEE, J. 1978. Friction of rocks. *Pure and Applied Geophysics*, **116**, 615–626.

CASSINIS, G., TOUTIN-MORIN, N. & VIRGILI, C. 1995. A general outline of the Permian continental basins in southwestern Europe. *In*: SCHOLLE, P.A., PERYT, T.M. & ULMER-SCHOLLE, D.S. (eds) *Sedimentary Basins and Economic Resources*. Springer, Berlin, 137–157.

CEMEN, I., WRIGHT, L.A., DRAKE, R.E. & JOHNSON, F.C. 1985. Cenozoic sedimentation and sequence of deformational events at the southeastern end of the Furnace Creek strike-slip zone, Death Valley region, California. *In*: BIDDLE, K.T. & CHRISTIE-BIDDLE, N. (eds) *Strike-Slip Deformation, Basin Formation, and Sedimentation*. Society of Economic Paleontologists and Mineralogists, Special Publications, **37**, 127–142.

CLEMENS, J.D. & MAWER, C.K. 1992. Granitic magma transport by fracture propagation. *Tectonophysics*, **204**, 339–360.

CLEMENS, J.D. & PETFORD, N. 1999. Granite melt viscosity and silicic magma dynamics in contrasting tectonic settings. *Journal of the Geological Society, London*, 6, **156**, 1057–1060.

COLLINS, W.J. & SAWYER, E.W. 1996. Pervasive granitoid magma transfer through the lower–middle crust during non-coaxial compressional deformation. *Journal of Metamorphic Geology*, **14**, 565–579.

COLLINS, W.J., VAN KRANKENDONK, M.J. & TEYSSIER, C. 1998. Partial convective overturn of Archean crust in the east Pilbara Craton, Western Australia: driving mechanisms and tectonic implications. *Journal of Structural Geology*, 9–10, **20**, 1405–1424.

COOPER, R.F. & KOHLSTEDT, D.L. 1986. Rheology and structure of olivine-basalt partial melts. *Journal of Geophysical Research*, B9, **91**, 9315–9323.

COSGROVE, J. 1997. The influence of mechanical anisotropy on the behaviour of the lower crust. *Tectonophysics*, **280**, 1–2, 1–14.

COX, S.F. 1999. Deformational controls on the dynamics of fluid flow in mesothermal gold systems. *In*: MCCAFFREY, K.J.W., LONERGAN, L. & WILKINSON, J.J. (eds) *Fractures, Flow and Mineralization*. Geological Society, London, Special Publications, **155**, 123–140.

CRITTENDEN, M.D., CONEY, P.J. & DAVIS, G.H. (eds) 1980. *Cordilleran Metamorphic Core Complexes*. Geological Society of America, Memoirs, **153**.

CRUDEN, A.C. 1990. Flow and fabric development during the diapiric rise of magma. *Journal of Geology*, **98**, 681–698.

DANIEL, J.M. & JOLIVET, L. 1995. Detachment faults and pluton emplacement: Elba Island (Tyrrhenian Sea). *Bulletin de la Societé Géologique de France*, **4**, 341–354.

DAVIDSON, C., HOLLISTER, L.S. & SCHMID, A.M. 1992. Role of melt in the formation of a deep crustal compressive shear zone: the McLaren Glacier metamorphic belt, South Central Alaska. *Tectonics*, **11**, 2, 348–359.

DAVIDSON, C., ROSENBERG, C. & SCHMID, S.M. 1996. Synmagmatic folding of the base of the Bergell pluton, Central Alps. *Tectonophysics*, **265**, 213–238.

DAVIDSON, C., SCHMID, S.M. & HOLLISTER, L.S. 1994. Role of melt during deformation in the deep crust. *Terra Nova*, **6**, 133–142.

DELL'ANGELO, L.N. & TULLIS, J. 1988. Experimental deformation of partially melted granitic aggregates. *Journal of Metamorphic Geology*, **6**, 495–515.

DE SAINT BLAQUAT, M., TIKOFF, B., TEYSSIER, C. & VIGNERESSE, J.L. 1998. Transpressional kinematics and magmatic arcs. *In*: HOLDSWORTH, R.E., STRACHAN, R.A. & DEWEY, J.F. (eds) *Continental Transpressional and Transtensional Tectonics*. Geological Society, London, Special Publications, **135**, 327–340.

DIETRICH, V.J. 1976. Plattentektonik in den Ostalpen. Eine Arbeitshypothese. *Geotektonische Forschung*, **50**, 1–84.

D'LEMOS, R.S., BROWN, M. & STRACHAN, R.A. 1992. Granite magma generation, ascent and emplacement within a transpressional orogen. *Journal of the Geological Society, London*, **149**, 487–490.

ENGLAND, P. & MOLNAR, P. 1993. Cause and effect among thrust and normal faulting, anatectic melting and exhumation in the Himalaya. *In*: TRELOAR, P.J. & SEARLE, P. (eds) *Himalayan Tectonics*. Geological Society, London, Special Publications, **74**, 401–411.

ESCHER, A., JACK, S. & WATTERSON, J. 1976. Tectonics of the North Atlantic Proterozoic dyke swarm. *Philosophical Transactions of the Royal Society, London, Series A*, **280**, 529–539.

ETHERIDGE, M.A. 1983. Differential stress magnitudes during regional deformation and metamorphism: upper bound imposed by tensile fracturing. *Geology*, **11**, 232–234.

ETHERIDGE, M.A., WALL, V.T. & VERNON, R.H. 1983. The role of the fluid phase during regional metamorphism and deformation. *Journal of Metamorphic Geology*, 3, **1**, 205–226.

FERREIRO MÄHLMANN, R. 1996. The pattern of diagenesis and metamorphism by vitrinite reflectance and illite 'crystallinity' in Mittelbünden and in the Oberhalbstein Part 2: Correlation of coal petrographical and of mineralogical parameters. *Schweizerische Mineralogische und Petrographische Mitteilungen*, **76**, 23–46.

FISCHER, G.J. & PATERSON, M.S. 1992. Measurements of permeability and storage capacity in rocks during deformation at high temperature and pressure. *In*: EVANS, B. & WONG, T.-F. (eds) *Fault Mechanics and Transport Properties of Rocks*. Academic Press, San Diego CA, 213–252.

GAPAIS, D. 1989. Shear structures within deformed granites: mechanical and thermal indicators. *Geology*, **17**, 144–147.

HANDY, M.R. 1989. Deformation regimes and the rheological evolution of fault zones in the lithosphere: the effects of pressure, temperature, grain size, and time. *Tectonophysics*, **163**, 119–152.

HANDY, M.R. 1990. The exhumation of cross sections of the continental crust: structure, kinematics and rheology. *In*: SALISBURY, M.H. & FOUNTAIN, D.M. (eds) *Exposed Cross-Sections of the Continental Crust*. NATO-ASI Series C, **317**, 485–507.

HANDY, M.R. 1994. Flow laws for rocks containing two nonlinear viscous phases: a phenomenological approach. *Journal of Structural Geology*, 3, **16**, 287–301.

HANDY, M.R. 1998. Fault rocks from an exhumed intracrustal extensional detachment: the Pogallo ductile fault zone. *In*: SNOKE, A.W., TULLIS, J. & TODD, V.R. (eds) *Fault-related Rocks: a Photographic Atlas*. Princeton University Press, Princeton, NJ.

HANDY, M.R. & ROSENBERG, C.L. 1999. The role of large shear zones in generating and transporting melts, and in changing the rheology of Variscan continental crust. *Terra Nostra*, **99**, 93–94.

HANDY, M.R. & STREIT, J.E. 1999. Mechanisms and mechanics of magmatic underplating: inferences from syntectonic mafic veins in deep crustal mylonite. *Earth and Planetary Science Letters*, **165**, 271–286.

HANDY, M.R. & ZINGG, A. 1991. The tectonic and rheological evolution of an attenuated cross section of the continental crust: Ivrea crustal section, southern Alps, northwestern Italy and southern Switzerland. *Geological Society of America Bulletin*, **103**, 236–253.

HANDY, M.R., FRANZ, L., HELLER, F., JANOTT, B. & ZURBRIGGEN, R. 1999*a*. Multistage accretion, orogenesis and exhumation of continental crust (southern Alps, Italy and Switzerland). *Tectonics*, **16**, 1154–1177.

HANDY, M.R., WISSING, S. & STREIT, J.E. 1999*b*. Strength and structure of mylonite with combined frictional–viscous rheology and varied bimineralic composition. *Tectonophysics*, 1–4, **303**, 175–192.

HANMER, S., MENGEL, F., CONNELLY, J. & VAN GOOL, J. 1997. Significance of crustal scale shear zones and synkinematic mafic dykes in the Nagssugtoqidian orogen, SW Greenland: a re-examination. *Journal of Structural Geology*, **19**, 59–75.

HIRTH, G. & KOHLSTEDT, D.L. 1995*a*. Experimental constraints on the dynamics of the partially molten upper mantle 1: Deformation in the diffusion creep regime. *Journal of Geophysical Research*, B2, **100**, 1981–2001.

HIRTH, G. & KOHLSTEDT, D.L. 1995*b*. Experimental constraints on the dynamics of the partially molten upper mantle 2. Deformation in the dislocation creep regime. *Journal of Geophysical Research*, B8, **100**, 15441–15449.

HOBBS, B.E., MULHAUS, H.-B. & ORD, A. 1990. Instability, softening, and localization of deformation. *In*: KNIPE, R.J. & RUTTER, E.H. (eds) *Deformation Mechanisms, Rheology and Tectonics*. Geological Society, London, Special Publications, **54**, 143–166.

HOGAN, J.P., PRICE, J.D. & CHARLES GILBERT, M. 1998. Magma traps and driving pressure: consequences for pluton shape and emplacement in an extensional regime. *Journal of Structural Geology*, **20**, 1155–1168.

HOLLISTER, L.S. & CRAWFORD, L.M. 1986. Melt-enhanced deformation: a major tectonic process. *Geology*, **14**, 558–561.

HUPPERT, H.E. & SPARKS, S.J. 1988. The generation of granitic magmas by intrusion of basalt into continental crust. *Journal of Petrology*, **29**, 599–624.

HUTTON, D.H.W. 1982. A tectonic model for the emplacement of the Main Donegal Granite, NW Ireland. *Journal of the Geological Society, London*, **139**, 615–631.

HUTTON, D.H.W. 1988*a*. Granite emplacement mechanisms and tectonic controls: inferences from deformation studies. *Transactions of the Royal Society of Edinburgh: Earth Sciences*, **79**, 245–255.

HUTTON, D.H.W. 1988*b*. Igneous emplacement in a shear zone termination: the biotite granite at Strontian, Scotland. *Geological Society of America Bulletin*, **100**, 1392–11399.

HUTTON, D.H.W. 1998. Syntectonic granites and the principle of effective stress: a general solution to the space problem? *In*: BOUCHEZ, J.L., HUTTON, D.H.W. & STEPHENS, W.E. (eds) *Granite: From Segregation of Melt to Emplacement Fabrics*. Kluwer, Dordrecht.

JIN, Z.M., BAI, Q. & KOHLSTEDT, D.L. 1994. High-temperature creep of olivine crystals from four localities. *Physics of the Earth and Planetary Interiors*, **82**, 55–64.

JOHANNES, W. & HOLTZ, F. 1996. *Petrogenesis and Experimental Petrology of Granitic Rocks*. Springer, Heidelberg.

KARATO, S. 1986. Does partial melting reduce the creep strength of the upper mantle? *Nature*, **318**, 309–310.

KELEMEN, P.B., HIRTH, G., SHIMIZU, N., SPIEGELMAN, M. & DICK, H.J.B. 1997. A review of melt migration processes in the adiabatically upwelling mantle beneath oceanic spreading ridges. *Philosophical Transactions of the Royal Society of London, Series A*, **355**, 283–318.

KNIPE, R.J. 1989. Deformation mechanisms—recognition from natural tectonites. *Journal of Stuctural Geology*, **11**, 127–146.

KOHLSTEDT, D.L., EVANS, B. & MACKWELL, S.J. 1995. Strength of the lithosphere: constraints imposed by laboratory experiments. *Journal of Geophysical Research*, B9, **100**, 17587–17602.

KÖPPEL, V. 1974. Isotopic U–Pb ages of monazites and zircons from the crust–mantle transition and adjacent units of the Ivrea and Ceneri Zones (Southern Alps, Italy). *Contributions to Mineralogy and Petrology*, **43**, 55–70.

KRIENS, B. & WERNICKE, B. 1990. Characteristics of a continental margin magmatic arc as a function of depth: the Skagit–Methow crustal section. *In*: SALISBURY, M.H. & FOUNTAIN, D.M. (eds) *Exposed Cross-Sections of the Continental Crust*. NATO-ASI Series, C, **317**, 159–174.

LACHENBRUCH, A.H., SASS, J.H. & GALANIS, S.P. 1985. Heat flow in southernmost California and the origin of the Salton Trough. *Journal of Geophysical Research*, **90**, 6709–6736.

LAMOUREUX, G., ILDEFONSE, B. & MAINPRICE, D. 1999. Modeling the seismic properties of fast-spreading ridge crustal low-velocity zones: insights from Oman Gabbro textures. *Tectonophysics*, **312**, 283–302.

LAPORTE, D., RAPAILLE, C. & PROVOST, A. 1997. Wetting angles, equilibrium melt geometry, and the permeability threshold of partially molten crustal protholiths. *In*: BOUCHEZ, J.L., HUTTON, D.H.W. & STEPHENS, W.E. (eds) *Granite: From Segregation of Melt to Emplacement Fabrics*. Kluwer, Dordrecht, 31–54.

LEJEUNE, A., RICHET, P. 1995. Rheology of crystal-bearing silicate melts: an experimental study at high viscosities. *Journal of Geophysical Research*, **100**, 4215–4229.

LINK, M.H., ROBERTS, M.T. & NEWTON, M.S. 1985. Walker Lake Basin, Nevada: an example of Later Tertiary (?) to Recent sedimentation in a basin adjacent to an active strike-slip fault. *In*: BIDDLE, K.T. & CHRISTIE-BIDDLE, N. (eds) *Strike-Slip Deformation, Basin Formation, and Sedimentation*. Society of Economic Paleontologists and Mineralogists, Special Publications, **37**, 105–126.

LISTER, G.S. & BALDWIN, S.L. 1993. Plutonism and the origin of metamorphic core complexes. *Geology*, **21**, 607–610.

MAGGI, A., JACKSON, J.A., MCKENZIE, D. & PRIESTLEY, K. 2000. Earthquake focal depths, effective elastic thickness, and the strength of the continental lithosphere. *Geology*, **28**, 495–498.

MCCAFFREY, K.J.W. 1992. Igneous emplacement in a transpressive shear zone: Ox Mountains igneous complex. *Journal of the Geological Society, London*, **149**, 221–235.

MCCAFFREY, K.J.W. 1994. Magmatic and solid state deformation partitioning in the Ox Mountains Granodiorite. *Geological Magazine*, **131**, 639–652.

MCKENZIE, D., NIMMO, F., JACKSON, J., GANS, P.B. & MILLER, E.L. 2000. Characteristics and consequences of flow in the lower crust. *Journal of Geophysical Research*, B5, **105**, 11029–11046.

MEANS, W.D. 1995. Shear zones and rock history. *Tectonophysics*, **247**, 157–160.

MILLER, C.F., WATSON, E.B. & HARRISON, T.M. 1988. Perspectives on source, segregation and transport of granitoid magmas. *Transactions of the Royal Society of Edinburgh: Earth Sciences*, **79**, 135–156.

MULCH, A., ROSENAU, M.R., DÖRR, W., HAACK, U. & HANDY, M.R. 1999. U–Pb ages of syn-mylonitic mafic and granitoid veins in the middle to lower continental crust (Ivrea–Verbano Zone, CH and I). *Terra Nova*, **99**, 150.

NELSON, K.D. & 27 OTHERS 1996. Partially molten middle crust beneath Southern Tibet: synthesis of Project INDEPTH results. *Science*, **274**, 1684–1687.

NICOLAS, A. & ILDEFONSE, B. 1996. Flow mechanism and viscosity in basaltic magma chambers. *Geophysical Research Letters*, **23**, 2013–2016.

NICOLAS, A. & JACKSON, M. 1982. High temperature dikes in peridotites: origin by hydraulic fracturing. *Journal of Petrology*, **23**, 568–582.

NIEVERGELT, P., LINIGER, M., FROITZHEIM, N. & FERREIRO MÄHLMANN, R. 1996. Early to mid Tertiary crustal extension in the Central Alps:

the Turba Mylonite Zone (Eastern Switzerland). *Tectonics*, **15**, 329–340.

OLSEN, K.H., BALDRIDGE, W.S. & CALLENDER, J.F. 1987. Rio Grande rift: an overview. *Tectonophysics*, **143**, 119–139.

PAQUET, J., FRANÇOIS, P. & NÉDÉLEC, A. 1981. Effect of partial melting on rock deformation: experimental and natural evidences on rocks of granitic compositions. *Tectonophysics*, **78**, 545–565.

PASSCHIER, C.W. & URAI, J.L. 1988. Vorticity and strain analysis using Mohr diagrams. *Journal of Structural Geology*, **10**, 755–763.

PATERSON, M.S. 1978. *Experimental Rock Deformation—the Brittle Field*. Springer, Berlin.

PATERSON, M.S. 1987. Problems in the extrapolation of laboratory rheological data. *Tectonophysics*, **133**, 33–43.

PATERSON, M.S. 1995. A theory for granular flow accommodated by material transfer via an intergranular fluid. *Tectonophysics*, **245**, 135–151.

PATERSON, M.S. 2001. A granular flow theory for the deformation of partially molten rock. *Tectonophysics*, in press.

PATERSON, S.R. & SCHMIDT, K.L. 1999. Is there a close spatial relationship between faults and plutons? *Journal of Strcutural Geology*, 8–9, **21**, 1131–1142.

PATERSON, S.R. & TOBISCH, A.T. 1992. Rates of processes in magmatic arcs: implications for the timing and nature of pluton emplacement and wall rock deformation. *Journal of Structural Geology*, **14**, 291–300.

PATERSON, S.R., VERNON, R.H. & TOBISCH, O.T. 1989. A review of criteria for the identification of magmatic and tectonic foliations in granitoids. *Journal of Structural Geology*, 3, **11**, 349–363.

PINARELLI, L., DEL MORO, A. & BORIANI, A. 1988. Rb–Sr geochronology of lower Permian plutonism in Massiccio dei Laghi, southern Alps (NW Italy). *Rendiconti Società Italiana di Mineralogia et Petrographia*, 2, **43**, 411–428.

PRICHARD, H.M., ALABASTER, T., HARRIS, N.B.W. & NEARY, C.R. (eds) 1993. *Magmatic Processes and Plate Tectonics*. Geological Society, London, Special Publications, **76**.

QUICK, J.E., SINIGOI, S. & MAYER, A. 1994. Emplacement dynamics of a large mafic intrusion in the lower crust, Ivrea–Verbano Zone, northern Italy. *Journal of Geophysical Research*, B11, **99**, 21559–21573.

RENNER, J., EVANS, B. & HIRTH, G. 2000. On the rheologically critical melt percentage. *Earth and Planetary Science Letters*, **181**, 585–594.

REY, P., BURG, J.P. & CARON, J.M. 1992. Middle and Late Carboniferous extension in the Variscan Belt: structural and petrological evidences from the Vosges massif (eastern France). *Geodinamica Acta*, **5**, 17–36.

RIVALENTI, G., ROSSI, A., SIENA, F. & SINIGOI, S. 1984. The Layered Series of the Ivrea–Verbano Igneous Complex, Western Alps, Italy. *Tschermaks Mineralogische und Petrographische Mitteilungen*, **33**, 77–99.

ROSENBERG, C.L. 2001. Deformation of partially molten granite: a review and comparison of experimental and natural case studies. *International Journal of Earth Sciences*, **90**, in press.

ROSENBERG, C.L. & HANDY, M.R. 2000. Syntectonic melt pathways during simple shearing of an anatectic rock analogue (norcamphor–benzamide). *Journal of Geophysical Research*, **105**, 3135–3149.

ROSENBERG, C.L. & HANDY, M.R. 2001. Melt migration during pure shear deformation of a partially molten rock analogue (norcamphor–benzamide). *Journal of Structural Geology*, in press.

ROSENBERG, C.L. & RILLER, U. 2000. Partial-melt topology in statically and dynamically recrystallized granite. *Geology*, **28**, 7–10.

ROSENBERG, C., BERGER, A. & SCHMID, S.M. 1995. Observations from the floor of a granitoid pluton: inferences on the driving force of final emplacement. *Geology*, **23**, 443–446.

RUBIN, A.M. 1998. Dike ascent in partially molten rock. *Journal of Geophysical Research*, B9, **103**, 20901–20919.

RUTTER, E.H. 1997. The influence of deformation on the extraction of crustal melts: a consideration of the role of melt-assisted granular flow. *In*: HOLNESS, M.B. (ed.) *Deformation-enhanced Fluid Transport in the Earth's Crust and Mantle*. Mineralogical Society Series, **8**, 82–110.

RUTTER, E.H. & MECKLENBURGH, J. 2001. The extraction of crustal melts from their protoliths and the flow behaviour of partially molten crustal rocks. *In*: BROWN, M. & RUSHMER, T. (eds) *Evolution and Differentiation of the Continental Crust*. Cambridge University Press, Cambridge.

RUTTER, E.H. & NEUMANN, D.H.K. 1995. Experimental deformation of partially molten Westerly granite under fluid-absent conditions, with implications for the extraction of granitic magmas. *Journal of Geophysical Research*, B8, **100**, 15697–15715.

RUTTER, E.H. & WHITE, S.H. 1979. The microstructures and rheology of fault gouges produced experimentally under wet and dry conditions at temperatures up to 400 °C. *Bulletin de Minéralogie*, 2–3, **102**, 101–109.

RUTTER, E.H., BRODIE, K.H. & EVANS, P.J. 1993. Structural geometry, lower crustal magmatic underplating and lithospheric stretching in the Ivrea–Verbano Zone, northern Italy. *Journal of Structural Geology*, **15**, 647–662.

SAWYER, E.W. 1991. Disequilibrium melting and the rate of melt–residuum separation during migmatization of mafic rocks from the Grenville Front, Quebec. *Journal of Petrology*, **32**, 701–738.

SAWYER, E.W. 1994. Melt segregation in the continental crust. *Geology*, **22**, 1019–1022.

SCHÄFER, A. & KORSCH, R.J. 1998. Formation and sediment fill of the Saar–Nahe Basin (Permo-

Carboniferous, Germany). *Zeitschrift der deutschen geologischen Gesellschaft*, 2, **149**, 233–269.
SCHEUBER, E. & REUTTER, K.J. 1992. Magmatic arc tectonics in the Central Andes between 21° and 25°S. *Tectonophysics*, **205**, 127–140.
SCHLUNEGGER, F. 1999. Controls of surface erosion on the evolution of the Alps: constraints from the stratigraphies of the adjacent foreland basins. *International Journal of Earth Sciences*, **88**, 285–304.
SCHMID, S.M. & HANDY, M.R. 1991. Towards a genetic classification of fault rocks: geological usage and tectonophysical implications. *In*: HSÜ, K.J., MACKENZIE, J. & MÜLLER, D. (eds) *Controversies in Modern Geology*. Academic Press, London, 339–361.
SCHMID, S.M., BOLAND, J.N. & PATERSON, M.S. 1977. Superplastic flow in fine grained limestone. *Tectonophysics*, **43**, 257–291.
SCHMID, S.M., PFIFFNER, O.A., FROITZHEIM, N., SCHÖNBORN, G. & KISSLING, E. 1996. Geophysical–geological transect and tectonic evolution of the Swiss–Italian Alps. *Tectonics*, **15**, 1036–1064.
SCHMID, S.M., RÜCK, P. & SCHREUERS, G. 1990. The significance of the Schams nappes for the reconstruction of the paleotectonic and orogenic evolution of the Penninic zone along the NFP-20 traverse (Grisons, eastern Switzerland). *Mémoire de la Société Géologique de France*, **156**, 263–287.
SCHOLZ, C.H. 1990. Mechanics of faulting. *Annual Review of Earth and Planetary Sciences*, **17**, 309–334.
SCHOLZ, C.H. 1998. Earthquakes and friction laws. *Nature*, **391**, 37–42.
SCHÖNBORN, G. & SCHUMACHER, M.E. 1994. Controls on thrust tectonics along basement–cover detachment. *Schweizerische Mineralogische und Petrographische Mitteilungen*, **74**, 421–436.
SECOR, D.T. 1965. Role of fluid pressure in jointing. *American Journal of Science*, **263**, 633–646.
SHAW, H.R. 1980. Fracture mechanisms of magma transport from the surface. *In*: HARGRAVES, R.B. (ed.) *Physics of Magmatic Processes*. Princeton University Press, Princeton, NJ, 201–264.
SIBSON, R.H. 1977. Fault rocks and fault mechanisms. *Journal of the Geological Society, London*, **133**, 191–213.
SIBSON, R.H. 1985. A note on fault reactivation. *Journal of Structural Geology*, **7**, 751–754.
SIBSON, R.H. 1986. Earthquakes and rock deformation in crustal fault zones. *Annual Review of Earth and Planetary Sciences*, **14**, 149–175.
SINGH, J. & JOHANNES, W. 1996. Dehydration melting of tonalites. Part I. Beginning of melting. *Contributions to Mineralogy and Petrology*, **125**, 16–25.
SINIGOI, S., QUICK, J., CLEMENS-KNOTT, D., MAYER, A. & DEMARCHI, G. 1994. Chemical evolution of a large mafic intrusion in the lower crust, Ivrea–Verbano Zone. *Journal of Geophysical Research*, B11, **99**, 21575–21590.
SMITH, J.V. 1997. Shear thickening dilatancy in crystal-rich flows. *Journal of Volcanological and Geothermal Research*, **79**, 1–8.
SNOKE, A.W., KALAKAY, T.J., QUICK, J.E. & SINIGOI, S. 1999. Development of a deep-crustal shear zone in response to syntectonic intrusion of mafic magma into the lower crust, Ivrea–Verbano zone, Italy. *Earth and Planetary Science Letters*, **166**, 31–45.
SPEER, J.A., MCSWEEN, H.Y. JR & GATES, A.E. 1994. Generation, segregation, ascent, and emplacement of Alleghanian plutons in the southern Appalachians. *Journal of Geology*, **102**, 249–267.
STREIT, J.E. 1997. Low frictional strength of upper crustal faults: a model. *Journal of Geophysical Research*, B11, **102**, 24619–24626.
TALBOT, C.J. 1999. Can field data constrain rock viscosities? *Journal of Structural Geology*, **21**, 949–958.
TROMMSDORFF, V. & CONNOLLY, J.A.D. 1996. The ultramafic contact aureole about the Bregaglia (Bergell) tonalite: isograds and a thermal model. *Schweizerische Mineralogische und Petrographische Mitteilungen*, **76**, 537–548.
TROMMSDORFF, V. & NIEVERGELT, P. 1983. The Bregaglia Iorio intrusive and its field relations. *Memorie della Società Geologica Italiana*, **26**, 55–68.
VAN DER MOLEN, I. & PATERSON, M.S. 1979. Experimental deformation of partially-melted granite. *Contributions to Mineralogy and Petrology*, **70**, 299–318.
VAUCHEZ, A. & EGYDIO DA SILVA, M. 1992. Termination of a continental-scale strike-slip fault in partially melted crust: the West Pernambuco shear zone, northeast Brazil. *Geology*, **20**, 1007–1010.
VIGNERESSE, J.L. 1995. Control of granite emplacement by regional deformation. *Tectonophysics*, **249**, 173–186.
VIGNERESSE, J.L., BARBEY, P. & CUNEY, M. 1996. Rheological transitions during partial melting and crystallization with application to felsic magma segregation and transfer. *Journal of Petrology*, **37**, 1579–1600.
VON BLANCKENBURG, F. & DAVIES, J.H. 1995. Slab breakoff: a model for syncollisional magmatism and tectonics in the Alps. *Tectonics*, **14**, 120–131.
WEERTMAN, J. 1968. Dislocation climb theory of steady state creep. *American Society of Metallurgy Transactions*, **61**, 681–694.
WICKHAM, S.M. 1987. The segregation and emplacement of granitic magmas. *Journal of the Geological Society, London*, **144**, 281–297.
ZHANG, S., COX, S.F. & PATERSON, M.S. 1994. The influence of room temperature deformation on porosity and permeability in calcite aggregates. *Journal of Geophysical Research*, **99**, 15761–15775.

Index

Note: Page numbers in **bold** type refer to tables, those in *italic* type to illustrations